Design, Simulation, and Optimization of Adsorptive and Chromatographic Separations

Design, Simulation, and Optimization of Adsorptive and Chromatographic Separations

A Hands-On Approach

Kevin R. Wood, Y.A. Liu, and Yueying Yu

WILEY-VCH

Authors

Dr. Kevin R. Wood
Virginia Polytech Institute and State University
Department of Chemical Engineering
635 Prices Fork Road
245 Goodwin Hall
Blacksburg VA 24061
United States

Prof. Y. A. Liu
Virginia Polytechnic Institute and State University
Department of Chemical Engineering
635 Prices Fork Road
245 Goodwin Hall
Blacksburg VA 24061
United States

Dr. Yueying Yu
Virginia Polytech Institute and State University
Department of Chemical Engineering
635 Prices Fork Road
245 Goodwin Hall
Blacksburg VA 24061
United States

Library of Congress Card No.: applied for

British Library Cataloguing-in-Publication Data
A catalogue record for this book is available from the British Library.

Bibliographic information published by the Deutsche Nationalbibliothek
The Deutsche Nationalbibliothek lists this publication in the Deutsche Nationalbibliografie; detailed bibliographic data are available on the Internet at <http://dnb.d-nb.de>.

Print ISBN: 978-3-527-34469-7
ePDF ISBN: 978-3-527-81499-2
ePub ISBN: 978-3-527-81501-2
Mobi ISBN: 978-3-527-81500-5
oBook ISBN: 978-3-527-81502-9

Cover Design Schulz Grafik-Design, Fußgönheim, Germany

Typesetting SPi Global, Chennai, India
Printing and Binding Strauss GmbH, Mörlenbach Germany

Printed on acid-free paper

10 9 8 7 6 5 4 3 2 1

Contents

Foreword by George E. Keller, II

This is a note that I too have lauded the relevance and completeness of the other documents regarding the contents of this tome. I know of no other publication in its class which even comes close. So I can rise and shout "Amen" to its authors.

I have always become both amused and frustrated by the reality that there are many more research papers in publication than there are specific pathways to get from a given starting point to a solution. Economics is many times not in touch with relevant problems, or alternative solutions are not evaluated for relevance. This book teaches us how to approach such problems more thoroughly.

It usually turns out that there is more than one solution to a problem and classical music is a good example. The grand music emanating from the brains of the Bach family, Mozart, Beethoven, Brahms, Handel, and many others has stood the test of time to the delight of millions of music lovers from around the world. But how did these masters create it all from tonality, notes, pitch, and dynamics to express their feelings so beautifully? After all, they had a myriad of instruments to choose from.

I wish to close this contribution on a sad note – for me, yet an inspiration for our readers who are developing new separation science and technology:

Last May my wife died. We had been married for 59 absolutely wonderful years. We met at State College, Pennsylvania. Judy, a smart, pretty, and vivacious co-ed and I a shy graduate student. The marriage was idyllic but about three years before her death she began to show serious signs of impending problems: heart, liver, kidney, melanoma, et al, including sepsis. Despite constant medical surveillance, nothing seemed to be helping, and nothing seemed to stop the onslaught. During this time, she used more and more oxygen which was fed to her as she needed. Mercifully, the home oxygen unit that I had developed helped to ease the pain, especially at night. But it didn't work for long and she succumbed. She seemed to suffer in no major way. But we had done our best and she seemed to be without any real pain at the end.

At best my efforts in oxygen feeding gave her a few months of pain-free existence. In all of the psychological pain that I suffered, Judy's death has catalyzed a desire to find ways to bring about longer, pain-free lives. Perhaps separation science and micro-technology will eventually help us prolong lives. I believe so.

Publisher's note: Dr. George E. Keller, II, is Chief Engineer, Mid-Atlantic Technology, Research and Innovation Center, South Charleston, West Virginia. He was formerly leader of the Separations and Process Fundamentals Skill Center at

Union Carbide and later DOW, arguably the highest profile industrial separation group in the United States. He is widely known for his pioneering work in separation science and technology, particularly for modernizing pressure swing adsorption (PSA) for use in medical oxygen generation. Dr. Keller was elected to the National Academy of Engineering in 1988, and has received three national awards from the American Institute of Chemical Engineers.

Preface

Motivation and Goals

Modern adsorptive and chromatographic technologies have been leaning heavily on cyclic processes, as opposed to traditional steady-state unit operations. A number of such technologies, including temperature swing adsorption (TSA), pressure swing adsorption (PSA), and simulated moving bed (SMB) chromatography, are becoming more common in laboratory-, pilot-, and industrial-scale processes. Engineers are grappling with a wide variety of methods for characterizing the cyclic operations of these technologies, and researchers are still inventing new heuristics and approximations for design and optimization.

Simulated moving bed technologies in particular are beginning to fulfill their earlier promise of creating efficient continuous chromatographic alternatives to older batch separations. They are finding widespread applications in separating and purifying enantiomers, petrochemicals, pharmaceuticals, and biochemicals with higher yield and lower solvent consumption. Because of this, there are a large number of engineers who need actionable methods for the selection, design, simulation, and optimization of these processes.

Simulated moving beds are new enough that there is a dearth of heuristics and shortcut design methods for practicing engineers to use. Indeed, even the ones that do exist are developed for narrow classes of separations and frequently do not map well to the separations of interest. This has led to a significant amount of duplicated work and a reliance on large over-design of SMB systems.

Engineers working on gas-adsorptive separations have similar needs to those who do SMB chromatography but are slightly better off since there are numerous sources of information. There are several well-established textbooks on the theory and design of TSA and PSA units, and resources are available for both practicing engineers and academic researchers. However, for systems as complex as TSA and PSA, being informed of the theory is necessary, but not sufficient, to perform the kind of simulation and optimization required in this competitive industry.

Often, experienced engineers and operators can make good estimates of unit performance. However, the growing tide of retiring professionals and loss of experience throughout the industry makes this task difficult or impossible. An engineer who wants to actually implement the modern theories to build computer models of his adsorptive and chromatographic processes will not find much help in current resources. There are software packages, which can put the finite-element analyses within reach of most engineers, but those without previous computer modeling experience will be hard pressed to use them effectively. This is because there is very little educational material that is devoted to the mapping between the theories of adsorptive and chromatographic separations and their implementation in such tools. Also, these software packages often omit intermediate methods (such as standing wave design) that allow the engineer to incorporate a wider set of physical effects (such as mass transfer resistance) than the idealized design equations without performing a rigorous, but time-consuming, finite-element simulation.

We prepare this text in order to teach undergraduate and graduate students, engineers, and scientists what they need to know to effectively design, simulate, and optimize industrial adsorptive and chromatographic separations. We do this by presenting a unified approach to the development of the ideal and intermediate design equations, while simultaneously teaching the reader to use the rigorous simulation packages, Aspen Adsorption and Aspen Chromatography (Aspen Technology, Inc., Bedford, Massachusetts). We explain the physical phenomena involved in adsorptive and chromatographic processes, and then perform the derivations of the various design equations using the computer algebra system, Mathematica (Wolfram Research, Inc., Campaign, Illinois). In this way, the reader can see the path from problem specification to solution, without getting bogged down in the details required to solve the complex systems of differential equations.

This book has a number of secondary goals, which all serve the main goals. These include the following:

1) Describe new software techniques, which can dramatically speed up simulations, and simplify model development. These include techniques to convert dynamic simulations into a simulation that attempts to directly solve for a cyclic steady state. We also describe techniques for storing results to avoid unnecessarily duplicating calculations.
2) Introduce the modeling of adsorptive separations, such as temperature and pressure swing absorbers (TSA/PSA). Because liquid and gas adsorption systems have extensive mathematical similarities, a comparison between these two types of separations can help those engineers who are already familiar with gas-phase adsorption understand the SMB processes.
3) Introduce some premade tools for performing the shortcut design of SMB systems, as well as instructions for modifying them to apply to systems outside the coverage of this book. This has largely been lacking in the field so far, and while the simplest methods can be easily implemented in a tool such as Excel, the majority of the methods require more advanced coding to solve.
4) Teach practicing engineers how to do parameter regression in adsorptive and chromatographic separations. Regression requires a solid understanding of the underlying physical processes and is frequently required during scale-up. While regression against steady-state models is well understood, the cyclic nature of the SMB and gas adsorption systems introduces some additional

nuances that must be taken into account. Moreover, regressing against dynamic models can require careful planning to be successful.

Approaches

To accomplish these goals, this text employs the following methodical approach:

- Thorough descriptions of adsorptive and chromatographic processes that highlight the underlying physical phenomena to be captured by our models.
- Description of how these phenomena manifest themselves in the ultimate models, or which assumptions cause them to be omitted.
- Best practices for using the developed models, including workflows that use multiple software packages.
- Strategies for improving the model performance, including parameter regression as well as customizing model code.
- Case studies of real processes to help the reader understand the details of TSA/PSA and SMB operations.
- Step-by-step, *hands-on workshops* to ensure that the reader is able to construct models with the complexity necessary to handle an immense variety of adsorptive and chromatographic separations.
- Instructions on how to develop new models in commercial software packages.
- *Hands-on practice problems* with hints and solutions.

Contents

In Chapter 1, we introduce the pressure and temperature swing strategies, and then learn the detailed design considerations by building a model of a PSA hydrogen separator, the classic Skarstrom process for producing an oxygen-enriched stream from air, and a simulation of TSA for air drying. In Chapter 2, we transfer our knowledge of Aspen Adsorption to Aspen Chromatography, and learn the differences between these software packages by building a model of an SMB process. We also discuss the differences in the assumptions that our models are allowed to make and the nuances of liquid- versus gas-phase isotherms. Because of the expensive calculations involved, we also discuss ways to improve the Aspen simulation performance. Chapter 3 covers the shortcut design methods for SMB processes. Some detailed derivation is performed using Mathematica, and the results are generalized to systems of arbitrary isotherms. Specifically, we provide a derivation and implementation of both triangle theory and standing wave design, two of the modern design techniques of SMBs. The Mathematica tools are used in a workshop to find the optimum operating conditions for the system modeled in Chapter 3. In Chapter 4, we extend our consideration to the so-called "Operational Modes" of SMB chromatography, which can yield large improvements in product purity, recovery, and other performance indices. Based on our research experience, we teach the reader how to simulate and optimize SMB operations by varying the feed flow rates ("PowerFeed" and "Partial Feeding"), varying the feed concentrations ("ModiCon"), and shifting asynchronously the inlet/outlet ports ("Varicol"), within the switching interval of the SMB operation. In Chapter 5, we cover methods for performing parameter estimation and regression using the models we built in Chapters 1 and 2, along with insights we gained from the analysis in Chapter 3.

Our review of currently available materials failed to find any text that encompassed the range of modeling techniques laid out here that emphasize *our approach of integrating fundamental principles, industrial applications, and hands-on workshops and practice problems of commercial software tools for design, simulation, and optimization.*

Since 1993, the senior author and the instructors he trained have taught industrial training courses to over 7,000 practicing engineers in 2017 FORTUNE global top one and two refining/chemical companies (SINOPEC and PetroChina) about hands-on applications of commercial software tools to the design, simulation, and optimization of petrochemical processes. These include the use of Aspen Adsorption for the TSA/PSA of hydrogen production, and of Aspen Chromatography for the petrochemical purification. Based on our 25 years of industrial training experience, we are fully convinced of the importance and effectiveness *of our integrating and hands-on approach* emphasized in this text.

Software Selection and Copyright Notice

Aspen Adsorption and Aspen Chromatography (Version 9.0 or new version) are available from Aspen Technology, Cambridge, MA (http.//www.aspentech .com/).

Screenshots of input information and output results from Aspen Adsorption® and Aspen Chromatography® are printed with permission by Aspen Technology, Inc. AspenTech®, aspen ONE®, Aspen Adsorption®, Aspen Chromatography®, and the Aspen leaf logo are trademarks of Aspen Technology, Inc.

Microsoft Excel and Visual Basic for Applications (VBA) for available as part of Microsoft's Office software package (http.//office.microsoft.com/en-us/default .aspx).

Mathematica and its Computable Document File (CDF) reader are available for user license from Wolfram Research, Champion, IL (http://www.wolfram.com/). We are allowed to offer CDFs freely in the workshop portion of the textbook.

Acknowledgments

It is a pleasure to thank a number of very special persons and organizations that contributed to the preparation of this book.

We would like to express our sincere appreciation to the following senior leaders of Aspen Technology, Inc., for their strong support for the Center of Excellence in Process System Engineering at Virginia Tech since 2002: Antonio Pietri, President and CEO; Willie Chan, Senior Vice President and Chief Technology Officer; Filipe Soares-Pinto, Senior Vice President, Asia-Pacific; Andy Lui, Vice President , R&D; Vikas Dhole, Vice President, Engineering Product Management; and Steven Qi, Vice President, Customer Support and Training; David Tremblay, Senior Director, Engineering Product Management; Fran Royer, Senior Manager, University Programs; and Theresa Foley, Senior Customer Care Specialist. We thank the following process modeling experts at Aspen Technology for teaching us the art and practice of adsorptive and chromatographic process modeling: Ashok Bhakta, Ben Hauser, Nickhil Rokkam, Theresa Foley, and Tan Zhang.

We would like to thank the China Petroleum and Chemical Corporation (SINOPEC) and China National Petroleum Corporation (CNPC or PetroChina) for supporting the Process System Engineering education and research at Virginia Tech.

We are very grateful to Mr. Cao Xianghong for his strong support for this work during his tenure as Senior Vice President and Chief Technology Officer of SINOPEC and to Mr. He Shengbao, Mr. Du Jizhou, Mr. Xu Yingjun, and Mr. Chen Yuanpeng, senior executives for research and development at PetroChina for their strong support in petrochemical technology development and engineering training.

We thank Dr. George E. Keller, II, Chief Engineer at Mid-Atlantic Technology and Innovation Center (MATRIX), South Charleston, West Virginia, who kindly took time to write the FOREWORD for our text.

The senior author would like to thank his wife, Hing-Har Liu, for her support through the laborious process of this book writing. Kevin Wood would like to dedicate this book to his parents, Stuart and Vicki Wood, and thank his wife Jessica for her patience and support.

About the Authors

Y.A. Liu, the Alumni Distinguished Professor and the Frank C. Vilbrandt Endowed Professor of Chemical Engineering at Virginia Tech, received his BS, MS, and PhD degrees from National Taiwan University, Tufts University, and Princeton University, respectively.

Professor Liu taught at Auburn University from 1974 to 1981, where his last position was as alumni associate professor endowed by the Auburn Alumni Association. He joined Virginia Tech as a professor of chemical engineering in 1982. In 1983, he was appointed the Vilbrandt Professor. In 2012, he was appointed the alumni distinguished professor. He has published numerous papers and nine books. These include six pioneering chemical engineering textbooks: (1) artificial intelligence in chemical engineering (Academic Press, 1991); (2) neural networks in bioprocessing and chemical engineering (Academic Press, 1995); (3) industrial water reuse and wastewater minimization (McGraw-Hill, 1999); (4) step-growth polymerization process modeling and product design (Wiley, 2008); (5) refinery engineering-integrated process modeling and optimization (Wiley-VCH, 2012); and (6) Petroleum Refining Process Modeling: Optimization Tools and Applications (Wiley-VCH, 2018).

Professor Liu's contributions to chemical engineering teaching, research, and industrial outreach have been recognized by over 25 university, national, and international awards. He is a Fellow of the American Institute of Chemical Engineers, a member of Virginia Tech's Academy of Teaching Excellence, and a recipient of the 1996 AspenTech International Award for University Teaching Excellence in computer-aided design. He has received three awards from the American Society of Engineering Education (ASEE): the Fred Merryfield Design Award (1993) for creativity and excellence in teaching and research of engineering design; the George Westinghouse Award (1990), ASEE's highest honor for an engineering educator under age 45 for outstanding achievements in both teaching and scholarship; and the Western Electric Award (1984) for excellence in instruction of engineering students. In 1986, he received the National Catalyst Award for excellence in chemical education from the Chemical Manufacturers Association. He received the Distinguished Alumni Award in 1990, and the Outstanding Career Achievement Award in 2010, both from Tufts University. In 2015, he received the Virginia Professor of the Year Award from the U.S. Professors of the Year Awards Program, the only U.S. national awards that recognize and celebrate extraordinary dedication to undergraduate teaching, sponsored by the Carnegie Foundation for the Advancement of Teaching and the Council for Advancement and Support of Education.

Over the past 25 years, Professor Liu devoted his school breaks helping petrochemical industries in developing countries and chemical industries in Virginia with technology development and engineering training. He has taught intensive training courses on computer-aided design, advanced process control, energy and water savings, and refinery and polymerization process modeling to over 7000 practicing engineers in China, Taiwan, and the United States. For his contributions to teaching, research, and industrial outreach, he received the Virginia Outstanding Faculty Award from Governor Jim Gilmore and the National Friendship Award from China's Premier Zhu Ronjie in 2000.

Kevin Wood received his PhD from the Department of Chemical Engineering at Virginia Polytechnic Institute and State University ("Virginia Tech") in June, 2016. He received his BS in chemical engineering from Purdue University in 2010. He completed his doctoral dissertation on the development of computational tools for the design of SMB systems, and design of software to enable Aspen simulations to be run in parallel. The tools he developed, and simulations he ran using them, form the basis of this textbook. He has worked on industrial modeling projects, including biodiesel, CO_2 capture, and petrochemical purification. These projects were collaborative efforts between Virginia Tech, Aspen Technology, and industrial partners. He is now an application engineer with ExxonMobil Corporation.

Yueying Yu received her PhD from the Department of Chemical Engineering at Virginia Tech in December, 2015. She received her BS in chemical engineering from Dalian University of Technology in China in 2009. She completed her doctoral dissertation on the simulation and comparison of operational modes in simulated moving bed (SMB) chromatography and gas-phase adsorptive separation. This is a part of the basis of this textbook. She is now engaged in her first love of educating young students in STEM (science, technology, engineering, and mathematics).

List of Computer Files

Accessible through Wiley-VCH website, wiley-vch.de/en/→shop→bookfinder→ design, simulation and optimization of adsorptive and chromatographic separations→student materials.

Direct link:

http://www.wiley-vch.de/en?option=com_eshop&view=product&isbn=9783527344697

Workshops are numbered according to the section of the text where they appear. For example, Workshop 1-7 appears in Section 1.7.

- Workshop 1-7 (Air Separation)
  - Workshop1A.ada
  - Workshop1B.ada
  - Workshop1C.ada
  - Workshop1D.ada
  - Workshop1E.ada
  - Workshop1F.ada
  - Workshop1G.ada
- Workshop 1-8 (Single-Bed H2)
  - H2Workshop1.ada
  - H2Workshop2.ada
  - H2Workshop3.ada
  - H2Workshop4.ada
  - H2Workshop5.ada
  - H2Workshop6.ada
  - H2Workshop7.ada
  - H2Workshop7e.ada
  - H2Workshop8.ada
- Workshop 1-9 (Dual-Bed H2)
  - Workshop3A.ada
  - Workshop3B.ada
  - Workshop3C.ada
  - Workshop3D.ada
  - Workshop3E.ada
- Workshop 1-10 (TSA for air drying)
  - TSAworkshop1.ada

- Workshop 2-4 (4-zone for Enantioseparation)
  - 4-zone basic SMB model.cra
- Workshop 2-5 (Tandem Insulin)
  - TandemSMBSizeEx.cra
- Workshop 3-3 (Triangle Theory Workshop)
  - TriangleTheoryWorkshop.cra
- Workshop 3-5 (Fixed Bed Transitions)
  - Fixed Bed Transitions.cdf
- Workshop 3-6 (Triangle Theory Constraints)
  - Langmuir Triangle Theory.nb
- Workshop 3-7 (SWD)
  - SWD Ideal Langmuir.nb
  - SWD Nonideal Langmuir.nb
- Workshop 4-3 (Varicol)
  - 4-zone Basic SMB model.cra
  - Workshop 4-3 Varicol.cra
- Workshop 4-4 (PowerFeed)
  - Workshop 4-4 PowerFeed.cra
- Workshop 4-5 wodicon)
  - Workshop 4-5 Modicon.cra
- Workshop 4-6 (Combined Modes)
  - Workshop 4-6 PF&MC.cra
- Workshop 4-7 (Parallel 2-zone)
  - Workshop 4-7 Train1.cra
  - Workshop 4-7 Train2.cra
  - Workshop 4-7 Combined.cra
  - Workshop 4-7 Final.cra
- Workshop 5-2 (Steady State Regression)
  - AspenRegression.cra
  - MathematicaRegression.nb
  - ExcelRegression.xlsx
- Workshop 5-3 (Dynamic Regression)
  - Workshop 5-3 Regression.cra
- Practice Problems
  - Practice Problem 1-1 (air separation2).ada
  - Practice Problem 1-2 (Naptha Upgrading).ada
  - Practice Problem 2-1 (Bad Steady State).cra
  - Practice Problem 2-1 (Steady State).cra
  - Practice Problem 2-2 (xylene).cra
  - Practice Problem 2-3 (5-zone).cra
  - Practice Problem 3-1 (TriangleTheory).cdf
  - Practice Problem 3-1 (Triangle Theory).cra
  - Practice Problem 4-1 (5-zone ModiCon).cra
  - Practice Problem 4-2 (Comparison).cra
  - Practice Problem 5-1 (Estimation).ada
  - Practice Problem 5-2 (Sensitivity).ada

1

Simulation of Adsorption Processes

This chapter introduces four key concepts inherent to all adsorptive separations, before presenting how these concepts are applied in rigorous simulations. Sections 1.1 and 1.2 provide a high-level overview of adsorption covering the history and driving forces involved. Section 1.3 discusses adsorption equilibrium and summarizes the theoretical assumptions behind the various mathematical models that have been used to describe it. Section 1.4 explains a number of terms and concepts inherent to working with packed beds. Section 1.5 covers the "classic" industrial adsorber configurations, along with commonly used industrial adsorbents. Section 1.6 onward discusses how to apply these concepts in Aspen Adsorption, specifically to pressure swing adsorption (PSA) for oxygen isolation from air (Section 1.7) and hydrogen separation from light hydrocarbons (Sections 1.8 and 1.9), and temperature swing adsorption (TSA) for air drying (Section 1.10). Section 1.11 provides the conclusions of the chapter. Section 1.12 presents practice problems to allow readers to apply the concepts described in this chapter. Section 1.13 includes the nomenclature, and Section 1.14 gives the bibliography.

1.1 Introduction to Gas-phase Adsorption Technologies

Gas-solid separations are incredibly important industrial problems, but have a short history compared to processes such as distillation. The difficulties associated with handling solid adsorbents require both creative process design and detailed understanding of the adsorptive properties of various solids. The investigation of solid-gas absorptivity began in the late 1700s, but it was not until the early 1900s that any industrial application was found. By the end of the 1950s, engineers had already filed patents on the major gas-adsorptive separations we know today, such as hydrogen sulfide removal [1], oxygen/nitrogen separation [2], and some drying applications [3, 4] as well as numerous short-chain hydrocarbon separations [5–7].

Over the years, refinements to these technologies have appeared, and numerous authoritative textbooks have consolidated the scientific knowledge of the subject. It is fair to say that at this point, adsorption technologies are fairly well understood, and there are a variety of resources available for the interested reader.

Design, Simulation, and Optimization of Adsorptive and Chromatographic Separations: A Hands-On Approach, First Edition. Kevin R. Wood, Y. A. Liu, and Yueying Yu.

Rather than rehashing the details of adsorptive separations, this textbook aims to make the reader proficient at software tools that bring all the knowledge contained in those textbooks to bear on adsorptive problems. Toward that end, the bulk of this chapter will consist of workshops on Aspen Adsorption. This approach systematically introduces the reader to all the relevant adsorption concepts by their implementation through the software tools.

In this chapter, we learn about both the theory of adsorptive separations as well as their implementation in Aspen Adsorption.

1.2 Core Concepts in Gas Adsorption

Gas adsorption occurs when gas molecules interact with and become retained by a solid material. The reasons for this adhesion are numerous, and may include charge differences, chemical reactions, and size or shape effects. These interactions frequently differ between gas molecules, and therefore serve as a useful basis for their separation.

Continuous adsorptive separation processes, which manipulate these effects, must always involve at least two steps:

1) The *adsorption* step, during which the adsorbent selectively retains the adsorbed species from the feed.
2) The *regeneration* or *desorption* step, during which the adsorbent releases the retained species, thus regenerating the adsorbent for use in the next cycle.

As there are only a limited number of methods that accomplish the release in the regeneration phase, these methods characterize gas adsorption technologies. When regeneration is accomplished by changing the pressure, the process is called PSA. The alternative, which is accomplishing the regeneration through temperature change, is called TSA. Of course, these approaches can be combined, and there are some subtypes of each; for example, when a PSA process involves a vacuum, it is referred to as a vacuum pressure swing adsorption, or a VPSA, process.

1.2.1 The Adsorption Process

Adsorption is a multistep process by which a molecule moves from the bulk liquid phase and becomes attached to the surface of a solid particle. For our purpose, we treat this process as involving three distinct regions, as shown in Figure 1.1. Specifically, the adsorption involves:

1) the diffusion of a component from the bulk fluid into the pores of a solid particle;
2) the binding of that component to the solid surface inside a pore.

The driving force for the migration of the chemical species into the pore phase is the difference in concentration. This is a very intuitive concept. If the pore phase is poor in a component that the bulk phase is rich in, then that component will be driven to diffuse into the pores.

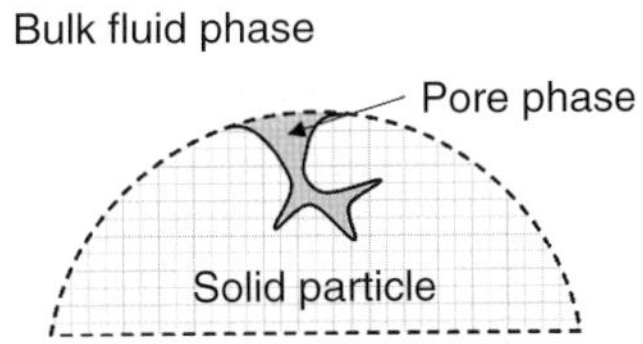

Figure 1.1 The phases involved in the adsorption process.

More detailed models of this process distinguish between the bulk phase and the laminar boundary layer surrounding the solid particle, and may also attempt to characterize the concentration gradients that form inside the pores. We could also attempt to describe the motion of the adsorbed components once they have been adsorbed onto the solid surface. That much detail, however, is not usually needed for the purpose of modeling the adsorptive unit operations we are interested in.

1.2.2 How the Driving Forces Achieve Separation

Figure 1.2 shows schematically how changing the pressure allows us to accomplish the regeneration, and therefore the separation. In this figure, the solid adsorbent is able to hold more of the retained component at high pressure. Therefore, when a bed containing this adsorbent is pressurized with the feed gas, the solid will be "loaded" with the adsorbed component along the line labeled "Uptake." Once the pressure is allowed to drop, the adsorbed molecules will be released along the curve labeled "Release." The curve labeled "strongly adsorbed gas isotherm" describes the maximum capacity of the solid for each pressure. The name isotherm comes from the fact that this entire cycle occurs at roughly constant temperature.

The "working capacity" of the adsorbent loading is defined as the difference between the isotherm point at P_{low} and the isotherm point at P_{High} in the figure. Since the y axis is loading, given by (mass adsorbed component)/(mass adsorbent), the difference represents the maximum mass of the adsorbed gas that can be recovered, per cycle, per mass of adsorbent.

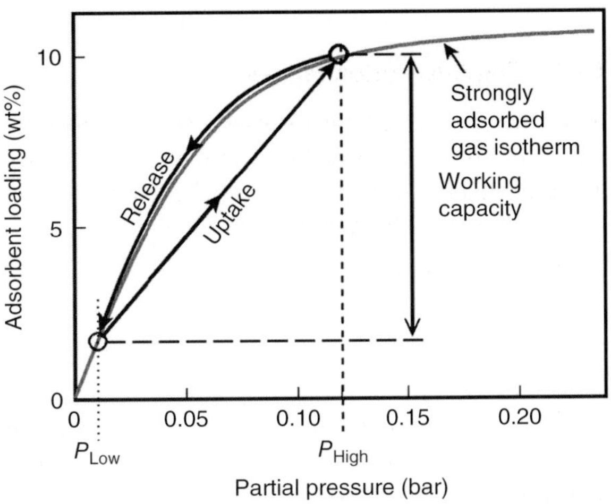

Figure 1.2 The basic schematic isotherms in a PSA operation for an adsorption process.

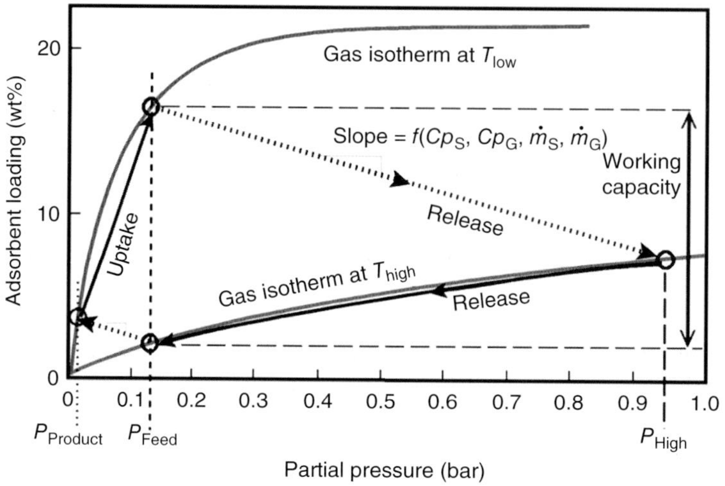

Figure 1.3 The basic schematic isotherms in a TSA operation for an adsorption process.

Temperature swing technology works in much the same way. Instead of simply alternating between high and low pressures, the bed also alternates between high and low temperatures. A higher temperature stream typically consisting of steam typically drives the temperature rise, although other methods such as microwave radiation are sometimes used. Figure 1.3 shows the analogous cycle for a process that cycles temperature and pressure. Notice the difference between the two gas isotherms; they represent isotherms at different temperatures. At higher temperatures, the solid phase typically has lower capacity, which is analogous to gas–liquid systems.

In Figure 1.3, we see that the cycle for pressure and temperature swing appears much more complex than that for PSA. The important distinction to be made is that the heating and cooling steps also change the pressure (dashed lines). The slope of these lines is a function of the heat capacities of the adsorbent and gas phases (Cp_s, Cp_g), along with the masses of the solid and adsorbent (m_s, m_g). Once again, we define a working capacity as the difference between the highest and lowest loading in the cycle.

PSA is, overall, more common than TSA. The significant reason for this difference is that the pressure swing can be accomplished much more quickly than the temperature swing. TSA is beneficial when the adsorbed component is very strongly adsorbed. In such a situation, the increase in temperature allows a much more effective regeneration step, since the pressure swing alone would be unable to release sufficient quantities of the adsorbed components.

1.3 Isotherms

In Section 1.2.2, we discussed how isotherms define a working capacity for an adsorption system. Here, we look at some specific isotherms and discuss what the different isotherms assume about the physical properties of the system.

Isotherms generally have a form as shown in Eq. (1.1). The total amount of a gaseous species (the adsorbate) adsorbed by the solid adsorbent is given as a function of the partial pressures of all the species in the gaseous phase (the nomenclature for this chapter appears in Section 1.13).

$$\frac{\text{mass adsorbate}}{\text{mass adsorbent}} = f(p_1, p_2, \dots, p_n). \tag{1.1}$$

A large number of equations can provide this relationship between the fluid-phase concentration and the adsorption capacity on the solid surface. Many of these relationships are essentially empirical; the equations were originally derived as convenient best-fits of experimental data. Some, most notably the Langmuir isotherm, are backed by physical models for describing how molecules transition from a solvated state to an adsorbed state. Not surprisingly, the model used to derive the Langmuir isotherm is called the Langmuir Adsorption Model, and serves as the basis for several isotherm variants.

1.3.1 The Langmuir Isotherm [8] (1918)

The Langmuir isotherm, Eq. (1.4), is the most popular isotherm in simulated moving bed applications. Although it was originally derived for ideal gasses (and, in fact, has the ideal gas behavior as an assumption), it turns out that it works well with liquid adsorption. There are a handful of Langmuir variations that are also used; for example, the "Dual Site" or "Bi-" Langmuir isotherms consist of two basic Langmuir equations summed together, see Eq. (1.3) (the nomenclature for this chapter appears in Section 1.12).

$$q_i = \frac{a_i c_i}{1 + \sum_{j=1}^{NC} b_j c_j} \tag{1.2}$$

$$q_i = \frac{a_i c_i}{1 + \sum_{j=1}^{NC} b_j c_j} + \frac{\alpha_i c_i}{1 + \sum_{j=1}^{NC} \beta_j c_j} \tag{1.3}$$

1.3.2 The Linear Isotherm

The linear isotherm is essentially a Henry's law model. It is considered useful in cases of low concentrations and loadings.

$$q_i = H_i c_i \tag{1.4}$$

1.3.3 The Brunauer–Emmett–Teller (BET) Isotherm [9] (1938)

This isotherm extends the Langmuir model to work in cases where adsorbed molecules can form layers on the surface of the adsorbent [2]. Relatively recent work finds that the appropriate form of the Brunauer–Emmett–Teller (BET) equation for use in liquid systems is given by

$$q_i = \frac{q_m K_i^S c_i^{eq}}{(1 - K_L c_i^{eq})(1 - K_L c_i^{eq} + K_i^S c_i^{eq})} \tag{1.5}$$

1.3.4 The Freundlich Isotherm [10] (1906)

This empirical equation was originally used on gasses, Eq. (1.6). It has applications in protein and other bioseparations. The model is almost entirely empirical.

$$q_i = h_i c_i^{1/n} \tag{1.6}$$

1.3.5 The Sips (Langmuir–Freundlich) Isotherm [11] (1948)

This model is an attempt to put the Freundlich isotherm into a more physically meaningful form. At low concentrations, it mimics the Freundlich behavior, while at high concentrations, it predicts a monolayer that is similar to the Langmuir isotherm.

$$q_i = \frac{a_i c_i^n}{1 + \sum_{j=1}^{\mathrm{NC}} b_j c_j^n} \tag{1.7}$$

1.3.6 The Toth Isotherm [12] (1971)

This empirical isotherm is primarily used in gas-phase adsorption, and is similar in form to the Sips equation.

$$q_i = \frac{h_i c_i}{(a_i + c_i)^{1/t}} \tag{1.8}$$

1.3.7 Summary

Figure 1.4 shows what these isotherms look like in the single-component case.

1.4 The Properties of Packed Beds

In Section 1.3 on isotherms, notice that the sample plots in Figure 1.4 show the relationship between fluid phase concentration and *adsorbent volume* rather than adsorbent mass. Of course, we can convert between mass and volume with information about the adsorbent density; but in packed beds, the concept of "density" is less well defined than in the case of gases and liquids.

In order to describe any adsorptive separation, it is important to think clearly about how to characterize the particles and voids within the adsorbent bed. As with most systems involving solid particles, the terms and conventions can be confusing or misleading because many different disciplines have developed their own partially overlapping naming schemes.

In adsorptive separations, the questions that we typically ask ourselves about the solid phase is essentially this: "Given some bed that is filled with both fluid and solid particles, what volume of the bed can be said to belong to the solid phase and what volume to the liquid phase?" It turns out that we can answer that question in several different ways, and which answer is "correct" depends on the details of the system. However, despite the variety of possible answers, the answer always involves reference to two physical properties: density and void fraction.

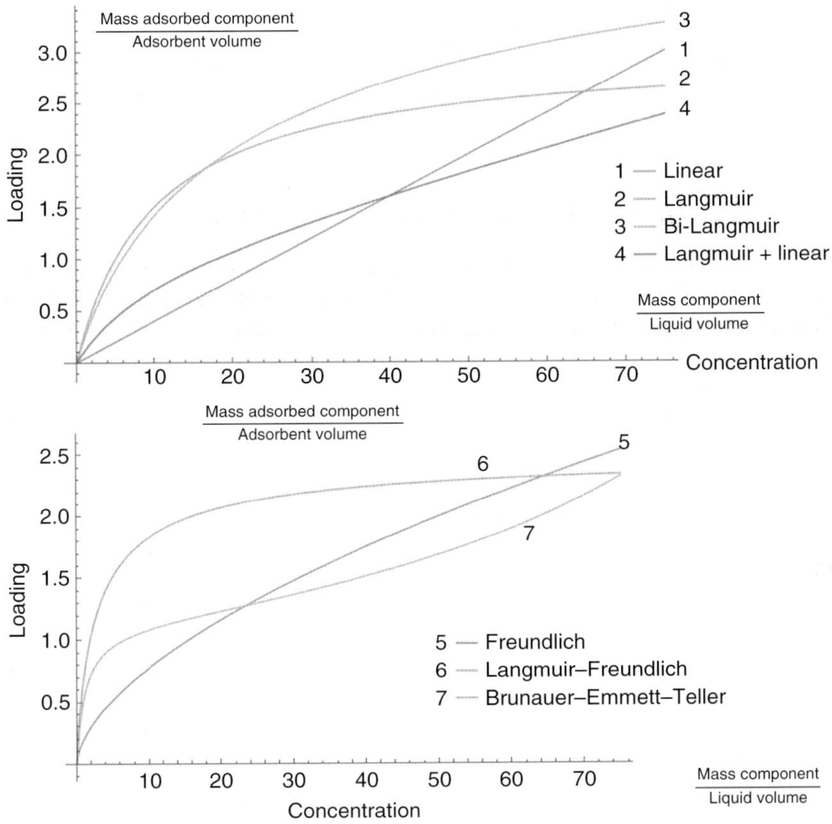

Figure 1.4 Sample isotherm curves.

1.4.1 Void Fractions

The first set of important parameters, which are intimately tied to density, are the void fractions (or porosities) of a packed bed. Figure 1.5 depicts the three voids that are most relevant to adsorptive separations. Of course, the drawings in Figure 1.5 represent the voids themselves, but we want to know about the void fractions, that is the size of the voids relative to some reference volume.

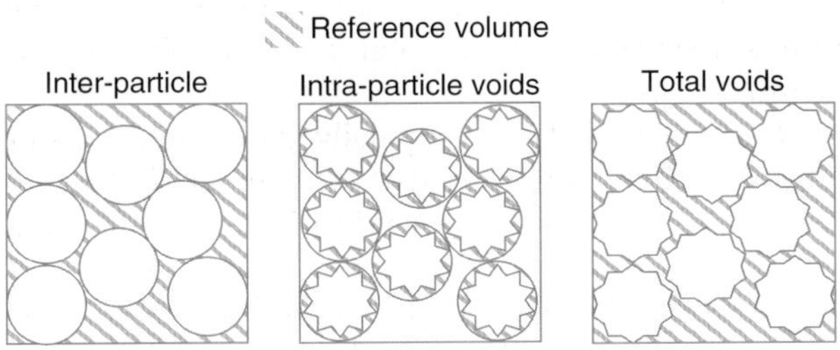

Figure 1.5 The locations of the various reference volumes.

1.4.2 External Voids

The first and most intuitive void fraction is the ratio of the volume of space *not* occupied by particles in the bed, to the total bed volume. We refer to this ratio as the **interparticle, external,** or **interstitial** void fraction. While this seems simple at first, it is somewhat difficult to exactly determine what space is "inside" a particle and what space is "outside." Particles can have very irregular geometry, which can make the boundaries very fuzzy for both theoretical definition and practical measurement.

1.4.3 Internal Voids

The second void fraction follows naturally from the first; since the external porosity defines the volume *outside* the particles, it is natural to want a ratio to describe the voids *inside* a particle. Unfortunately, toward this end, two conventions have arisen. *In chromatographic practice,* it is common to define the **internal** or **intraparticle** void fraction as the ratio of voids in the particles to the total column volume. However, *in chemical engineering circles,* it is more common to use **internal** or **intraparticle** void fraction to refer to the ratio of voids in the particles to the total volume of *particles* instead of the *column*. To see this in mathematical terms, refer Table 1.1.

The final void fraction is simply derived from the sum of both internal and external voids. The **total** void fraction is therefore the ratio of all voids in the system, to the total system volume.

1.4.4 Densities

Density is simply characterized by dividing a mass of particles by a reference volume, and as you might have guessed, there are several meaningful ways to characterize the density of a group of particles. Figure 1.6 illustrates the different ways that can be used to create reference volumes to divide our particle mass.

1.4.4.1 Bulk Density

The simplest and most intuitive is the density of many particles in aggregate. This is the density found by pouring the particles in a beaker and measuring their mass and volume. It is commonly referred to as the **bulk** density.

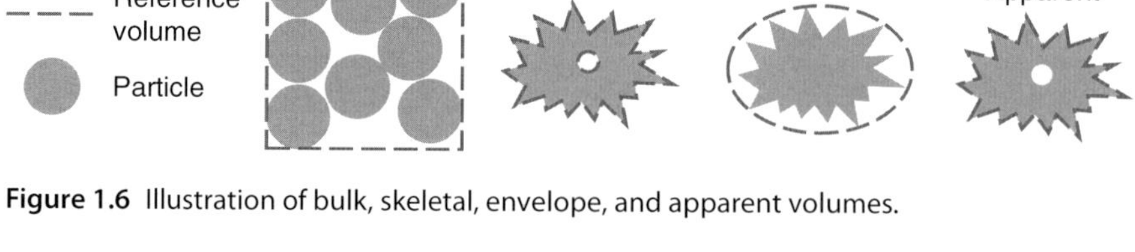

Figure 1.6 Illustration of bulk, skeletal, envelope, and apparent volumes.

Table 1.1 Definitions of different voids, densities, and porosities.

Quantity	Symbol	Units	Example relationships
Total system volume	V_T	Volume	For example, in a cylindrical column $V_T = \pi R^2 L$
External void volume	V_X	Volume	$V_X = \varepsilon_i V_T$
Internal void volume (pore volume)	V_P	Volume	Chromatographic convention: $V_P = \varepsilon_p V_T$ Chemical engineering convention: $V_P = \varepsilon_p (1 - \varepsilon_i) V_T$
Bulk density	ρ_b	$\frac{\text{Mass}}{\text{Volume}}$	$\rho_b = \frac{M_{\text{particles}}}{V_T} = (1 - \varepsilon_i)\rho_e = (1 - \varepsilon_t)\rho_s$
Skeletal or solid density	ρ_s	$\frac{\text{Mass}}{\text{Volume}}$	$\rho_s = \frac{M_{\text{particles}}}{V_T - V_X - V_P} = \frac{\rho_b}{1 - \varepsilon_t}$
Envelope density	ρ_e	$\frac{\text{Mass}}{\text{Volume}}$	$\rho_e = \frac{M_{\text{particles}}}{V_T - V_X} = \frac{\rho_b}{1 - \varepsilon_i}$
External or interparticle porosity	ε_i	$\frac{\text{Volume}}{\text{Volume}}$	$\varepsilon_i = \frac{V_X}{V_T} = 1 - \frac{\rho_b}{\rho_e}$
Internal or intraparticle porosity	ε_p	$\frac{\text{Volume}}{\text{Volume}}$	Chromatographic convention: $\varepsilon_p = \frac{V_P}{V_T} = (1 - \varepsilon_i) - \frac{\rho_b}{\rho_s}$ Chemical engineering convention: $\varepsilon_p = \frac{V_P}{V_T - V_X} = 1 - \frac{\rho_b}{\rho_s}$ Conversion: $\varepsilon_{p,\text{eng}} = \varepsilon_{p,\text{cro}} + \varepsilon_i$
Total porosity	ε_t	$\frac{\text{Volume}}{\text{Volume}}$	$\varepsilon_t = \frac{V_P + V_X}{V_T}$ Chromatographic convention: $\varepsilon_t = \varepsilon_i + \varepsilon_p$ Chemical engineering convention: $\varepsilon_t = \varepsilon_i + \varepsilon_p (1 - \varepsilon_i)$

1.4.4.2 Skeletal or Solid Density

The next relevant density is that of the pure solid. This is the density you would get if there were a perfect cube of a solid (i.e. with no holes or voids) and its mass was divided by the volume of the cube. Since it is impractical to make such a perfect solid, this density is often measured by forcing gas into all the pores of the particles. When the volume of the gas is known, and the volume of the gas-particles mix is known, the volume of the pure solid "skeleton" is simply their difference. The density is therefore the mass of the particles divided by the "skeleton" volume. This density is referred to as the **skeletal** or **solid** density.

The skeletal density is very similar to another density, which includes a particle's internal voids. When the reference volume includes internal voids, the density is referred to as the **apparent** density.

1.4.4.3 Envelope or Particle Density

The final density is less intuitive and falls between the bulk and skeletal densities. This density is found by defining a hypothetical "envelope" around each particle; the envelope contains both the solid mass as well as the particle's internal and external pores. The density is then found by dividing the mass of the particles by the volume of their envelopes. However, arriving this density value is difficult thanks to the difficulty in selecting and measuring an appropriate envelope; what constitutes appropriate frequently depends on the application. There are some techniques, such as mercury porosimetry, which allow measurement of envelope volumes. The density thus obtained is usually referred to as the **envelope** or **particle** density.

1.4.4.4 Caveats

Some authors use **true** density to mean **skeletal** density, while others use it to refer to **apparent** density. Some authors use **apparent** density to refer to **bulk** density, while others use it to refer to **envelope** density. When gathering data from the literature, vendors, or even your own experiments, you should be careful to ensure that you know exactly what your density is referring to.

1.4.5 Relationships

As might be expected, all these parameters are related to each other. We summarize their relationships in Table 1.1. This text will exclusively use *the chemical engineering convention,* but the chromatographic convention is presented here for completeness. The source of the difference between the engineering and chromatographic conventions is the definition of ε_p, where the engineering convention defines the fraction in terms of **the solid volume**, while the chromatographic convention defines it in terms of **the total system volume**.

1.4.6 Gas-phase Behavior

The behavior of gasses in packed beds is the final piece of the puzzle we need before we can begin to discuss actual absorber configurations, and construction of Aspen models. As a gas flows through a packed bed, its properties change, even in the absence of adsorption.

1.4.6.1 Pressure Drop

Gas flowing through a packed bed suffers a decrease in pressure. In an empty tube, the no-slip boundary condition at the walls represents the only frictional force; but in a packed bed, each particle represents a source of additional friction. Just like flow through a pipe, this frictional force depends on whether or not the flow is laminar, as well as on some properties of the gas.

The simplest packed-bed pressure drop equation is the Carman–Kozeny equation (1.9), which dates back to 1927 [13]. This equation calculates pressure drop by assuming that the space between the particles behaves like many small pipes governed by Poiseuille's law [14]. As such, this equation is only valid for laminar flow.

$$\frac{\partial P}{\partial z} = -1.5^{*}10^{-3}\frac{\mu(1-\varepsilon_{\mathrm{i}})^2}{(2r_{\mathrm{p}}\psi)^2\varepsilon_{\mathrm{i}}^3}v_{\mathrm{g}} \tag{1.9}$$

Here, ψ is a *shape factor* that describes how spherical the particles are, and v_g is the interstitial velocity of the gas phase. Refer to Section 1.13 for nomenclature used throughout this chapter.

The Burke–Plummer equation (1.10) describes the pressure drop in the case of turbulent flow (Re > 2000) [15]. MW in this equation refers to the molecular weight of the gas.

$$\frac{\partial P}{\partial z} = -1.75^{*}10^{-3}\frac{\rho_{\mathrm{g}}\,\mathrm{MW}\,(1-\varepsilon_{\mathrm{i}})}{2r_{\mathrm{p}}\psi\varepsilon_{\mathrm{i}}^3}v_{\mathrm{g}}^2 \tag{1.10}$$

The most commonly used equation, the Ergun equation [16], is able to handle both turbulent and laminar flows. It does this by combining the Carman–Kozeny and the Burke–Plummer equations.

$$\frac{\partial P}{\partial z} = -\left(1.5^{*}10^{-3}\frac{\mu(1-\varepsilon_{\mathrm{i}})^2}{(2r_{\mathrm{p}}\psi)^2\varepsilon_{\mathrm{i}}^3}v_{\mathrm{g}} + 1.75^{*}10^{-3}\frac{\rho_{\mathrm{g}}\mathrm{MW}\,(1-\varepsilon_{\mathrm{i}})}{2r_{\mathrm{p}}\psi\varepsilon_{\mathrm{i}}^3}v_{\mathrm{g}}^2\right) \tag{1.11}$$

1.4.6.2 Compressibility

As a gas flows through the pressure gradients in a packed bed, it changes volume. This volume change affects the gas velocity and temperature, which in turn influences the pressure drop. Therefore, it is important to be able to select the correct thermodynamic model of the gas phase, so that our calculations generate correct volumetric flow rates. It is also important for nonisothermal systems.

There are a large number of options when it comes to selecting these thermodynamic models. The simplest and most idealized is simply to use the ideal gas law. While this works reasonably well for low-pressure systems (less than a few bar), higher pressure systems require better models.

One such model is the **Redlich–Kwong–Soave (RK–Soave)** [17] equation of state. It is a cubic equation of state that works well for most vapor systems up to around 10 bar. It was designed for use with hydrocarbons, and works well.

A second detailed model is the **Peng–Robinson** [17] equation of state, which works well with moderately nonideal gas phases over a slightly wider range of pressures than RK–Soave. It works very well with gas phases containing CO, CO_2, H_2 and O_2, and is commonly used in refinery processes.

1.5 PSA and TSA Implementation Details

1.5.1 Common Adsorbent Characteristics

The most common adsorbent for PSA systems are zeolites, because they are selective for a number of industrially important molecules. Table 1.2 [18] shows several common zeolite types, along with a molecular diameter that describes the kinds of molecules such an adsorbent would be able to separate.

Another common feature of PSA systems is the use of two adsorbent beds. While it is possible to operate a PSA separation with one bed, some separations use six or more. The two-bed configuration allows for important energy-saving schemes that easily outperform the one-bed process, while the additional complexity of more beds is typically only worthwhile for larger-scale separations. We discuss the energy-saving schemes in the first workshop in Section 1.7.

1.5.2 Common Process Configurations

The classic two-bed PSA process was patented in the 1960s by Charles Skarstrom [5], and has come to be known as *the Skarstrom cycle*. The basic Skarstrom cycle utilizes two packed adsorbent beds, as shown schematically in Figure 1.7. The "core" cycle consists of four steps: ***pressurization, adsorption, countercurrent blowdown, and countercurrent purge***. It is possible to add an additional **pressure equilibration** step that allows to save energy. In step 1, feed flows into bed 1, and raises its pressure, while the fully loaded second bed is connected to the extract product. Once the desired pressure has been reached, the system enters step 2, where the more strongly adsorbed component is retained in bed 2 and the gas outlet stream is enriched in the less strongly adsorbed component. A portion of the raffinate product is diverted to the second bed, to help fully regenerate it. Once the first bed has been loaded, and the second bed regenerated, a pressure equalization step may take place. This allows for energy saving by reducing the quantity of gas that needs to be compressed. After the equilibration step, the process repeats, but with the roles of the two beds reversed.

Figure 1.7 shows a common PSA scheme. This is only one possible configuration; Table 1.3 lists several additional options for each step, as well as the rationale behind them.

Table 1.2 Zeolites and their adsorptive properties.

Molecular sieve zeolite	Effective molecular diameter cut-off (molecules smaller than this will be adsorbed; larger molecules will be rejected) (Å)
3A	3
4A	4
5A	5
10X	8
13X	10

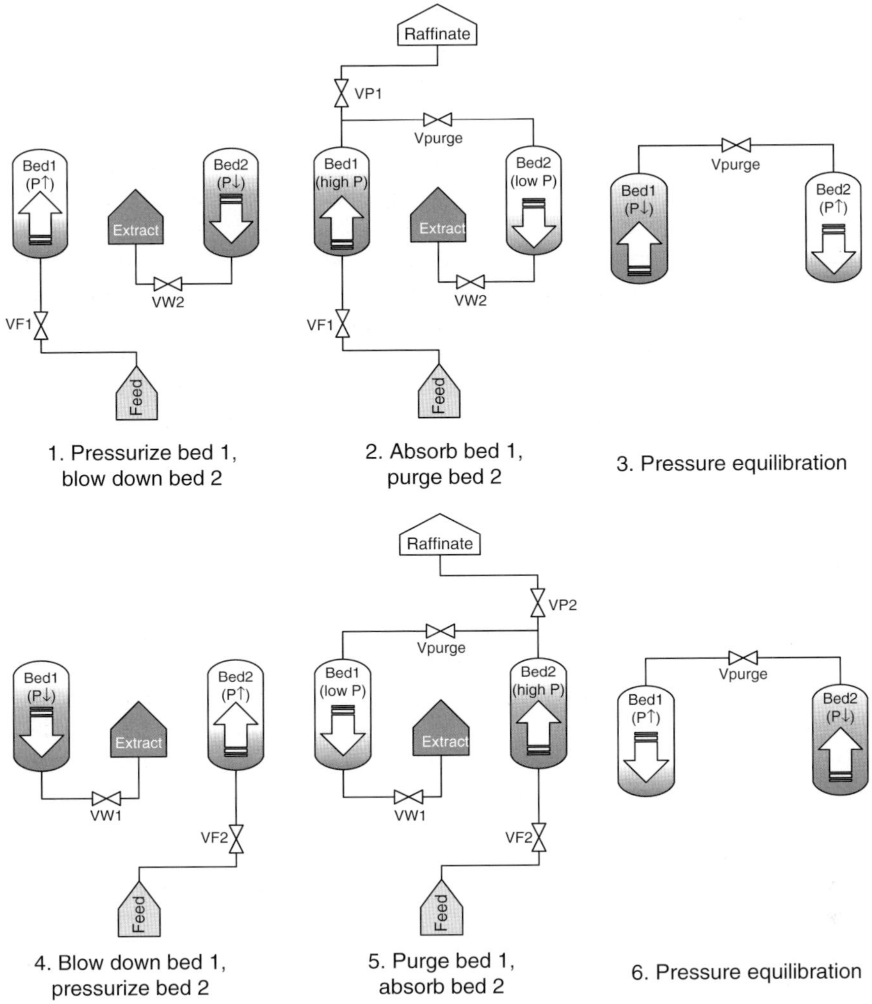

Figure 1.7 The basic principle of the Skarstrom cycle in the PSA process.

1.6 Introduction to Aspen Adsorption

Given the above introduction to the basics of TSA/PSA systems, we learn the details of their design through a hands-on workshop using Aspen Technology's software, Aspen Adsorption.

Aspen Adsorption is a comprehensive flowsheet simulator for the optimal analysis, design, simulation, and optimization of industrial gas- and liquid-adsorption processes. This software simulates gas processes with adsorption only, or reactive gas adsorption processes where both reaction and adsorption occur simultaneously.

For gas processes, Aspen is capable of handing the following conditions "out of the box":

- Isothermal or nonisothermal operation.

Table 1.3 Common steps used in PSA.

Step name	Description	Motivation
Pressurization	In this step, the bed is pressurized, typically with material from the feed stream	This step forces any enriched gas still present in the bed toward the product stream, while raising the bed pressure in preparation for the adsorption step
Adsorption	This step is operated at high pressure; the gas and solid interact to extract the strongly adsorbed component. The raffinate is collected during this step, and its withdrawal may or may not lower the bed pressure.	Allow maximum extract adsorption and recovery of the less-adsorbed component
Blowdown	One end of the bed is connected to a low-pressure sink. If the blowdown is counter-current, the raffinate will have higher purity. Co-current blowdown is sometimes used to improve extract purity (see also: equilibration)	Purge the bed of the remaining raffinate gas and prepare the bed for low pressure desorption
Desorption	The solid phase is allowed to release the adsorbed component, and the gas is collected as the extract. May be done as in vacuum, and with or without a purge	Regenerate the solid for the next Adsorption cycle. The use of a purge may involve a trade-off between the extract purity and raffinate recovery
Equilibration/ equalization	The high pressure bed is connected to the low pressure bed, typically at the raffinate end	Lower energy consumption (less pressurization required)

- Plug flow or plug flow with axial dispersion.
- The system may or may not be well-mixed in the radial direction.
- The gas phase is ideal or nonideal.
- Gas-phase pressure may be constant, or the pressure may vary according to momentum balances.
- Mass transfer is described using a lumped overall resistance, or by a model that accounts separately for micro- and macropore effects. The driving force is based on a liquid or solid film, and is either linear, quadratic, or user-specified. Mass-transfer coefficients are either constant, or vary with local conditions.
- Adsorption isotherms are available for single or competitive multicomponent adsorption.
- Reactions may occur at any point.

We cover the details of these conditions as we encounter them in our workshops.

1.7 PSA Workshop: Aspen Adsorption Modeling for Air Separation

Having briefly introduced the concept of PSA for gas-phase separation, we now cover the details required to specify a complete PSA operation, specifically the use of PSA for air separation. Since we have already established that the Skarstrom process is considered to be the classic PSA implementation, and because the original Skarstrom patent dealt with oxygen isolation [5], this is the process we now model in Aspen Adsorption.

The Skarstrom process for oxygen purification has been commercialized at scales ranging from a few liters per minutes for small-scale units to tens of tons per day for industrial systems. The processes originally used a zeolite adsorbent, generally 5A or 13X, while modern processes employ a LiLSX adsorbent [19,20]. Most small-scale units use a two-bed system, sometimes with the addition of pressure equalization steps. For this model, we assume a binary air separation system of 79% N_2 and 21% O_2 using a 5A zeolite as the adsorbent.

In this PSA system, the zeolite preferentially adsorbs O_2 at higher pressure in the adsorption step, and releases O_2 when lthe pressure during the desorption step is lowered. We can find the information required to describe this adsorption in the literature [2,5,19,21–23].

Figure 1.8 shows the cyclic PSA unit we build with two beds for a binary separation. The process is inherently cyclic, because the beds will alternate

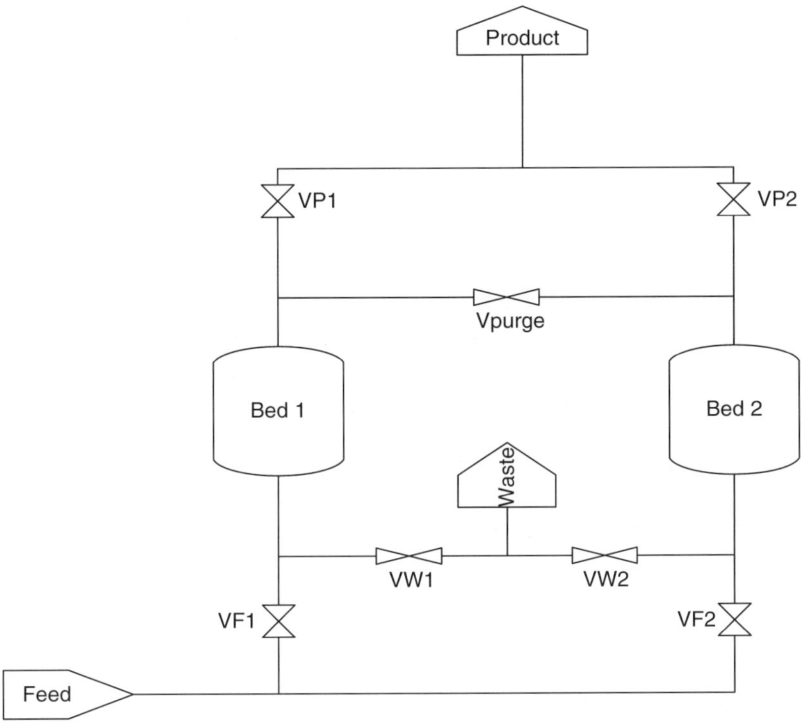

Figure 1.8 Diagram of a basic two-bed pressure swing adsorption separator.

between adsorption and desorption. While bed 1 is in the adsorption stage, bed 2 is in the desorption stage. During the adsorption stage, at a relatively high pressure, the solid adsorbent in the bed will selectively adsorb the oxygen from the binary gas mixture. The leftover nitrogen produced in this step is referred to as the **raffinate**, since it has been refined by having the oxygen removed. Since this product is less strongly adsorbed, and is removed from the top of the column, it is sometimes referred to as **the light product (LP)**. In the following desorption step, by lowering the gas-phase partial pressure inside the bed, the adsorbent releases the adsorbed components, in this case oxygen, to produce a gas stream called the **extract** (because it was *extracted* from the main gas stream.) Some sources refer to this as **the heavy product (HP)** because it is enriched in the more strongly adsorbed components. From here, the cycle repeats.

To accomplish this kind of switching, we need to carefully schedule when to open and close the valves. To do this, we construct a diagram like the one shown in Figure 1.8, which shows the four steps required to cycle between adsorption and desorption. The steps for bed 1 are specifically as follows:

1) Under high pressure, the feed gas is fed into the bed, and adsorption takes place.
2) Once the bed is adequately pressurized, we begin collecting the least adsorbed component in the raffinate, while the bed is being loaded. We take a portion of the raffinate and feed it into the second bed to help drive out the more strongly adsorbed component.
3) When the bed is fully loaded, we depressurize it in the "blow down" step. The bed is connected.
4) Once the bed has been completely cleared of the HP, the LP is used to repressurize the bed for the feed stream.

If we look carefully at the schedule, as well as the valve layout described in Figure 1.8, we can easily come up with the schedule for valve opening and closing listed in Table 1.4. Table 1.5 explains the valve specification numbers 0, 1, 2, and 3 appearing in Table 1.4.

Now that we have an idea of the layout of our system, as well as the system behavior, let us begin creating the flowsheet in Aspen Adsorption.

Table 1.4 Valve schedule for a two-bed four-step process.

Step	Function	VF1	VF2	VW1	VW2	VP1	VP2	VPurge
Step 1	Adsorb bed 1, purge bed 2	1	0	0	1	1	0	2
Step 2	Blowdown bed 1, pressurize bed 2	0	1	1	0	0	1	0
Step 3	Purge bed 1, adsorb bed 2	0	1	1	0	0	1	2
Step 4	Pressurize bed 1, blowdown bed 2	1	0	0	1	1	0	0

Table 1.5 Valve specification meanings.

0	The valve is fully closed (the flow rate through the valve always be zero)
1	The valve is fully open (the flow rate through the valve will be determined by mass balance)
2	The valve's flow rate will have a linear relationship with pressure drop
3	The valve will have a fixed flow rate

1.7.1 Adding Components to an Aspen Adsorption Simulation

Step 1: If you already have Aspen Adsorption open, select "new" from the file menu to create a new blank simulation (Figure 1.9)

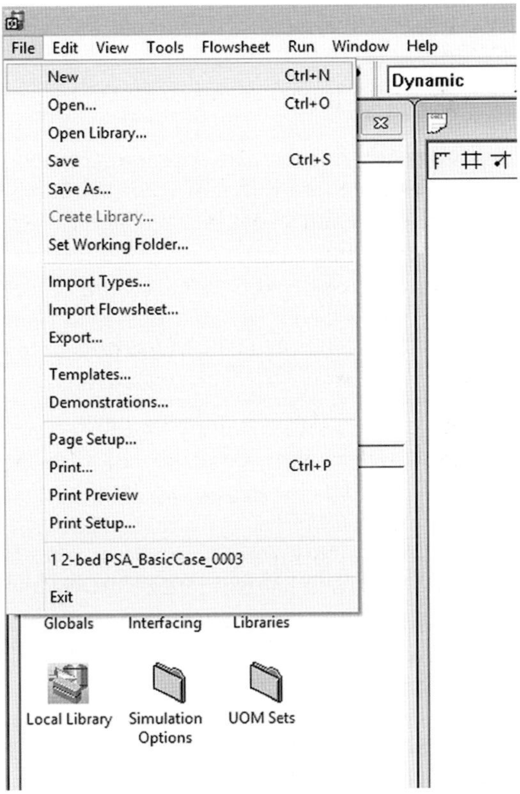

Figure 1.9 A new file in Aspen Adsorption.

Step 2: It is important to save Aspen simulations frequently

Saving guards against two undesirable outcomes:

- Computer crashes
- Irreversible changes to the Aspen flowsheet.

The numeric solvers Aspen Adsorption uses frequently store temporary, initial, and intermediate values while you are creating and running simulations. When the values the solvers have stored correspond well to your simulation, it can make your simulation run faster. However, making certain changes to the flowsheet can cause these values to become unhelpful, and your simulation might not run at all; in some cases, the simulation may continue to fail even after you ostensibly revert the simulation to a state that previously worked. In such a scenario, the best thing to do is to open a saved file from before you made the changes.

Save the simulation as **Workshop1A.ada** (Figure 1.10)

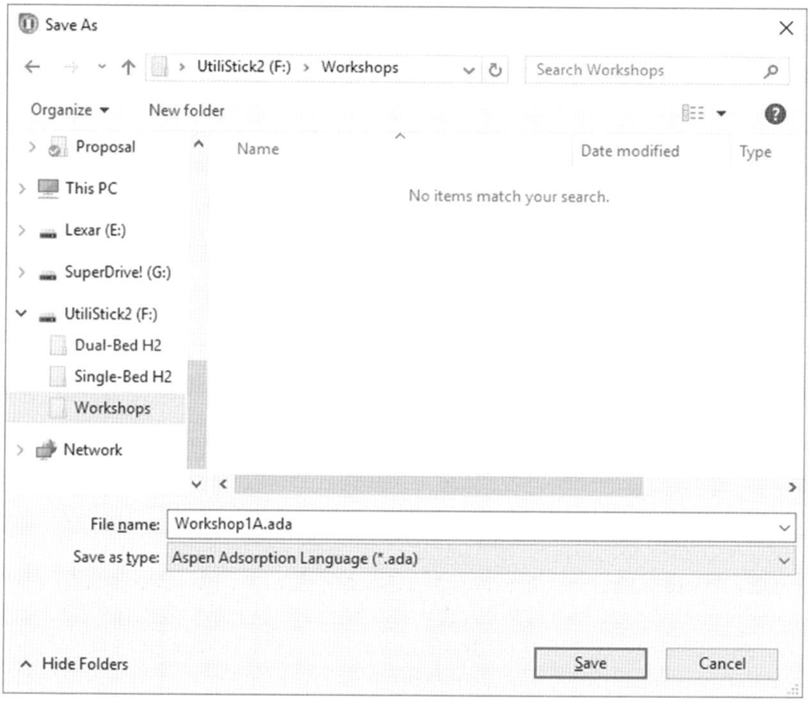

Figure 1.10 Saving the PSA process with four steps for air separation.

Step 3: Add chemical components to Aspen's component list

There are two methods of adding chemical components into the model: component list and component set.

A Component List is the simplest way to add components. All that is required is a list of chemical names. However, a component list contains no information about the physical properties of these chemical components, and so all required properties must be provided by the user.

On the other hand, a Component Set allows for much more detailed simulations because each component is associated with a list of optional parameters and their values. These values provide the information required for the Aspen property system to calculate the component's physical properties automatically, without any additional input from the user. When we have access to a nice collection of physical property data (as we do in Aspen Properties) using a component set is the natural choice.

When we use Aspen Properties for physical properties, we can invoke the Aspen Properties user interface directly if we have it installed. In order to edit physical property options, we need to follow the steps in Figure 1.11:

1) In the Simulation Explorer, click "**Component Lists**".
2) In the **Explorer of the Component Lists folder**, double-click the "**Configure Properties" node**.
3) In the **Physical Properties Configuration** dialog box, choose "Use Aspen property system" option, and then click "**Edit using Aspen Properties" button to launch the user interface**.

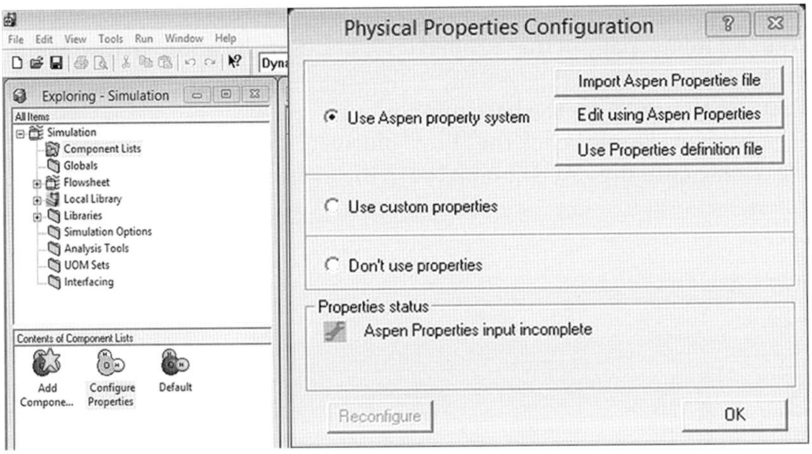

Figure 1.11 The physical properties configuration pane.

Step 4: Add component names in Aspen Properties

Once "**Edit using Aspen Properties**" in the **Physical Properties Configuration** pane is clicked, it automatically opens the Aspen Properties for us to add component names and edit their physical and chemical properties. Figure 1.12 displays that we are trying to separate nitrogen from the nitrogen and oxygen mixture.

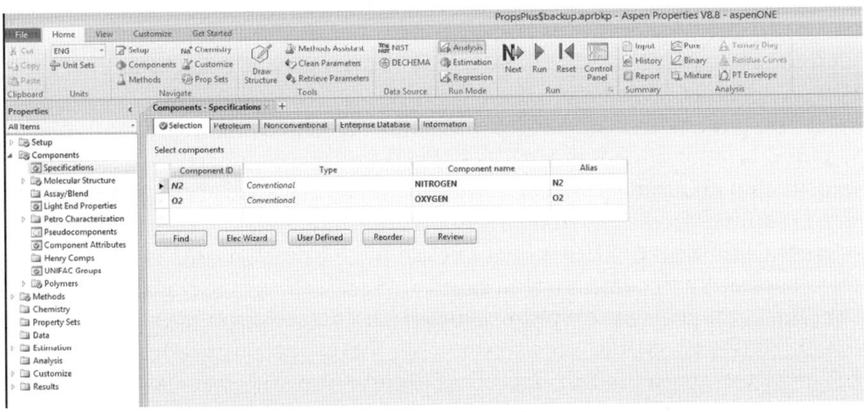

Figure 1.12 The Aspen Properties interface.

Step 5: Select the Appropriate Property Method

In this step, we specify the methods that calculate the thermodynamic properties for our simulation. The "Base-method" and "Method name" drop-down menus in Figure 1.13 are essentially redundant, although the "Base method" option allows you to filter the long list of methods by application. For example, PENG-ROB shows up when the method-filter is set to COMMON; however, specialized variants of the Peng–Robinson method (such as PR-BM, the Peng–Robinson model with Boston–Mathias alpha-function) do not show up unless the filter is set to "GasProc"

To edit the components' physical properties in Aspen Properties, we follow the steps in Figure 1.13:

1) In the "All Items" pane of the "Properties", click "Specification" in the "Methods" folder.
2) Select "PENG-ROB" as the base method.
3) In the pane of "Home", click the "Run" button (Figure 1.14).
4) Save the file until the run is completed in the Control Panel. See Figure 1.14.

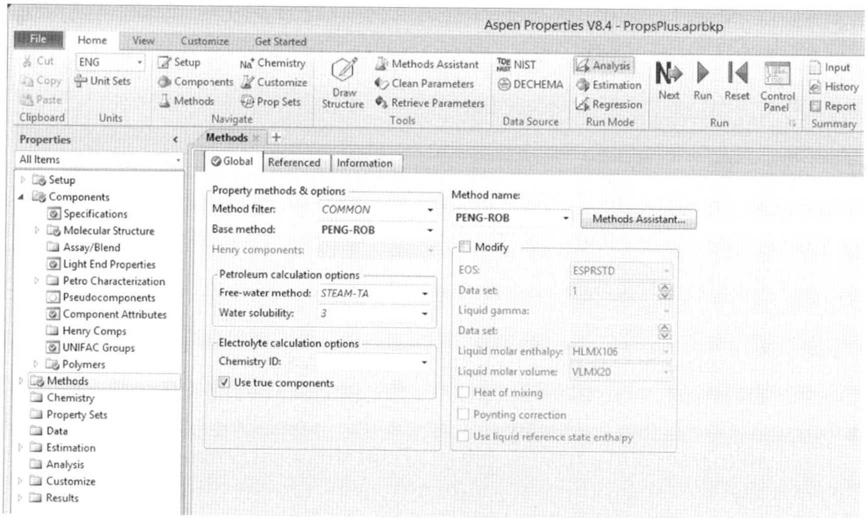

Figure 1.13 The specification of the components in Aspen Properties.

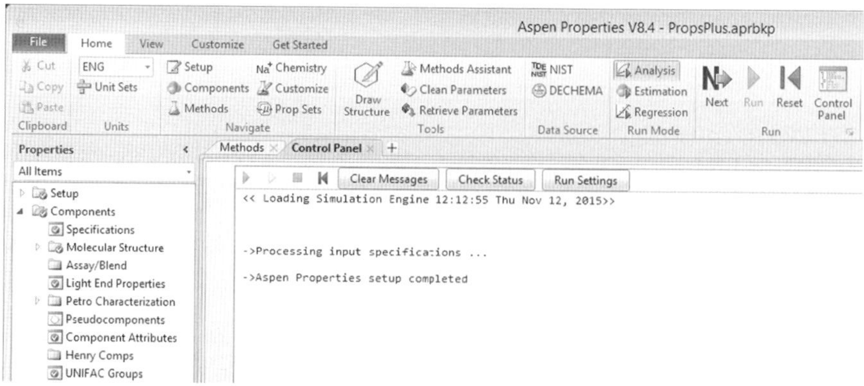

Figure 1.14 The completion in control panel in Aspen Properties.

Step 6: Tell Aspen Adsorption to use the Aspen Properties file that is created

Once the Aspen Property file completes the run, you may close Aspen Properties. Make sure that you choose "yes" to save the document as **PropsPlus.aprbkp** while closing! (Figure 1.15)

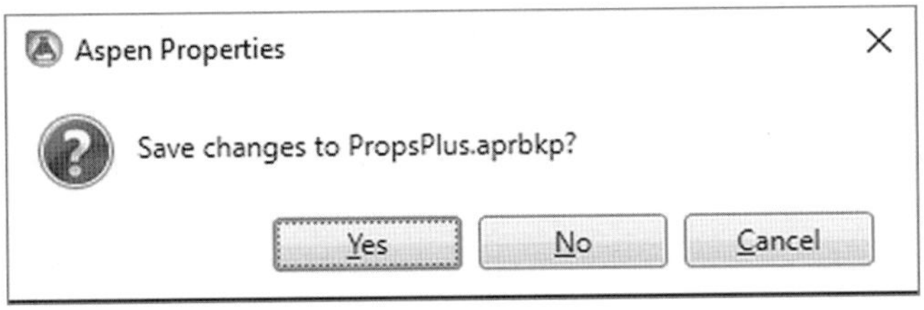

Figure 1.15 Ensure the properties document is saved.

The saved properties document will be included in our Aspen flowsheet automatically. We can tell that the properties are successfully configured by looking for a green square next to "**Configured using Embedded Aspen Properties**" in the properties configuration form (Figure 1.16).

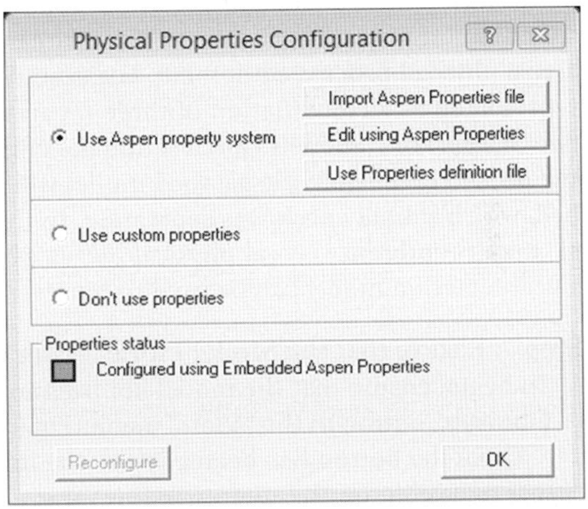

Figure 1.16 The completion of the physical properties configuration.

Once we press "OK" on this form, we must add the components we choose in Aspen Properties to our flowsheet. To do this, double-click on "Default" in the Explorer to open up the component list builder. Move components N2 and O2 from Available Components on the left to Components on the right (Figure 1.17).

This distinction between the components available in Aspen Properties and those available in the flowsheet is useful in situations where we are building a chemically complex model, but want to start with only a few components.

Once you have added the components, press **ctrl+S** to save the Workshop1A file with the same name.

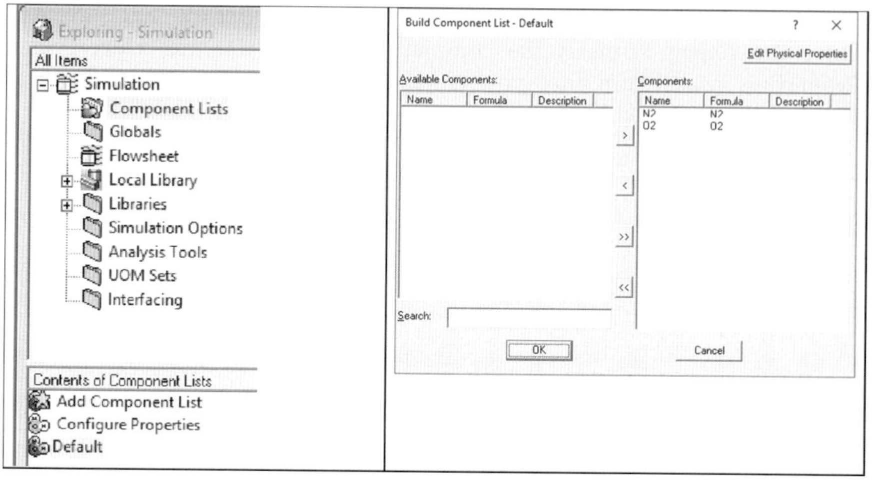

Figure 1.17 Adding the defined components to the flowsheet.

1.7.2 Creating a Flowsheet in Aspen Adsorption

Now that we have successfully added several chemical components to Aspen, it is time to begin specifying the layout of our process. Aspen provides a large number of "Models" that are contained in "Model Libraries." These models roughly correspond to common unit operations, but take care not to confuse these models with actual physical process equipment. For example, a physical packed bed might be represented as a combination of three separate "Models" in Aspen. Specifically, two "empty void" models can describe the empty voids at the top and bottom of the packed bed, while the gas-bed model will describe the region of the packed bed, which actually performs separation. In the same way, a single Aspen "Model" might represent several physical pieces of equipment. For example, a single gas_interaction model can be used to represent several identical packed beds.

Step 7: Ensure that the Model Libraries are visible

To begin, ensure that the model libraries are visible by clicking on the "Model Libraries" button in the "View" menu (Figure 1.18).

Once the button has been clicked, the following area shown in Figure 1.19 will be visible on the main window. This area displays all the prebuilt models available in Aspen Adsorption, and are colloquially referred to as "blocks." Each model or block roughly corresponds to a unit operation, such as a packed bed, tank, or valve. However, some models have subtler roles, which we cover as we get to them.

Step 8: Understand an overview of the flowsheet and of implementing the forward-reverse flow reversibility through the pressure setters and flow setters in Aspen Adsorption

Let us first take a look at the final flowsheet of Figure 1.20 that we are to draw and understand the concepts of pressure setters and reversible flow setters within the Aspen Adsorption's dynamic simulation environment.

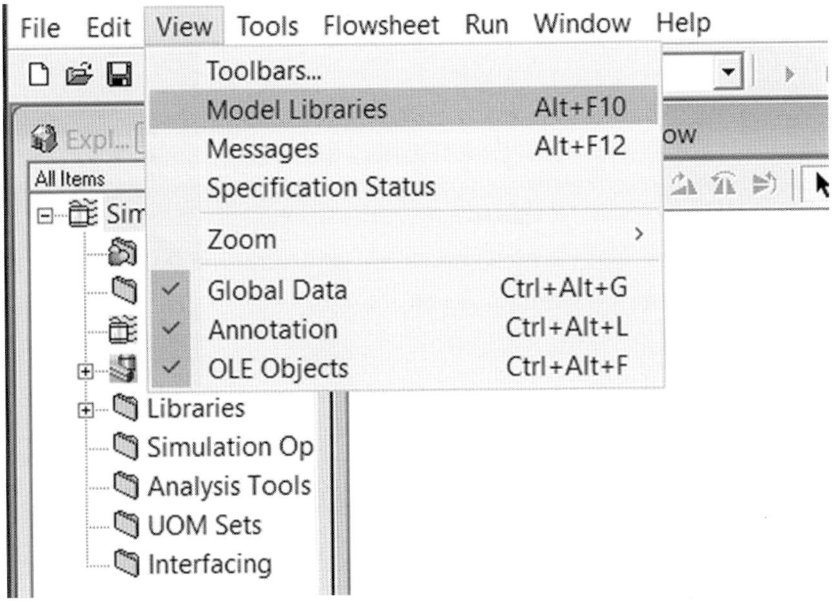

Figure 1.18 Display of the model libraries.

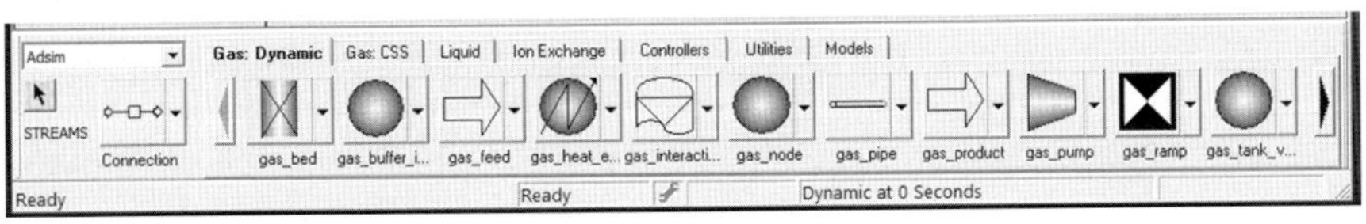

Figure 1.19 The model libraries.

Within the flowsheet, we have the following feed and product "streams":

1) Feed F1 (an air stream; a mixture of N2 and O2)
2) Product P1 (a high-purity O2 product)
3) Waste W1 (a waste mixture of a high-purity N2 and a low-purity O2).

Each stream is defined by $(C + 2)$ specifications, where C is the number of components, and the remaining two specifications are typically temperature and pressure.

The flowsheet also includes the following "blocks," which are models for unit operations, such as mixer, splitter, valve, and adsorption column:

1) Gas_valve model, which serves as a control valve for closing or opening the valve, and as **a "reversible" flow setter** that sets the flow rate and **the forward-reverse direction of flow** through the valve: feed valves VF, VF1, and VF2; waste valves VW1 and VW2; purge valve VPurge; and product valves VP, VP1, and VP2.

   In Aspen Adsorption, valves operate under the assumption that the molar flow rate is linearly related to pressure drop across the valve, as shown in Eq. (1.12):

$$F = \mathrm{Cv}\Delta P \tag{1.12}$$

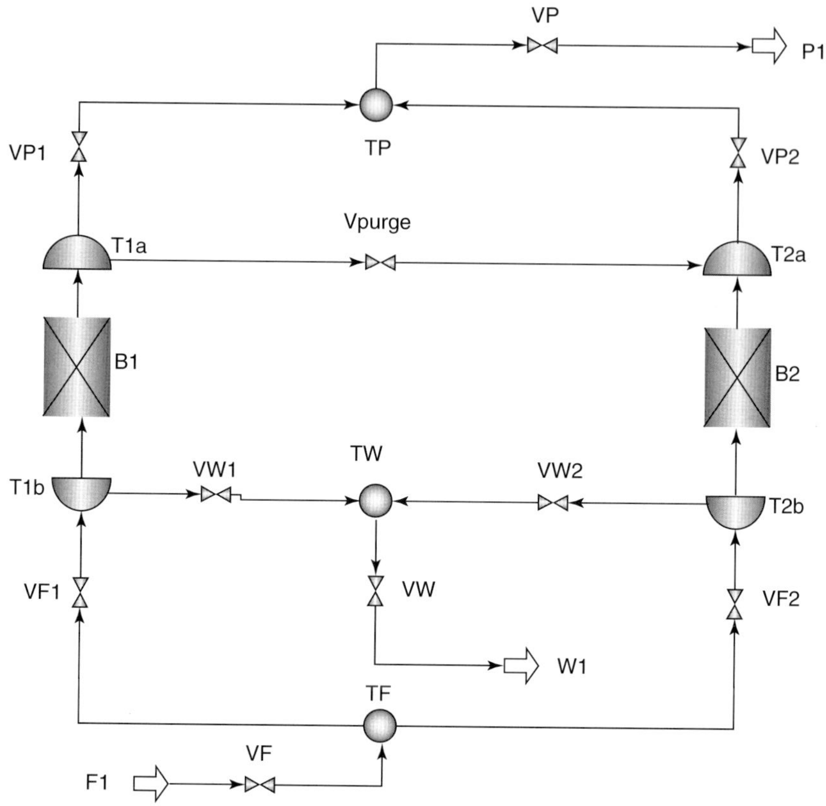

Figure 1.20 An overview of the final flowsheet.

In the equation, F is the molar flow rate in kmol s^{-1} and ΔP is the pressure drop across the valve in bar.This means that the Cv value in Aspen *does not* correspond to the conventional form of the Cv equation shown in Eq. (1.13):

$$\mathrm{Cv} = q\sqrt{\frac{G_\mathrm{f}}{\Delta P}}. \tag{1.13}$$

In the equation, q is the volumetric flow rate in GPM (gallon per minute) and G_f is the specific gravity of the fluid. In Eq. (1.13), ΔP is still the pressure drop across the valve, but it has units of psi.

2) The Gas_tank_void model, serves as a mixer or a splitter, and as **a pressure setter** that sets the pressure to enable the calculation of flow rate through the valve equation based on the pressure drop across the valve, and of the pressure drop across an adsorber bed. We set the initial pressure of each gas_tank_void model through the Presets/Initial form within the block. In the flowsheet, we have the following gas_tank_void blocks: TF, TW, T1a, T1b, T2a, T2b, and TP.
3) Gas_bed model for the absorber, including beds B1 and B2.

The online "Help" section of Aspen Adsorption explains in detail the concept that implements **the forward-reverse flow directions** corresponding to the cyclic adsorption under a dynamic environment. Basically, in the overall flowsheet of Figure 1.20, we do the following:

1) Add a gas_tank_void block TF as a pressure setter to fix the pressure between valves VF and VF1, and between valves VF and VF2. Add a gas_tank_void block TW as a pressure setter to fix the pressure between valves VW1 and VW, and between valves VW2 and VW. Also, add a gas_tank_void block TP as a pressure setter to fix the pressure between valves VP1 and VP, and between valves VP2 and VP.
2) Add gas_tank_void blocks T1a and T1b as pressure setters to fix the pressures at the inlet and outlet of adsorber bed B1. By doing so, we also fix the pressure between valves VF1 and VW1, and between valves VPurge and VP1. Do the same for adsorber bed B2 with gas_tank_void blocks T2a and T2b.

Step 9: Place your first blocks: the adsorber beds

To select the adsorber bed block, click on the first icon in Figure 1.19, labeled gas_bed. To place it in the flowsheet, merely click anywhere in the flowsheet. Click a second time to place a second gas block, then right-click to stop adding repeatedly the same blocks. Once the gas_beds have been placed, an icon will appear that represents a new packed bed. See Figure 1.21.

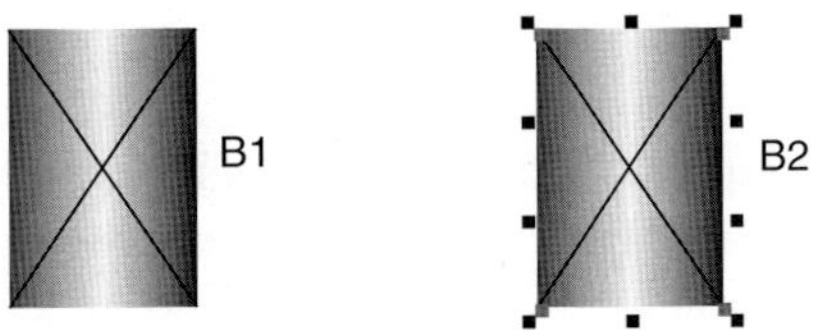

Figure 1.21 Freshly placed gas_beds in the flowsheet.

Step 10: Place the gas_tank blocks

We can repeat this process to add the rest of the blocks to the flowsheet, but there is one additional consideration. Each model has multiple icons that can be used to improve the intelligibility of the flowsheet. To view the extra icons, click on the small arrow beside the button in the model library. Use this arrow to add seven different tanks (gas_tank_void) to the flowsheet, as shown in Figure 1.22. It is important to note that the icons do not change the behavior of the block in any way, and they are purely cosmetic.

Step 11: Place the gas_valves

The gas_valve models are the very last on the list in the Gas: Dynamic tab. Therefore, if you are working on a small screen, you may need to use the arrow on the far right-hand side to scroll the gas_valve button into view. Once you have, place the valves as shown in Figure 1.23.

Step 12: Add the Feed and Product blocks

The final blocks to add to the flowsheet are the "feed" and "product" blocks. These represent points where material may enter or exit the flowsheet. They are not required in all circumstances, but they are needed here. We cover situations where they are not needed in later exercises. Add one feed block (B19) at the lower left of the flowsheet and two product blocks (B20, B21), as shown in Figure 1.24.

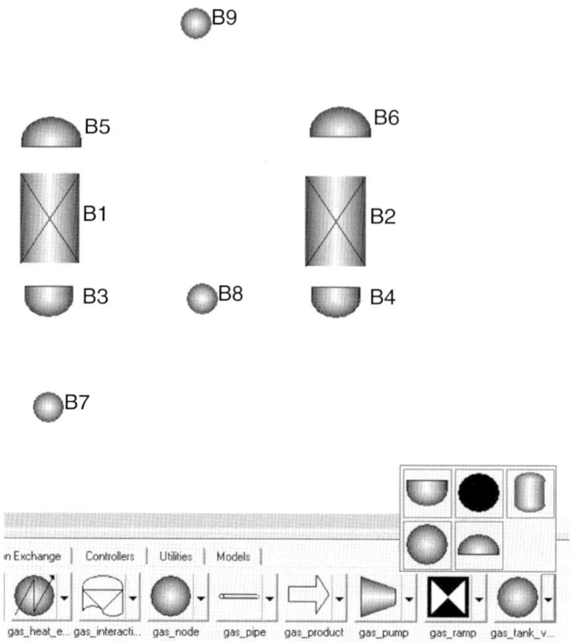

Figure 1.22 Placing gas_tank_voids using the various icon options.

Step 13: Connect the blocks using gas_material_connections

Similarly to the procedure when placing models with different icons, use the small arrow by the "connection" icon to select gas_material_connection. Unlike with the models, the different connection types do have different behaviors, and so it does matter which is selected. After selecting the connection icon, we see that the blocks in the flowsheet show blue and red arrows, indicating points where they can be connected together. You can connect the blocks in any order by clicking on the arrow of one block, and an arrow of a second block. Blocks may accept more than one connection to the same point. For example, a gas_tank_void can accept many inlet and outlet streams simultaneously. If you want to change the location of a connection for esthetics, hold down the mouse button while clicking on the blue or red arrow, and drag it to the desired location before releasing. See Figure 1.25 for the connection options and arrows.

One additional feature that can improve the legibility of flowsheet diagrams is the rotate button at the top of the flowsheet window, which allows blocks (and therefore their connection points) to be rotated (Figure 1.26). Use this feature, along with the gas_material_connection to connect the flowsheet as shown in Figure 1.27. Pay close attention to the direction of the arrows, which indicate the expected direction of fluid flow. These directions are reversible for some, but not all, connection types. While the gas_material_connection is reversible, having the streams in the correct order facilitates initialization.

If the rotation options are not visible, you may enable them by right-clicking on a block in the flowsheet to see "Rotate Icon." Figure 1.26 shows the right-click menu with the option that enables the rotation buttons highlighted.

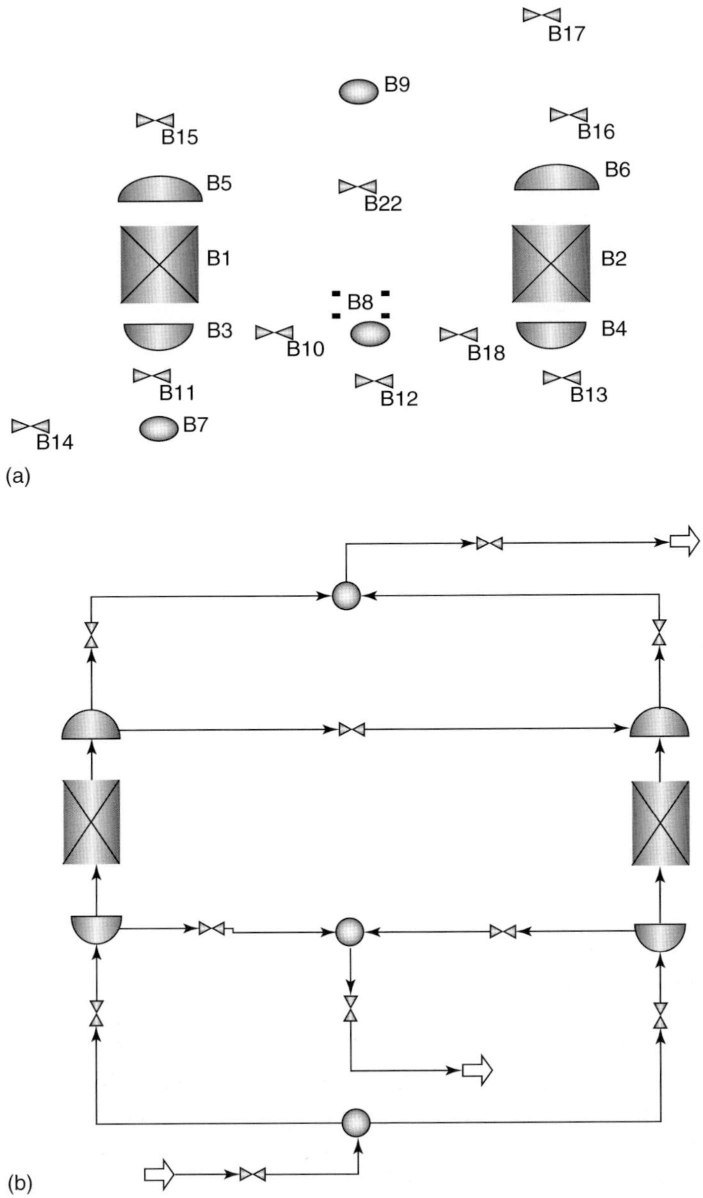

Figure 1.23 The flowsheet with valves added. (a) The valves after initial placement and (b) the valves after "clean up" in step 13.

Step 14: Clean Up

The flowsheet you created may not appear as tidy as that in Figure 1.27. In this step, we cover three tips for cleaning up a new flowsheet, and nearly every model you build will need these.

First: Align blocks and redraw the streams.

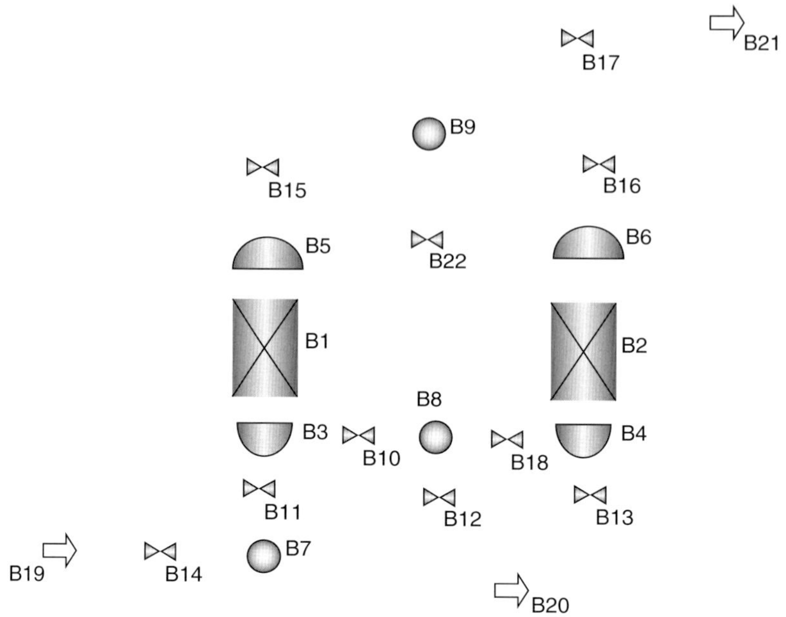

Figure 1.24 The flowsheet with all blocks added.

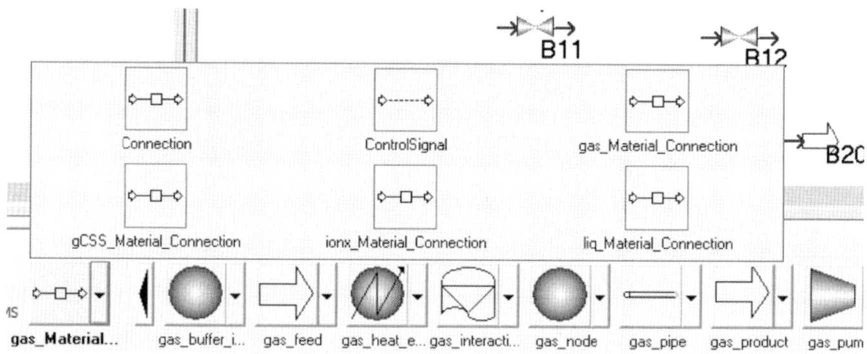

Figure 1.25 Connection options and arrows.

There are two simple ways to align and redraw the blocks in the flowsheet. First, select the stream connecting the two blocks you wish to align, and press **ctrl+B** to align them. Then, press **ctrl+J** to redraw the stream into a more efficient path. You can do this all at once by first using **ctrl+A** to select all the streams, and alternating between **ctrl+B** and **ctrl+J**, although this will not resolve all esthetic issues, and you may need to manually move some blocks and iterate.

Second: Rename the important blocks and streams.

To rename a block or stream, first click on it, then press ctrl+m and enter the new name. Use this process to rename the blocks as shown in Figure 1.28.

Third: Hide the names of unimportant streams or blocks.

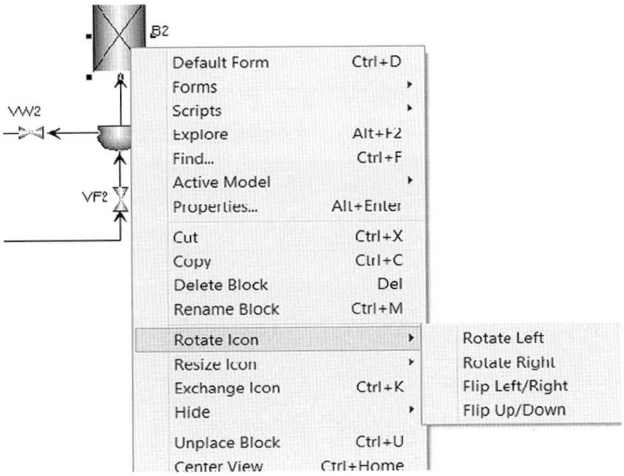

Figure 1.26 The flowsheet right-click menu.

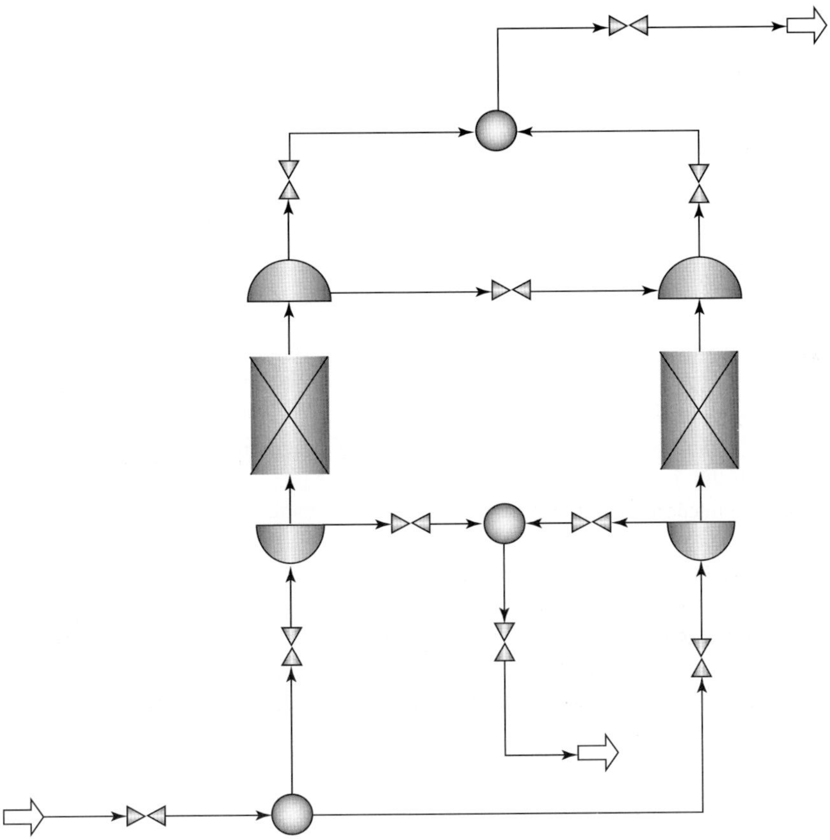

Figure 1.27 The connected flowsheet.

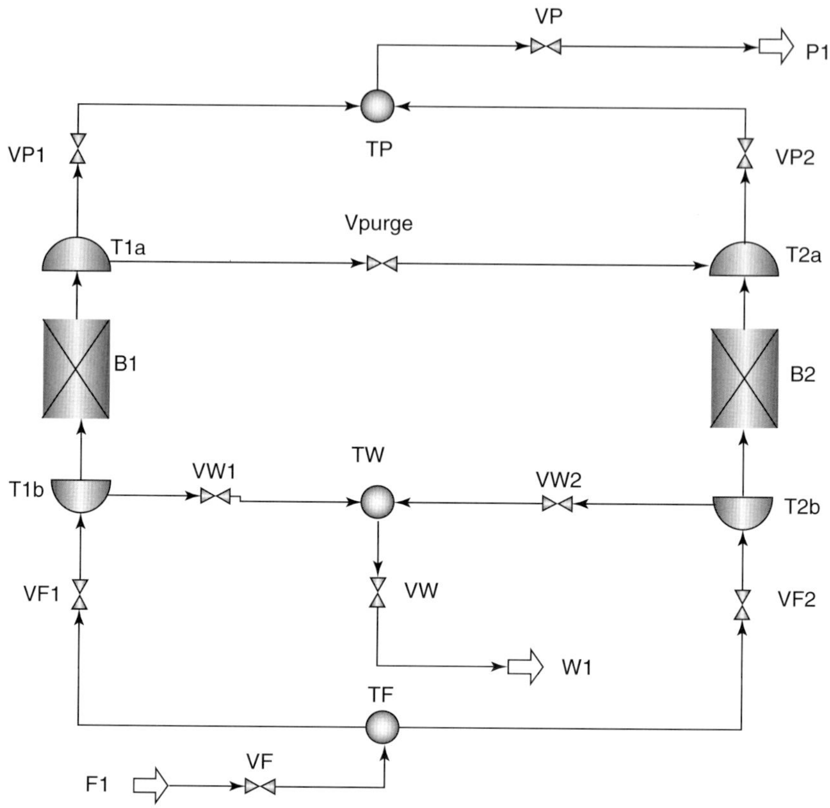

Figure 1.28 The completely labeled and connected flowsheet.

In this flowsheet, the names of the streams will not end up being important. Therefore, we can hide their labels. Do this by selecting the stream, then pressing ctrl+H.

Once you have completed these three "clean up" operations, the flowsheet will look like Figure 1.28.

Once you are satisfied with the layout of the flowsheet, save the file as **Workshop1B.ada**

Step 14a: An alternative approach to drawing the process flowsheet quickly

Now that we have understood how to use the relevant streams and blocks within Aspen Adsorption, and practiced applying various tools in drawing the flowsheet, we can proceed to demonstrate an alternative approach to drawing the flowsheet that is straightforward. We show this in a few steps below.

1) Cancel the automatic name generation for blocks and streams: Tools → Settings → unclick the automatic name generation for blocks and streams → Apply → OK (Figure 1.29).
2) Draw stream F1, valve VF, pressure setter (gas_tank_void) TF, valve VF1, and valve VF2 together with the gas_material_connections S1 to S4. Rotate the blocks appropriately if needed. (Figure 1.30).

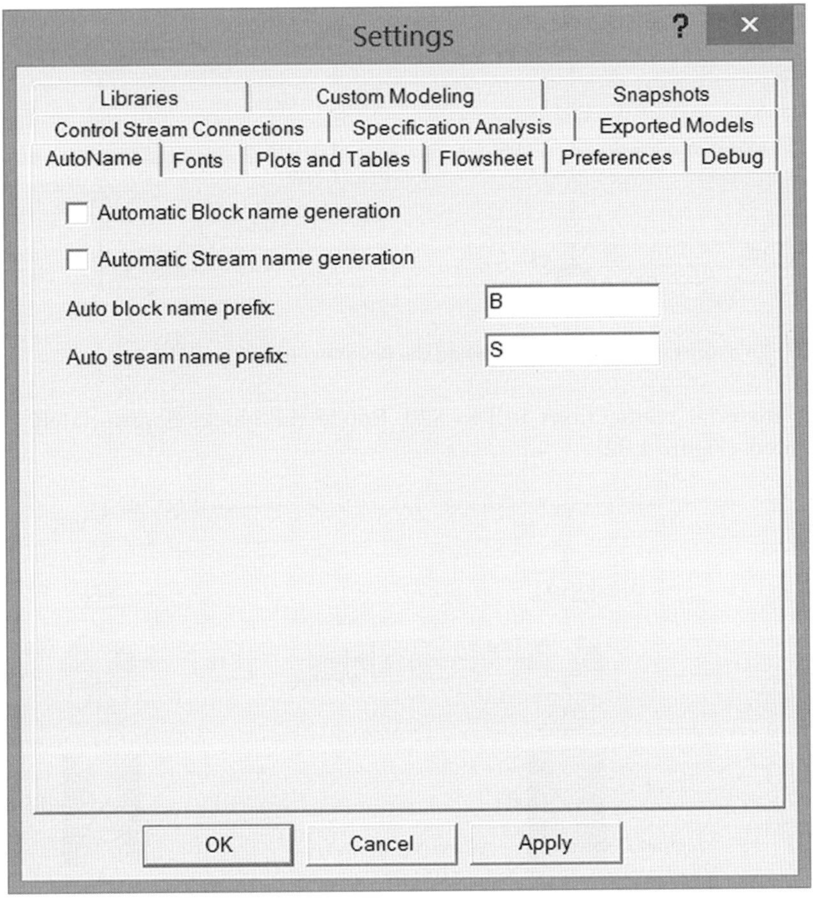

Figure 1.29 Cancellation of automatic name generation for blocks and streams.

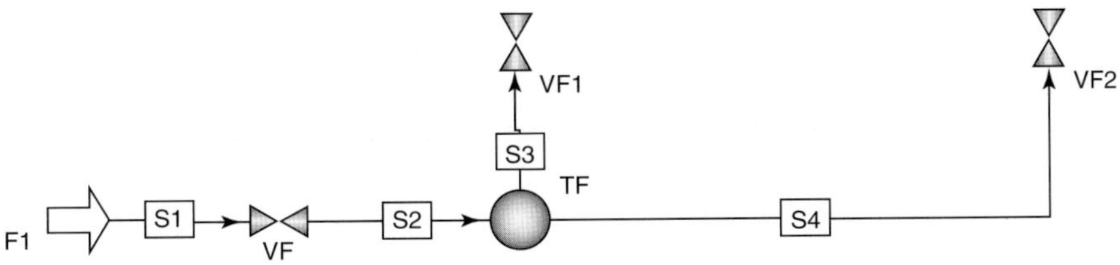

Figure 1.30 An alternative approach to drawing the flowsheet (step 1).

3) Draw the pressure setters (gas_tank_voids) T1b, T2b, and TW, valves VW1, VW2, and VW, and product stream VW, as well as the gas_material_ connections S5 to S12. Rotate the blocks appropriately if needed. (Figure 1.31)
4) Draw the adsorber beds B1 and B2, pressure setters (gas_tank_voids) T1a and T1b, and valves VPurge, VP1, and VP2, together with the

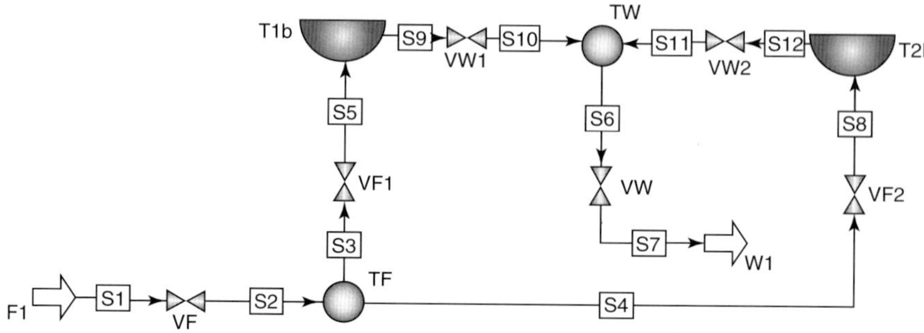

Figure 1.31 An alternative approach to drawing the flowsheet (step 2).

gas_material_connections S13 to S20. Rotate the blocks appropriately if needed. (Figure 1.32)

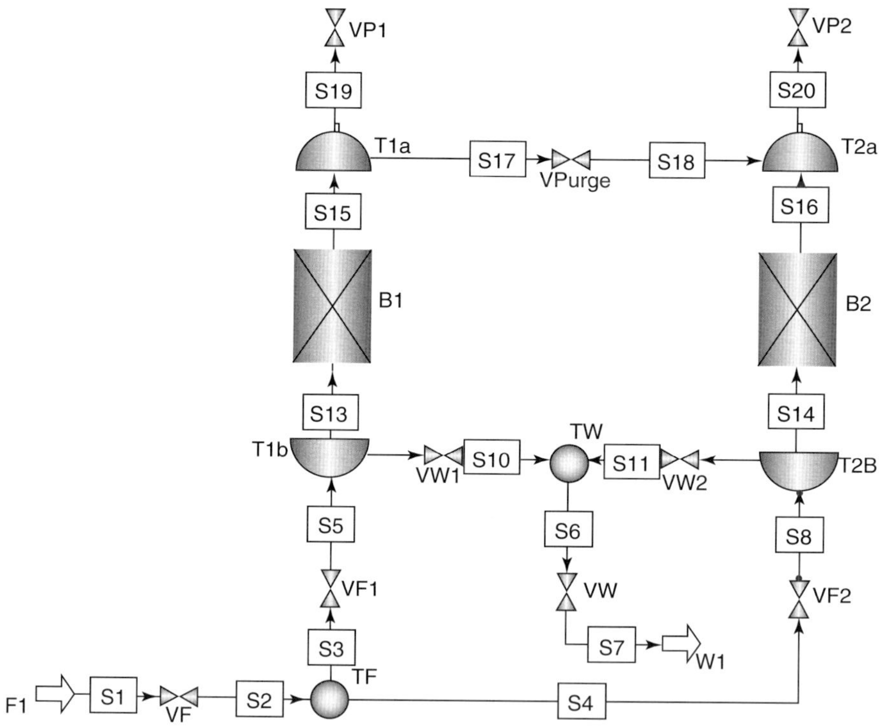

Figure 1.32 An alternative approach to drawing the flowsheet (step 3).

5) Draw the pressure setters (gas_tank_void) TP, and valve VP, product P1, together with the gas_material_connections S21 to S24. Rotate the blocks appropriately if needed. (Figure 1.33)
6) Hide the names of all streams (Figure 1.34). Save the file as **Workshop1B.ada**.

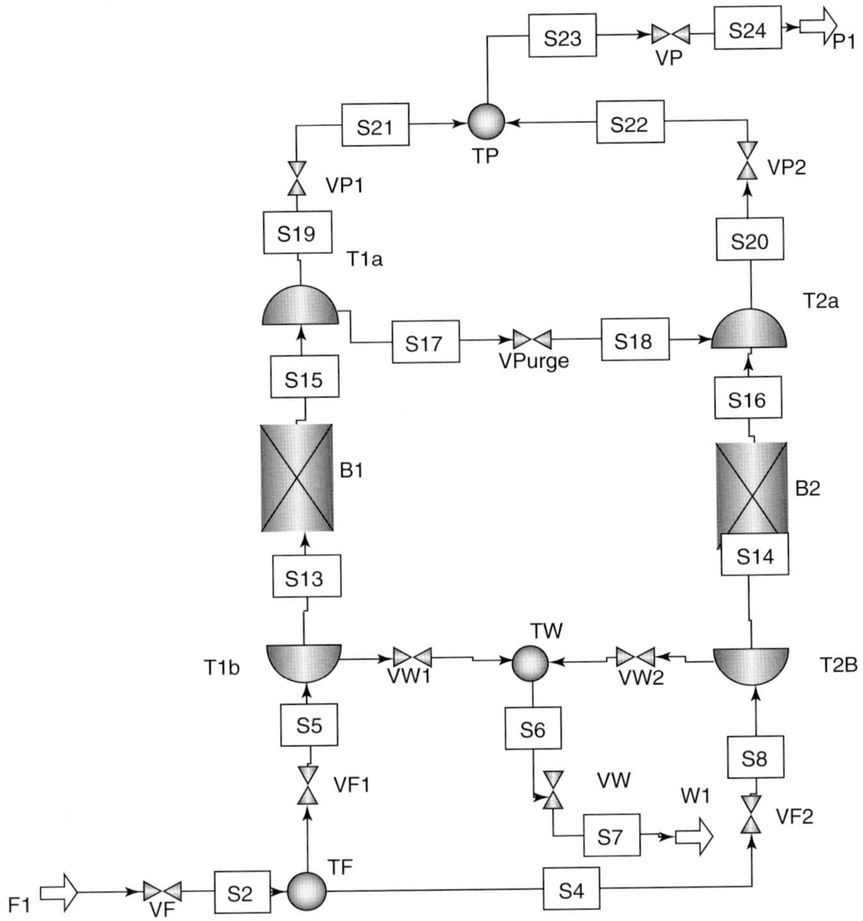

Figure 1.33 An alternative approach to drawing the flowsheet (step 4).

1.7.3 Specifying Operating Conditions: Tables and Forms

There are two main methods for entering values into an Aspen simulation: Tables and Forms, with Tables being the most common. Tables are exactly what you would expect: a window dedicated to tabulating variables and their values. Forms are similar, except Aspen Tech has applied formatting to organize and present the data. To open the default form for a process model, simply double-click on it. It is ***critical*** to be aware that the default form does not always include entries to all the variables you might need to provide data for.

Step 15: Specify the beds B1 and B2

Figure 1.35 shows the specification of the column in the PSA model. It has only one vertical adsorbent layer within the bed with the same packed adsorbent molecular sieve 5A. We choose 1-D (one-dimensional) as the spatial dimensions, since we do not expect there to be any issues with 2-D (two-dimensional)

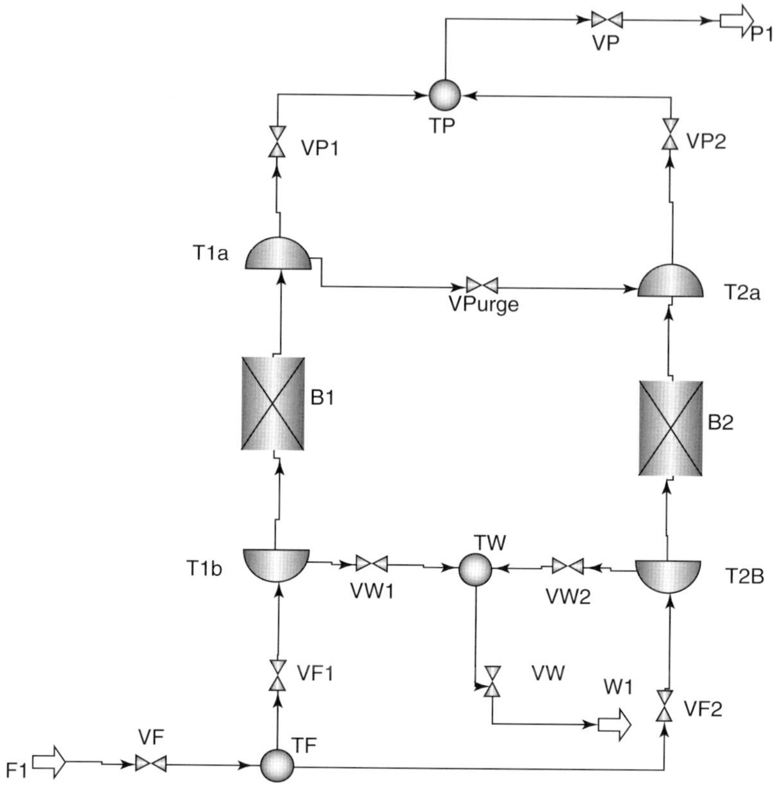

Figure 1.34 An alternative approach to drawing the flowsheet (step 5).

radial mixing. There is no heat exchanger within the adsorbent layer, so we may leave that option alone.

Step 16: Specify the discretization method to be used

From this point, we want to specify more detailed properties of the adsorption bed. To do this, click on the "Configure" button to the right of "Layer 1." Once this has been clicked, Aspen will show the General tab (Figure 1.36), which allows us to specify the numerical options for solving the partial differential equations (PDEs), and to select the gas model assumption.

Aspen Adsorption uses a set of PDEs, ordinary differential equations (ODEs) and algebraic equations, together with the appropriate initial and boundary conditions, to fully describe the adsorption column.

The *upwind differencing scheme 1* (UDS1) is the preferred option in most cases. We prefer to use UDS1 for this dynamic PSA model because it saves the simulation time and the results are reasonably accurate. Therefore, we use the USD1 with 60 nodes.

Step 17: Provide the Material/Momentum Balance Assumptions

We use the Material/Momentum Balance tab in Figure 1.37 to specify the material and momentum balances, and the dispersive properties. For simplicity, we choose convection only as the mass balance assumption. The

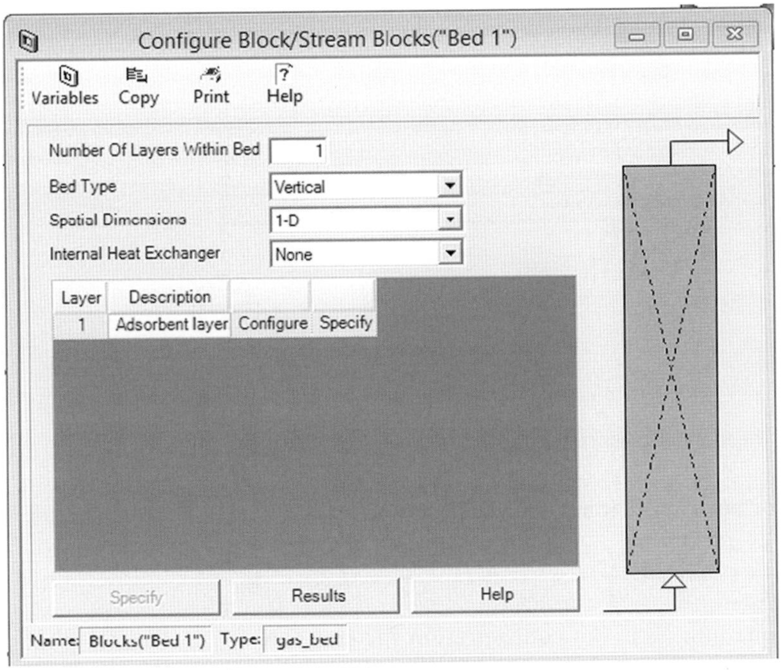

Figure 1.35 The column configuration specification.

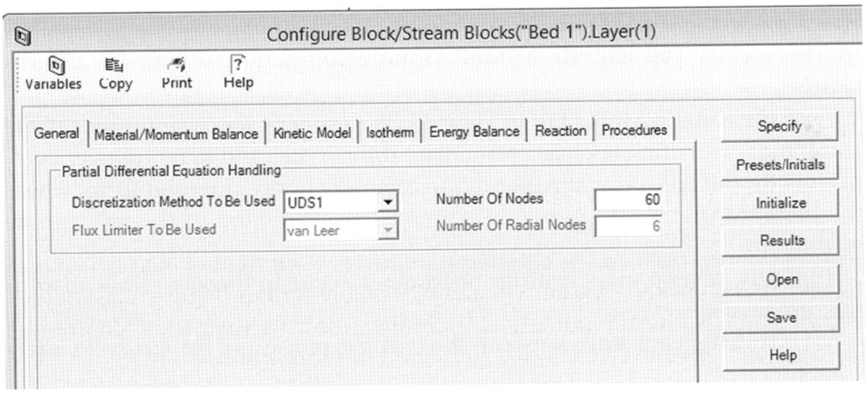

Figure 1.36 The discretization method in the PSA model.

Convection Only option drops the dispersion term from the material balance in Eq. (1.14), so the model represents the plug flow with a zero dispersion coefficient. Local equilibrium is achieved instantaneously between the adsorbent and adsorbates at each axial location.

We use the Momentum Balance Assumption box to specify how the adsorption bed layer model treats gas velocity and pressure. For the calculation of the pressure drops in the PSA system, we use the Ergun equation, Eq. (1.11), which

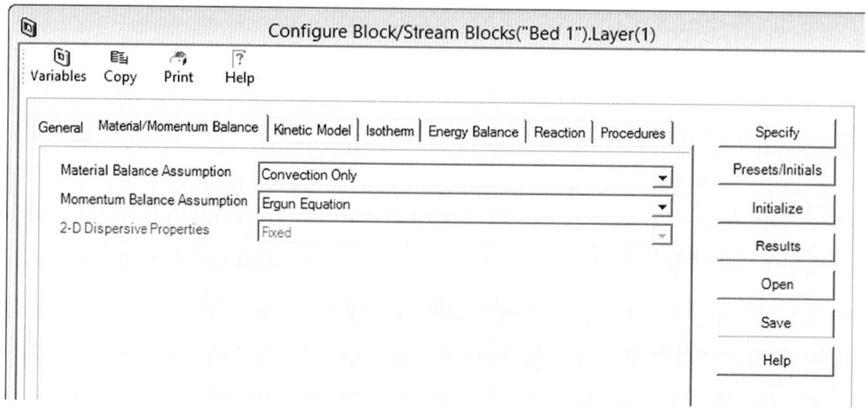

Figure 1.37 The material balance assumption in the PSA model.

is valid for both laminar and turbulent flows and is the most popular option.

$$\frac{\partial C_i}{\partial t} + \frac{(1-\varepsilon)}{\varepsilon}\frac{\partial q_i}{\partial t} + u\frac{\partial C_i}{\partial z} = D_{\mathrm{L}}\frac{\partial^2 C_i}{\partial z^2} = 0. \tag{1.14}$$

In the equation, C_i is the concentration for component i in the gas phase, q_i is the concentration for component i in the solid phase, and ε represents the overall bed voidage, z is the axial distance through the column, and D_{L} is the axial dispersion coefficient. P is the pressure drop within the column, u is the superficial velocity of the gas flow, μ = fluid viscosity, r_{p} = particle radius, and ψ = particle shape factor.

Step 18: Specify the Kinetic Model Assumption & Mass-Transfer Coefficients

We use the Kinetic Model tab in Figure 1.38 to specify the assumptions made about model kinetics, such as resistances, diffusivities and mass transfer coefficients. In the Film Model Assumption box, we choose the solid form, where the mass transfer driving force is expressed as a function of the solid-phase loading.

Typically, several mass transfer resistances occur in the gas-phase adsorption processes:

- Mass transfer resistance between the bulk gas phase and the gas-solid interface.
- Mass transfer resistance due to the porous structure of the adsorbent.

In our case, we consider the mass transfer resistance as *a lumped and linear resistance* or *a linear driving force (LDF) assumption* in the Kinetic Model Assumption box. It means that we lump these mass transfer resistances as a single overall factor, or one resistance dominates all others. The mass transfer coefficient for each component is constant throughout the bed. The mass transfer driving force for component i is a linear function of the solid-phase loading (solid film) as shown in Eq. (1.15):

$$\frac{\partial q_i}{\partial t} = k_i(q_i^* - q_i), \tag{1.15}$$

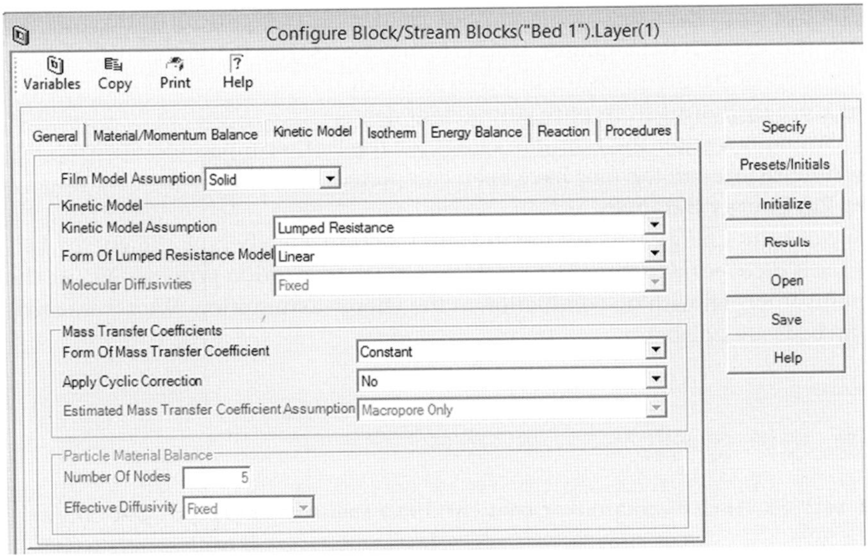

Figure 1.38 The kinetic method in the PSA model.

where k_i is *the linear lumped mass transfer coefficient*, q_i is solute concentrations of the solid phase, and q_i^* represents the adsorbed phase concentration at equilibrium with the gas phase.

Step 19: Specify the isotherm assumptions in the bed

We use the Isotherm tab in Figure 1.39 to choose the assumptions regarding the isotherm of the system.

In an adsorber design, we are usually interested in the adsorption equilibria of mixtures, rather than those of pure components. This is because adsorbed gas components interact on the solid surface, so individual gas components adsorb differently when mixed with other components. Mixture adsorption equilibrium data are not readily available. Although measurements can be made, they are tedious and time consuming to perform, so it is a common

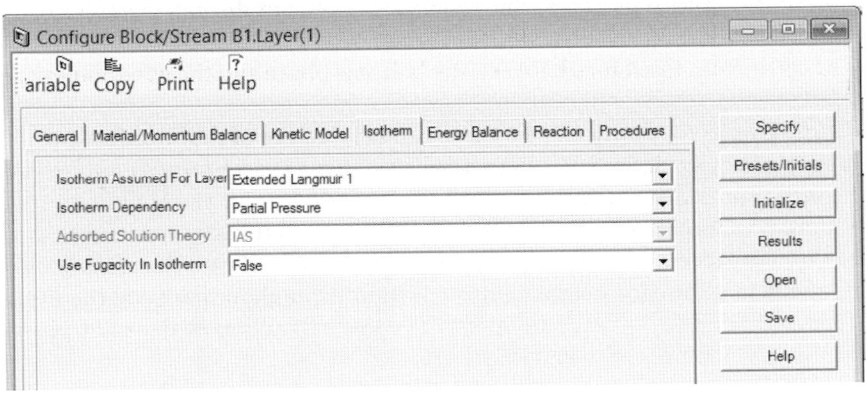

Figure 1.39 The isotherm method in the PSA model.

practice to predict mixture isotherms from pure component isotherms. We use the extended Langmuir isotherm as the isotherm model, which is a function of partial pressure in the system. In this approach, we can predict mixture isotherms from pure component data.

We do not enter the numeric values for the isotherm parameters here, but wait until after all the tabs have been completed

Step 20: Energy Balances

We use the Energy Balance tab in Figure 1.40 to specify how the energy balance is incorporated into the model for this gas adsorption process. We use the nonisothermal with no conduction as the energy assumption. We assume that the beds are adiabatic.

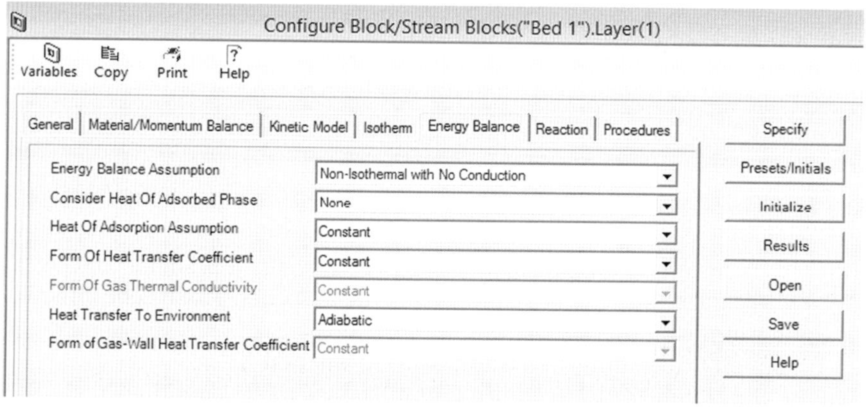

Figure 1.40 The energy balances in the PSA model.

Step 21: Reactions

There are no reactions in this PSA system, and we specify that here (Figure 1.41).

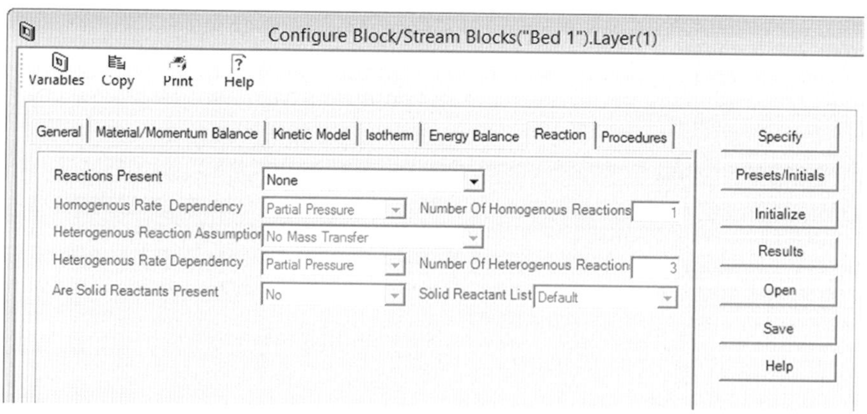

Figure 1.41 The reaction method in the PSA model.

Step 22: User procedures

We can use the Procedures tab in Figure 1.42 to view a list of user procedures within the current adsorption layer model. This is where our choice of Component List pays off. Had we chosen Component Set, we would need to provide our own FORTRAN code to calculate the quantities shown in this list for Aspen to be able to calculate the values and simulate the bed.

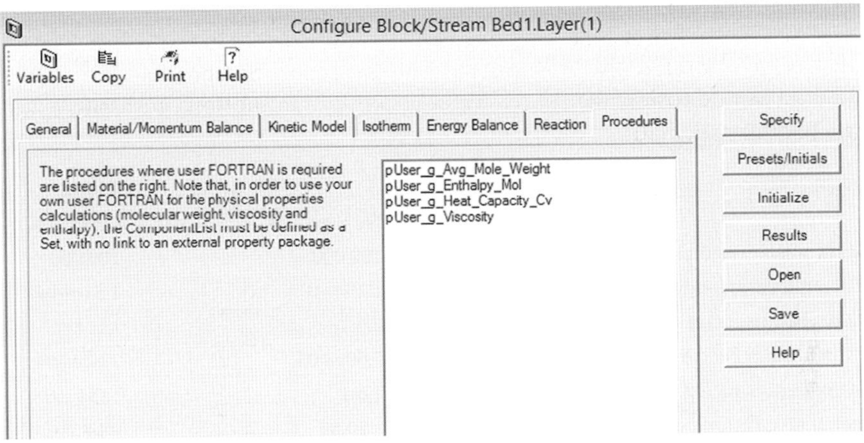

Figure 1.42 The available procedures in the PSA model.

Step 23: The "Specify" button

After defining all assumptions for the layer, we click the "Specify" button in the Configuration form to open the data table containing the required data for the layer. Table 1.6 contains the required data, and Figure 1.43 shows the filled table.

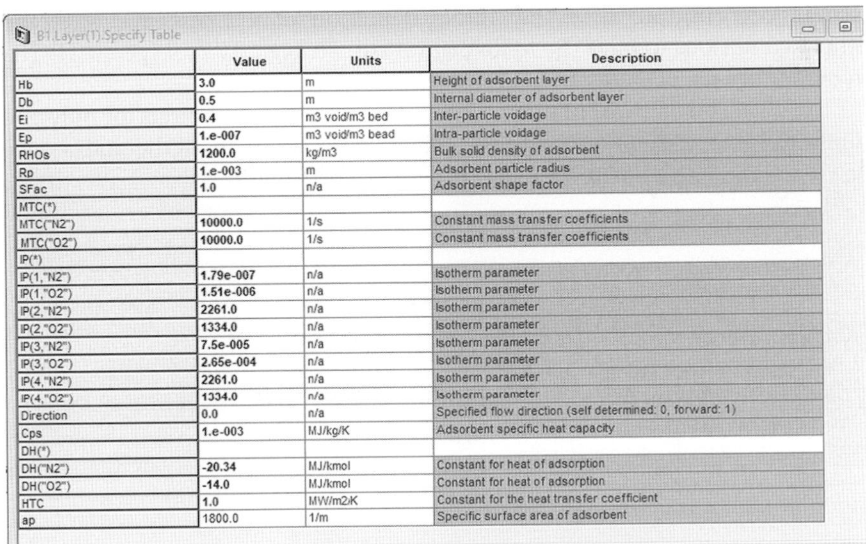
B1.Layer(1).Specify Table

	Value	Units	Description
Hb	3.0	m	Height of adsorbent layer
Db	0.5	m	Internal diameter of adsorbent layer
Ei	0.4	m3 void/m3 bed	Inter-particle voidage
Ep	1.e-007	m3 void/m3 bead	Intra-particle voidage
RHOs	1200.0	kg/m3	Bulk solid density of adsorbent
Rp	1.e-003	m	Adsorbent particle radius
SFac	1.0	n/a	Adsorbent shape factor
MTC(*)			
MTC("N2")	10000.0	1/s	Constant mass transfer coefficients
MTC("O2")	10000.0	1/s	Constant mass transfer coefficients
IP(*)			
IP(1,"N2")	1.79e-007	n/a	Isotherm parameter
IP(1,"O2")	1.51e-006	n/a	Isotherm parameter
IP(2,"N2")	2261.0	n/a	Isotherm parameter
IP(2,"O2")	1334.0	n/a	Isotherm parameter
IP(3,"N2")	7.5e-005	n/a	Isotherm parameter
IP(3,"O2")	2.65e-004	n/a	Isotherm parameter
IP(4,"N2")	2261.0	n/a	Isotherm parameter
IP(4,"O2")	1334.0	n/a	Isotherm parameter
Direction	0.0	n/a	Specified flow direction (self determined: 0, forward: 1)
Cps	1.e-003	MJ/kg/K	Adsorbent specific heat capacity
DH(*)			
DH("N2")	-20.34	MJ/kmol	Constant for heat of adsorption
DH("O2")	-14.0	MJ/kmol	Constant for heat of adsorption
HTC	1.0	MW/m2/K	Constant for the heat transfer coefficient
ap	1800.0	1/m	Specific surface area of adsorbent

Figure 1.43 The specify form of the adsorbent layer in the PSA model.

Table 1.6 The specifications of the adsorbent layer.

Column (adsorbent layer) height (m)		3
Internal diameter of the column (m)		0.5
Interparticle voidage (m^3 void/m^3 bed)		0.4
Intraparticle voidage (m^3 void/m^3 bead)		1.00e−07
Solid density of the adsorbent (kg m^{-3})		1200
Adsorbent particle radius (mm)		1.00
Adsorbent shape factor (n/a)		1
Specific surface area (1 m^{-1})		1800
	Nitrogen (N_2)	Oxygen (O_2)
Mass transfer coefficient (1 s^{-1})	100 000	100 000
IP1 equilibrium loading	1.79e−07	1.51e−06
IP2 Arrhenius factor	2261	1334
IP3 saturation Factor	7.50e−05	2.65e−04
IP4 Arrhenius factor	2261	1334
Heat of adsorption constant (kJ mol^{-1})	−20.34	−14

When these numbers are typed into the column, they look like Figure 1.43.

Once the values have all been entered into this table, it may be closed.

Step 24: Set the initial conditions inside the bed

The final step before the bed is completely specified is to press the "Presets/initials" button to set the initial condition of the bed. We want the initial conditions to have the following ratios: nitrogen 0.79, oxygen 0.21. The temperature for both gas and solid phases is 25 °C.

The form shown in Figure 1.44 represents the initial conditions for only one finite-element node in the bed. We will apply this value to all the nodes once we reach the "initialization" step.

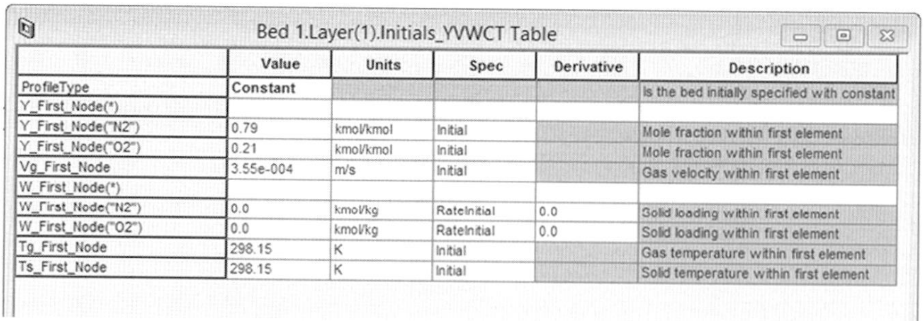
Bed 1.Layer(1).Initials_YVWCT Table

	Value	Units	Spec	Derivative	Description
ProfileType	Constant				Is the bed initially specified with constant
Y_First_Node(*)					
Y_First_Node("N2")	0.79	kmol/kmol	Initial		Mole fraction within first element
Y_First_Node("O2")	0.21	kmol/kmol	Initial		Mole fraction within first element
Vg_First_Node	3.55e-004	m/s	Initial		Gas velocity within first element
W_First_Node(*)					
W_First_Node("N2")	0.0	kmol/kg	RateInitial	0.0	Solid loading within first element
W_First_Node("O2")	0.0	kmol/kg	RateInitial	0.0	Solid loading within first element
Tg_First_Node	298.15	K	Initial		Gas temperature within first element
Ts_First_Node	298.15	K	Initial		Solid temperature within first element

Figure 1.44 The initialization form in the gas_bed.

Step 25: Specify the feed and product blocks

In this PSA model, we set the feed-in flow rate in the "Feed" block as free, because we control the feed flow in the Cycle Organizer through the VF valve. The feed composition is fixed with 0.79 mole fraction nitrogen and 0.21 mole fraction oxygen. The temperature and pressure in the feed are both fixed, with 25 °C and 8.5 bar, respectively (Figure 1.45).

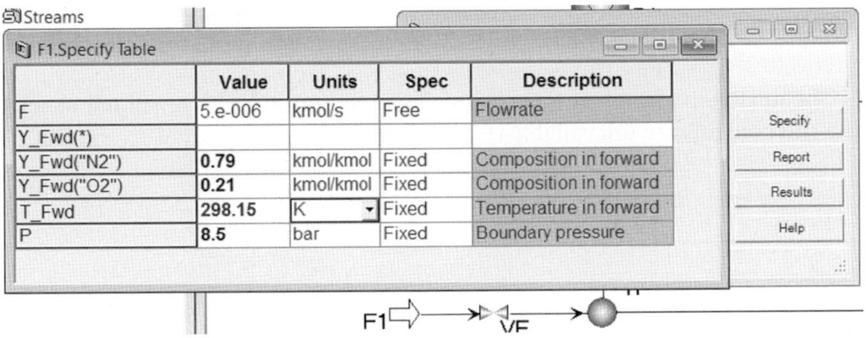

	Value	Units	Spec	Description
F	5.e-006	kmol/s	Free	Flowrate
Y_Fwd(*)				
Y_Fwd("N2")	0.79	kmol/kmol	Fixed	Composition in forward
Y_Fwd("O2")	0.21	kmol/kmol	Fixed	Composition in forward
T_Fwd	298.15	K	Fixed	Temperature in forward
P	8.5	bar	Fixed	Boundary pressure

Figure 1.45 The feed specification in the PSA model.

Just as with the feed block, the valves will decide the product block flow rates. However, we need to specify their pressure, as well as the "reverse" concentrations that get invoked if the flow rate of the product blocks ever becomes negative. Set both product blocks to have a pressure of 1 bar, and reverse concentration that matches the feed block (Figure 1.46).

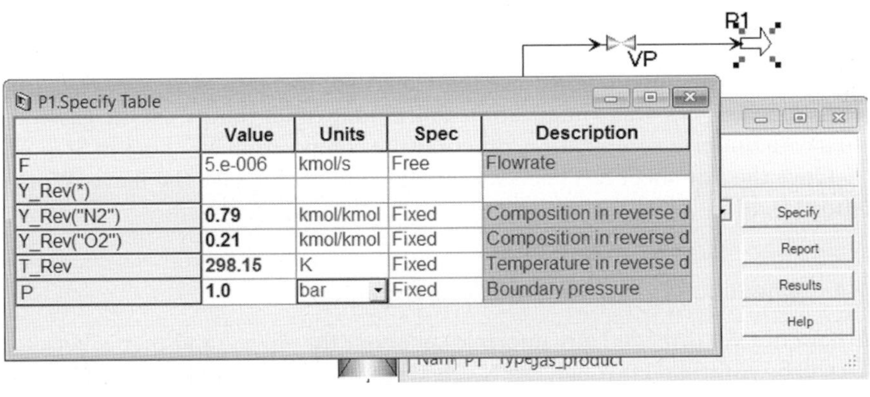

	Value	Units	Spec	Description
F	5.e-006	kmol/s	Free	Flowrate
Y_Rev(*)				
Y_Rev("N2")	0.79	kmol/kmol	Fixed	Composition in reverse d
Y_Rev("O2")	0.21	kmol/kmol	Fixed	Composition in reverse d
T_Rev	298.15	K	Fixed	Temperature in reverse d
P	1.0	bar	Fixed	Boundary pressure

Figure 1.46 Specifying a product block.

Once you have entered the data, save the flowsheet as **Workshop1C.ada.**

Step 26: Initializing the flowsheet

At this point, we have defined a valid flowsheet. We are not yet done entering information, but Aspen should be able to run what we have so far. Therefore, to check our work, initialize the simulation for a run. The first thing to do is to look at the run controls shown in Figure 1.47. If they are greyed out, Aspen considers our flowsheet to be incomplete (i.e. things are not connected properly) or misconfigured (i.e. we have too many or too few degrees of freedom.) At this stage, the buttons *should* be greyed out.

Figure 1.47 The Aspen run controls.

The next step to look at is the "Specification Status," which will tell us why the run controls are unavailable (Figure 1.48).

Figure 1.48 The Specification Status for an overspecified flowsheet (a) and a completely specified flowsheet (b).

If we have failed to connect the blocks correctly, the indicator will show a grey wrench icon. If we have done everything correctly, we should see an upward pointing triangle that indicates that we have an over-specified simulation. Why is our simulation overspecified? Because we have not yet *initialized* the gas beds and tanks. Initialization changes various internal variables according to how the flowsheet is connected and configured (dynamic/steady state). To perform the initialization, use the Flowsheet menu to pick "Check & Initial" (Figure 1.49). This initialization will, among other things, set the gas beds to have a uniform composition, precisely the composition we chose in Step 24:

Once the flowsheet has been successfully initialized, save the file as **Workshop1D.ada.**

1.7.4 Scheduling Events with the Cycle Organizer

We have not yet provided any information about the valves, and that is because they will be controlled through the "Cycle Organizer." This block makes it easy to program schedules into our simulation, so that we can achieve the dynamic valve opening and closing required to simulate our PSA process.

The way we have placed the valves in our flowsheet, Figure 1.28, will allow us to employ the following strategies:

- Flow rates are set by the VF, VP, and VW valves, and they will not close during any step. That is to say, they will behave as valves with constant flow coefficient, so that the flow rate through them is a function of the pressure drop across them.

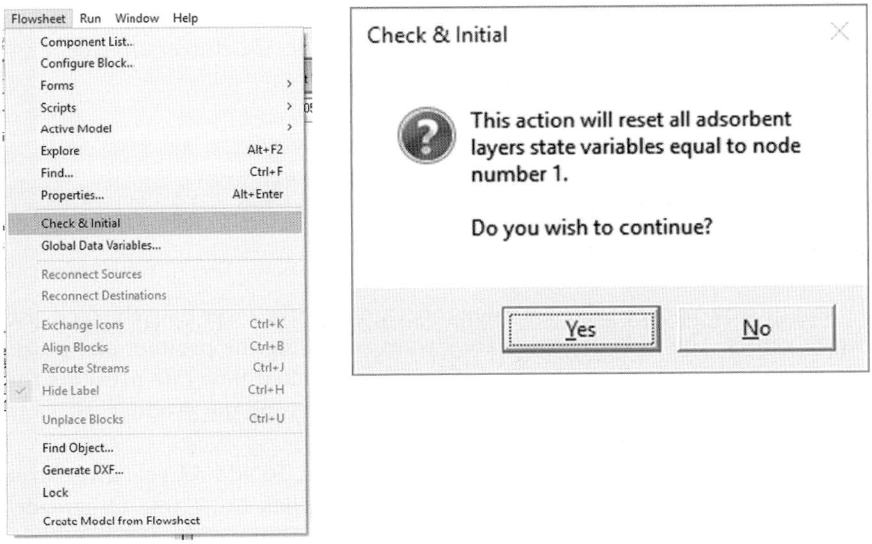

Figure 1.49 The Check and Initial function.

- The VPurge valve will also function as a Cv valve, but will only be open during purge steps.
- All other valves (VF1, VF2, VW1, VW2, VP1, and VP2) are either completely open or completely closed, and simply select which bed the gas will flow through.

Step 27: Specify the VF, VW, VPurge, and VP valves

We have previously explained, in Tables 1.4 and 1.5, the four active specifications (ASs) for valves within Aspen Adsorption: (1) AS = 0, valve is fully closed; (2) AS = 1, valve is fully open; (3) AS = 3, the gas flow rate through the valve is linearly dependent on the pressure drop across the valve according to Eq (1.12); and (4) AS = 3, the valve has a fixed flow rate.

Since the Cv values of these valves do not change over time, we can specify them before working with the Cycle Organizer. To change the Cv value of a valve, double-click on it, then press the "Specify" button (Figure 1.50).

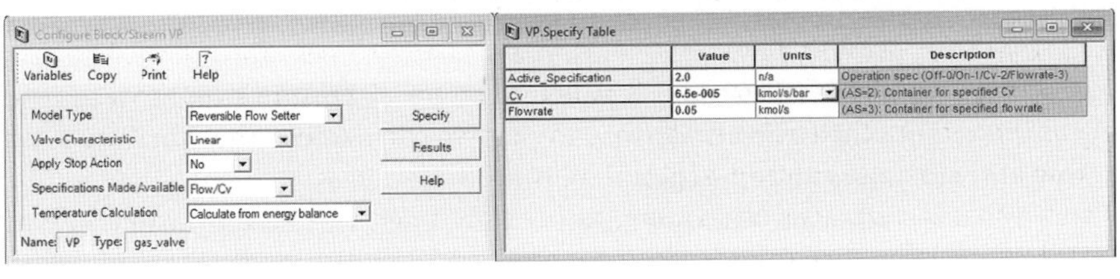

Figure 1.50 Specifying the VP valve.

Set the valve Cv values according to Table 1.7. Note that the valve acts as **a flow setter** according to Eqs. (1.12) and (1.13) presented previously in Step 8.

Table 1.7 Cv valve settings.

Valve	Cv setting (kmol/(s bar^{-1}))
VF	0.0095
VP	6.5e−6
Vpurge	4.5e−5
VW	0.065

The VP and Vpurge valve Cv values are significantly lower than the feed and waste valves. This is because we are always collecting the product and purge from the high-pressure tank. Having a more restricted flow rate prevents the bed from depressurizing too quickly and contaminating the product.

Step 28: Add the Cycle Organizer

It is now time to think about the valve scheduling. To add a Cycle Organizer to the flowsheet, press the "tools" button, then click on "Cycle Organizer" (Figure 1.51).

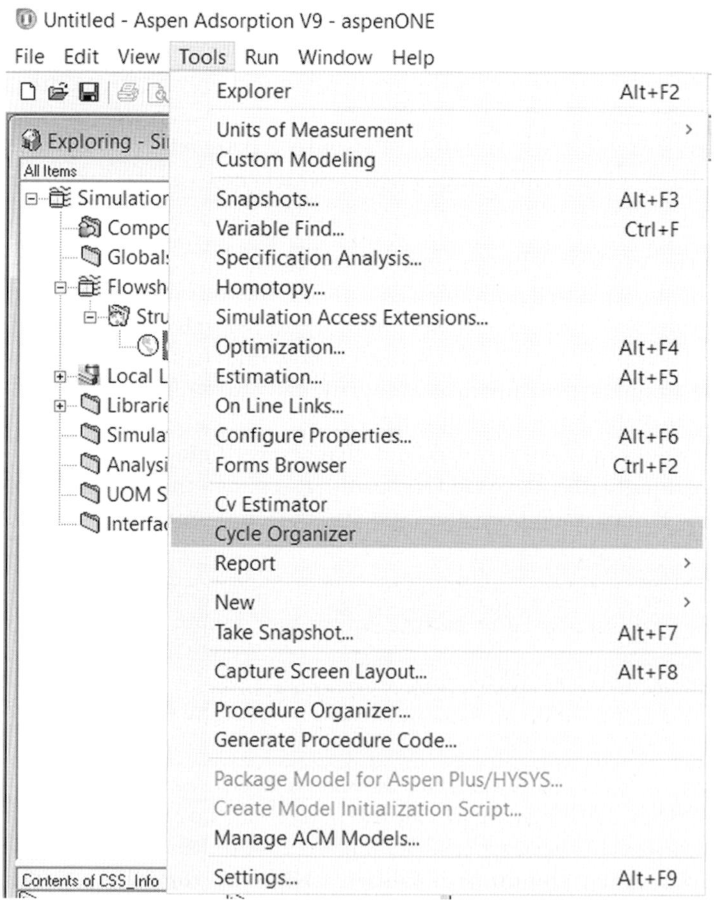

Figure 1.51 Adding the Cycle Organizer.

Step 29: Specifying a step in the Cycle Organizer

After adding a Cycle Organizer, we need to specify the various steps that the Cycle Organizer will go through. The input form for the Cycle Organizer presents one undefined step. There are many options to control these steps. They can be triggered by events that occur in the flowsheet, time, or even other cycles. To specify the first step, we put in a description and indicate that the step will simply run for 40 s, as shown in Figure 1.52.

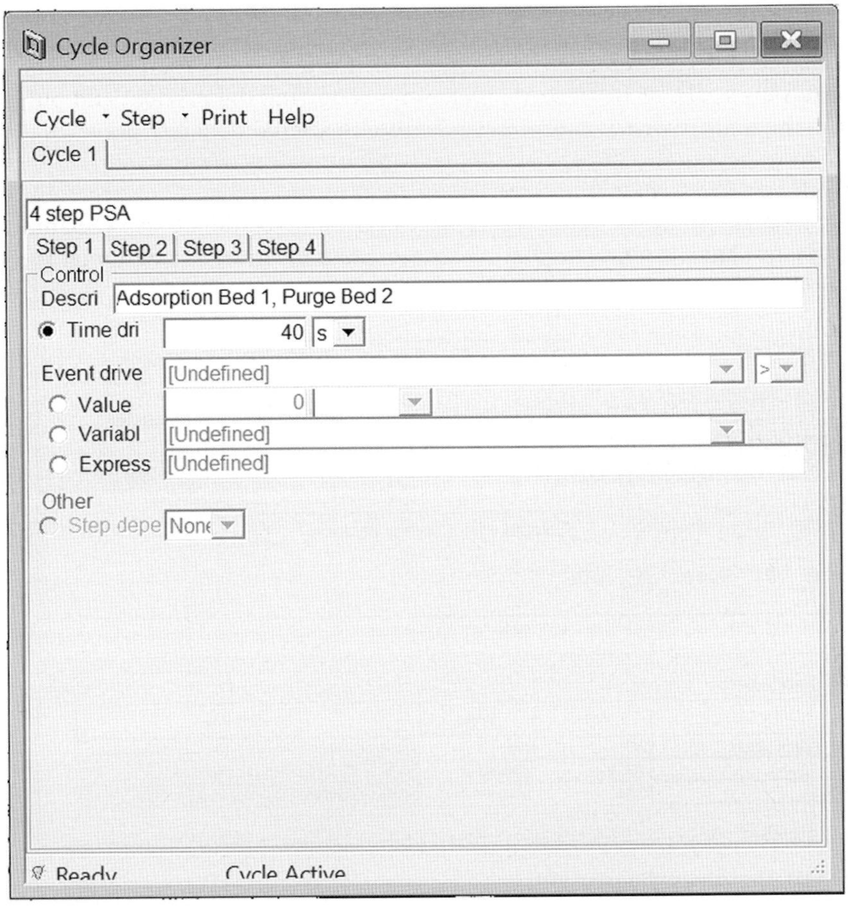

Figure 1.52 Specifying the first step (adsorption bed 1, purge bed 2).

Of course, we also need to specify which valves will be open and closed during this step. To do this, click on the "step" button and select "Manipulated" (Figure 1.53).

Now we need to know what variables to manipulate. Double-click on any valve icon to see what settings are available. Figure 1.54 shows the high-level options, and common numerical variables that define the valve. For the Cycle Organizer, we will try to configure the flowsheet so that we only need to modify the Active_Specification for simplicity's sake.

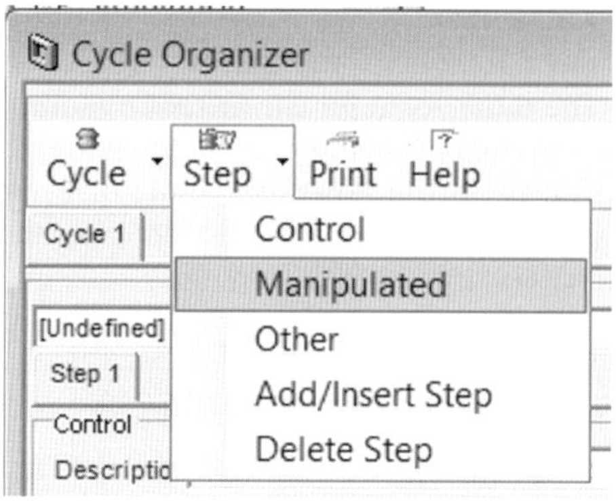

Figure 1.53 Viewing manipulated variables in the cycle organizes.

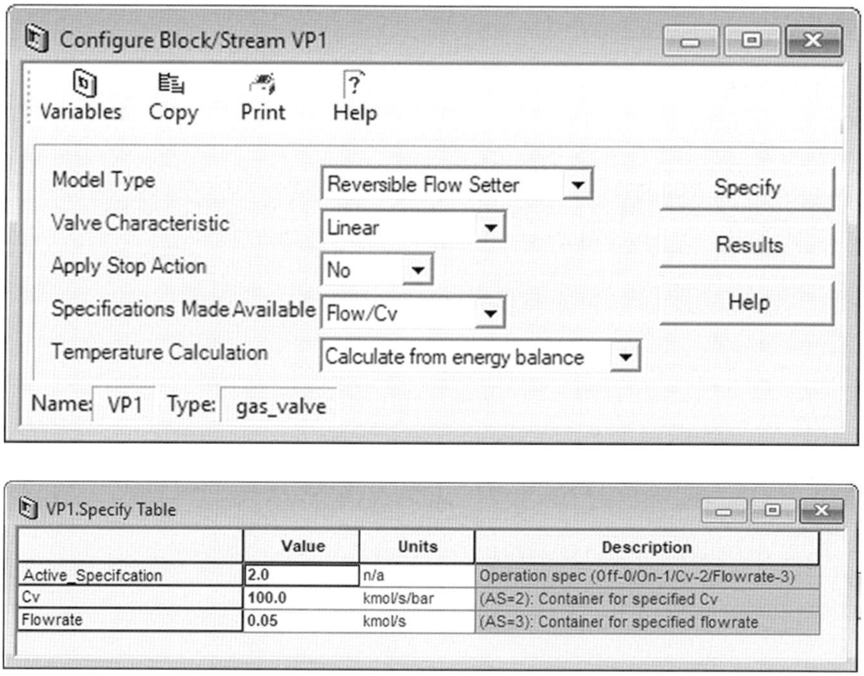

Figure 1.54 The valve settings for VP1.

To add variables, right-click and choose "Add" (Figure 1.55).

Following Figure 1.7 and Table 1.4 in Section 1.7, we specify the variables and values for Step 1 (adsorption bed 1, purge bed 2) of the PSA cycle (Figure 1.56).

What do these "Active_Specification" variables mean? We have defined these previously in Table 1.4. They tell Aspen that the valves will behave according to Table 1.4.

Step 30: Adding steps to the Cycle Organizer

To add more steps, use the step button and select "Add/Insert Step" (Figure 1.57).

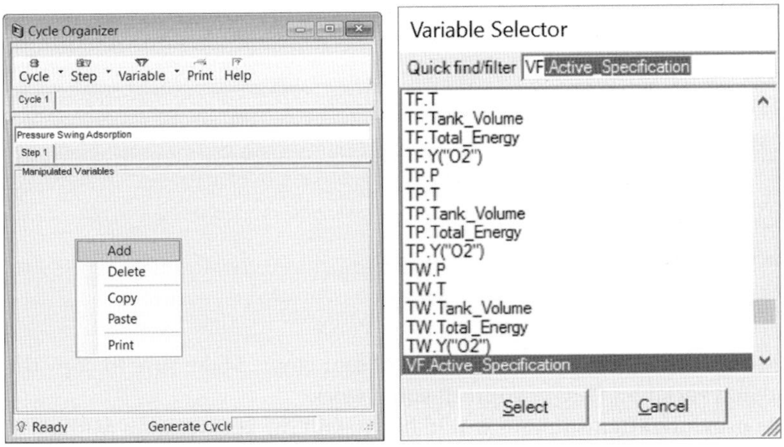

Figure 1.55 Adding manipulated variables.

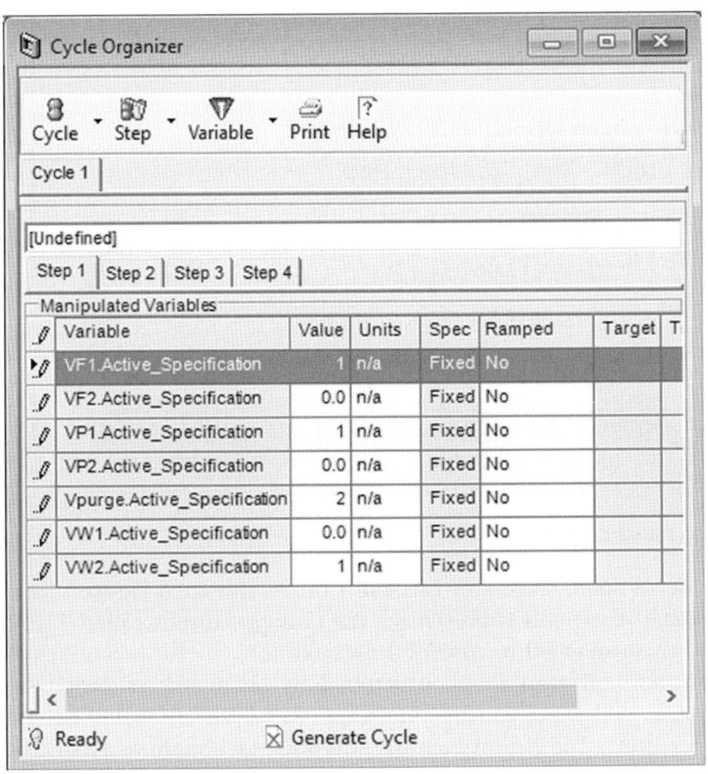

Figure 1.56 Variables manipulated during step 1(adsorption bed 1, purge bed 2).

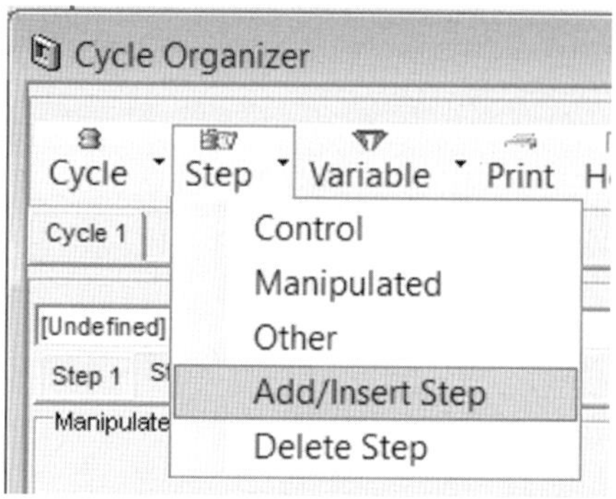

Figure 1.57 Adding steps to the cycle.

Add three additional steps for the total of four steps. Make sure to copy the manipulated variables to save yourself some work.

In this pressurization step, we would like to end the step as soon as the bed pressure reaches some fraction of the maximum pressure (8.5 bar). To do this, we can create an event that will trigger the step change once the void pressure at the top of the bed reaches 8.4 bar. To find the Aspen variable that represents tank pressure, we will use the variable finder. To open the variable finder, press ctrl+f while on the flowsheet, or use the tools menu and select "Variable Find…" as shown in Figure 1.58.

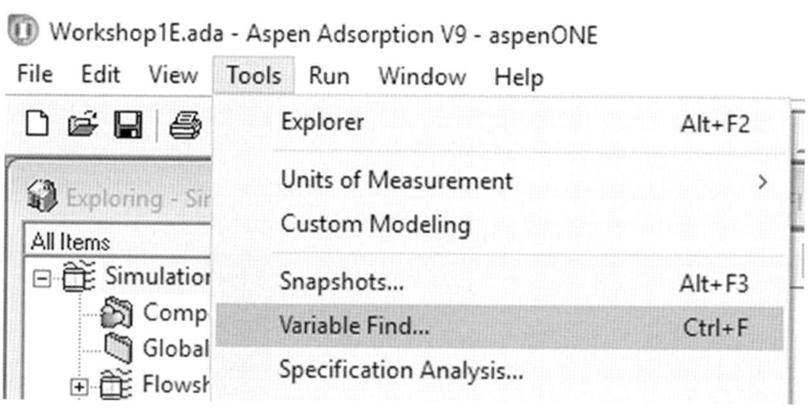

Figure 1.58 Opening the variable finder.

Once the finder is open, search for variable T1a.* as shown in Figure 1.59.

As you can see, the variable representing the tank pressure is called T1a.P, so we can now create an event to control the step time.

We make the second step to be controlled by an event, as shown in Figure 1.60. Aspen will warn you that the expression you enter will not be checked for correctness, so be very careful to enter the equations correctly. This expression indicates that step 2 will run until bed 2 has been pressurized to a pressure greater than 8.4 bar, and bed 1 has been depressurized to less than 1.1 bar.

For step 2, we choose the manipulated variables listed in Figure 1.61.

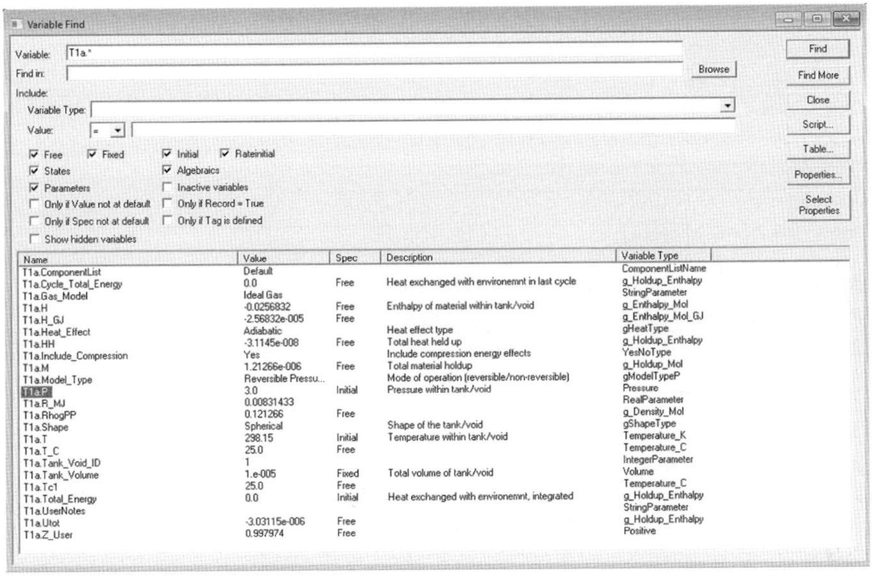

Figure 1.59 The variable finder.

Figure 1.60 The event-controlled step 2 (blowdown bed 1, pressurize bed 2).

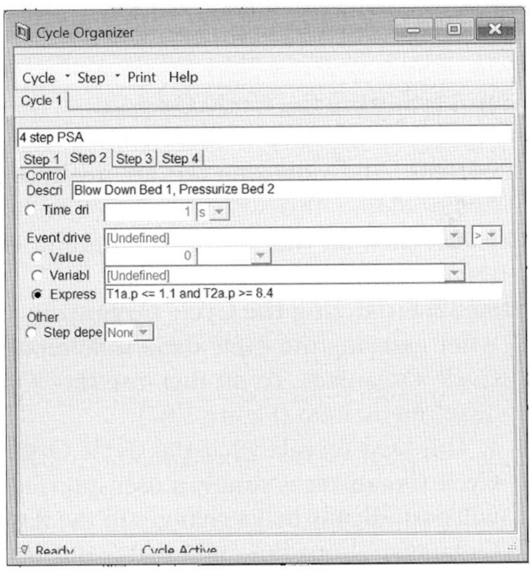

Step 3 is a mirror of step 1, which runs for 40 s (Figure 1.62).

Since we only need to change two valves, we can remove all the other variables, if we wish (Figure 1.63).

And step 4 is a mirror of step 2 (Figure 1.64).

The manipulated variables are also reversed (Figure 1.65).

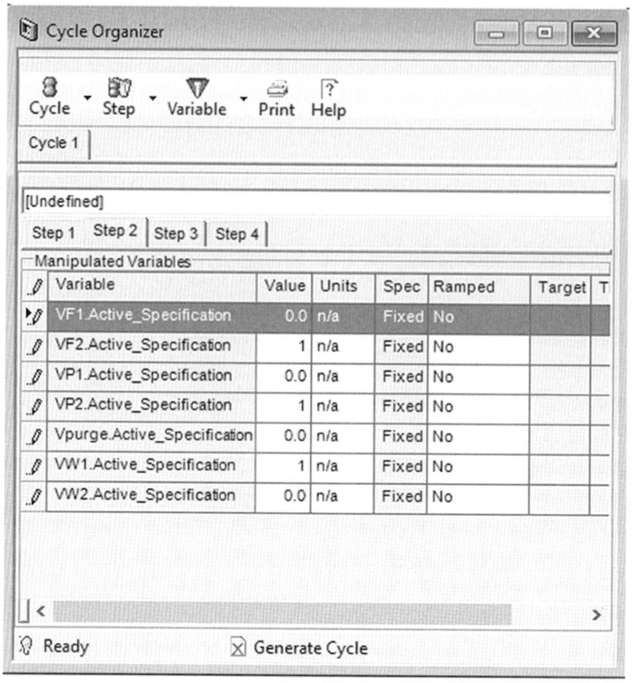

Figure 1.61 Manipulated variables in step 2 (blowdown bed 1, pressurize bed 2).

Step 31: Setting the Cycle Options

Now that we have defined all our steps, we will look at the overall cycle options. The only one we are really interested in is the maximum number of steps. This will impose an endpoint on our simulation. To set this value, click on the "Cycle" button and enter 10 in the Maximum cycles box (Figure 1.66)

Step 32: Finalizing the Cycle Organizer

After entering the cycle data, it is important to "generate" and "activate" the Cycle Organizer. To do this, use the "Cycle" button and press the "Generate Task" menu item (Figure 1.67).

This step reveals what the Cycle Organizer really is: a code generator. The Cycle Organizer is simply a user interface to make it easier to write tasks that will periodically make changes to the flowsheet. It is possible to skip the Cycle Organizer altogether and create your own task, which will perform the exact functions as the task generated by the Cycle Organizer.

After generating this task, navigate to the Flowsheet and look for the "Cycle1" task (Figure 1.68).

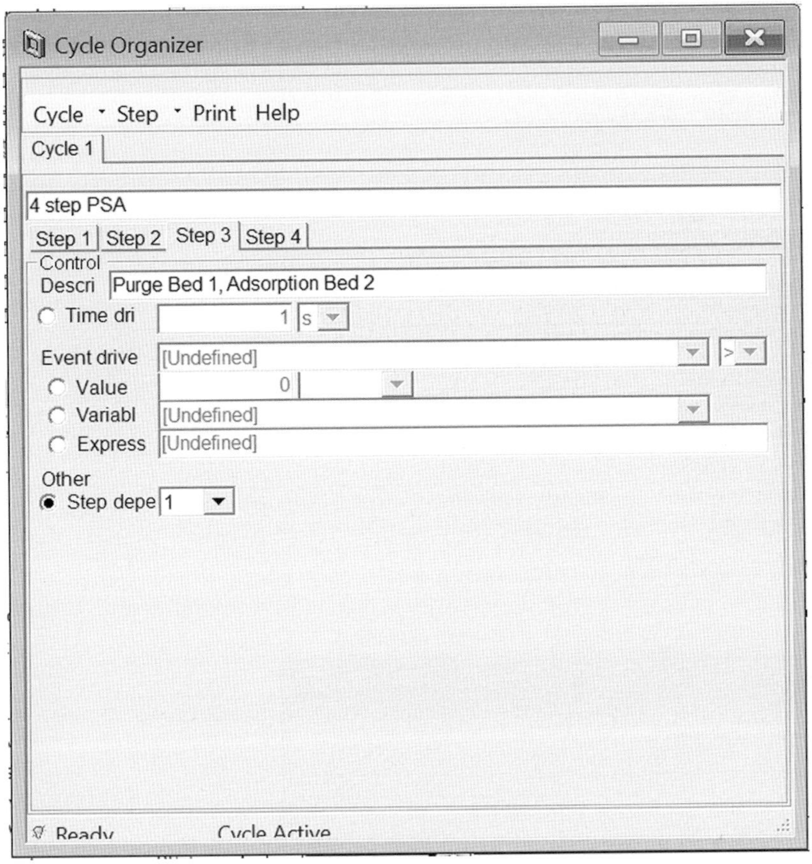

Figure 1.62 Step 3 (purge bed 1, adsorption bed 2).

1.7.5 Running an Aspen Simulation

Before you run the simulation, save the simulation as **Workshop1E.ada**.

Step 33: Run the simulation in Dynamic mode using the "play" button

To begin the simulation, ensure that the dropdown menu says "Dynamic," then press the solid blue "play" button (Figure 1.69). This will perform an initialization run followed by a dynamic simulation.

1.7.6 Viewing and Exporting Simulation Results

Once the simulation has completed, we want to view some results.

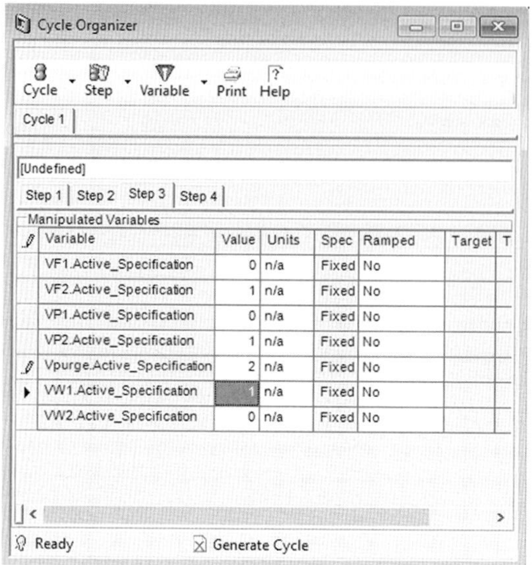

Figure 1.63 Manipulated variables in step 3 (purge bed 1, adsorption bed 2).

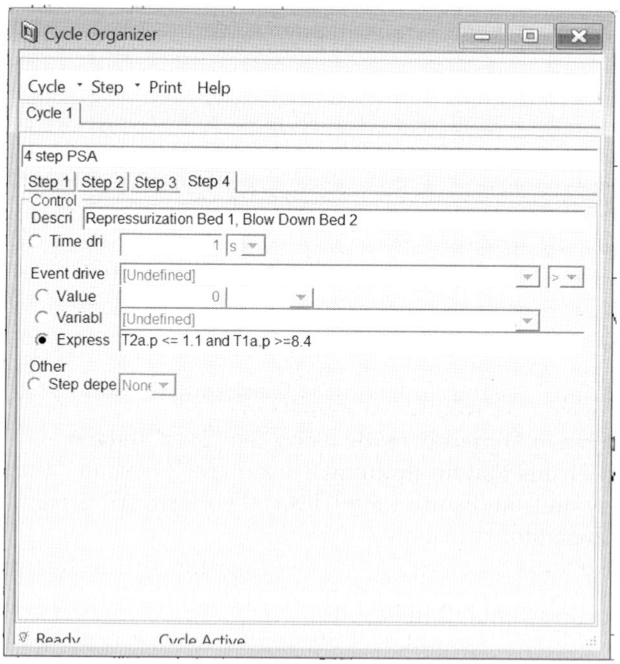

Figure 1.64 The event driving step 4 (pressurize bed 1, blowdown bed 2).

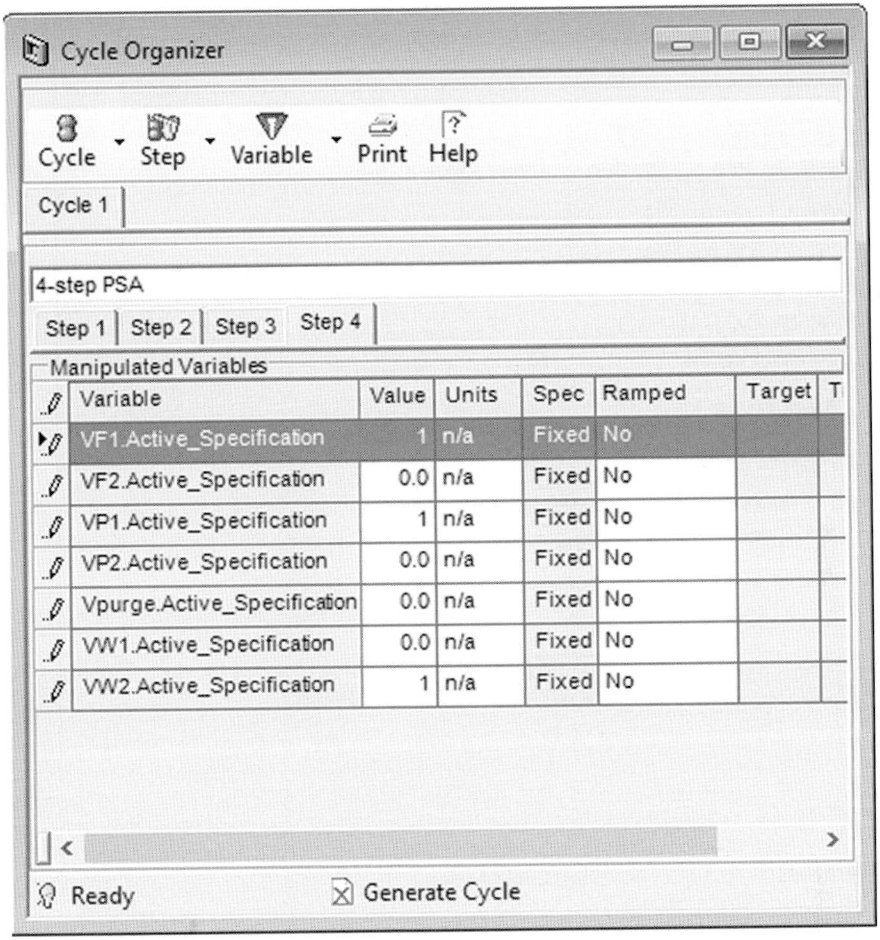

Figure 1.65 Step 4 manipulated variables (pressurize bed 1, blowdown bed 2).

Step 34: Add a plot of the oxygen product composition

To add a plot of the oxygen composition over time, open the flowsheet in the Explorer and click on the "Add Form" (Figure 1.70).

Once the form is created, drag the variables from the P1 block's "Results" table into the plot (Figure 1.71).

Right-click on the plot and "Zoom Full" in order to see the full time history of the oxygen product composition (Figure 1.72).

Step 35: Create an axial composition plot (Figure 1.73)

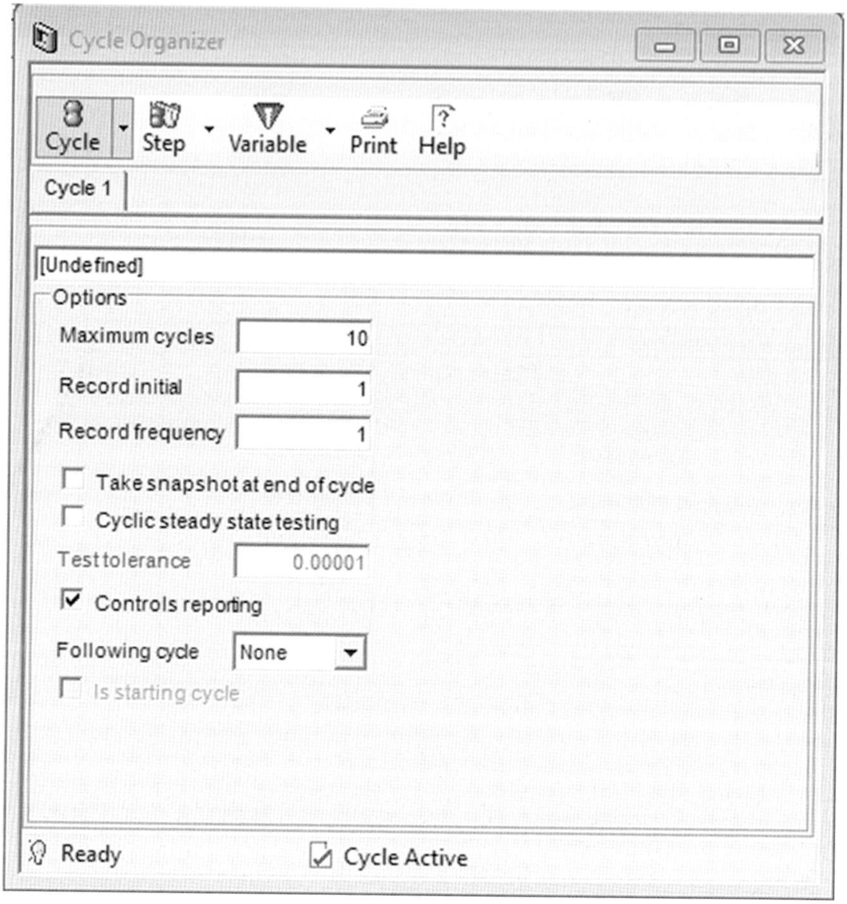

Figure 1.66 The Cycle Options.

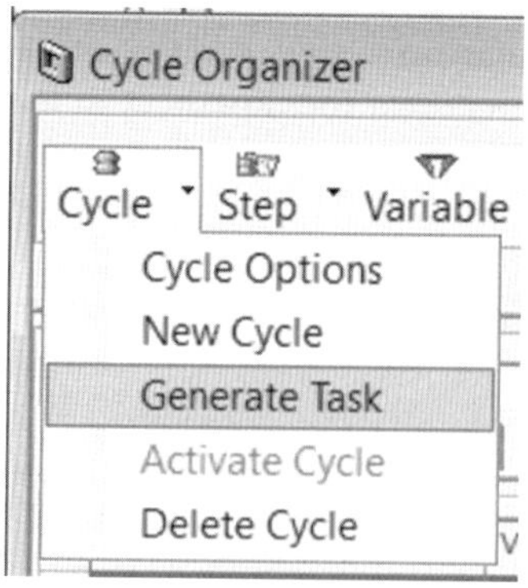

Figure 1.67 Generating the Cycle Organizer task.

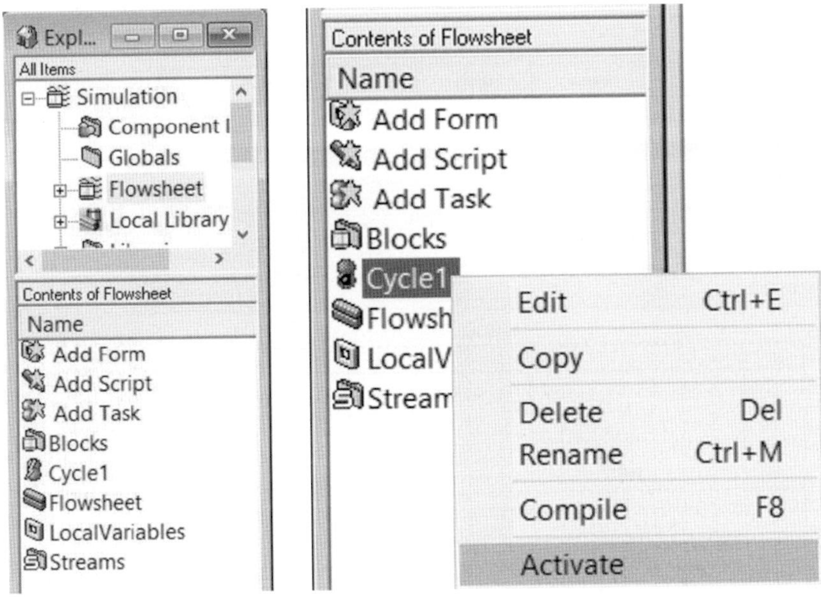

Figure 1.68 If the cycle task does not have a lightning bolt icon, use the right mouse button to activate it.

Figure 1.69 The run control buttons.

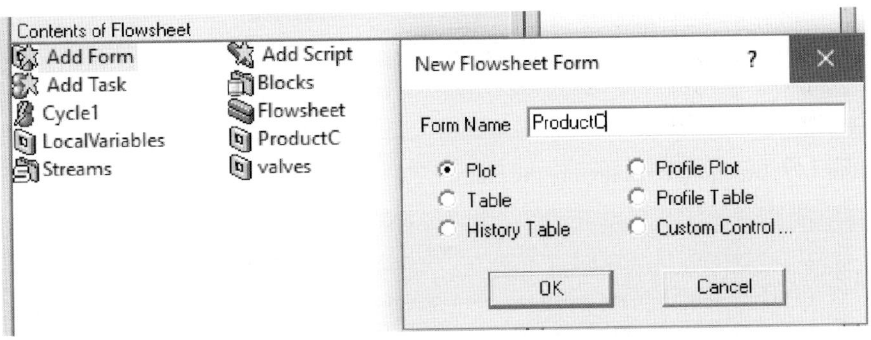

Figure 1.70 Adding a product composition plot.

Step 36: Profile Editor for the Axial_Composition profile plot

Position: B1. Layer(*).Axial_Distance(*)

N2 Partial Pressure: B1. Layer(*).Y (*,"N2")

O2 Partial Pressure: B1. Layer(*).Y (*,"O2")

Since our column has one layer, you could use Layer(1), instead of Layer(*); however, it is good practice to always use Layer(*) so that you always view the full column profile (Figure 1.74).

Step 37: Conclusion

Save the file as **Workshop1F.ada.**

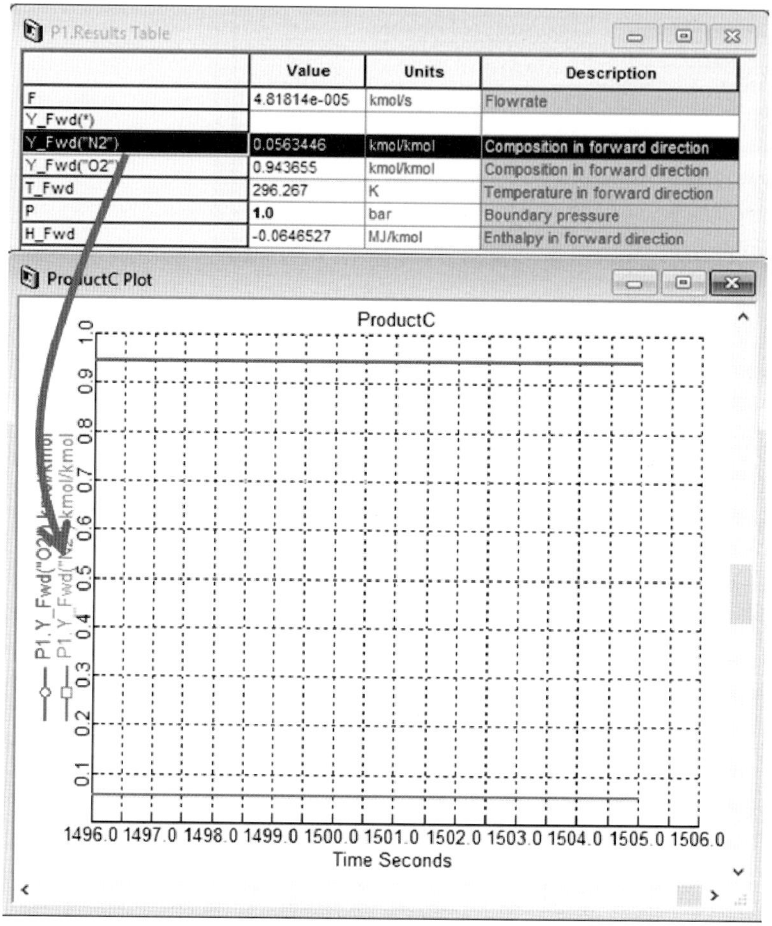

Figure 1.71 Dragging variables into the plot.

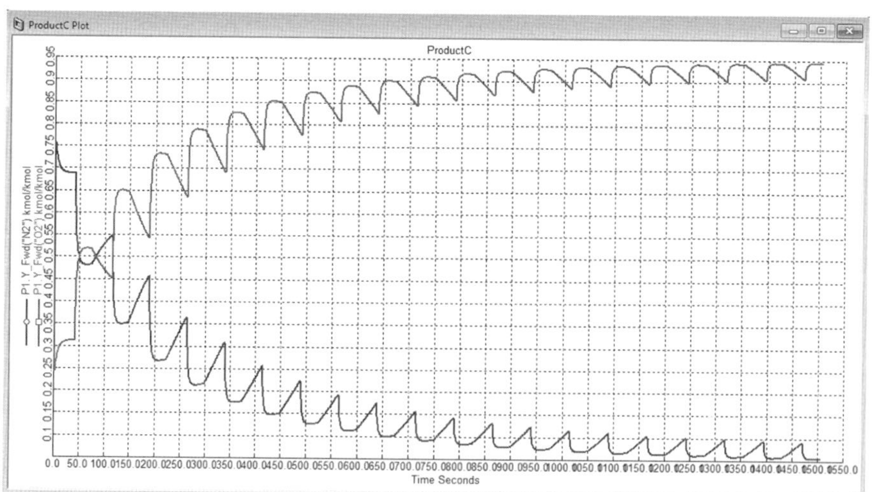

Figure 1.72 The composition plot in the light product stream (raffinate or "waste"stream or nitrogen-rich stream) over time in the PSA model.

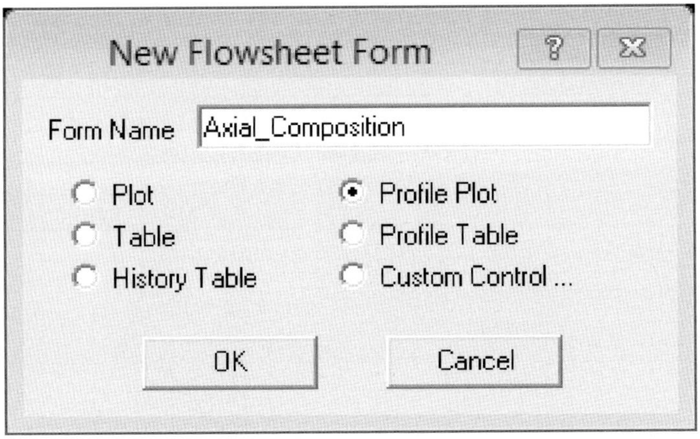

Figure 1.73 The generation of a concentration profile within the column.

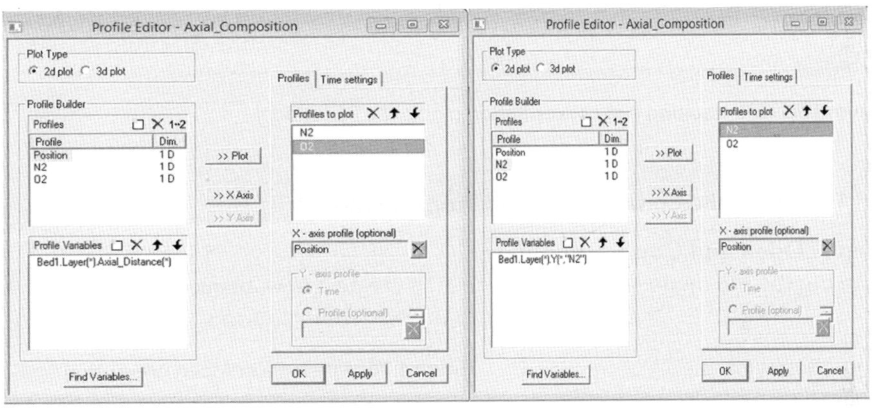

Figure 1.74 The axial composition profile plot in the PSA model.

In this workshop, we developed a two-bed PSA model for a binary separation of an O2/N2 mixture. Figure 1.75 shows how the composition of the gas phase changes during adsorption. In practice, you may consider also creating graphs of the axial pressure, or tank pressures over time.

1.8 PSA Workshop: Hydrogen Separation in Aspen Adsorption

Hydrogen isolation is an important component of several refinery processes, such as cleaning coke oven gas. In most cases, hydrogen is present alongside combustion products (CO/CO_2) and/or light hydrocarbons like methane. In this workshop, we will simulate a variety of hydrogen separation processes, and learn about Aspen Adsorption as we proceed.

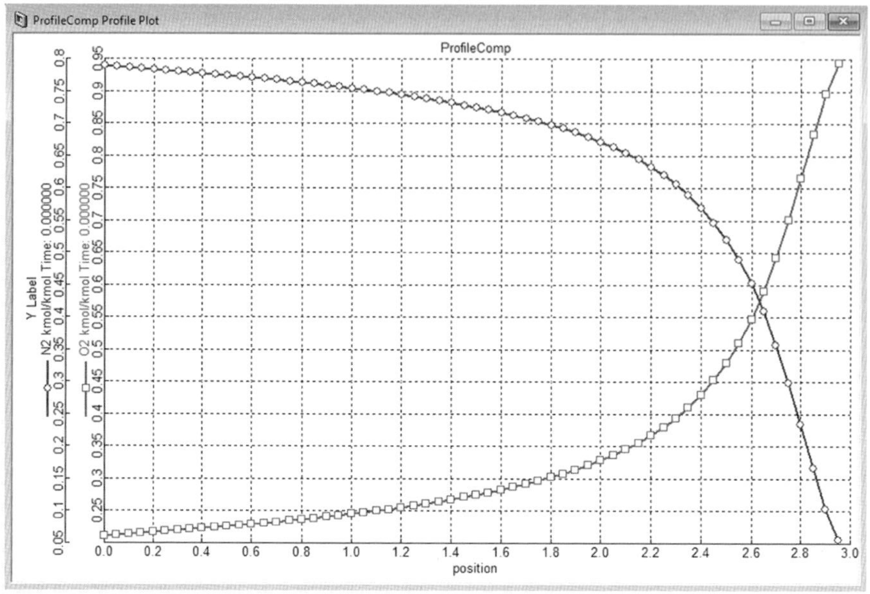

Figure 1.75 The resulting bed concentration plot in the PSA model.

1.8.1 Define the Components and Property Model

Step 1: Open and Configure Aspen

Using the file menu shown in Figure 1.76 create a new, blank flowsheet.

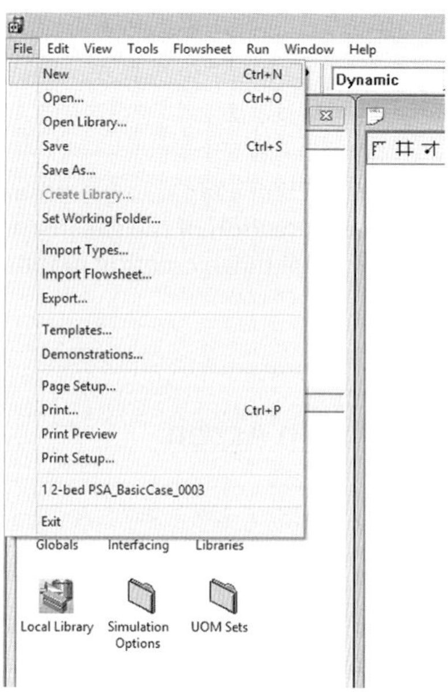

Figure 1.76 A new file in Aspen Adsorption.

Step 2: It is important to save Aspen simulations frequently

Save the file as `H2Workshop1.ada`. We will be saving the file after each major change; it is good practice to remember to save the file under a new name each time. That way, if a change you have made results in incorrect flowsheet behavior at a later time, you will not have to completely start over (Figure 1.77).

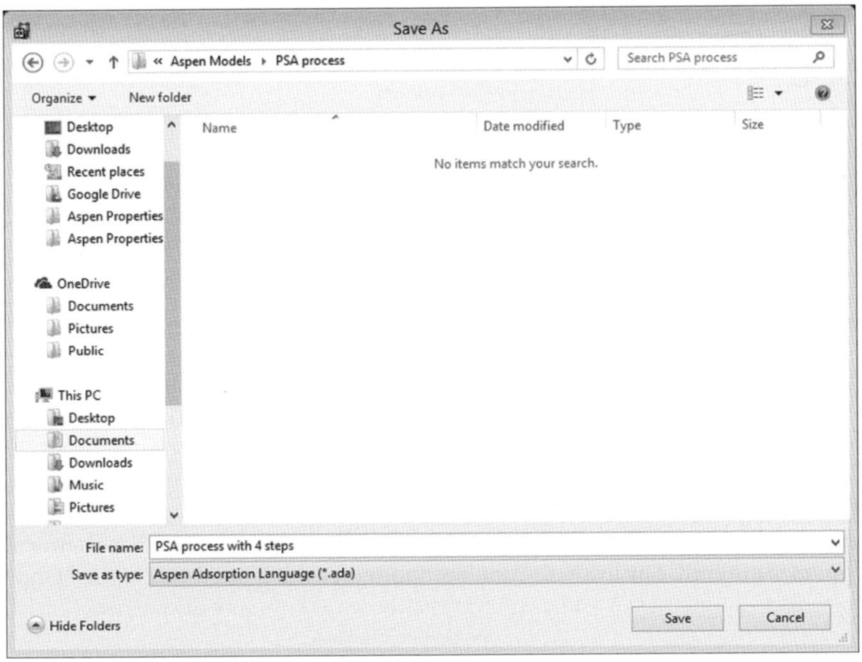

Figure 1.77 The PSA process with four steps for air separation.

Step 3: Add chemical components to Aspen's component list

The first thing to do in a simulation is to tell Aspen which chemicals will be present in our simulation. As we have seen in the previous workshop, Aspen Adsorption will automatically prompt the user to supply components as soon as a new flowsheet has been created. There are two methods for adding chemical components into the model: component list and component set, as discussed in Section 1.7.1.

When we use Aspen Properties for physical properties, we can invoke the Aspen Properties user interface directly if we have it installed. In order to edit physical property options, we need to follow the steps in Figure 1.78:

1) In the Simulation Explorer, click "**Component Lists**".
2) In the **Explorer of the Component Lists folder**, double-click the "**Configure Properties" node**.
3) In the **Physical Properties Configuration** dialog box, choose "Use Aspen property system" option, and then click "**Edit using Aspen Properties" button to launch the user interface**.
4) **Step 4: Add component names in Aspen Properties**

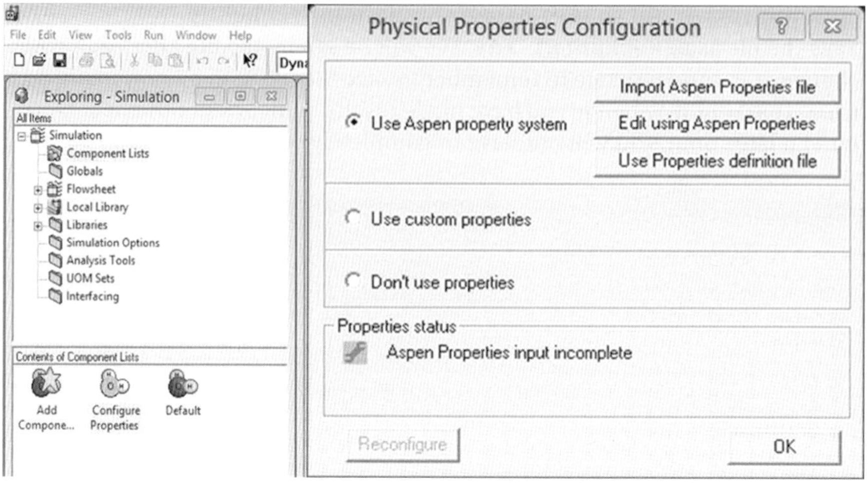

Figure 1.78 The physical properties configuration pane.

After we click "**Edit using Aspen Properties**" in the **Physical Properties Configuration** pane, it automatically opens the Aspen Properties for us to add component names and edit their physical and chemical properties. Figure 1.79 shows all the compounds that might be relevant for our hydrogen separation. To enter these yourself, you can type the names into the Component ID column. For most common chemicals, Aspen will know right away what chemical you have entered. For less common components, you'll need to use the "Find"

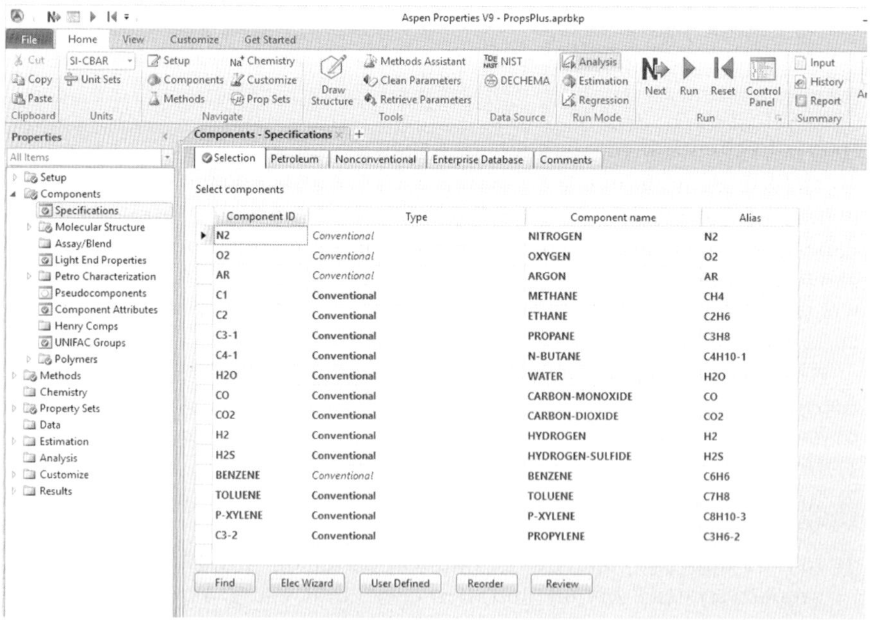

Figure 1.79 The Aspen Properties interface.

button to search. Figure 1.80 shows how to use the "Find" tool to get components from Aspen's databanks. Once you select propane in the list of matches, click on the "Add selected compounds" button to add it to the list. You may then search for more compounds if you wish. Use the "Find" tool to add all the components in Figure 1.79.

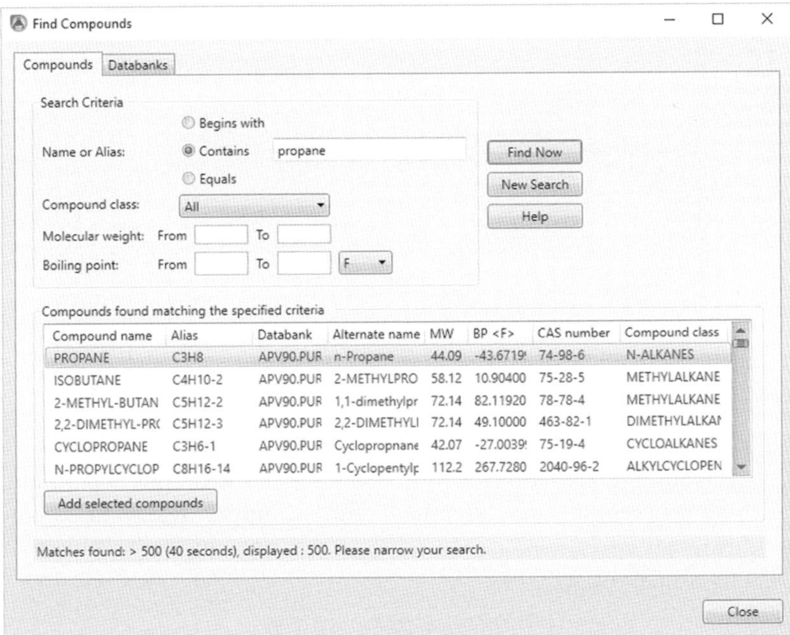

Figure 1.80 Searching the Aspen Property Databanks for Propane.

Step 5: Select the appropriate property method

In this step, we specify the methods that calculate the thermodynamic properties for our simulation. The "Base method" and "Method name" drop-down menus are essentially redundant, although the "Base method" option allows you to filter the long list of methods by application. For example, PENG-ROB shows up when the method-filter is set to COMMON; however, specialized variants of the Peng–Robinson method (such as PR-BM) do not show up unless the filter is set to "GasProc."

To edit components' physical properties in Aspen Properties, we follow the steps in Figure 1.81:

1) In the "All Items" pane of the "Properties", click "Specification" in the "Methods" folder.
2) Select "PENG-ROB" as the base method.
3) In the pane of "Home", click "Run" button.
4) Save the file until the run is completed in the Control Panel, as shown in Figure 1.82.

Once you have completed these steps, close the Aspen properties window and choose "yes" when asked to save the properties file.

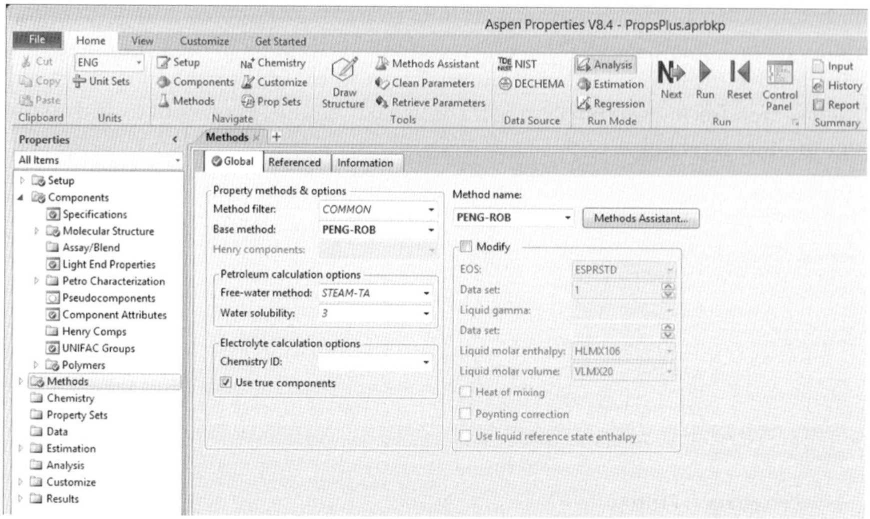

Figure 1.81 The specification of the components in Aspen Properties.

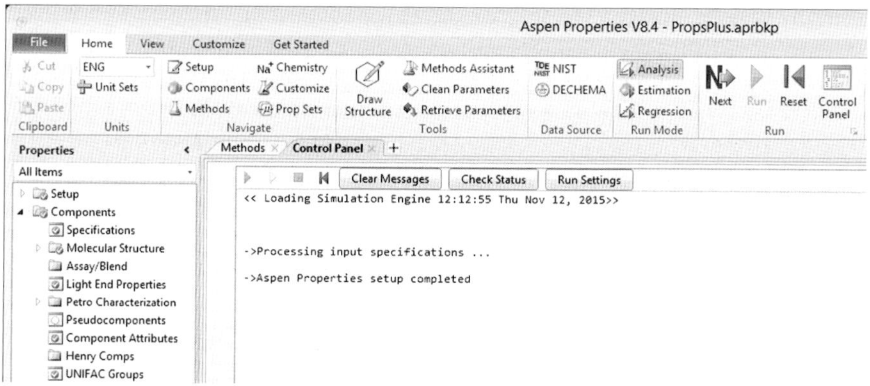

Figure 1.82 The completion in control panel in Aspen Properties.

Step 6: Tell Aspen Adsorption to use the Aspen Properties file we created

Once we have done the import in Aspen Properties, the Aspen Properties input file, **PropsPlus.aprbkp**, will be included in the file when we save the simulation file. Then go back to the Physical Properties Configuration panel, and the Properties status should show a green square next to "Configured using Embedded Aspen Properties" (Figure 1.83).

Click ok on the properties configuration window to be presented with the resulting component list. In order to make the components actually available in the flowsheet, they need to be moved into the right-hand column. To do this, click on the component you want, then press the right arrow, >. Add H2, CO and methane to the right-hand column, then click "ok".

Once you have completed this step, save the file as `H2Workshop2.ada`.

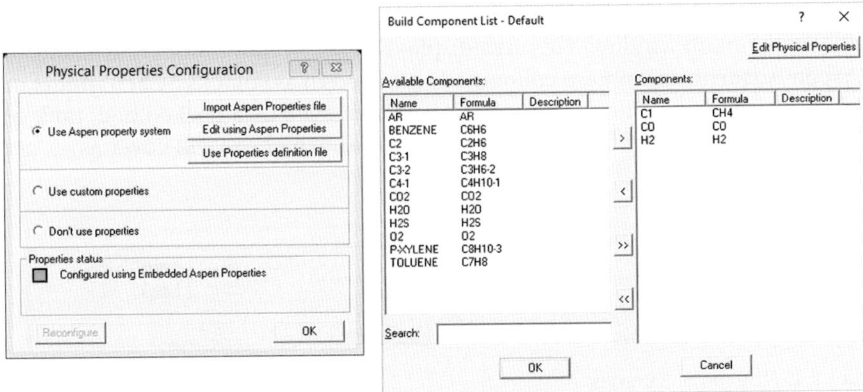

Figure 1.83 The completion of the physical properties configuration.

1.8.2 Creating a Flowsheet in Aspen Adsorption

Now that we have successfully added several chemical components to Aspen, it is time to begin specifying the layout of our process. Aspen provides a large number of "Models" that are contained in "Model Libraries." These models roughly correspond to common unit operations, but take care not to confuse these models with actual physical process equipment. For example, a physical packed bed might be represented as a combination of three separate "Models" in Aspen. Specifically, two gas_tank_void models can describe the empty voids at the top and bottom of the packed bed, and also serve as **pressure setters** to fix the pressure drop across the adsorption bed; while the gas-bed model will describe the region of the packed bed that actually performs the separation. In the same way, a single Aspen "Model" might represent several physical pieces of equipment. For example, a single gas_interaction model can be used to represent several identical packed beds.

Step 7: Ensure that the model libraries are visible

To begin, ensure that the model libraries are visible by clicking on the "Model Libraries" button in the "View" menu (Figure 1.84).

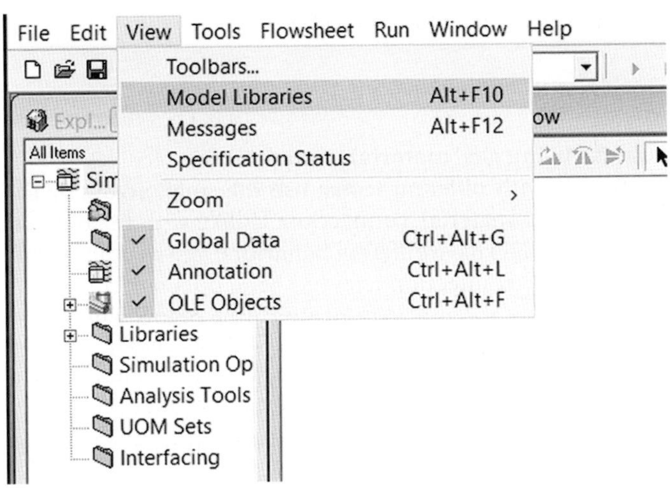

Figure 1.84 Display of the model libraries.

Once the button has been clicked, the area shown in Figure 1.85 will be visible on the main window. This area displays all the prebuilt models available in Aspen Adsorption, and are colloquially referred to as "blocks." Each model or block roughly corresponds to a unit operation, such as a packed bed, tank or valve. However, some models have subtler roles, which we will cover as we get to them.

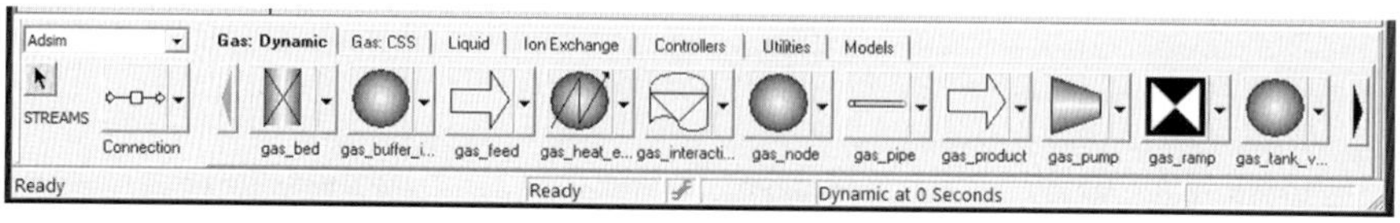

Figure 1.85 The model libraries.

Step 8: Place the first blocks on the flowsheet

To place these blocks on the flowsheet, first click on the block in the model library, then click on the flowsheet. You can add as many blocks of that type as you want by clicking multiple times. Click with the right mouse button to stop adding blocks. For our first step, we will add three blocks to the flowsheet, a gas_bed, a gas_feed, and a gas_product. Add them as shown in Figure 1.86. Note that you may want to rename the blocks; you can do this by clicking on a block, then pressing ctrl+m. This is especially useful here since there is no visual way to differentiate between the feed and product blocks.

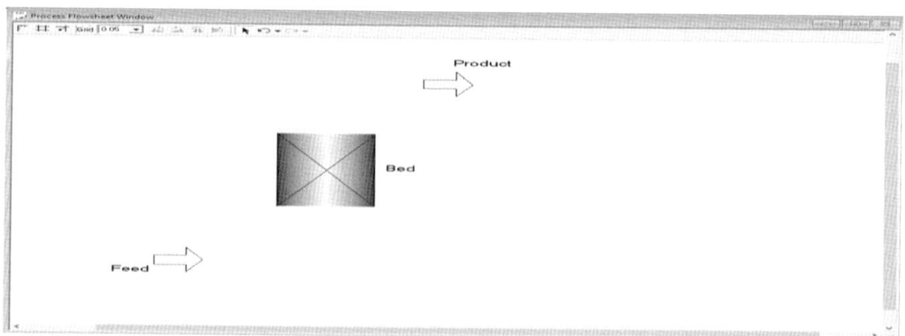

Figure 1.86 Placing the first blocks on the flowsheet.

Step 9: Connect the blocks using gas_material_connections

Similarly to placing models with different icons, use the small arrow by the "connection" icon to select gas_material_connection. Unlike with the models, the different connection types DO have different behaviors, and so it does matter which is selected. Once the connection icon has been selected, the blocks in the flowsheet will expose blue and red arrows, indicating points where they can be connected together. You can connect the blocks in any order by clicking on the arrow of one block, and an arrow of a second block. Blocks may accept more than one connection to the same point (for example, a gas_tank_void can accept many inlet and outlet streams simultaneously.) If you want to change the location of a connection for aesthetics, hold down the mouse button while clicking on the blue or red arrow, and drag it to the desired location before releasing. The connection options and arrows, are shown in Figure 1.87.

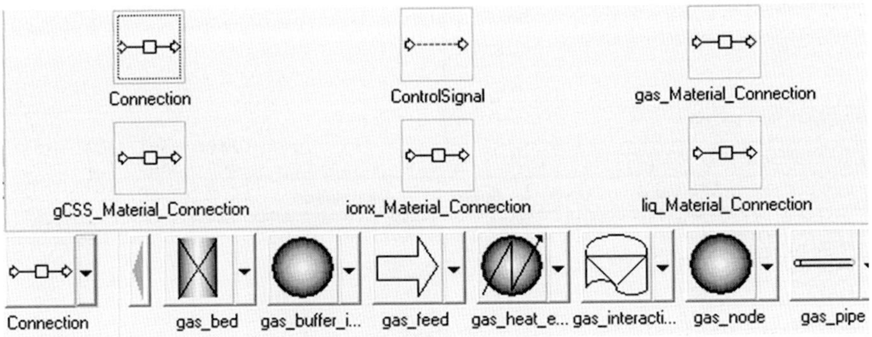

Figure 1.87 The methods for connecting blocks.

Connect the blocks as shown in Figure 1.88.

Figure 1.88 The connected blocks.

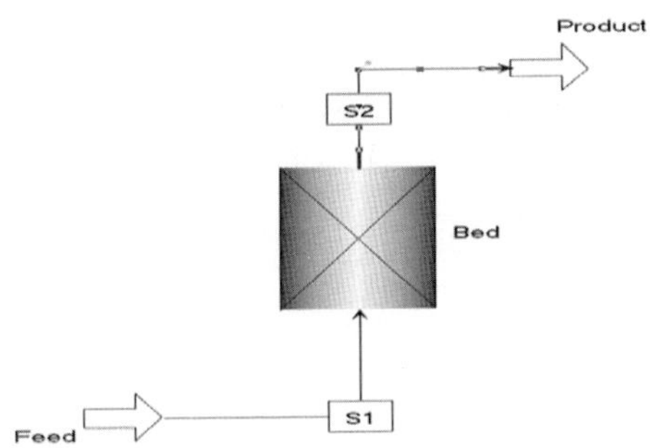

Save the resulting simulation as **`H2Workshop3.ada`**.

1.8.3 Run a Breakthrough Simulation

Step 10: Specify the feed block and the product block

What we have specified so far is clearly not a PSA process. All we have done is to create a model of a gas stream flowing through an adsorber bed. However, this is still a useful model. We can use this model to predict breakthrough curves. In order to do this, we need to first tell Aspen about the gas stream that will be flowing through the bed. Double-click on the "Feed" block to open the default form. This shows us some options for the feed block. For right now, all we want is to "Specify" the values using the "Specify" button (Figure 1.89). Note that the feed block also serves as a pressure setter to fix the inlet pressure of the adsorber bed.

Enter the values shown in Figure 1.89, then close the windows.

Specify the "Product" block in the same way. The Product block also sets the outlet pressure for the adsorber bed (Figure 1.90).

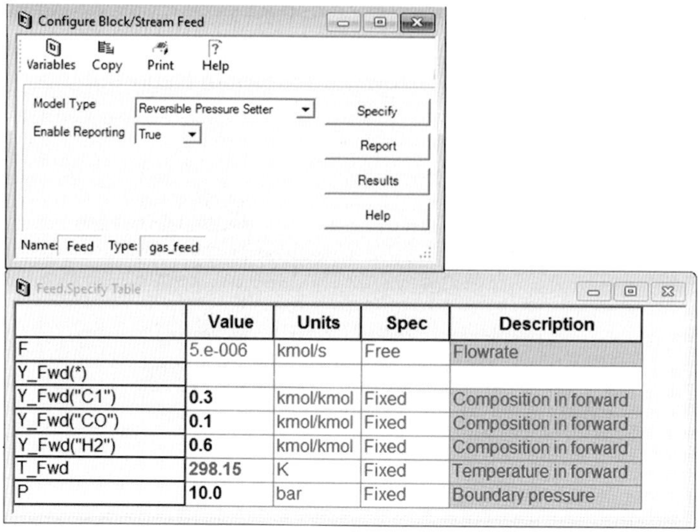

	Value	Units	Spec	Description
F	5.e-006	kmol/s	Free	Flowrate
Y_Fwd(*)				
Y_Fwd("C1")	0.3	kmol/kmol	Fixed	Composition in forward
Y_Fwd("CO")	0.1	kmol/kmol	Fixed	Composition in forward
Y_Fwd("H2")	0.6	kmol/kmol	Fixed	Composition in forward
T_Fwd	298.15	K	Fixed	Temperature in forward
P	10.0	bar	Fixed	Boundary pressure

Figure 1.89 Specifying the feed block.

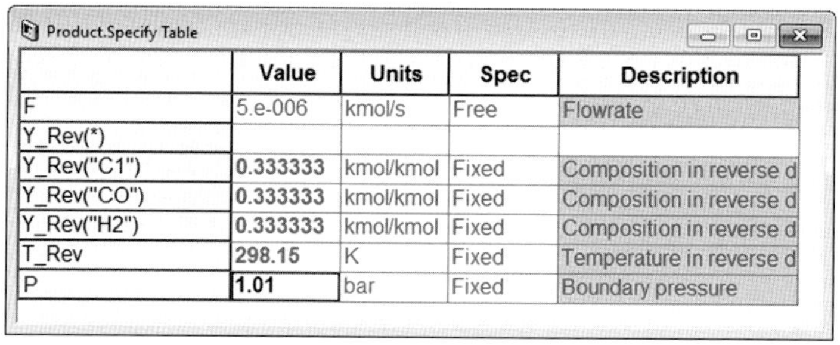

	Value	Units	Spec	Description
F	5.e-006	kmol/s	Free	Flowrate
Y_Rev(*)				
Y_Rev("C1")	0.333333	kmol/kmol	Fixed	Composition in reverse d
Y_Rev("CO")	0.333333	kmol/kmol	Fixed	Composition in reverse d
Y_Rev("H2")	0.333333	kmol/kmol	Fixed	Composition in reverse d
T_Rev	298.15	K	Fixed	Temperature in reverse d
P	1.01	bar	Fixed	Boundary pressure

Figure 1.90 Set the product pressure to 1.01 bar.

Step 11: Specify the Adsorber bed

Double-click on the gas_bed to open its default form. This form is different from the feed block, since the gas_bed model is more complex. The gas_bed form allows you to create packed beds consisting of any number of adsorbent layers. For this separation, we will use two layers, to represent the layered bed from Jee et al. [24], which used zeolite 5A and activated carbon. To create a two-layer bed, simply enter 2 in the Number of Layers box, as shown in Figure 1.91.

Now, each layer has its own "configure" and "specify" buttons. Generally speaking, the Configure tab is where you choose the assumptions you make about the physical behavior of the layer. The Specify tab is where you provide the numeric values for the layer's physical properties. Changing the assumptions in the configure tab can change what variables are present in the specify tab.

Step 12: Specify the first layer

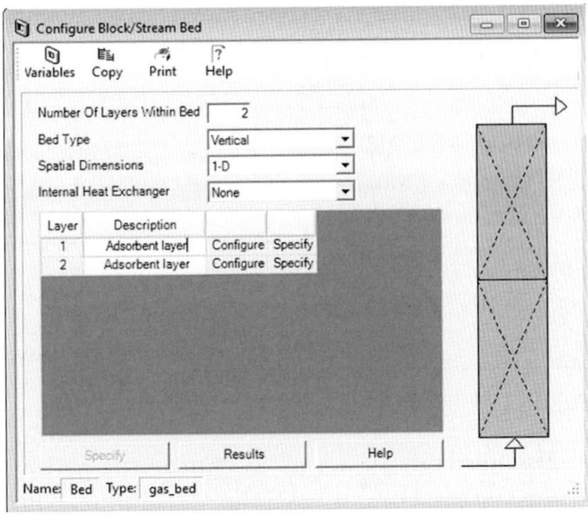

Figure 1.91 Configuring the gas_bed block.

Click on the "configure" button for Layer 1. The first tab that appears is the Finite Element discretization method options. In this tab, you can specify exactly how the PDE defined by the gas_bed will be solved. For this example, we will use *a Central Differencing Scheme, CDS2* with 16 nodes. This will allow for very fast simulation times when we include axial dispersion term in our momentum balance according to the online help file in Aspen Adsorption. Figure 1.92 shows this selection.

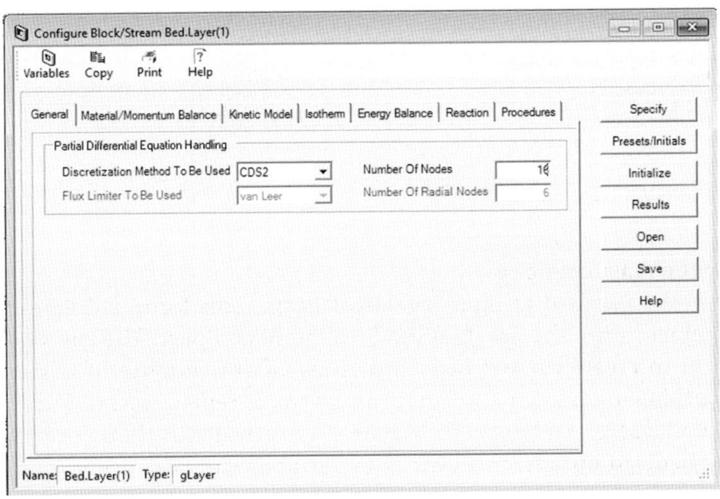

Figure 1.92 Choose the second-order central difference scheme (CDS2) to improve the simulation efficiency with axial dispersion effect.

The next tab allows us to choose the momentum and material balances we wish to calculate for the system. Both of our layers will use "convection with constant dispersion" and the Ergun Equation, Eq. (1.11), for pressure drop (Figure 1.93).

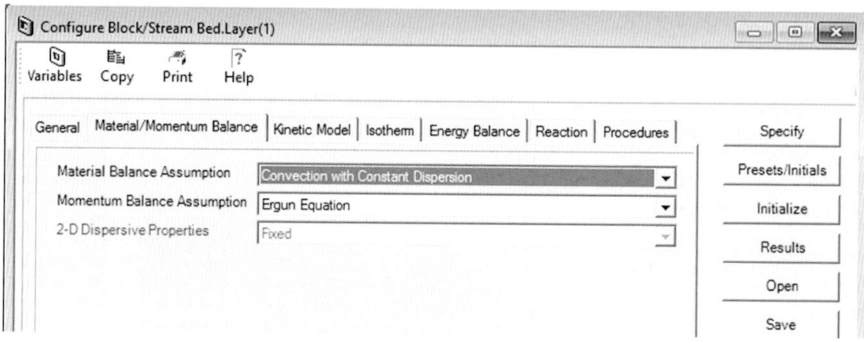

Figure 1.93 Specifying the material balances within the bed.

The next tab allows us to choose the assumptions regarding the kinetics of the system. For our case, the default linear-lumped resistance or LDF assumption is acceptable, and the resistance should be based on the solid-phase concentrations. Refer to Eq. (1.14) in Section 1.7.3 about the LDF assumption (Figure 1.94).

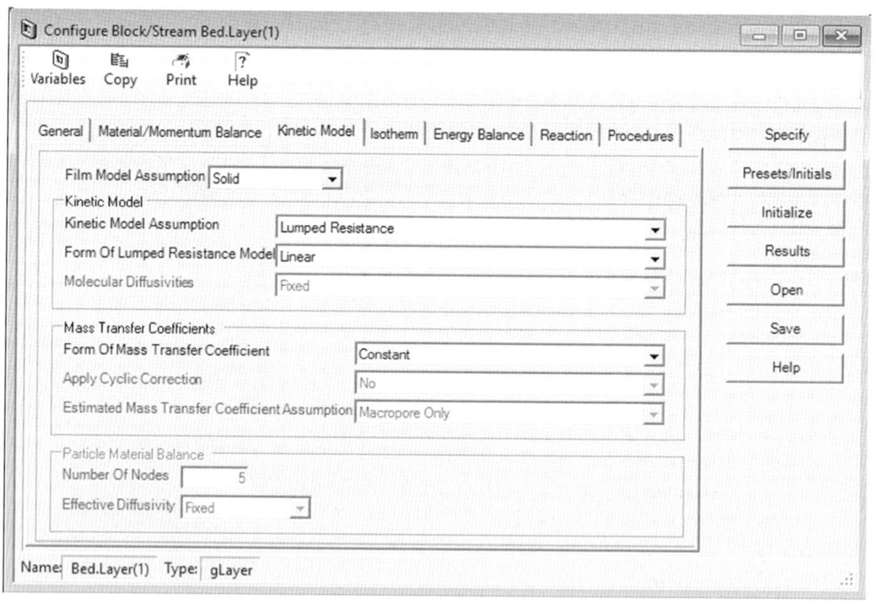

Figure 1.94 The kinetic assumptions.

Next, we choose the equilibrium behavior, the isotherm assumptions. For both layers, an Extended Langmuir 3 model based on partial pressure is appropriate. Note that this isotherm does not perfectly match that reported in Jee et al. [24], and we will revisit this discrepancy later. After choosing the isotherm, you should see Figure 1.95.

The final tab we will make changes to is the energy balance tab. This tab allows a wide range of assumptions about the heat transfer in the system. For

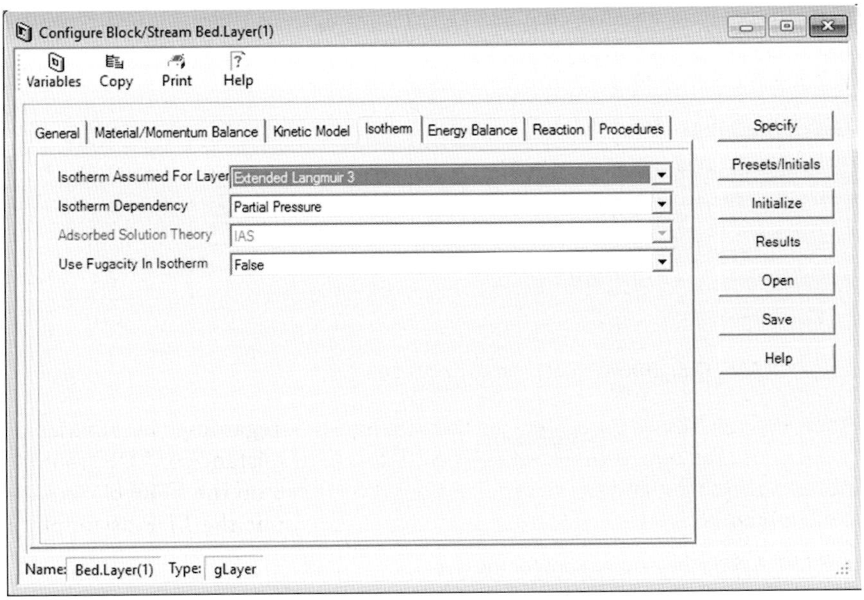

Figure 1.95 Isotherm selection.

this bed, the only heat transfer occurs by conduction in the gas phase. We will be performing a rigorous energy balance with the wall, but all heat capacities and heats of adsorption are constant. The thermal conductivity of the gas is a function of the axial dispersion. To make these assumptions, choose the options shown in Figure 1.96.

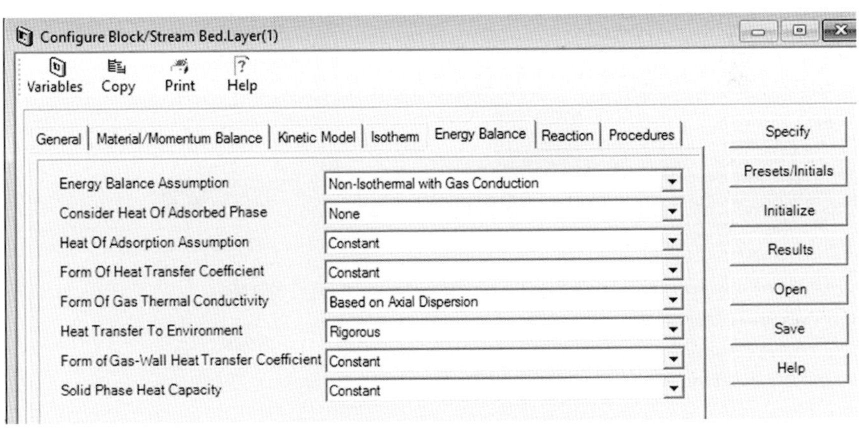

Figure 1.96 The energy balance and heat transfer assumptions.

With this, we have completed the assumption configuration for the first layer. Do not close the configure window yet though!

Step 13: Specify the assumptions for the second layer

The assumptions for this layer are identical to the first. To quickly copy the settings from one layer to the next, we will use the "Save" button in the right-hand panel to create a file that contains all the assumptions (Figure 1.97). Save the file as **HydrogenAssumptions.ada**.

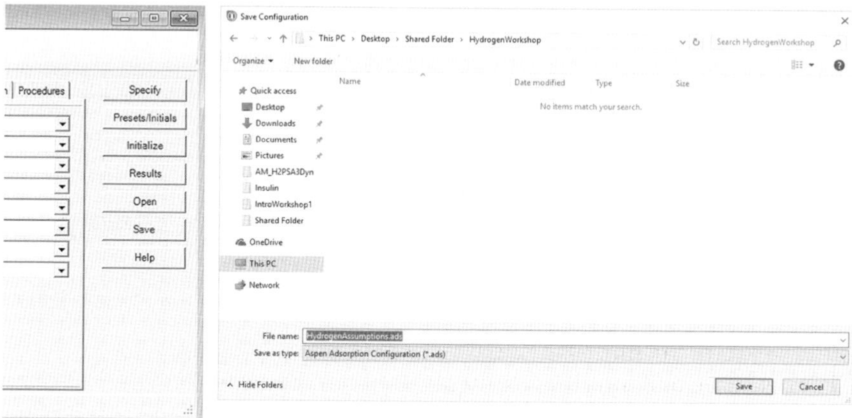

Figure 1.97 Saving the assumptions for the layer.

Once we have created the assumption file, we can open it in the second layer (Figures 1.98 and 1.99).

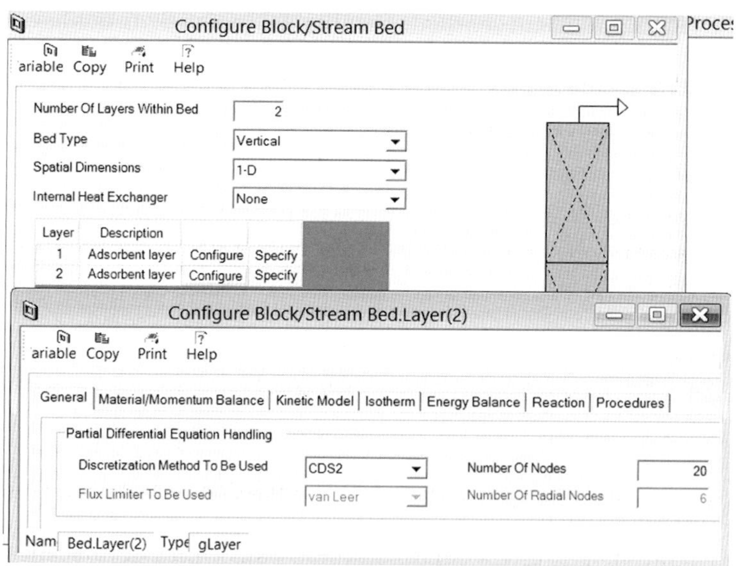

Figure 1.98 Opening the configure form for layer 2.

Once we have opened the file, confirm that all the parameters match what were specified in step 13. Pay attention to the number of nodes in the PDE discretization method.

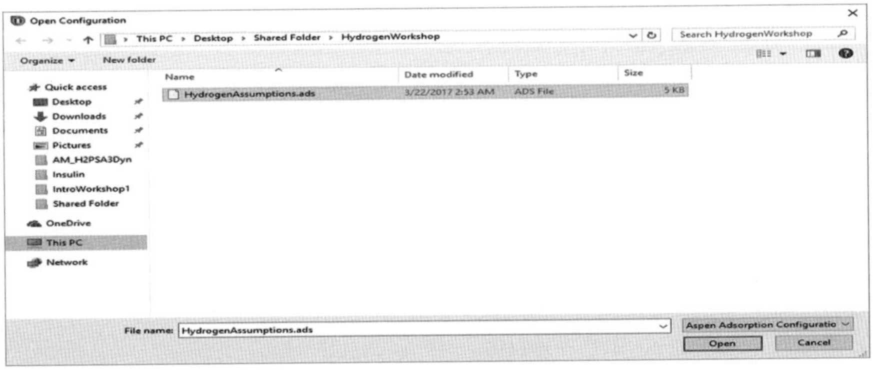

Figure 1.99 Selecting the configuration file from the first layer.

Step 14: Specify the numeric values

So far we have only specified assumptions about the layers, but it is now time to provide the numbers required for the various models we chose. To do this, click on the "specify" button for both layers. Enter the information in Tables 1.8 and 1.9 into the layer specification forms.

Table 1.8 Data shared between layers 1 and 2.

Bed height	0.5 m
Bed diameter	0.1855 m
Bed wall thickness	0.00134 m
Dispersion coefficient (all components)	1.5e−005
Heat of adsorption (CO)	−22.175 MJ $kmol^{-1}$
Heat of adsorption (C1)	−20.92 MJ $kmol^{-1}$
Heat of adsorption (H2)	−11.715 MJ $kmol^{-1}$
Fluid/solid heat transfer coefficient	1 MW/(m^2 K^{-1})
Fluid/wall heat transfer coefficient	3.851e−005 MW/(m^2 K^{-1})
Wall/Env heat transfer coefficient	1.423e−005 MW/(m^2 K^{-1})
Wall thermal conductivity	1e−20 MW/(m K^{-1})
Wall material heat capacity	5.0232e−4 MJ/(kg K^{-1})
Wall material density	7830 kg m^{-3}

Once these values have been entered, save the simulation as `H2Workshop4.ada` (Figure 1.100).

Double-click on the green–red–triangle button. We see the specification status indicating that the current simulation is underspecified by 12 initial variables. See Figure 1.101.

Step 15: Initialize the simulation

Now that the values have all been entered, the simulation should be ready to run. Unfortunately, there is one final impediment. Double-click on the green–red–triangle button. We see the specification status indicating that the

Table 1.9 Data that are distinct for each layer.

Variable	Layer 1	Layer 2	Units
Particle radius	0.001 15	0.001 57	m
Bed voidage	0.433	0.357	—
Intraparticle voidage	0.61	0.65	—
Surface area/volume ratio	2608.7	1910.83	$1\ m^{-1}$
Particle density	850	1160	$kg\ m^{-3}$
Mass transfer coefficient (CO)	0.15	0.063	$1\ s^{-1}$
Mass transfer coefficient (C1)	0.195	0.147	$1\ s^{-1}$
Mass transfer coefficient (H2)	0.7	0.7	$1\ s^{-1}$
Solid phase heat capacity	0.001 046 5	9.209e−004	$MJ/(kg\ K^{-1})^{-1}$
IP(1,"C1")	0.023 86	0.005 833	Isotherm parameter
IP(1,"CO")	0.033 85	0.011 845	Isotherm parameter
IP(1,"H2")	0.016 943	0.004 314	Isotherm parameter
IP(2,"C1")	−5.62e−05	−1.19e−05	Isotherm parameter
IP(2,"CO")	−9.07e−05	−3.13e−05	Isotherm parameter
IP(2,"H2")	−2.10e−05	−1.06e−05	Isotherm parameter
IP(3,"C1")	0.003 478	6.51e−04	Isotherm parameter
IP(3,"CO")	2.31e−04	0.020 2	Isotherm parameter
IP(3,"H2")	6.25e−05	0.002 515	Isotherm parameter
IP(4,"C1")	1159	1731	Isotherm parameter
IP(4,"CO")	1751	763	Isotherm parameter
IP(4,"H2")	1229	458	Isotherm parameter

current simulation is underspecified by 12 initial variables. See Figure 1.101. This message means that the degrees-of-freedom analysis has found that the model we have built needs additional information before Aspen can solve it.

Now that it certainly looks like we have provided all the required information, we need to see what is missing. The answer is that the gas_bed block needs to be "initialized" so that a few additional internal variables will automatically be set. To do this, use the "flowsheet" menu to click on "Check and Initial" Following this, the simulation should be ready to run, because the initialization logic will have fixed the underspecification problem (Figure 1.102).

Step 16: Run the simulation

Figure 1.103 shows the controls for running simulations. As it stands, our simulation has no defined "end" criteria. If we run the simulation using the solid blue "play" icon, it would run until we manually stop it with the "pause" button. So first, click on the "run options" button represented by the play button with red circles. Here, we tell the simulation to pause after 100 s.

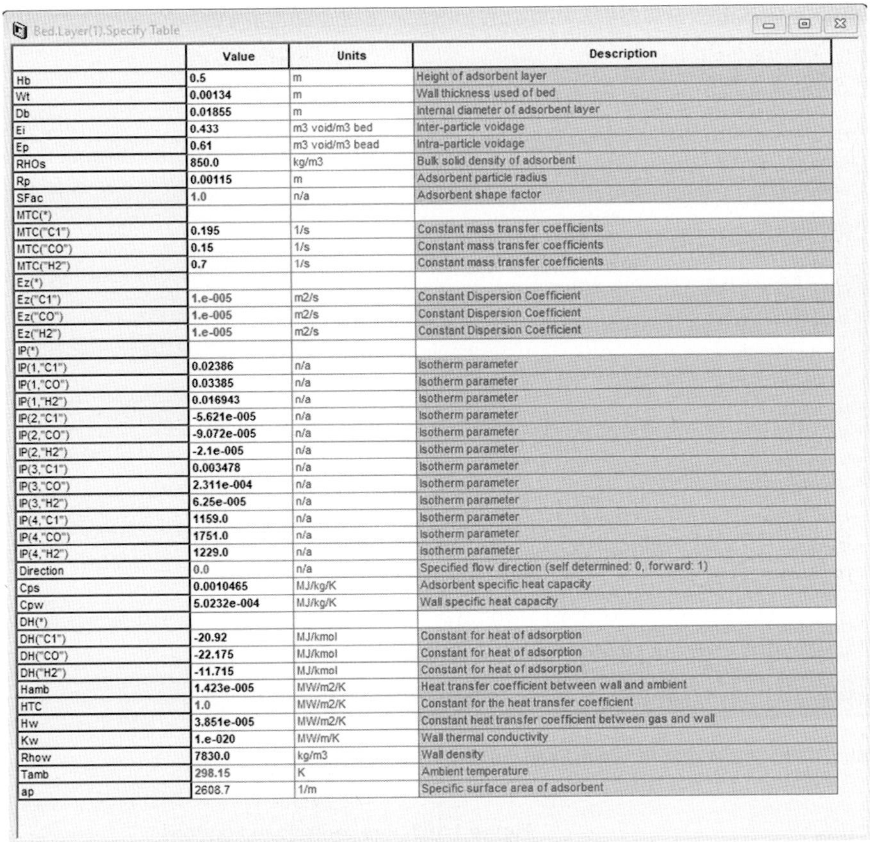

	Value	Units	Description
Hb	0.5	m	Height of adsorbent layer
Wt	0.00134	m	Wall thickness used of bed
Db	0.01855	m	Internal diameter of adsorbent layer
Ei	0.433	m3 void/m3 bed	Inter-particle voidage
Ep	0.61	m3 void/m3 bead	Intra-particle voidage
RHOs	850.0	kg/m3	Bulk solid density of adsorbent
Rp	0.00115	m	Adsorbent particle radius
SFac	1.0	n/a	Adsorbent shape factor
MTC(*)			
MTC("C1")	0.195	1/s	Constant mass transfer coefficients
MTC("CO")	0.15	1/s	Constant mass transfer coefficients
MTC("H2")	0.7	1/s	Constant mass transfer coefficients
Ez(*)			
Ez("C1")	1.e-005	m2/s	Constant Dispersion Coefficient
Ez("CO")	1.e-005	m2/s	Constant Dispersion Coefficient
Ez("H2")	1.e-005	m2/s	Constant Dispersion Coefficient
IP(*)			
IP(1,"C1")	0.02386	n/a	Isotherm parameter
IP(1,"CO")	0.03385	n/a	Isotherm parameter
IP(1,"H2")	0.016943	n/a	Isotherm parameter
IP(2,"C1")	-5.621e-005	n/a	Isotherm parameter
IP(2,"CO")	-9.072e-005	n/a	Isotherm parameter
IP(2,"H2")	-2.1e-005	n/a	Isotherm parameter
IP(3,"C1")	0.003478	n/a	Isotherm parameter
IP(3,"CO")	2.311e-004	n/a	Isotherm parameter
IP(3,"H2")	6.25e-005	n/a	Isotherm parameter
IP(4,"C1")	1159.0	n/a	Isotherm parameter
IP(4,"CO")	1751.0	n/a	Isotherm parameter
IP(4,"H2")	1229.0	n/a	Isotherm parameter
Direction	0.0	n/a	Specified flow direction (self determined: 0, forward: 1)
Cps	0.0010465	MJ/kg/K	Adsorbent specific heat capacity
Cpw	5.0232e-004	MJ/kg/K	Wall specific heat capacity
DH(*)			
DH("C1")	-20.92	MJ/kmol	Constant for heat of adsorption
DH("CO")	-22.175	MJ/kmol	Constant for heat of adsorption
DH("H2")	-11.715	MJ/kmol	Constant for heat of adsorption
Hamb	1.423e-005	MW/m2/K	Heat transfer coefficient between wall and ambient
HTC	1.0	MW/m2/K	Constant for the heat transfer coefficient
Hw	3.851e-005	MW/m2/K	Constant heat transfer coefficient between gas and wall
Kw	1.e-020	MW/m/K	Wall thermal conductivity
Rhow	7830.0	kg/m3	Wall density
Tamb	298.15	K	Ambient temperature
ap	2608.7	1/m	Specific surface area of adsorbent

Figure 1.100 The Layer(1) specify table.

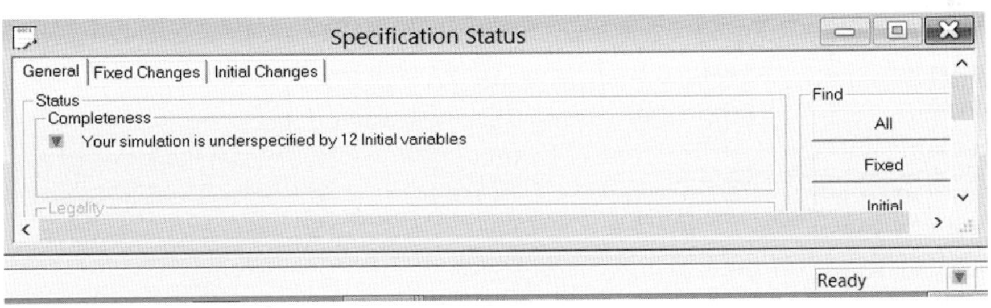

Figure 1.101 An underspecified simulation.

Before running the simulation, save the file as **`H2Workshop5.ada`**. Once you run the simulation, you should see messages like those shown in Figure 1.104.

Step 17: View breakthrough results

To view a plot of the concentration profile, we need to create a plot in the Flowsheet section of the Explorer window. Double-click on the "Add Form" tool

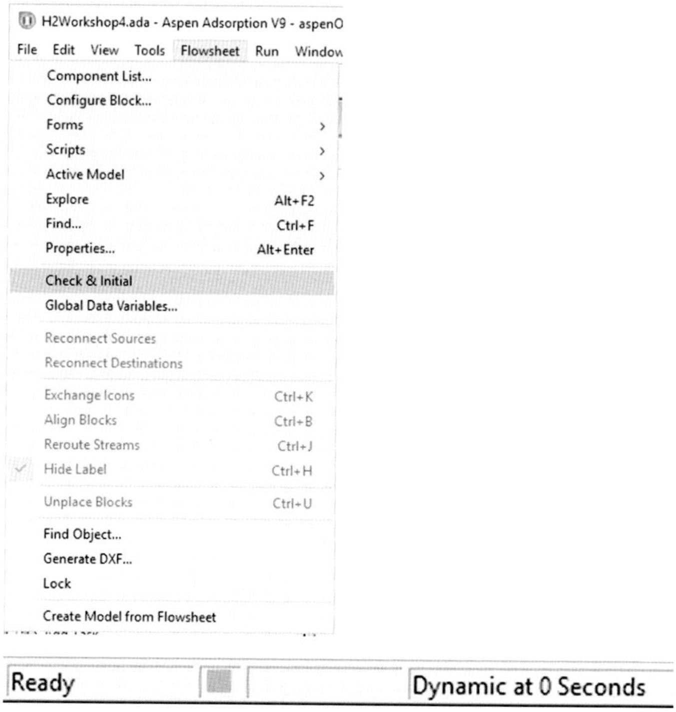

Figure 1.102 Initializing the flowsheet.

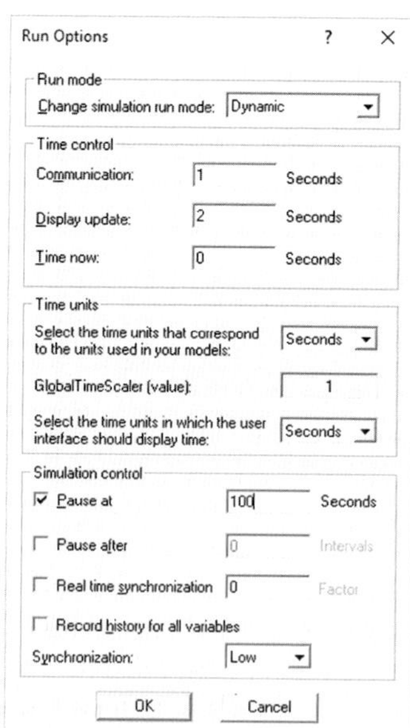

Figure 1.103 Run controls and options.

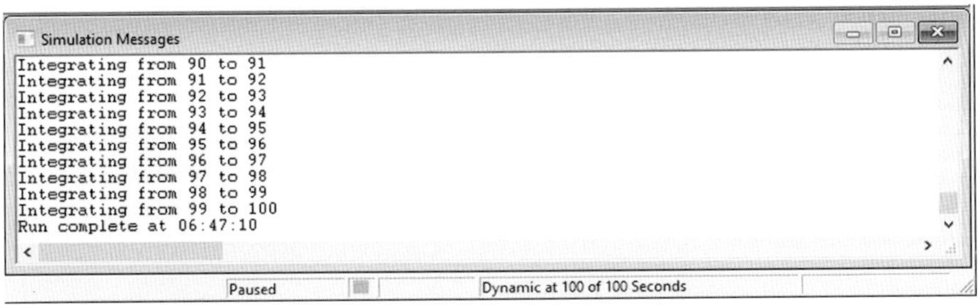

Figure 1.104 Messages from a successful simulation.

shown within Contents of Flowsheet in Figure 1.105 to begin the process of adding a plot.

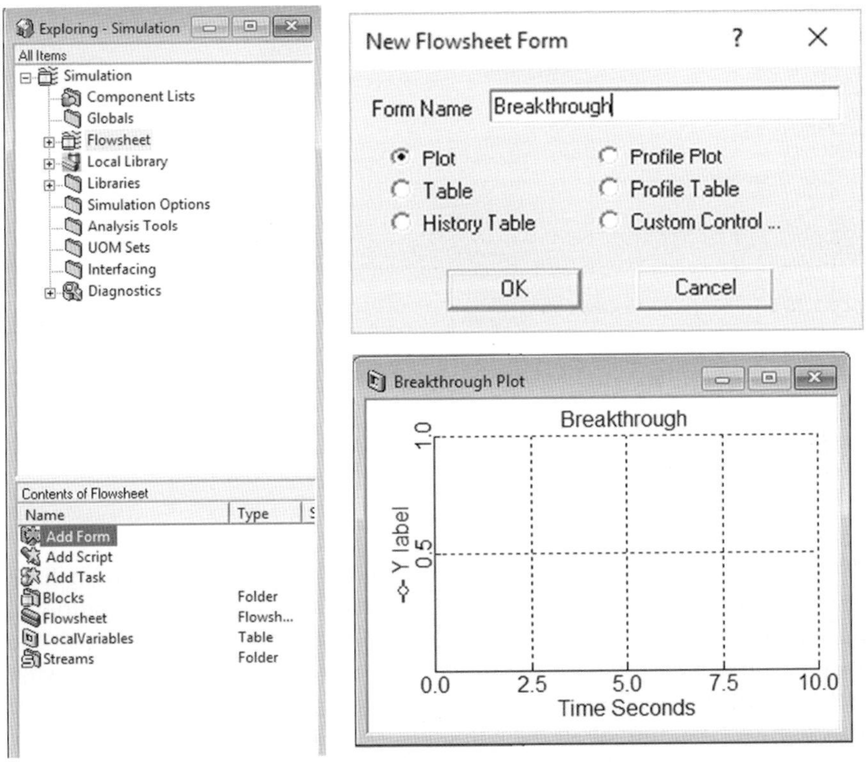

Figure 1.105 The "Add form" tool.

Name the new form "Breakthrough" and ensure that the "Plot" radio button is selected. The result will be a blank plot. In order to add data to this plot, we can click and drag variables from any other form. In this case, we want to see the "Results" from the product block. Double-click on the "Product" block, then click on "Results" to view the results table. Select the three variables shown in Figure 1.106 and drag them onto the y-axis of the plot that is created.

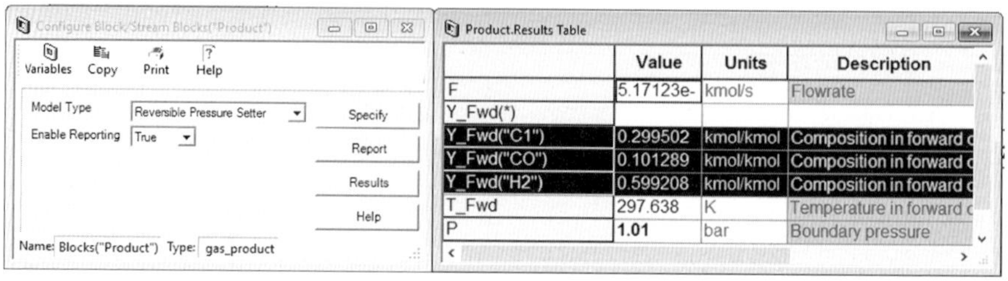

Figure 1.106 The product results.

Once the variables have been added, we need to show the full simulation history. To do this, right-click on the graph and choose "Zoom Full" as shown in Figure 1.107.

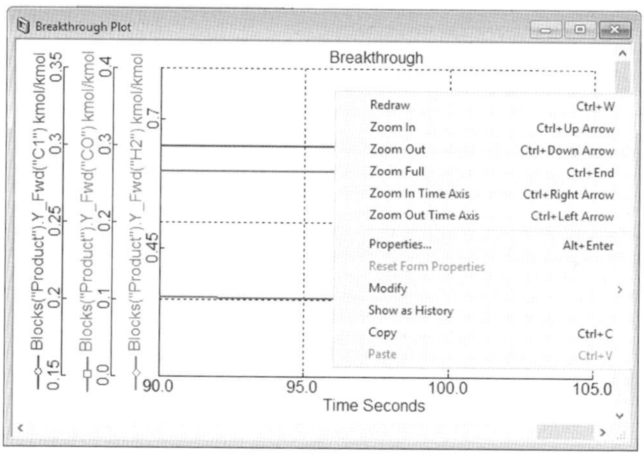

Figure 1.107 View the full history.

Now we see the full results, but they are not quite what we might have expected (Figure 1.108). Why does the result not look like a traditional breakthrough curve, with those sudden jumps in the first few seconds? The answer is that *we initialized the simulation "blindly"* without checking what the simulation would be initialized to. The gas_bed was initialized to have an equimolar mixture of all the components, instead of the composition we might have expected (i.e. equal to the feed)

Step 18: Rewind the simulation

There are two ways of resetting a simulation to its initial conditions. The first is the use of "Rewind using snapshots" and the second is the use of the "Restart" button. Snapshots are useful if you have several initial conditions, and want to switch between them. The restart button, on the other hand, simply resets the values to those used to initialize the simulation. Note that these buttons become inactive if you make *structural* changes to the simulation after a run. That is, if you run a simulation for 100 s and then add a new valve, you will

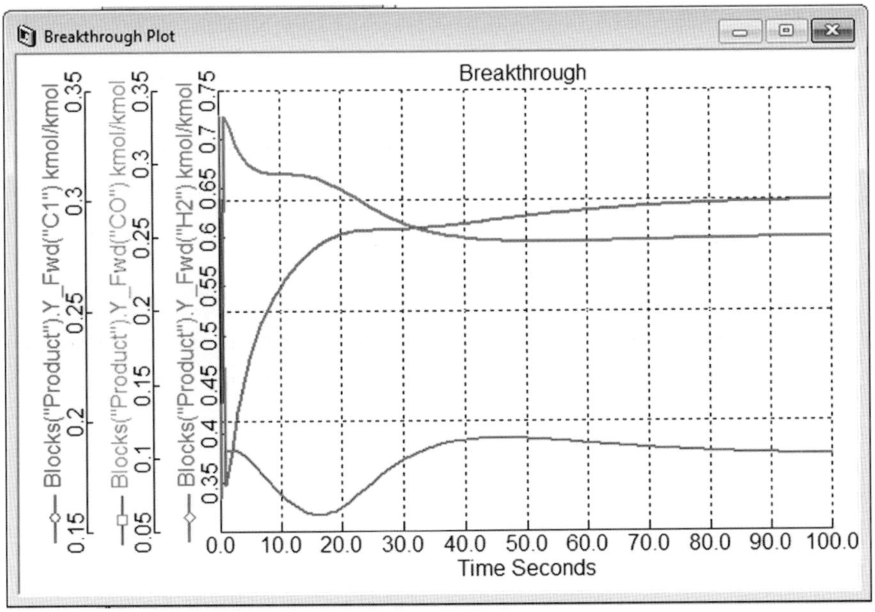

Figure 1.108 The simulation results plotted.

not be able to rewind to t = 0 s anymore. **Always remember to rewind before making changes after a run** (Figure 1.109).

Figure 1.109 The rewinding options in Aspen Adsorption.

Using the "restart" button, rewind the simulation.

Step 19: Set the initial conditions correctly

Open the configuration forms for both bed layers. Click on the "presets/initials" button and set the Y_First_Node(*) values to match the feed (Figure 1.110). Do this for both layers.

Once you have entered the values, repeat the "Check and Initial" process we used in Step 15.

Step 20: Rerun the simulation

This time, the breakthrough curve looks normal. CO and C1 get adsorbed initially, resulting in a higher fraction of H2 in the product stream. However, the bed eventually becomes saturated, and the yield matches the feed.

Rewind the simulation, then save it as `H2Workshop6.ada`.

1.8.4 Create the PSA Flowsheet

Now that we understand the key aspects of building an Aspen Adsorption simulation, let us move on to building the complete PSA process. Starting with `H2Workshop6.ada`, we will add additional blocks to the flowsheet.

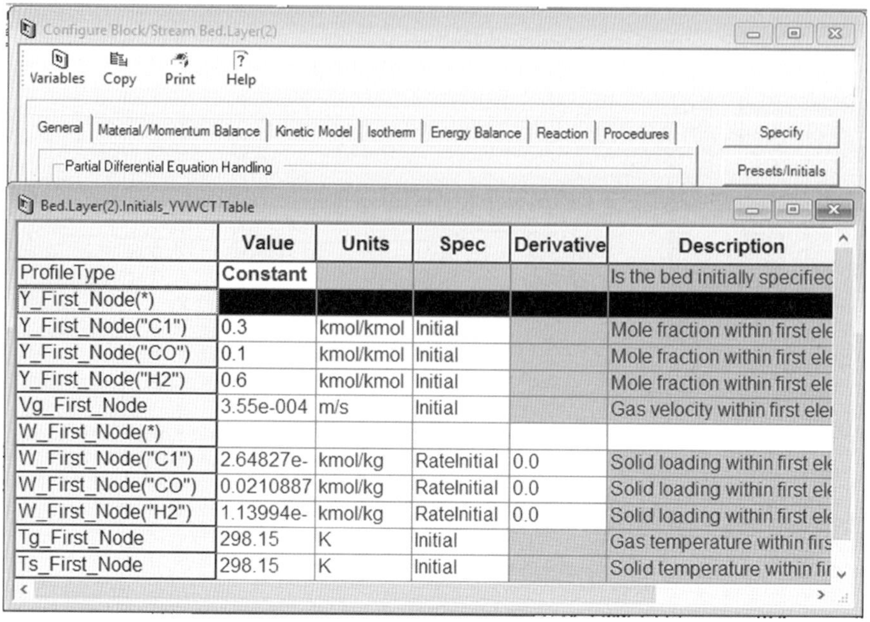

Figure 1.110 Specifying the initial bed composition.

Before we begin to describe the PSA process in Aspen, we should make a plan for what we want to do. In this workshop, we will be using a "single bed" approach for simulating a pressure swing adsorber. Specifically, we want to simulate the process shown in Figure 1.111. We will introduce the use of a gas_interaction block in this model.

Step 21: Add the blocks required for PSA

Following Step 8, Section 1.7.2, we add two gas_tank_void blocks as our pressure setters, and add four valves as our reversible flow setters. We also add an extra product block, and a gas_interaction block; we connect them as shown in Figure 1.112. We explain the gas_interaction block in the next step.

If we review Figure 1.111, we can now plan out how to schedule the valves in order to achieve the scheme we described. See Table 1.10.

Previously in Figure 1.7 and Table 1.5, we have explained the four ASs for valves within Aspen Adsorption: (1) AS = 0, valve is fully closed; (2) AS = 1, valve is fully open; (3) AS = 2 the gas flow rate through the valve is linearly dependent on the pressure drop across the valve according to Eq. (1.12); and (4) AS = 3, the valve has a fixed flow rate.

Step 22: Specify the gas_interaction block (D1)

We use the gas_interaction model as a part of the single-bed modeling approach, to record the profile of the material received, and later replay this profile to simulate the returned material. The model records: (1) molar flow rate, (2) mole fraction composition, (3) temperature, (4) upstream bed

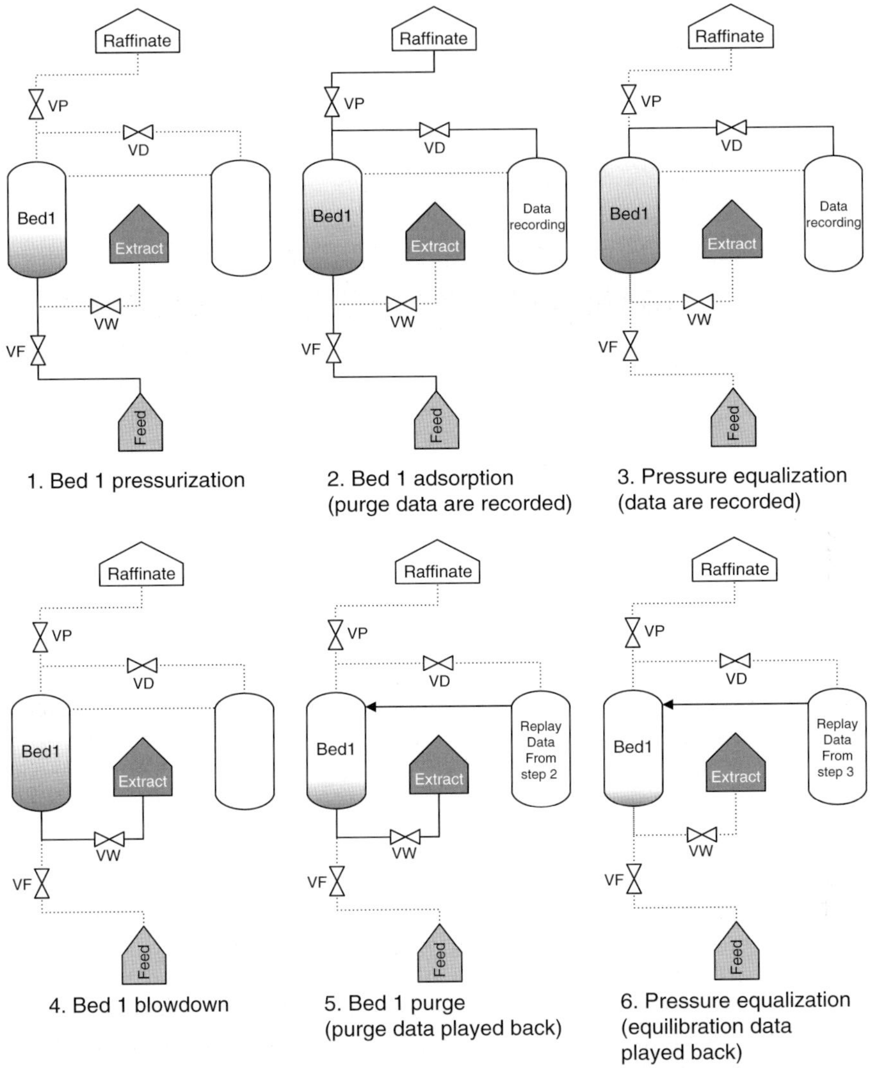

Figure 1.111 The steps of this H2 separation process (identical to PSA steps in Figure 1.7, except that the gas_intercation block replaces bed 2).

pressure, and (5) specific enthalpy. The gas_interaction block will automatically "replay" input from the "Bed" block based on our Cycle Organizer settings. Before we configure what data get recorded, however, we should specify a few properties of the block. Specifically, we should set the values that are used for initial guesses on the first iteration, before data have been recorded.

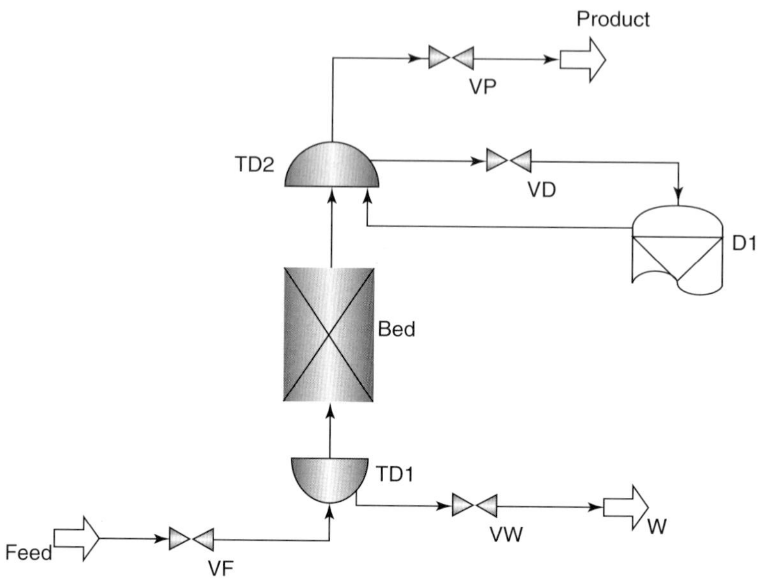

Figure 1.112 The additional blocks required for a PSA separation.

Table 1.10 Valve scheduling and active specifications (ASs) for the hydrogen PSA process.

Step	Function	VF	VD	VP	VW
Step 1	Pressurize the bed	2	0	0	0
Step 2	Adsorb bed 1 – fully load the bed	2	2	2	0
Step 3	Equalize the pressure of bed 1 and "data recorder" (gas_interaction_block)	0	2	0	0
Step 4	Depressurize (blowdown) bed 1	0	0	0	2
Step 5	Purge bed 1 to remove the strongly adsorbed component	0	0	0	2
Step 6	Repressurize bed 1 using pressure from the "data recorder" (gas_interaction_block)	0	0	0	0

Specifically, we should set the XFac number to adjust the effective volume of the interaction block. In our case, a XFac of 10 works well, since it makes the effective volume approximately equal to that of our bed. The pressure values are also important to set; in this case, we set them to 2 bar (Figure 1.113).

D1.Specify Table

	Value	Units	Description
Notional_Volume	**2.09966e-004**	m3	Notional bed volume for use in pressure estima
P_Stage_Start	2.0	bar	Estimated sink pressure
XFac	**10.0**	n/a	Notional bed volume correction factor
F_Initial_Reverse	**0.0**	kmol/s	Initial molar flowrate for reversed interaction
Y_Initial_Reverse(*)			
Y_initial_reverse("C1"	**0.3**	kmol/kmol	Initial mole fraction for reversed interaction
Y_initial_reverse("CO"	**0.1**	kmol/kmol	Initial mole fraction for reversed interaction
Y_initial_reverse("H2")	**0.6**	kmol/kmol	Initial mole fraction for reversed interaction
T_Initial_Reverse	**298.15**	K	Initial temperature for reversed interaction
P_Initial_Reverse	**2.0**	bar	Initial pressure for reversed interaction
P	2.0	bar	Actual pressure of the sink

Figure 1.113 Configuring the Gas_interaction block.

Step 23: Add a Cycle Organizer

Use the tools menu to add a Cycle Organizer (Figure 1.114).

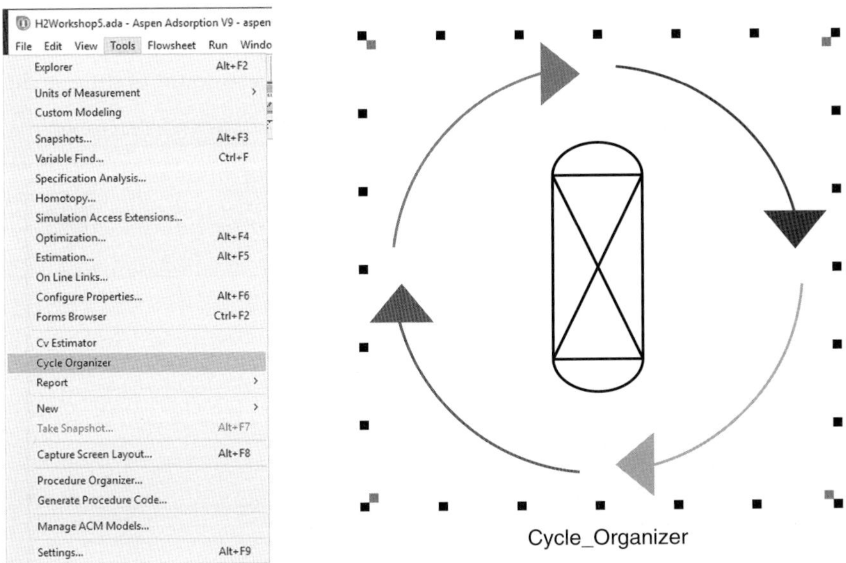

Figure 1.114 Adding a Cycle Organizer.

This is a special block, and only one is allowed per flowsheet. It presents an easy interface for scheduling changes (e.g. valve opening and closing) during a simulation.

Step 24: Specify the Cycle Organizer

Each cycle tracked by the Cycle Organizer comprises steps. These steps have three major components.

- Control: Manages when the step triggers
- Manipulated: Manages what variables the step changes

- Interactions: When using a gas_interaction bed, the interactions define how information is shared between the gas_interaction block and other parts of the flowsheet.

To begin with, we will define six steps and set their Control options. To add a new step, use the button indicated in Figure 1.115. Ensure that there are a total of six steps.

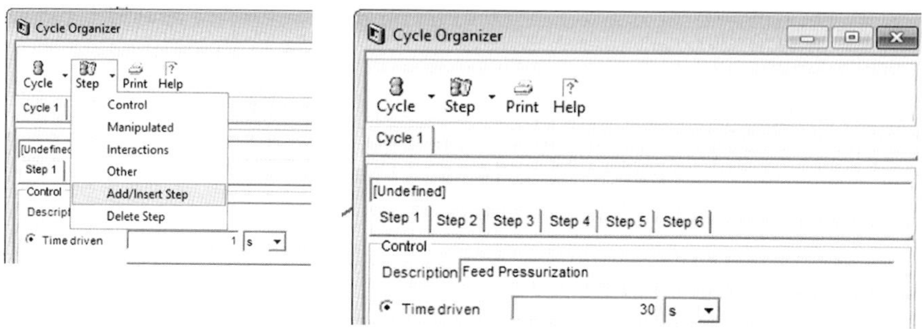

Figure 1.115 Add step 1 – feed pressurization.

We now set the control options for all six steps in figures. Step-dependency is a way to assert that one step will take exactly the same amount of time as the second step (Figures 1.116–1.120).

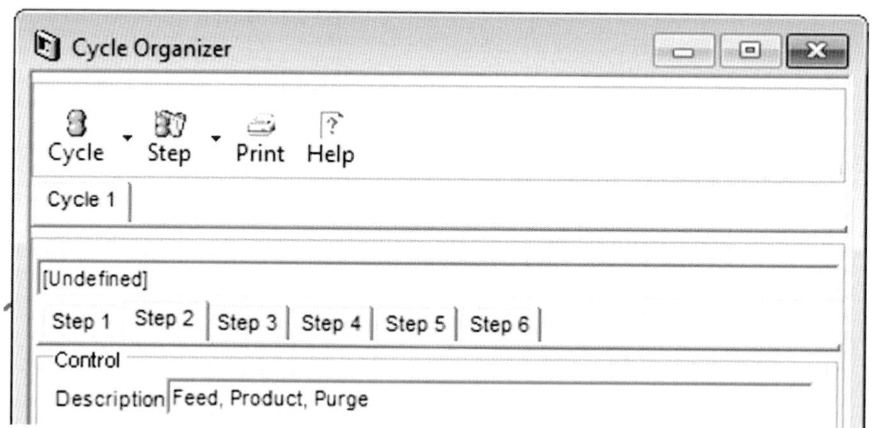

Figure 1.116 Step 2 – feed adsorption.

How do we specify steps 5–6 with the help of the gas_interaction block? We use the gas_interaction model as part of the single-bed modeling approach, to record the profile of material received, and later replay this profile to simulate the returned material. The model records: (1) molar flow rate, (2) mole fraction composition, (3) temperature, (4) upstream bed pressure, and (5) specific enthalpy.

We need to open their Interaction settings. Do this with the button highlighted in Figure 1.121.

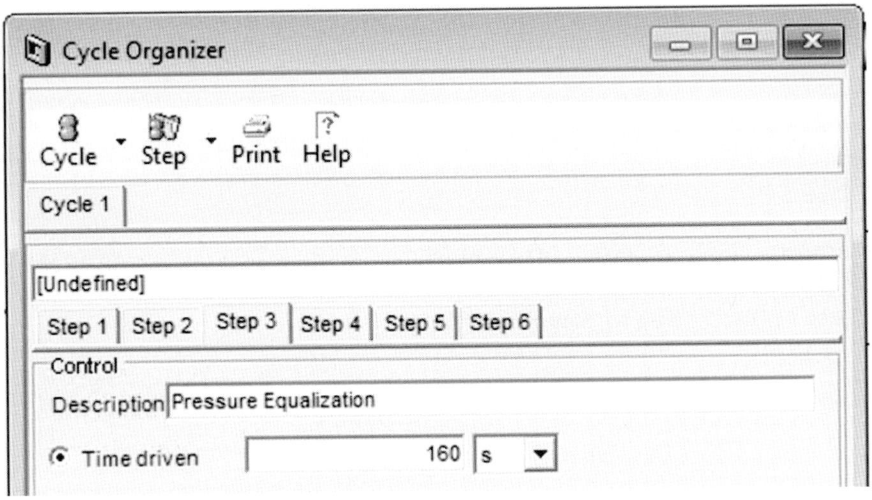

Figure 1.117 Step 3 – pressure equalization.

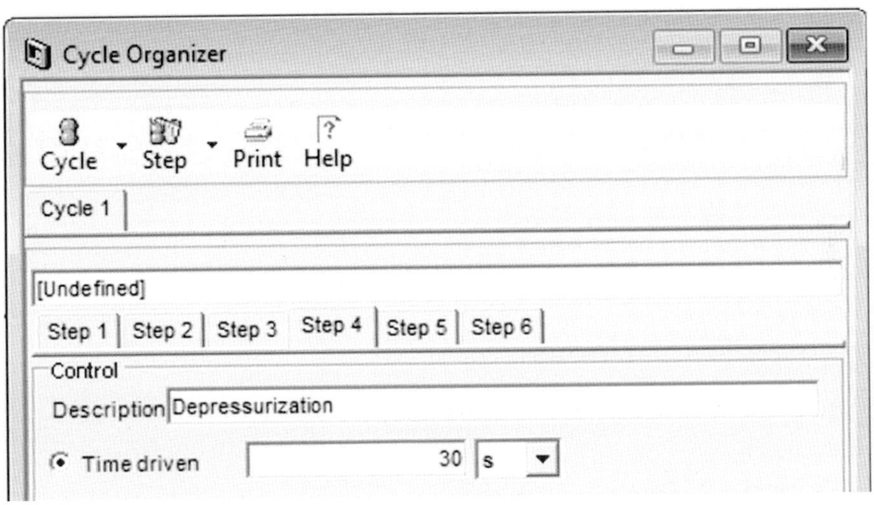

Figure 1.118 Step 4 – depressurization.

In the step interaction window, use the drop downs shown in Figure 1.121 to specify the interactions depicted in Figure 1.122. According to both Figures 1.7 and 1.111, step 5 and step 2, as well as step 6 and step 2, are closely related in a forward–reverse mode operationally (Figure 1.123).

Next, we move on to the manipulated variables. This is where we specify the variables that will change between each step of the cycle (Figures 1.124–1.129).

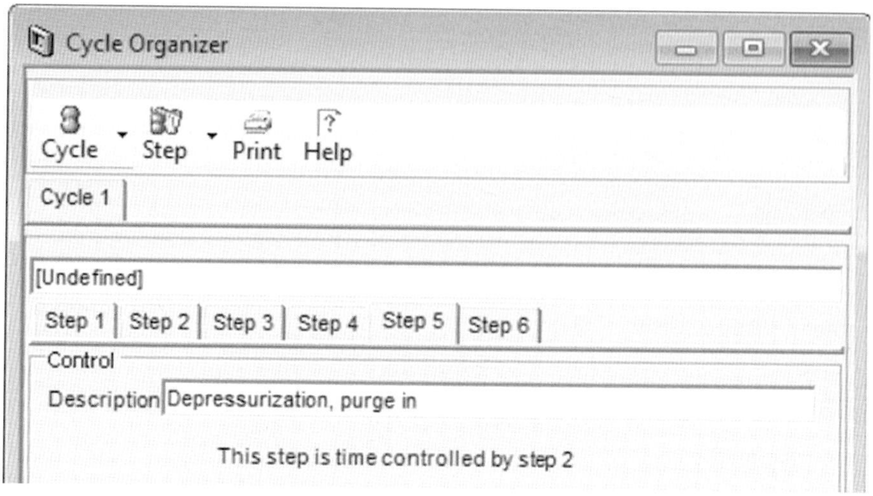

Figure 1.119 Step 5 – bed purging and depressurization.

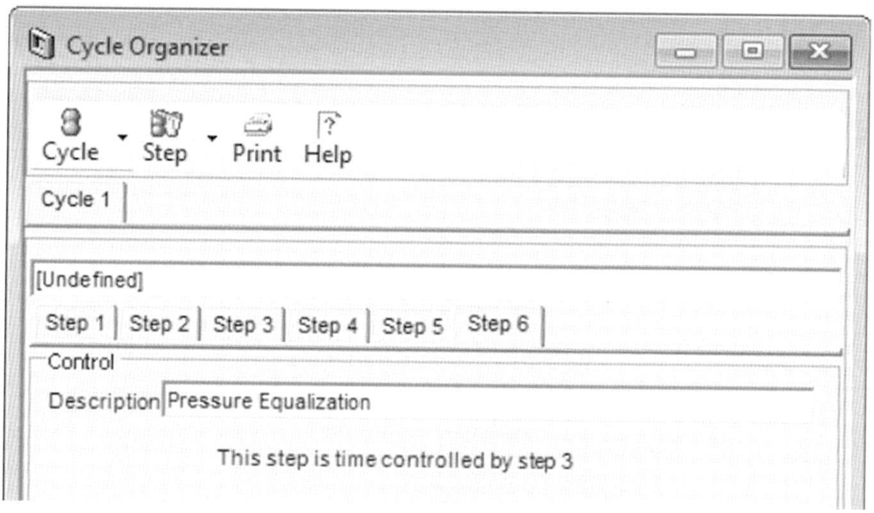

Figure 1.120 Step 6 – pressure equalization.

Step 25: Generate the cycle

Merely specifying the Cycle Organizer is not enough. We also must "Activate" it. Figure 1.130 illustrates this process.

We can also check to ensure that the cycle is active by looking in the Explorer. A block called Cycle1 with a lightning bolt indicates that the Cycle Organizer is ready to go. See Figure 1.131.

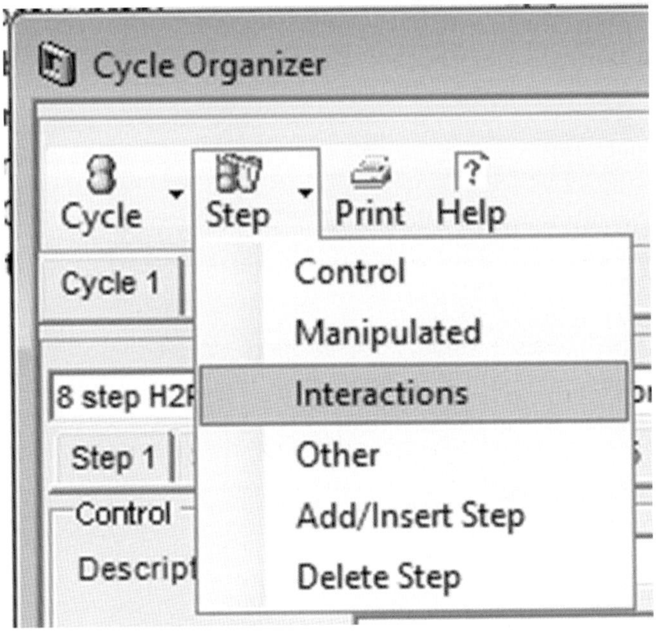

Figure 1.121 Step interactions.

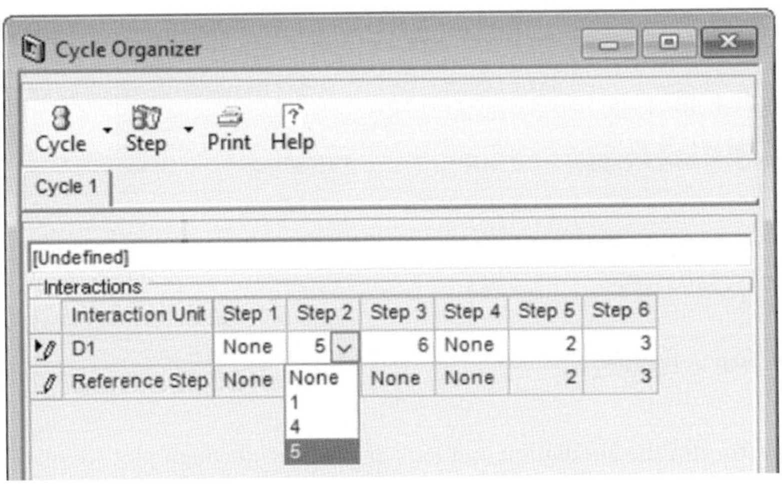

Figure 1.122 The step interaction setting.

Step 26: Run the simulation

Once again, use the "Check and Initial" to resolve the underspecified problem. The Initialization scripts check to see if a Cycle Organizer is present, and initialize the flowsheet differently, if it is. Set the Cycle Organizer to run for five cycles. See Figure 1.132.

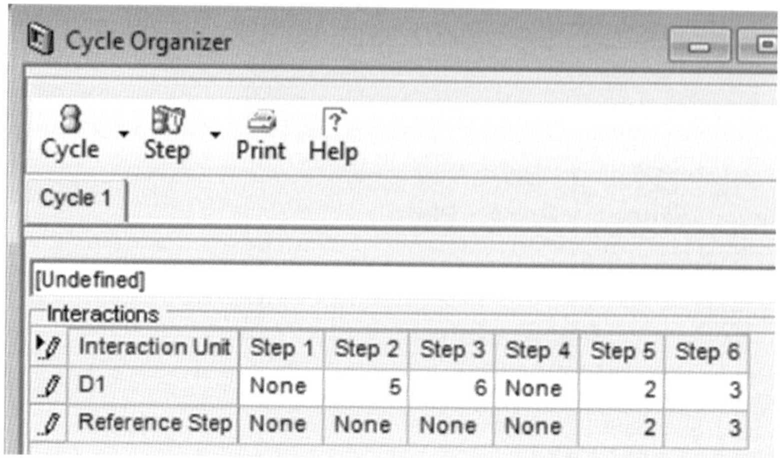

Figure 1.123 The completed interactions.

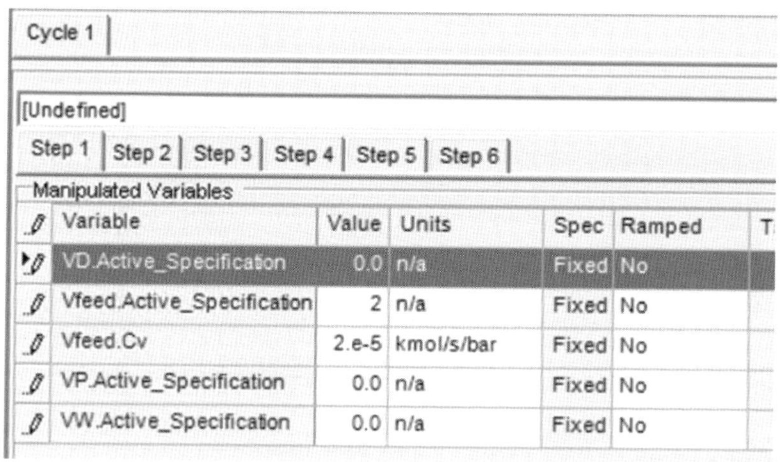

Figure 1.124 Step 1 – feed pressurization.

Afterward, run the simulation and look at the breakthrough plot we made earlier (Figure 1.133).

Save the finished simulation as `H2Workshop7.ada`.

As you can see, the H2 concentration (in red) is slowly increasing each cycle. After many cycles, the separation unit will reach a steady state at approximately 95% hydrogen. If you have a powerful computer, try running the simulation for 15 cycles.

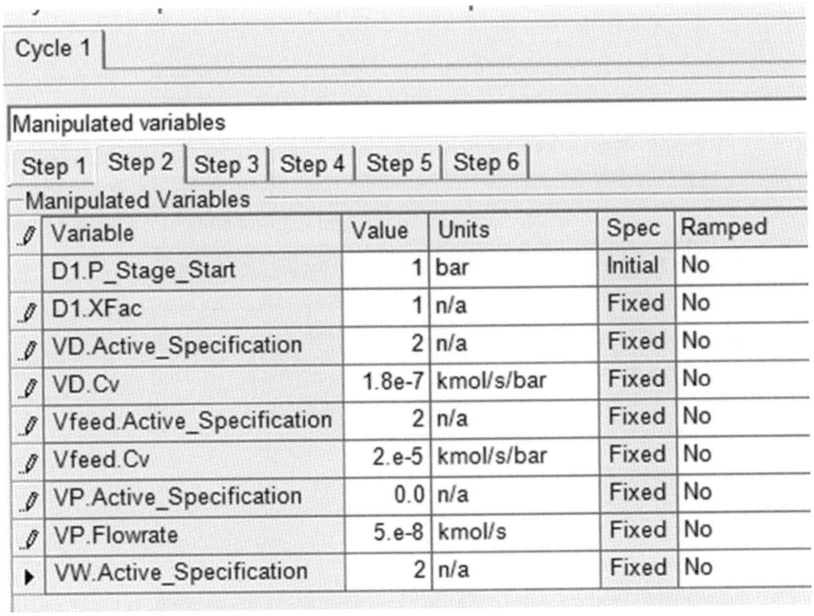

Cycle 1

Manipulated variables

Step 1 | Step 2 | Step 3 | Step 4 | Step 5 | Step 6

Manipulated Variables

Variable	Value	Units	Spec	Ramped
D1.P_Stage_Start	1	bar	Initial	No
D1.XFac	1	n/a	Fixed	No
VD.Active_Specification	2	n/a	Fixed	No
VD.Cv	1.8e-7	kmol/s/bar	Fixed	No
Vfeed.Active_Specification	2	n/a	Fixed	No
Vfeed.Cv	2.e-5	kmol/s/bar	Fixed	No
VP.Active_Specification	0.0	n/a	Fixed	No
VP.Flowrate	5.e-8	kmol/s	Fixed	No
VW.Active_Specification	2	n/a	Fixed	No

Figure 1.125 Step 2 – feed adsoprtion.

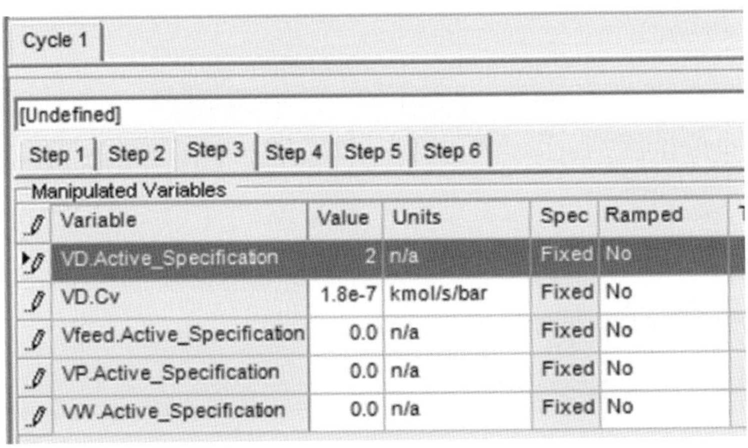

Cycle 1

[Undefined]

Step 1 | Step 2 | Step 3 | Step 4 | Step 5 | Step 6

Manipulated Variables

Variable	Value	Units	Spec	Ramped
VD.Active_Specification	2	n/a	Fixed	No
VD.Cv	1.8e-7	kmol/s/bar	Fixed	No
Vfeed.Active_Specification	0.0	n/a	Fixed	No
VP.Active_Specification	0.0	n/a	Fixed	No
VW.Active_Specification	0.0	n/a	Fixed	No

Figure 1.126 Step 3 – pressure equalization.

1.9 PSA Workshop: Modeling Hydrogen Separation using gCSS

In the previous workshop, we used the "one bed" approach to model hydrogen separation. It allows for faster simulations, but Aspen is certainly capable of

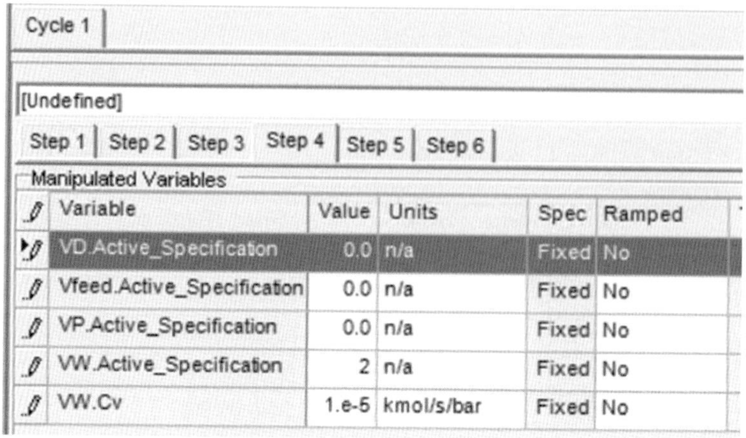

Cycle 1

[Undefined]

Step 1 | Step 2 | Step 3 | Step 4 | Step 5 | Step 6

Manipulated Variables

Variable	Value	Units	Spec	Ramped
VD.Active_Specification	0.0	n/a	Fixed	No
Vfeed.Active_Specification	0.0	n/a	Fixed	No
VP.Active_Specification	0.0	n/a	Fixed	No
VW.Active_Specification	2	n/a	Fixed	No
VW.Cv	1.e-5	kmol/s/bar	Fixed	No

Figure 1.127 Step 4 – depressurization.

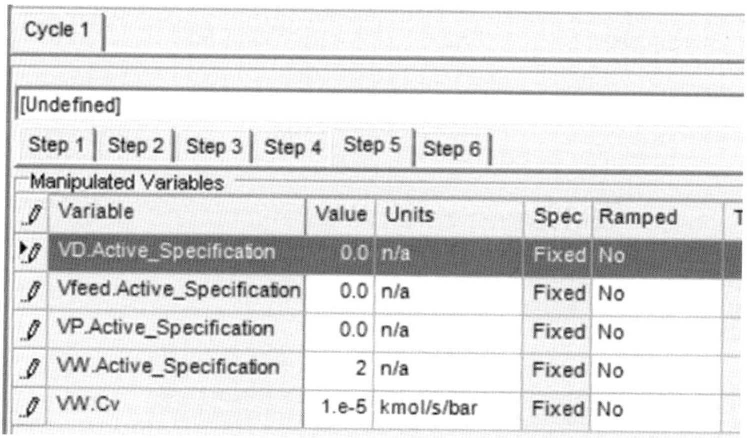

Cycle 1

[Undefined]

Step 1 | Step 2 | Step 3 | Step 4 | Step 5 | Step 6

Manipulated Variables

Variable	Value	Units	Spec	Ramped
VD.Active_Specification	0.0	n/a	Fixed	No
Vfeed.Active_Specification	0.0	n/a	Fixed	No
VP.Active_Specification	0.0	n/a	Fixed	No
VW.Active_Specification	2	n/a	Fixed	No
VW.Cv	1.e-5	kmol/s/bar	Fixed	No

Figure 1.128 Step 5 – bed purging and depressurization.

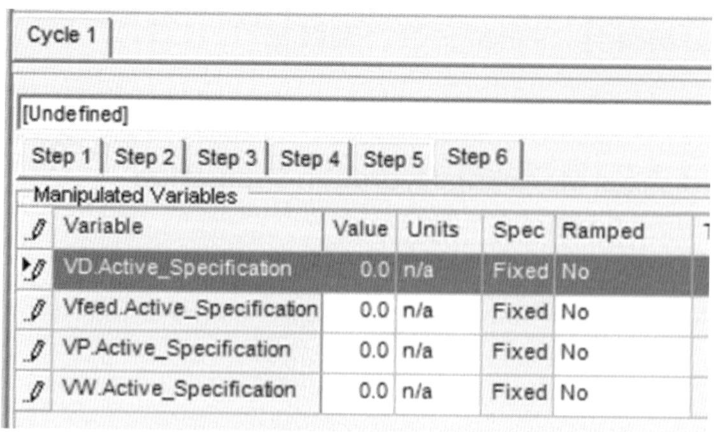

Cycle 1

[Undefined]

Step 1 | Step 2 | Step 3 | Step 4 | Step 5 | Step 6

Manipulated Variables

Variable	Value	Units	Spec	Ramped
VD.Active_Specification	0.0	n/a	Fixed	No
Vfeed.Active_Specification	0.0	n/a	Fixed	No
VP.Active_Specification	0.0	n/a	Fixed	No
VW.Active_Specification	0.0	n/a	Fixed	No

Figure 1.129 Step 6 – pressure equalization.

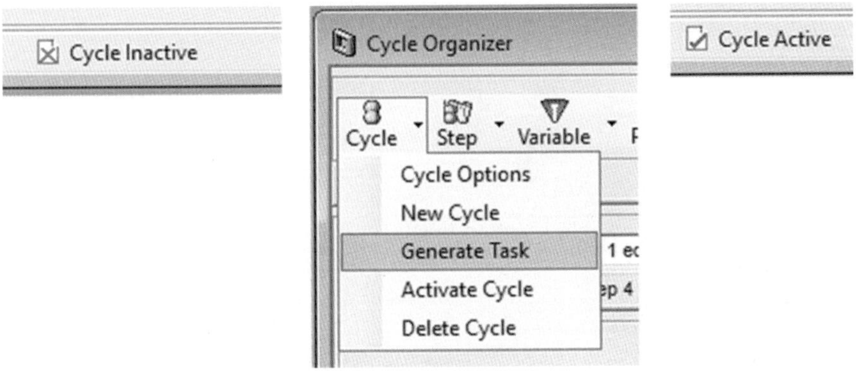

Figure 1.130 Activating a deactivated cycle.

Figure 1.131 Contents of flowsheet showing that cycle1 is ready to go.

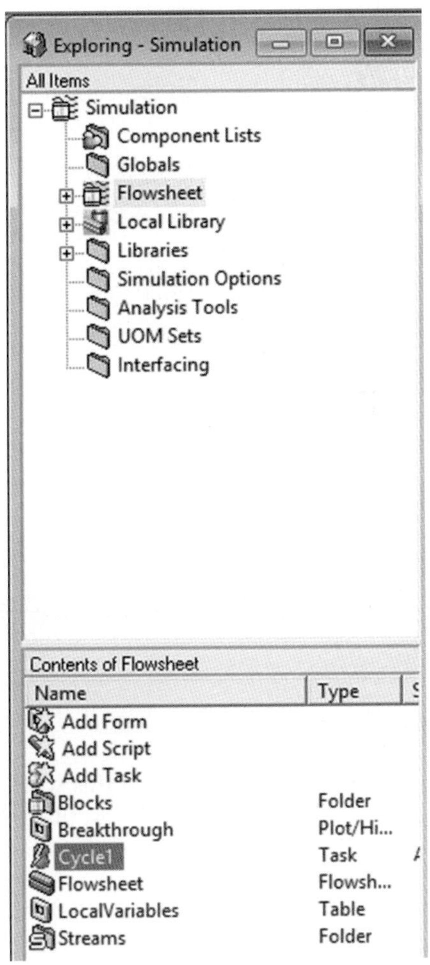

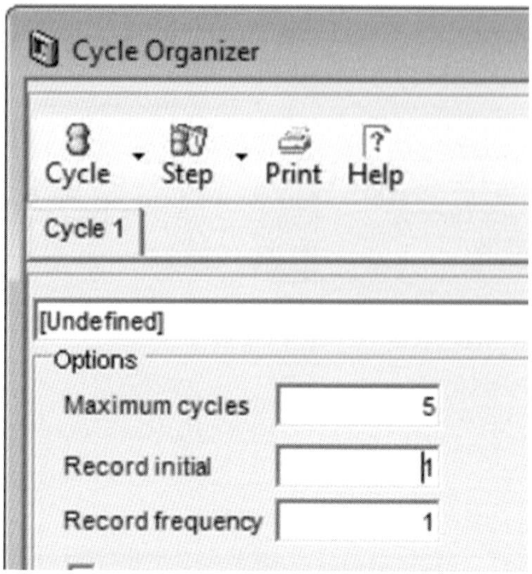

Figure 1.132 Setting the maximum cycles.

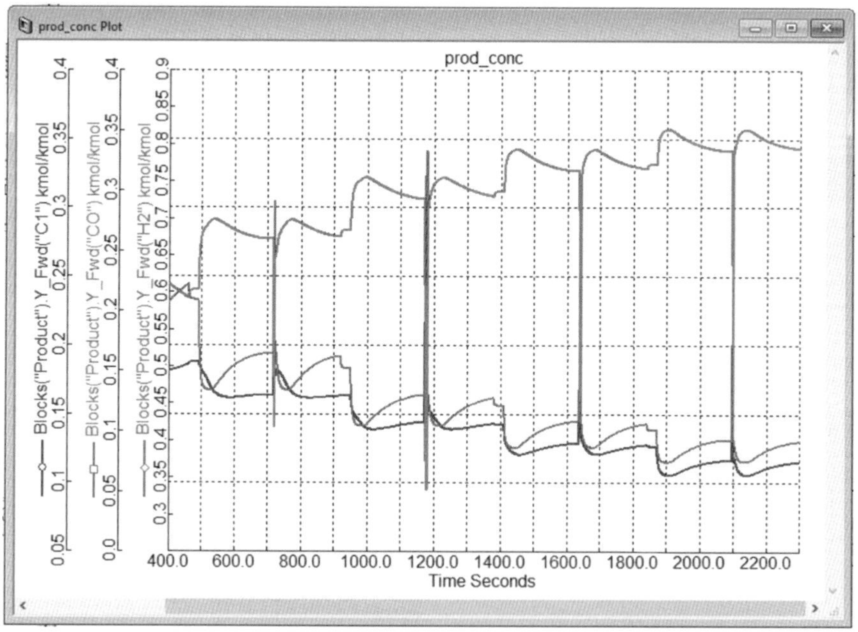

Figure 1.133 Product composition.

modeling multiple beds. In this workshop, we revisit the hydrogen separation, using the two-bed model described in Figure 1.134. We will also replace the "dynamic" blocks with gas cyclic steady state (gCSS) blocks.

The gCSS models enable the direct determination of the cyclic steady state, without carrying out a dynamic simulation over a large number of cycles. gCSS models result from complete discretization of both time and space, and present a periodic adsorption process as a steady-state problem. They offer an extremely

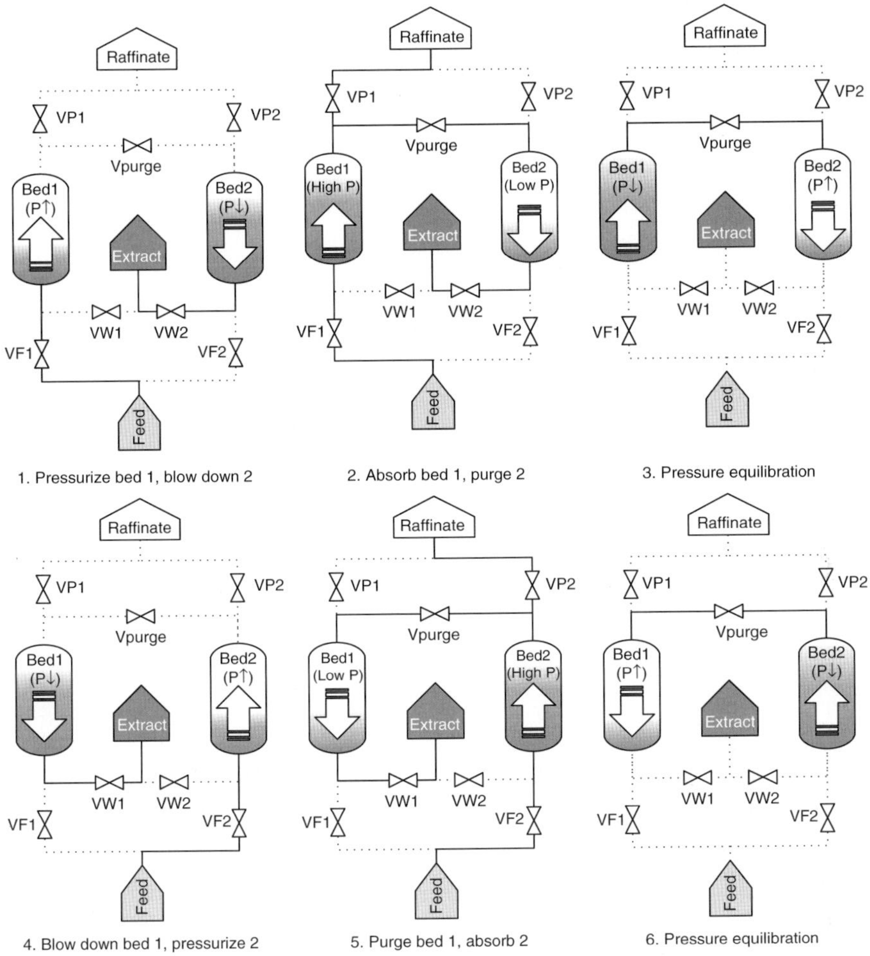

Figure 1.134 The configuration for hydrogen separation.

efficient design tool to optimize the design and operating conditions for an adsorption process. gCSS blocks must be configured differently than dynamic blocks, and so it is important that we know how to work with them.

1.9.1 Define the Components and Property Models

Step 1: Open Aspen

To start with, open a blank Aspen Adsorption document, and show the model libraries using the procedure described in workshop 1 (Figure 1.135). Save the file as **Workshop3a.ada.**

Step 2: Components

Just as before, we start creating our simulation by defining the components that will be present in the simulation. To do this, double-click on the "Default" component list (Figure 1.136) to bring up the "Physical Properties Configuration"

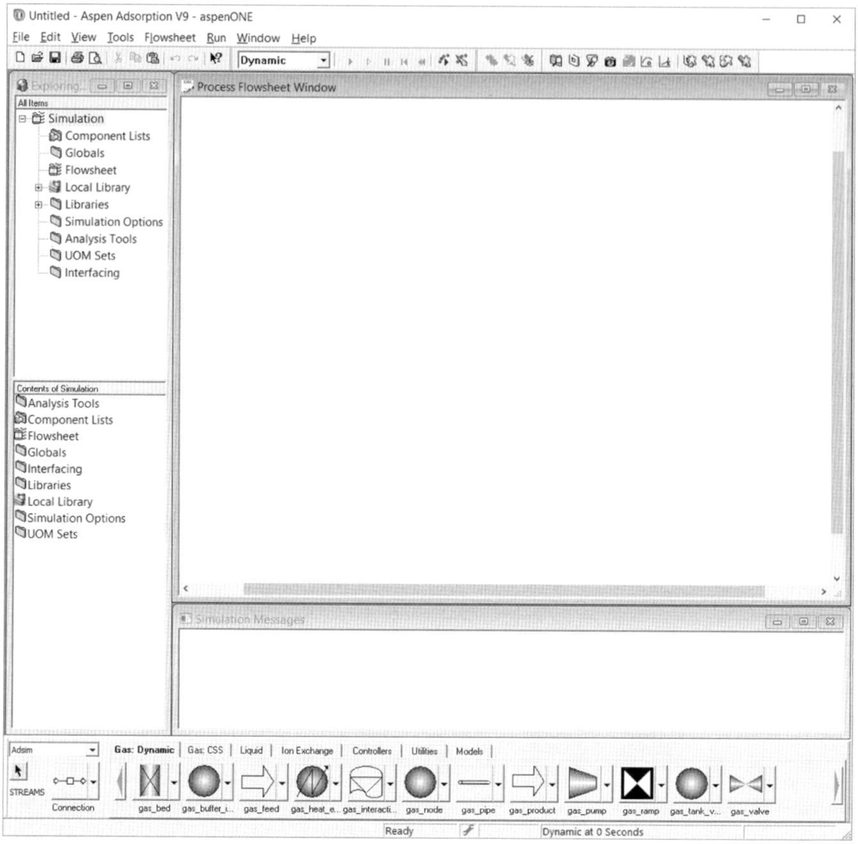

Figure 1.135 A blank adsorption flowsheet.

dialog (Figure 1.137). Configure Aspen Adsorption to use Aspen Properties, which will allow our simulation to calculate important physical properties at a variety of temperatures and conditions.

We need to request data for all of the components in the separation, so we enter all 5 components. Fortunately, we can easily find the simple components we are working with. All we need to do is to enter their names in the "Component ID" column, and Aspen will automatically retrieve the data (Figure 1.138).

After selecting the components, we choose the method for estimating their physical properties. For this separation, the Peng–Robinson method is appropriate (Figure 1.139). Set this value either by clicking on the "specifications" entry in the "Methods" folder on the left, or by clicking on the "Next" button in the ribbon.

Once this step is completed, we are ready to generate the property file that will be used by Aspen Adsorption. To do this, we click on the "Run" button in the ribbon. If the run button is inactive, it is likely that some parameters were not automatically retrieved from the databank. The simplest way to retrieve the parameters is to click on the "Next" button. It will automatically take you

Figure 1.136 Selecting the component list in the explorer window.

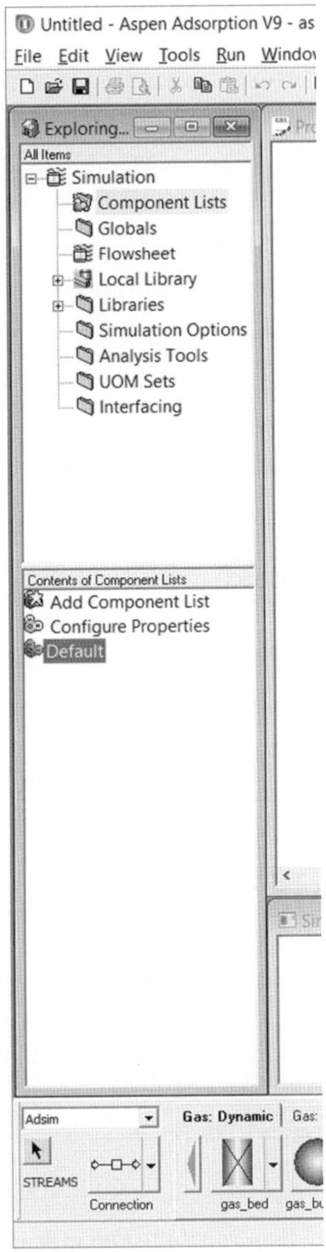

to the parameter page, and any parameters that were not previously retrieved will be updated. Once they have been filled in, you may then either click the run button, or "Next" button to run the "Property Analysis/Setup" as shown in Figures 1.140 and 1.141.

Once the property analysis is complete, close the Aspen Properties window, and save the document as **PropsPlus.aprbkp** file when prompted

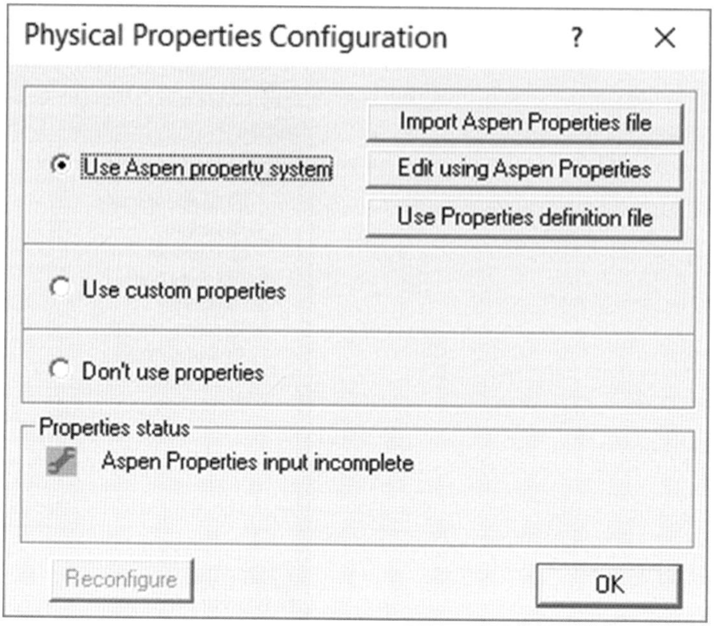

Figure 1.137 Configure the Aspen property system.

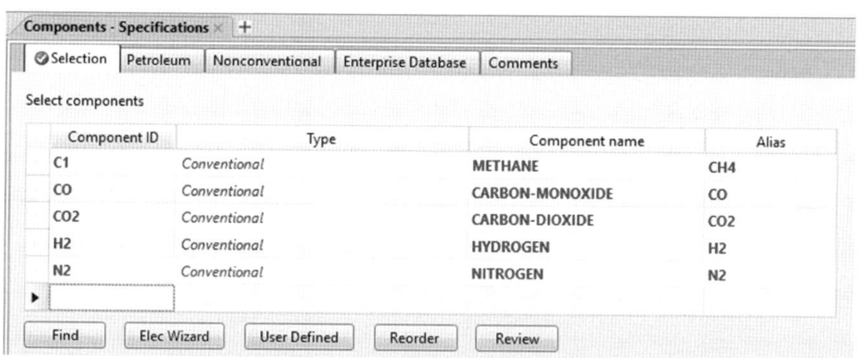

Figure 1.138 Retrieving components from the property databank.

(Figure 1.142). You may save the file as an Aspen Properties Document format (*.aprop file) if you want to, but we recommend that you select "no" and ask Aspen to not show the message again (Figure 1.143).

If the property file was saved correctly, Aspen Adsorption will report that the Physical Properties have been configured. Click "OK" (Figure 1.144) and then move all of the components from the "available" section to the right-hand column so that they will appear in the simulation (Figure 1.145). Save the file as **Workshop3b.ada**.

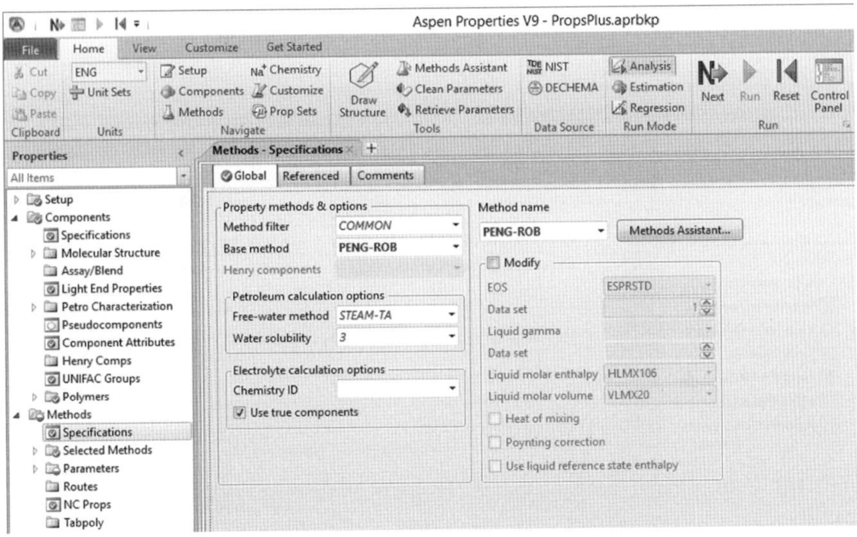

Figure 1.139 Selecting the property method.

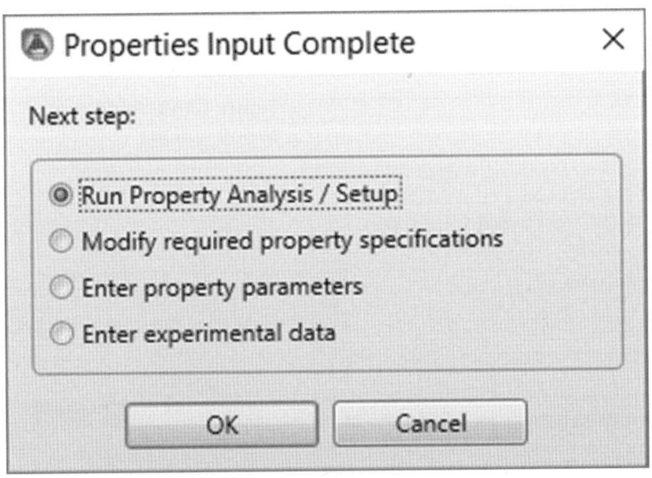

Figure 1.140 Running the Property Analysis.

1.9.2 Working with Model Libraries: Advanced Flowsheet Options

Step 3: Configure CSS_Info and add Cycle Organizer

Now that we have completed the component setup, we need to prepare the flowsheet. Because we use gCSS blocks in this simulation, there are some important steps we need to complete before we begin adding the blocks. The first new step is to create an instance of the CSS_Info structure (Figure 1.146). This object contains settings essential to gCSS operation, including:

- The step time
- The number of step time nodes in CSS simulations

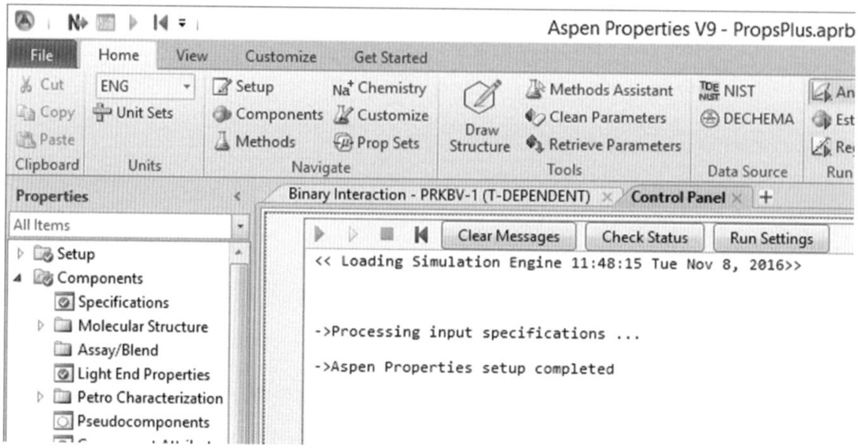

Figure 1.141 Control Panel reporting that the Property Analysis was successful.

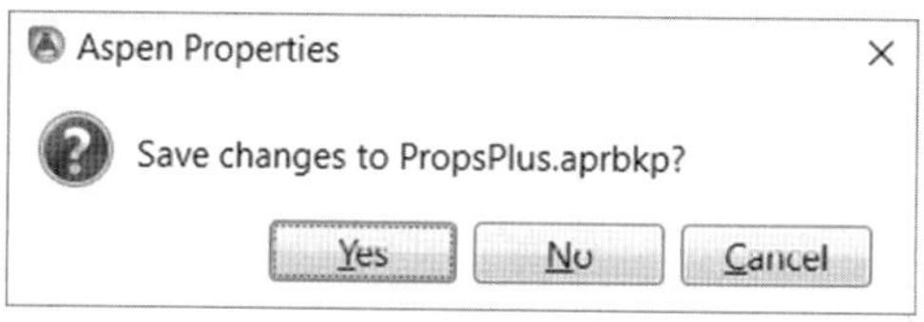

Figure 1.142 Saving the properties for use in Adsorption.

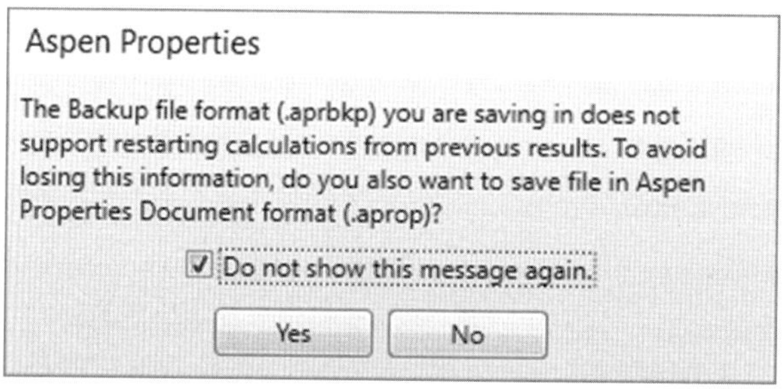

Figure 1.143 Additional saving options.

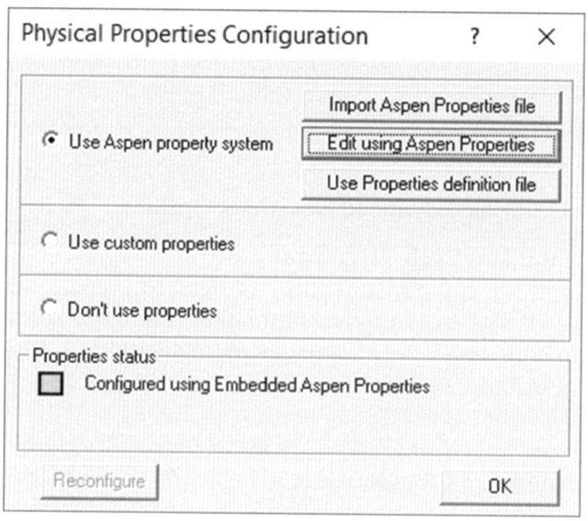

Figure 1.144 The property file is recognized.

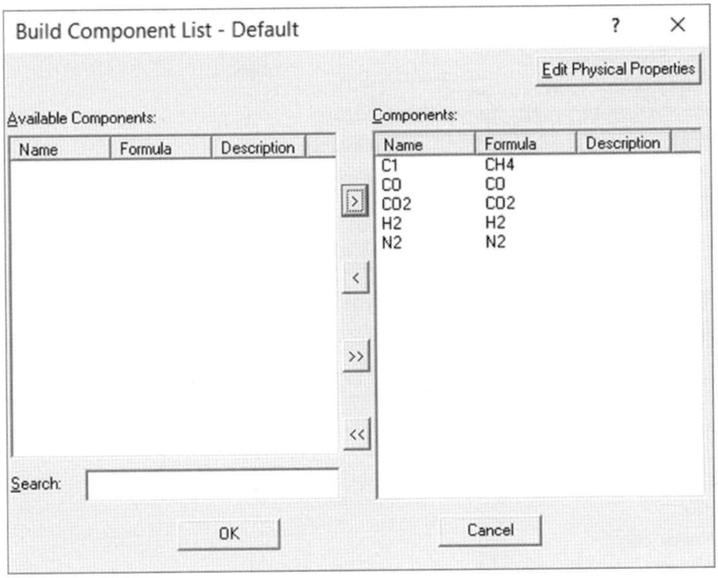

Figure 1.145 Adding all of the components to the simulation.

- Global tearing values
- Whether all gCSS blocks are isothermal or nonisothermal.

Note that it is mandatory to create this object ***before*** placing any blocks on the flowsheet. If you place blocks on the flowsheet before a **CSS_Info** object (Figures 1.147 and 1.148) exists, they will not be able to get the values they need, even if you add a CSS_Info object later. Also note that the name of the CSS_Info block matters. Most of the gCSS blocks we add later are correctly coded to be able to use a CSS_Info block of any name, but some, such as the gCSS_Tank, assume that the CSS_Info block will be named "**CSSInfo**" and will not work if you supply any other name.

Once we have added the structure, we also add a Cycle Organizer (Figure 1.149). At this time, we are not adding any information to the Cycle Organizer, so simply close the Cycle Organizer window that pops up. If you are presented with a "Store Data" dialog, click "yes" (Figure 1.150).

With this, we have performed all the necessary set-up steps, and may now begin laying out the H2 separation. Save as the file as **Workshop3b.ada**.

Step 4: Build flowsheet

We lay out the flowsheet in a very similar way to the air separation workshop, except we use blocks from the Gas: CSS tab in the model library. To begin with, add two gCSS_Adsorber blocks, using any of the icons you wish. We can see icon options by clicking on the small arrow in Figure 1.151.

At this point, the precise placement of the adsorber blocks (Figure 1.152) is not too important, we can align them more carefully once everything is in place.

Now add the void volumes for our simulation. We may use several different icons to help indicate the role of each tank. For example, we can place the

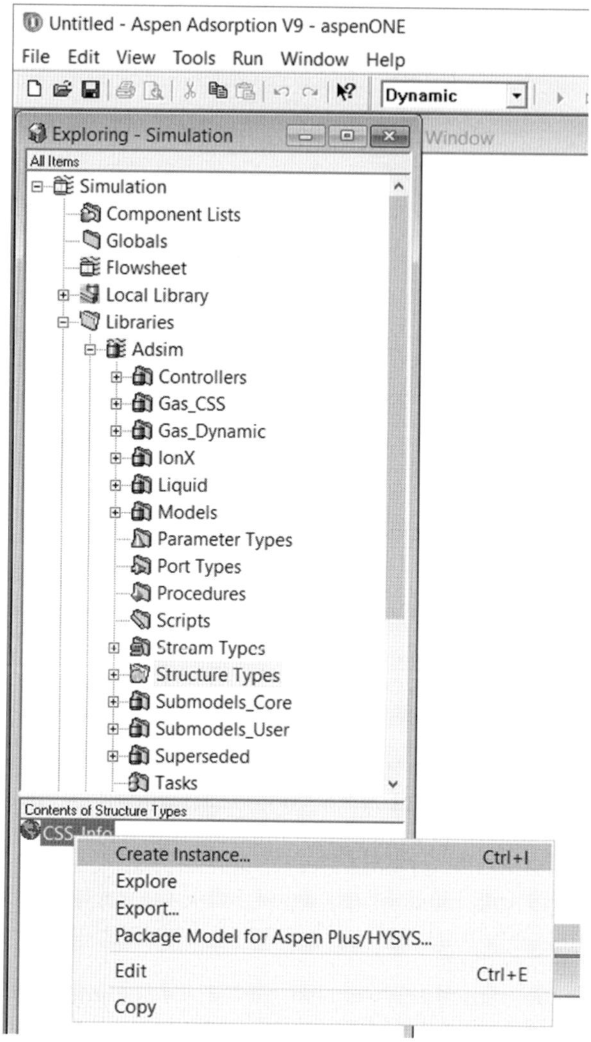

Figure 1.146 Creating an instance of the CSS_Info object.

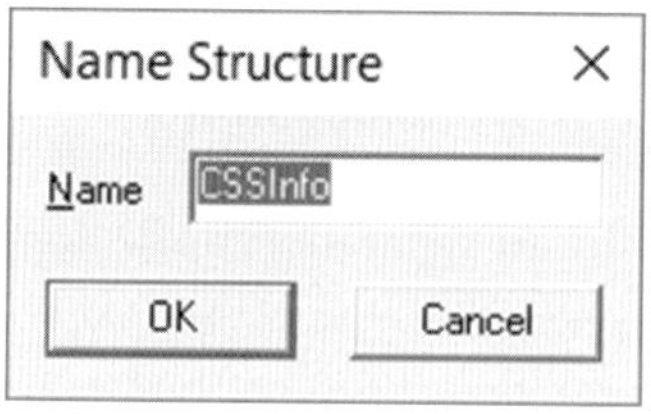

Figure 1.147 Naming the CSS_Info structure. This name will work correctly with the gCSS_Tank block.

top and bottom cap icons to indicate that the tanks represent voids at the top and bottom of the adsorption bed, respectively (Figures 1.153 and 1.154). As described previously in Step 8, Section 1.7.2, these gas_tank_void blocks also serve as **essential pressure setters** for the flowsheet.

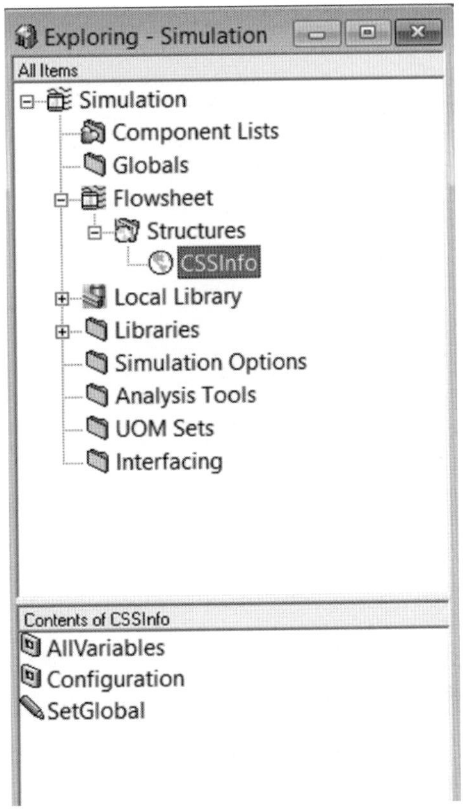

Figure 1.148 The CSS_Info structure has been successfully added to the flowsheet.

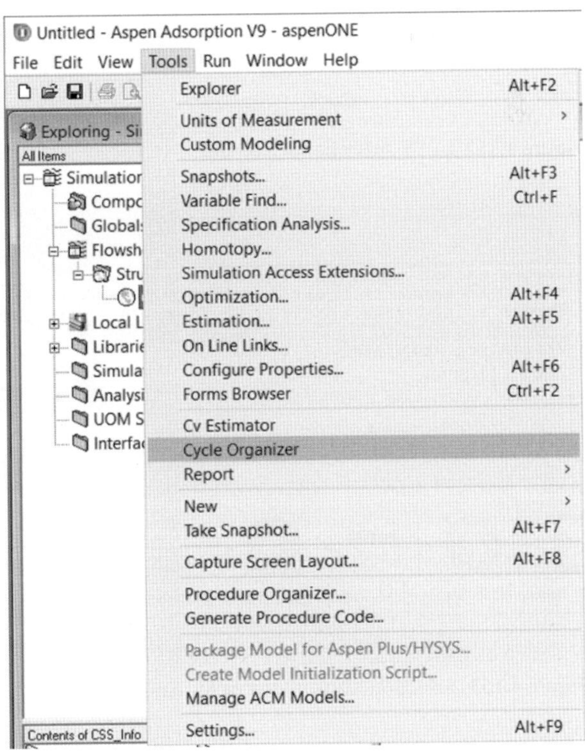

Figure 1.149 Adding the Cycle Organizer.

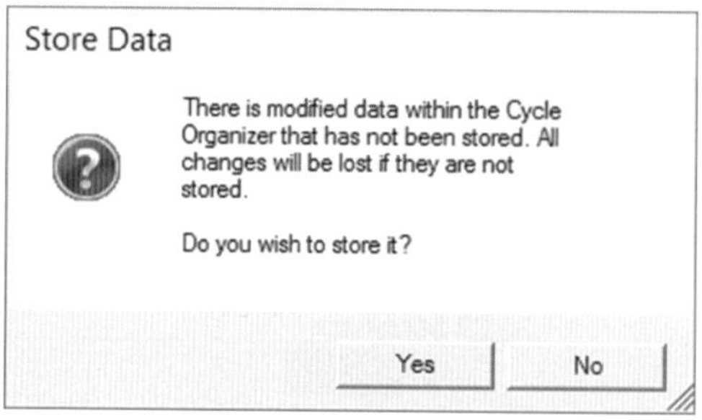

Figure 1.150 Always click "yes" when you see this dialog.

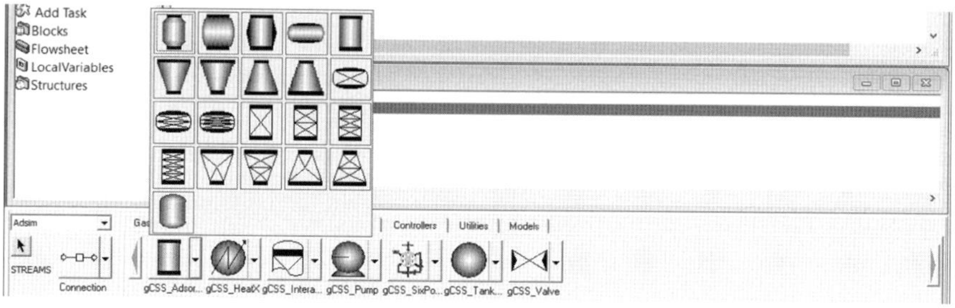

Figure 1.151 Selecting an icon for the gCSS_Adsorber.

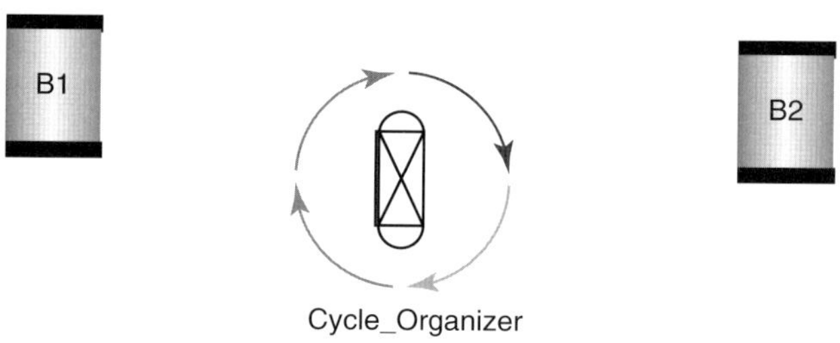

Figure 1.152 gCSS_Adsorber placement.

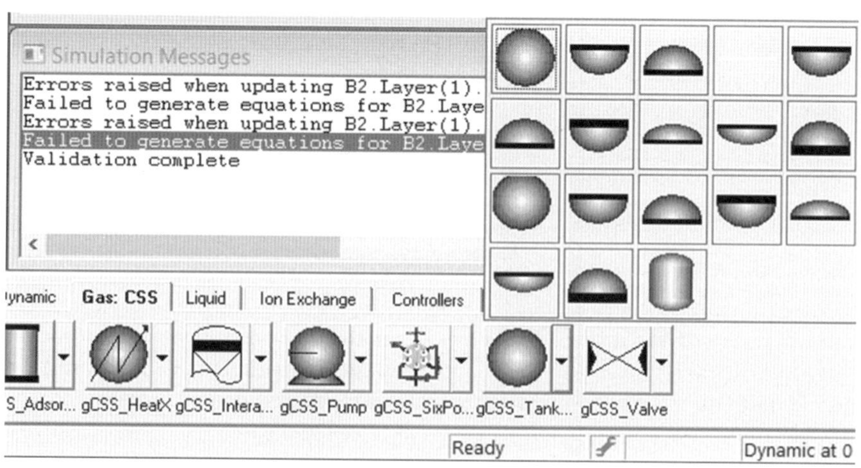

Figure 1.153 Selecting icons for the gCSS_TankVoid block.

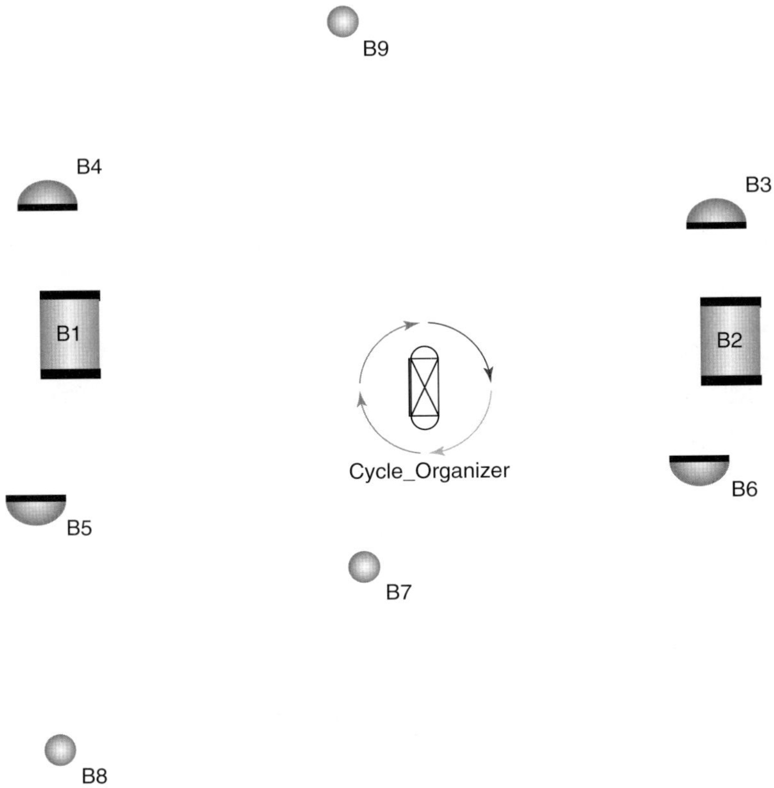

Figure 1.154 Placing tanks on the flowsheet, using three different icons.

Once again, do not spend too much time aligning blocks at this stage, the goal is to get the correct blocks into the flowsheet.

Finally, we add the 10 valves (Figures 1.155 and 1.156).

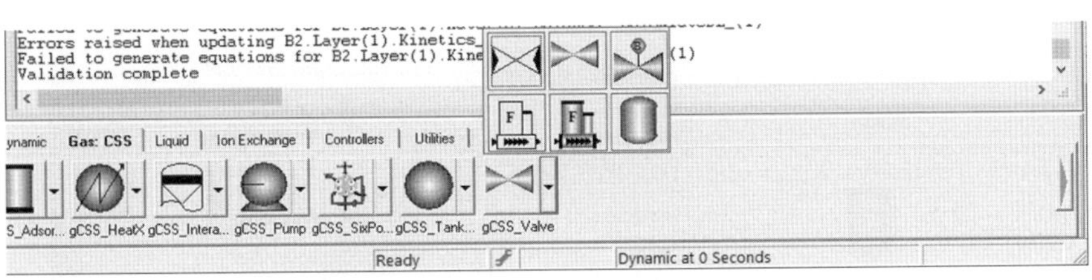

Figure 1.155 Valve icon selection.

Once again, there is no need for precise placement.

Now, in order to get ready for the connecting streams, let us rotate the valves on either side of the adsorbent beds. To do that, make sure that the Process Flowsheet ToolBar is visible. It can be toggled by right-clicking on the flowsheet and clicking on the Process Flowsheet ToolBar entry (Figure 1.157).

Now once the rotation icons are visible, rotate the four valves that are selected in Figure 1.158 to the left.

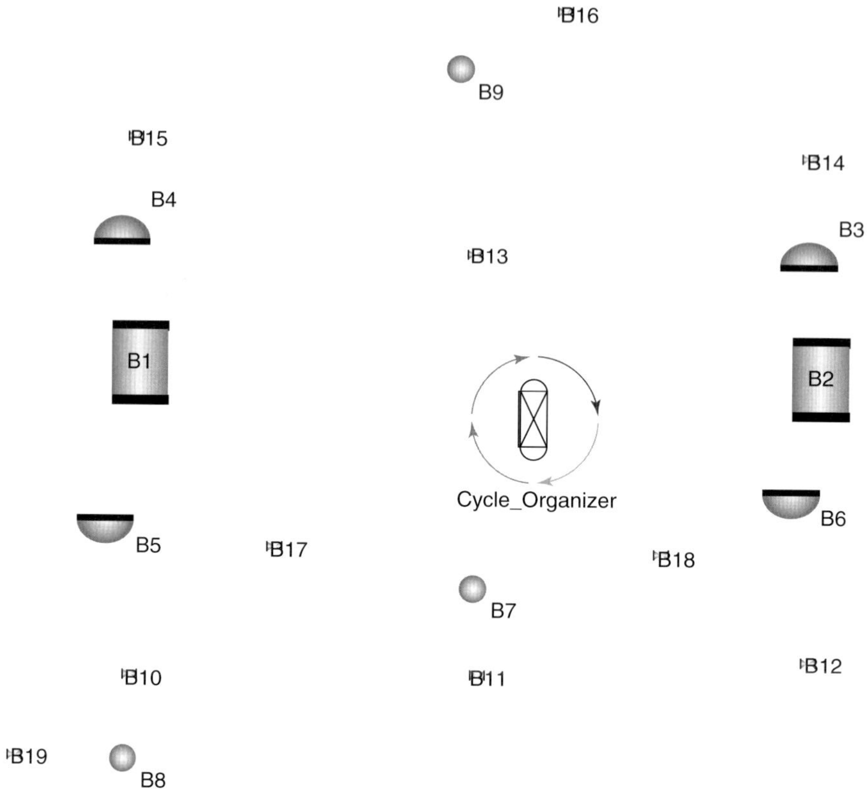

Figure 1.156 Valve placement.

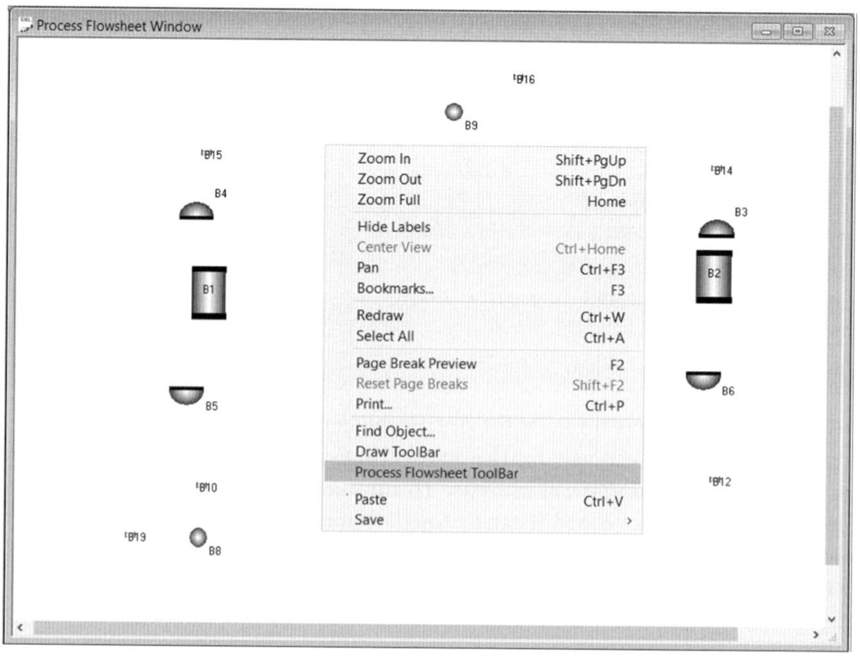

Figure 1.157 Toggling the Process Flowsheet ToolBar.

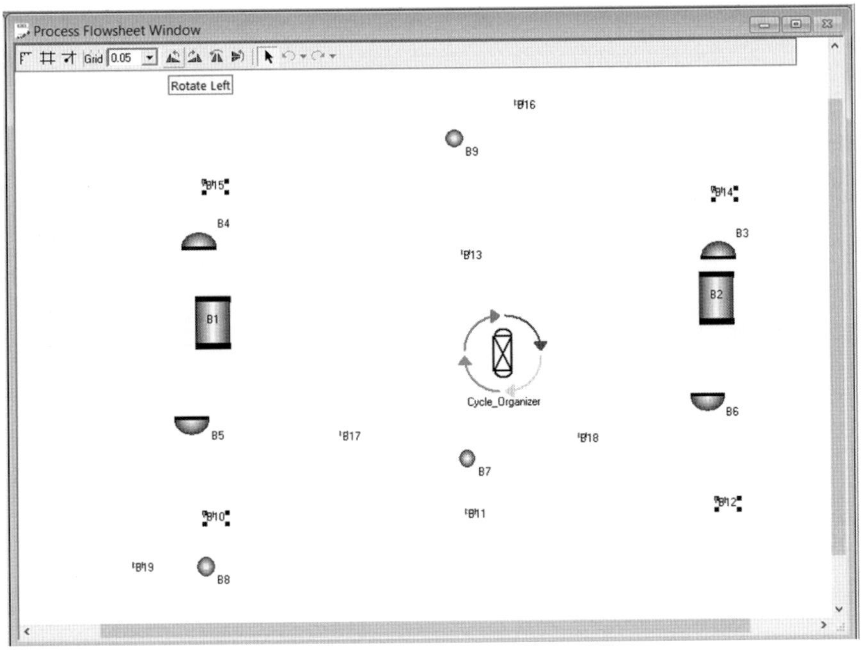

Figure 1.158 Rotating the adsorbent bed valves.

After rotating the valves, we begin connecting the blocks together. Because we are using gCSS blocks, we must select the gCSS_material_connection before connecting the blocks (Figure 1.159).

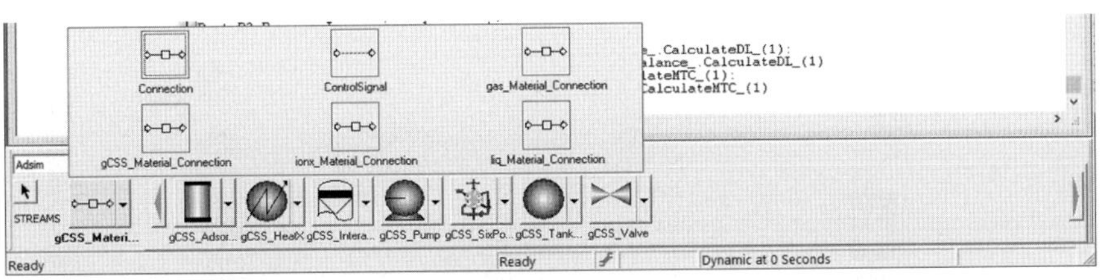

Figure 1.159 Selecting the gCSS_material_connection in order to define flowsheet connections.

Once we have selected the gCSS_material_connection, the connectivity arrows will appear on the flowsheet (Figure 1.160). Connect the streams as usual, and double-check that the streams are connected in the correct directions.

Once the connection streams have been laid out, we save the flowsheet. Also consider flipping the valve on the right-hand side so that the connections will be more linear (Figure 1.161).

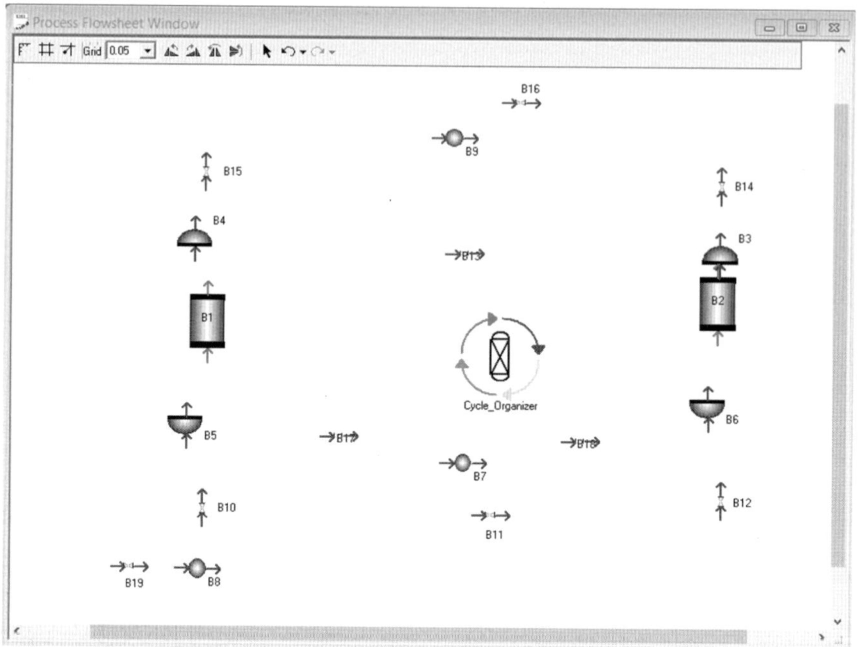

Figure 1.160 Flowsheet connection arrows.

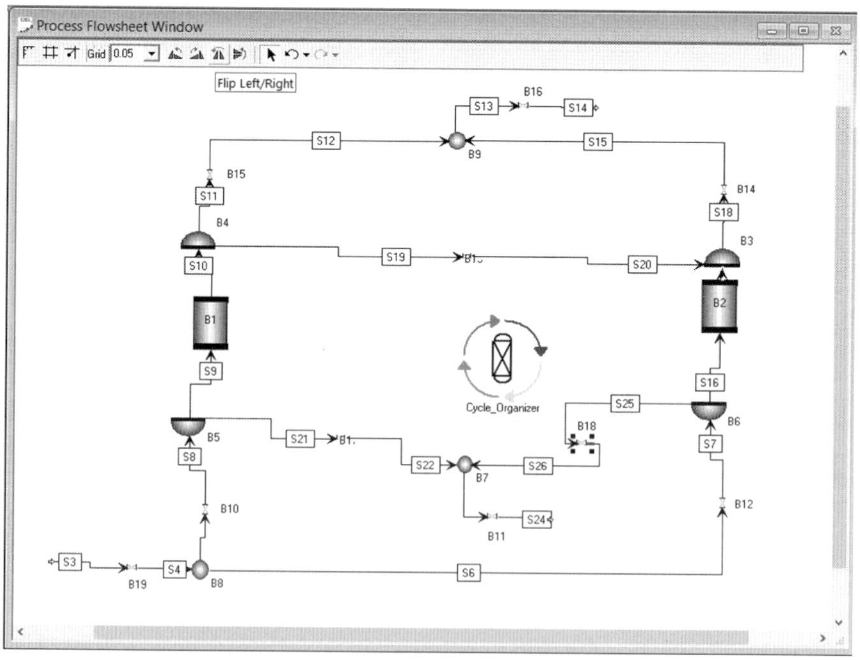

Figure 1.161 Flipping the B18 block left-to-right.

Step 5: Save the file as *Workshop3c.ada* (Figure 1.162)

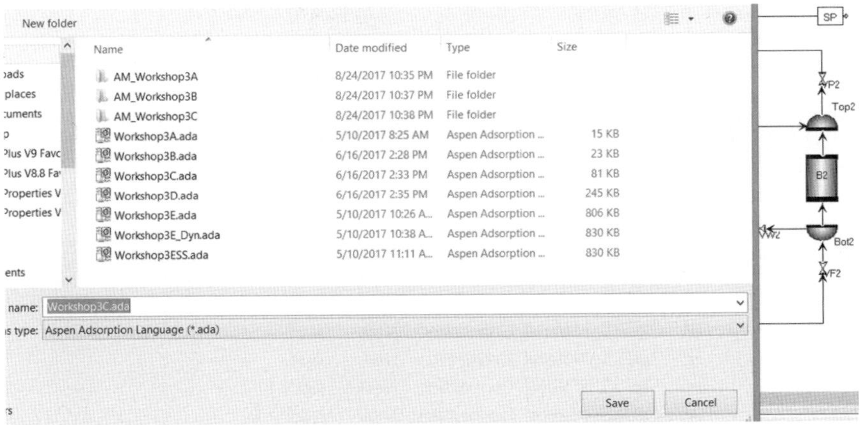

Figure 1.162 Save the document using the default file type as **Workshop3c.ada**.

Step 6: Clean up

Before we finalize the flowsheet, it is useful to check if the blocks and streams are correctly recognizing the CSS_Info structure that we created in step 1. A simple way to do this is to double-click on the feed stream, and look at the form which comes up. If the form has no variables (Figure 1.163), then the streams cannot locate the CSS_Info structure. The simplest fix for this miscommunication is to cut all the blocks out of the flowsheet, then paste them back in.

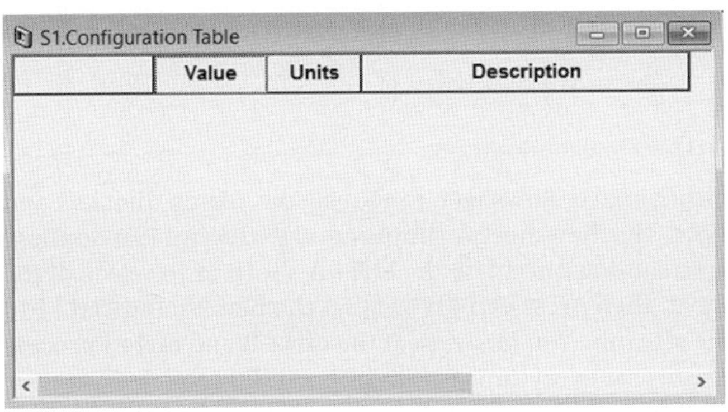

Figure 1.163 A stream that cannot communicate with the CSS_Info structure.

When you cut the objects, you will be presented with a delete dialog, and you must pick "OK" (Figure 1.164).

When you paste the objects, you will be presented with a dialog that allows you to specify which objects you wish to paste, and make changes to their names if you wish (Figure 1.165).

Once we have ensured that all the blocks are correctly communicating with the CSS_Info Structure, we can do some housekeeping. The simplest

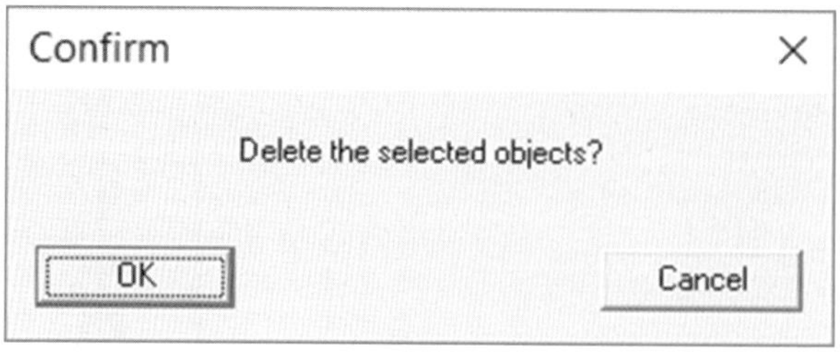

Figure 1.164 Cutting objects from the flowsheet.

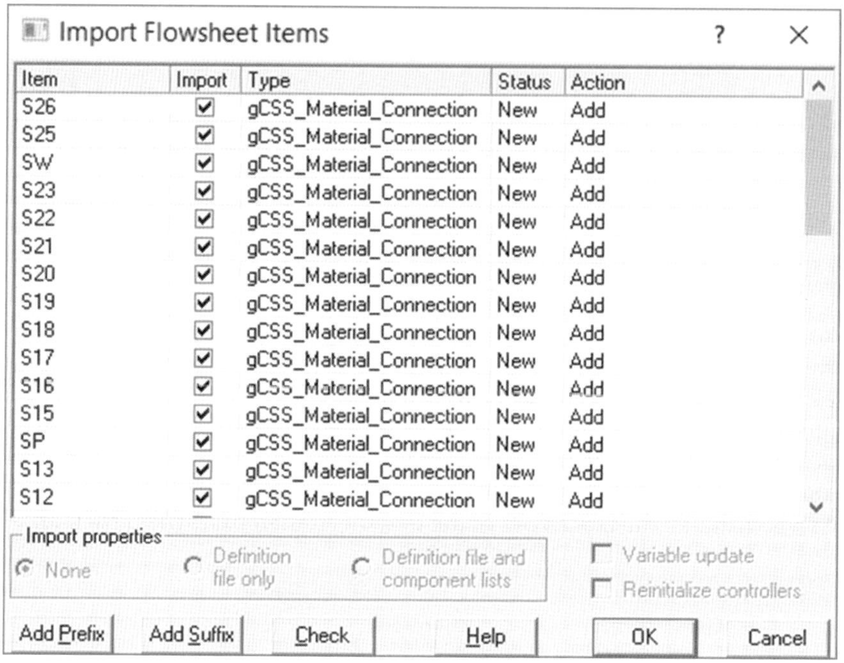

Figure 1.165 Pasting objects back into the flowsheet.

way to make a well-organized flowsheet is to use the "Align Blocks" and "Redraw streams" tools. This flowsheet is simple enough that we can do these operations on all the streams at once. Use the **ctrl+A** shortcut to select all the objects in the flowsheet, then press **ctrl+B** to align the blocks, and **ctrl+J** to update the path of the streams. You may repeat the **ctrl+B** and **ctrl+J** process a few times to make sure that everything is fully aligned (Figure 1.166). If your flowsheet is well laid out to begin with, you will quickly end up with a neat and organized final result. If your flowsheet is too disorganized to begin with, you may have to manually move blocks and streams around before you can get a good end-product.

Now that everything is organized, we can declutter the flowsheet by hiding the names of some streams. In this flowsheet, we do not need to know the names of any streams, except those of the feed and product streams, so we can hide everything else. Use ctrl+click to select all of the process streams, except the feed and products, then use ctrl+H to hide the labels containing the stream names (Figure 1.167).

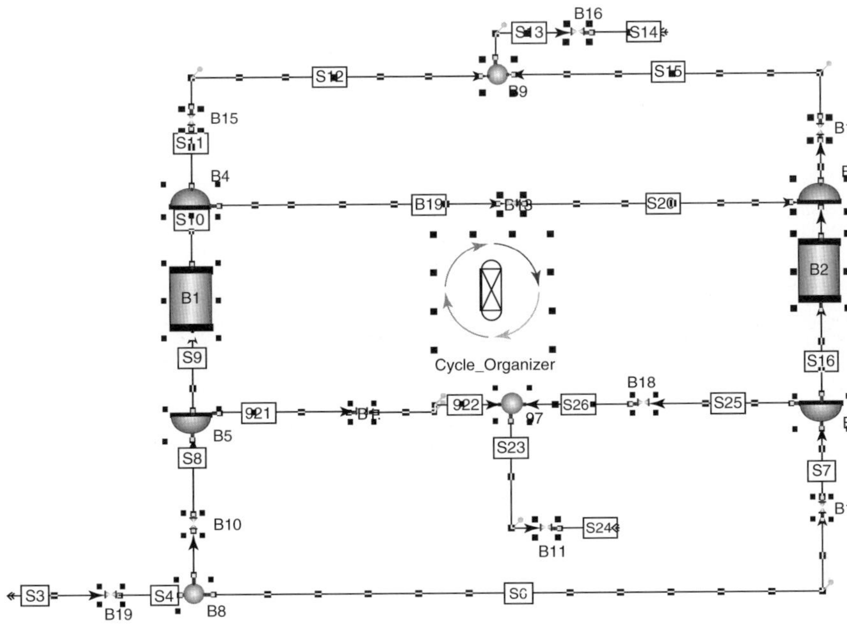

Figure 1.166 Flowsheet with all streams selected and aligned.

Finally, we can give more descriptive names to the blocks and streams in the flowsheet. To do this, click on the block you wish to rename, then press

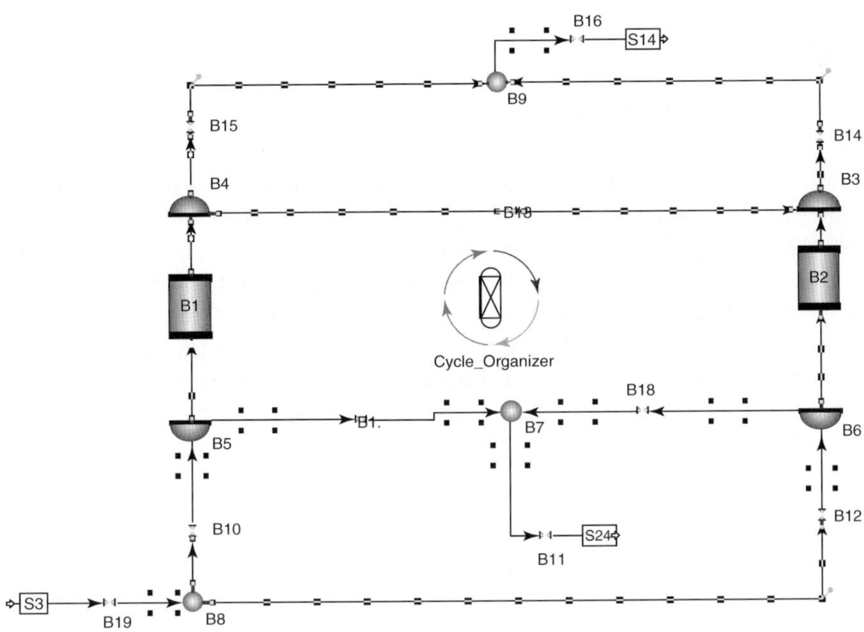

Figure 1.167 Flowsheet with stream names hidden.

ctrl+M to rename it. There are some restrictions on the names you may use, and Aspen will give you a helpful message if you accidentally enter a disallowed name. Rename all the tanks and valves in the flowsheet with helpful names, as shown in Figure 1.168.

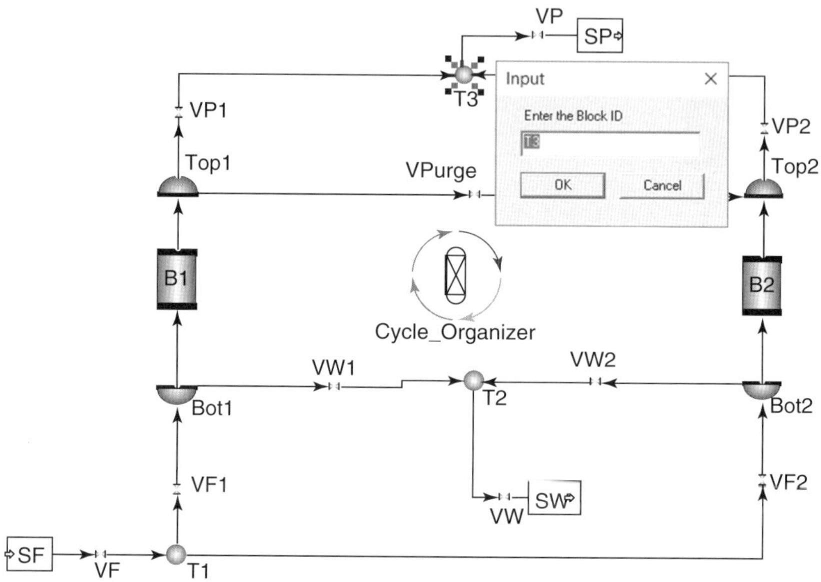

Figure 1.168 Renaming the tanks, valves, and streams.

Step 7: Specify Feed Streams

The first step to specifying the feed blocks is to make sure that the feed blocks will have all the variables we wish to specify. By default, the gCSS blocks are considered to be isothermal, which we do not want. To change this, we must edit the CSS_Info structure we created in step 3. Open the "Configuration" form from the Explorer, and set the "NonIsothermal" variable to "True" (Figure 1.169). This will update all the gCSS streams, tanks, and valves to have additional temperature-related options and variables.

Now we can specify the concentrations and temperatures of our feed, product, and waste streams (Figures 1.170–1.172). To do this, double-click on the stream whose values you wish to set. Be careful not to click close to the end of a stream, or you may end up moving the location of the end.

1.9.3 Introduction to Scripting: Set Repeated Values and Initialize Blocks

Step 8: Set all tanks to be nonadiabatic

Double-click on each tank to open their configuration form. Set the "NonAdiabaticTankVolume" to "True" (Figure 1.173). Doing this manually can be slow, but there is an alternative!

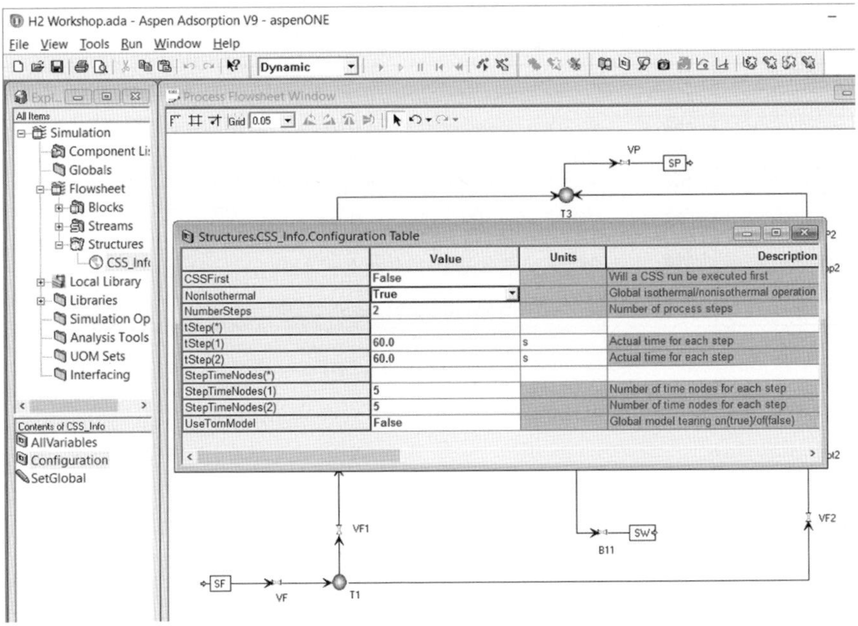

Figure 1.169 Configuring the CSSInfo structure to be nonisothermal.

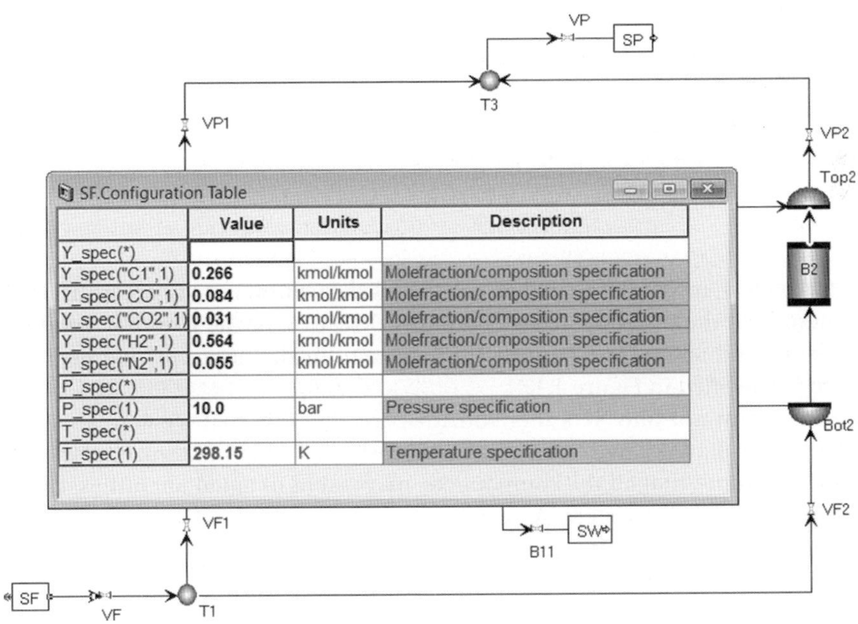

Figure 1.170 Specifying the Feed Stream.

SP.Configuration Table

	Value	Units	Description
Y_spec(*)			
Y_spec("C1",1)	0.266	kmol/kmol	Molefraction/composition specificatio
Y_spec("CO",1)	0.084	kmol/kmol	Molefraction/composition specificatio
Y_spec("CO2",1)	0.031	kmol/kmol	Molefraction/composition specificatio
Y_spec("H2",1)	0.564	kmol/kmol	Molefraction/composition specificatio
Y_spec("N2",1)	0.055	kmol/kmol	Molefraction/composition specificatio
P_spec(*)			
P_spec(1)	1.013	bar	Pressure specification
T_spec(*)			
T_spec(1)	298.15	K	Temperature specification

Figure 1.171 Specify the Product Stream.

SW.Configuration Table

	Value	Units	Description
Y_spec(*)			
Y_spec("C1",1)	0.266	kmol/kmol	Molefraction/composition specificatio
Y_spec("CO",1)	0.084	kmol/kmol	Molefraction/composition specificatio
Y_spec("CO2",1)	0.031	kmol/kmol	Molefraction/composition specificatio
Y_spec("H2",1)	0.564	kmol/kmol	Molefraction/composition specificatio
Y_spec("N2",1)	0.055	kmol/kmol	Molefraction/composition specificatio
P_spec(*)			
P_spec(1)	1.013	bar	Pressure specification
T_spec(*)			
T_spec(1)	298.15	K	Temperature specification

Figure 1.172 The Waste Stream.

If we wish, instead of manually setting all the values, we can create a *script* to automate the process. In the "Flowsheet" section of the Explorer window, click on "Add Script" and name it UtilityScript. Inside the script, add the lines of code specified in Figure 1.174.

This script not only sets the NonAdiabatic variable that we care about, it also sets an initial value for the gas composition inside the tank. By default, Aspen assumes that all the components are present in equal quantities, so for our five-component system, Aspen assumes a value of 0.2 for the mole fraction of each component. This script gives us a nice way to change that assumption in all the tanks at once.

To execute this script, you can either press ctrl+shift+I while the script is open, or simply double-click the UtilityScript object in the explorer window. Once you do that, you should see the following messages in the message window to confirm that the script set the values in each of the seven tanks (Figure 1.175).

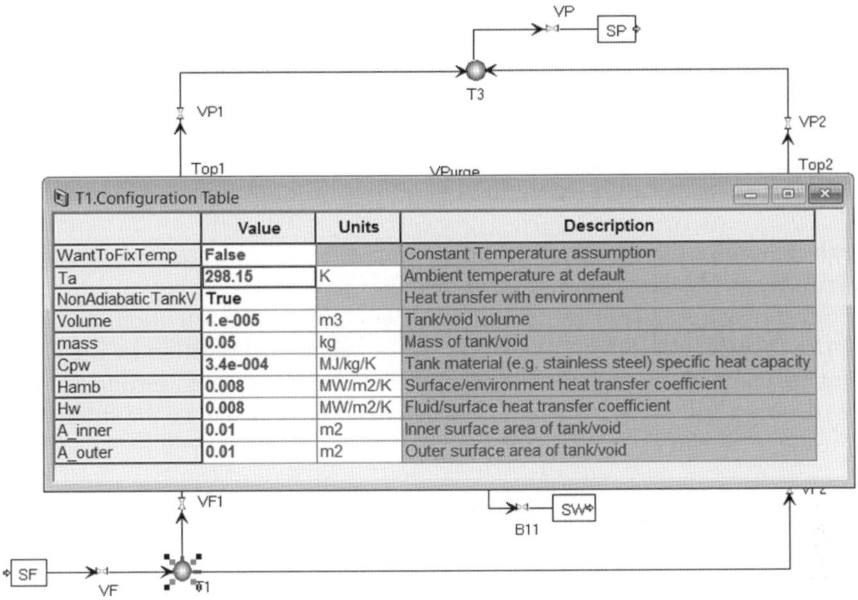

Figure 1.173 Specifying the tank to be nonadiabatic.

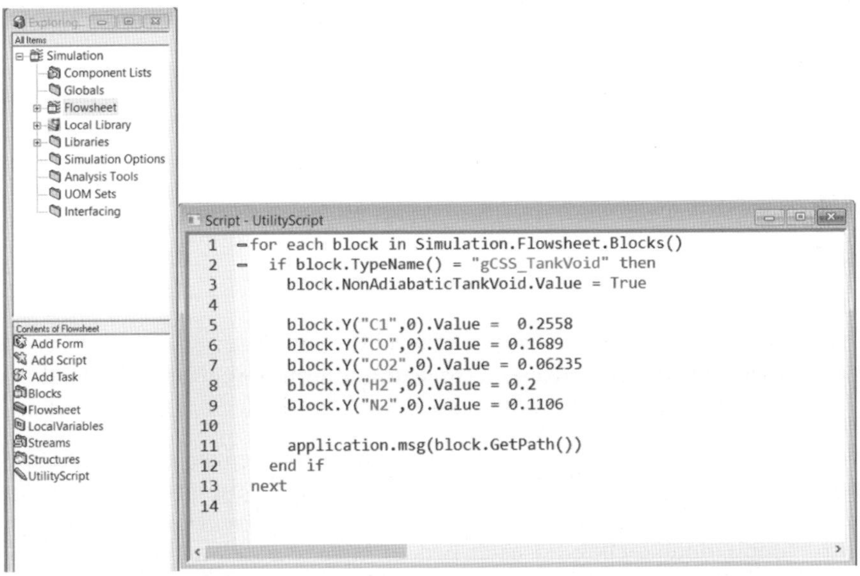

Figure 1.174 Creating the UtilityScript.

Step 9: Specify valves

We use the same strategy as the air separation workshop. VF, VP, VW, and Vpurge will all be implemented as control valves that have their flow rate set by the pressure drop across them, while the bed-specific valves (VF1, VF2, VP1, VP2, VW1, and VW2) will be either completely open or completely closed. As

Figure 1.175 The output of our utility script.

with the tanks, the names of the variables and their presentation is different between the "dynamic" and gCSS valves.

Set the values shown in Table 1.11. If you are paying attention, you may notice that the Use_Spec(1) variables will simply be overwritten by the Cycle Organizer, once we implement it. However, in this workshop, setting the values does actually serve a purpose, which will be explained once the simulation is ready to be run.

Table 1.11 Valve specifications for workshop 3.

Valve variable	#	Valve variable	kmol/(s bar^{-1})
VF1.Use_spec(1)	1	VF.Cv_spec(1)	2.00e−05
VF2.Use_spec(1)	1	VP.Cv_spec(1)	5.60e−09
VF.Use_spec(1)	2	Vpurge.Cv_spec(1)	1.80e−07
VP1.Use_spec(1)	1	VW.Cv_spec(1)	1.00e−05
VP2.Use_spec(1)	1		
VP.Use_spec(1)	2		
Vpurge.Use_spec(1)	0		
VW1.Use_spec(1)	1		
VW2.Use_spec(1)	1		
VW.Use_spec(1)	2		

1.9.4 Inspecting Blocks: Advanced Operating Conditions

Step 10: Specify separators

We have now set up the flowsheet, but we have two remaining tasks before we can run the simulation. First, we need to specify the properties of the adsorbent bed. Following this, we schedule the valve timings. To start with, simply double-click on each bed to open the geometry form. Specify that the adsorbent bed has two layers, and have the sizes specified in Figure 1.176.

Unlike with the "Gas: Dynamic" blocks, the gCSS blocks do not provide a nice user interface for specifying the rest of the property values. We have to find

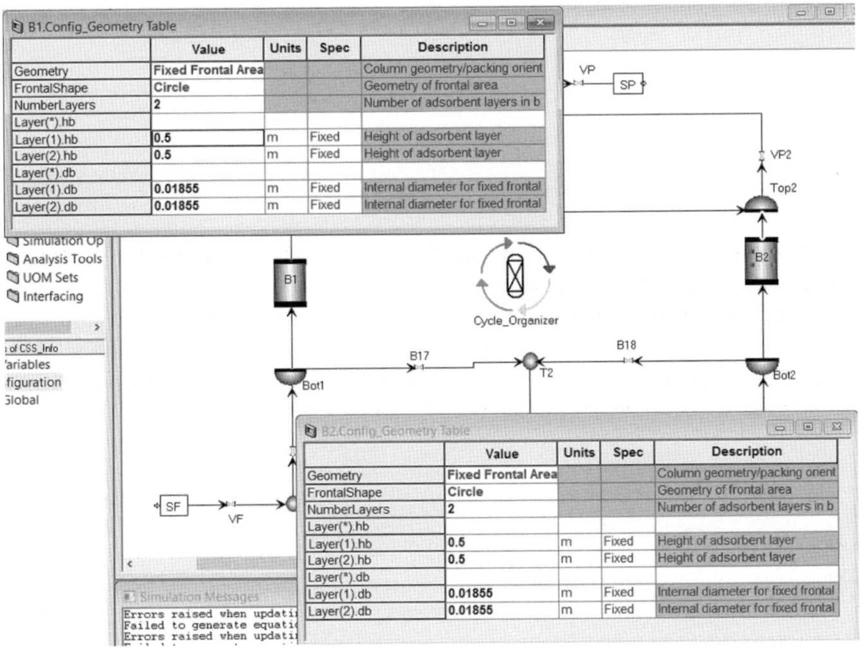

Figure 1.176 Identical bed geometries.

and specify the values manually. To begin, explore the first bed by right-clicking on it (Figure 1.177).

Once the explorer is open, we see a long list of "Config" forms, as shown in the left-hand window in Figure 1.178.

Step 11: Specify the adsorbent properties

Once the "Config" forms are visible, begin opening them one at a time in order to specify the adsorbent bed's properties. To begin with, open the Adsorbent Property form.

Note that each variable has two entries, because our adsorber has two layers. Remember that we are modeling an adsorber that uses two different adsorbents, so this is precisely the behavior we need, provided we have the data for both adsorbents.

Step 12: Make assumptions about heat transfer and specify energy balance values

Since we have specified that our gCSS simulation is nonisothermal, we need to provide information about the energy balances in the adsorber. Fortunately, the options in this block are well named, so it is easy to tell what options are available to you. Note that the exact mathematical forms of these assumptions are available in the help files, and in most cases, are identical to the Gas: Dynamic blocks from the first workshop.

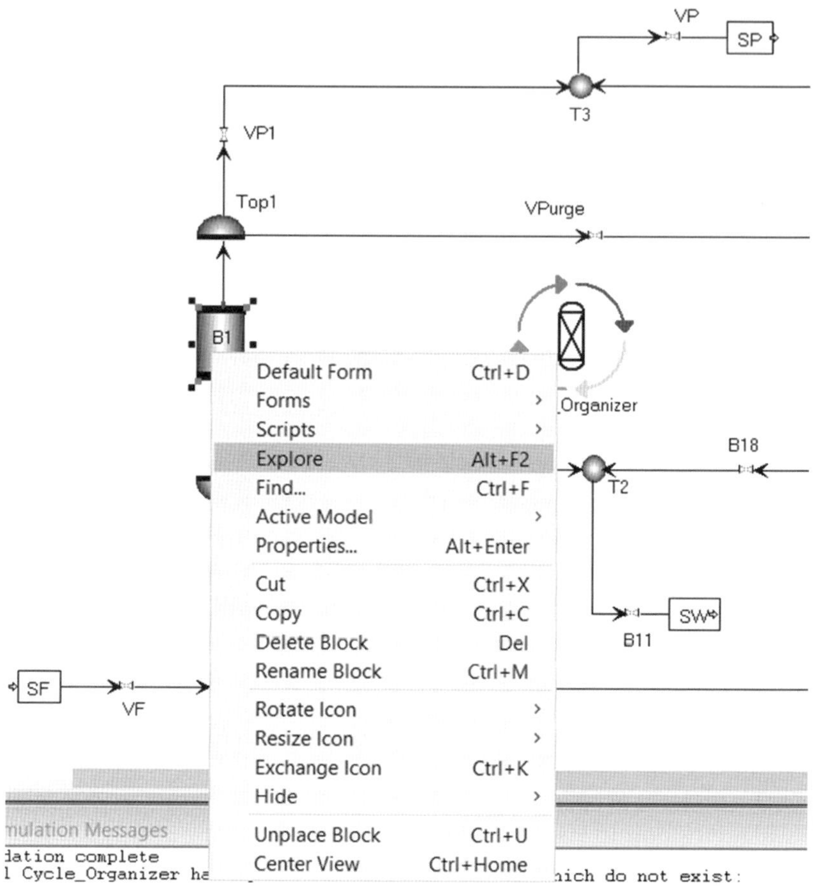

Figure 1.177 Opening the Explorer at the first bed.

The general form of the energy balance used here, when no reaction is present, is given in Eq. (1.16):

$$\overbrace{-k_{\mathrm{ga}}\varepsilon_i\frac{\partial^2 T_{\mathrm{g}}}{\partial z^2}}^{\text{Conduction}} + \overbrace{C_{\mathrm{vg}}v_{\mathrm{g}}\rho_{\mathrm{g}}\frac{\partial T}{\partial z}}^{\text{Convention}} + \overbrace{\varepsilon_{\mathrm{b}}C_{\mathrm{vg}}\rho_{\mathrm{g}}\frac{\partial T_{\mathrm{g}}}{\partial t}}^{\text{Accumulation}} + \overbrace{P\frac{\partial v_{\mathrm{g}}}{\partial z}}^{\text{Compression}} + \overbrace{\mathrm{HTC}\ a_{\mathrm{p}}(T_{\mathrm{g}}-T_{\mathrm{s}})}^{\text{Gas–Solid Transfer}} + \overbrace{\frac{4H_{\mathrm{w}}}{D_{\mathrm{B}}}(T_{\mathrm{g}}-T_{\mathrm{W}})}^{\text{Gas–Wall Transfer}} = 0. \tag{1.16}$$

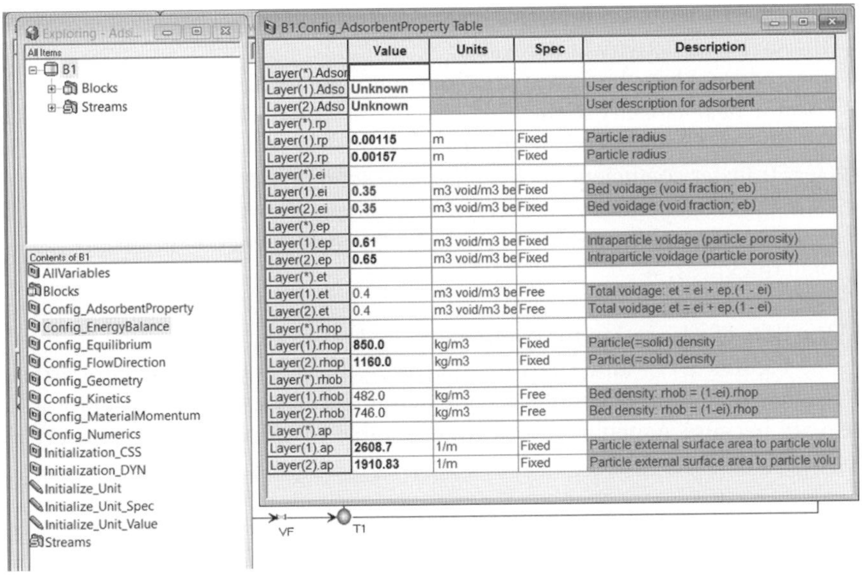

Figure 1.178 Using the explorer window to open the Adsorbent Properties form.

When we enable the rigorous wall balance, Aspen Adsorption calculates the change in wall temperatures according to Eq. (1.17):

$$\overbrace{-k_{\mathrm{w}}\frac{\partial^2 T_{\mathrm{w}}}{\partial z^2}}^{\text{Conduction}} + \overbrace{\rho_{\mathrm{w}} C_{\mathrm{pw}}\frac{\partial T_{\mathrm{w}}}{\partial t}}^{\text{Accumulation}} - \overbrace{H_{\mathrm{w}}\frac{4D_{\mathrm{B}}}{(D_{\mathrm{B}}+W_{\mathrm{T}})^2 - D_{\mathrm{B}}^2}(T_{\mathrm{g}} - T_{\mathrm{w}})}^{\text{Bed-Wall Transfer}} + \overbrace{H_{\mathrm{amb}}\frac{4(D_{\mathrm{B}}+W_{\mathrm{T}})^2}{(D_{\mathrm{B}}+W_{\mathrm{T}})^2 - D_{\mathrm{B}}^2}(T_{\mathrm{w}} - T_{\mathrm{amb}})}^{\text{Wall-Ambient Transfer}} = 0. \tag{1.17}$$

And finally, since we are assuming that the adsorption is nonisothermal, Aspen will use Eq. (1.18) to calculate the temperature of the adsorbent:

$$\overbrace{\rho_{\mathrm{s}} C_{\mathrm{ps}}\frac{\partial T_{\mathrm{s}}}{\partial t}}^{\text{Accumulation}} + \overbrace{\rho_{\mathrm{s}}\sum_{i=1}^{n}\Delta H_i\frac{\partial w_i}{\partial t}}^{\text{Heat of Adsorption}} + \overbrace{\mathrm{HTC}\, a_{\mathrm{p}}\,(T_{\mathrm{g}} - T_{\mathrm{s}})}^{\text{Gas-Solid Transfer}} = 0. \tag{1.18}$$

That means, by enabling the nonadiabatic and rigorous wall balances, we need to specify a large number of property parameters to describe these heating effects. Specifically, we need:

- Gas-phase thermal conductivity, k_{ga}.
- Thermal conductivity of the wall, k_w.
- Specific gas-phase heat capacity (constant volume) C_{vg}.
- Specific heat capacity of the wall, c_{pw}.
- Specific heat capacity of the adsorbent, c_{ps}.
- Heat transfer coefficient between gas and solid phase, HTC.
- Heat transfer coefficient between the gas phase and wall, H_w.
- Heat transfer coefficient between the wall and ambient conditions, H_{amb}.
- Wall thickness, W_T.
- Wall density, ρ_W.
- The heat of adsorption for component i, ΔH_i.

See Figures 1.179 and 1.180 for specifying the energy balance properties. Note that we do not need information about gas density, compressibility, and velocity because Aspen Properties will be able to calculate these properties for us.

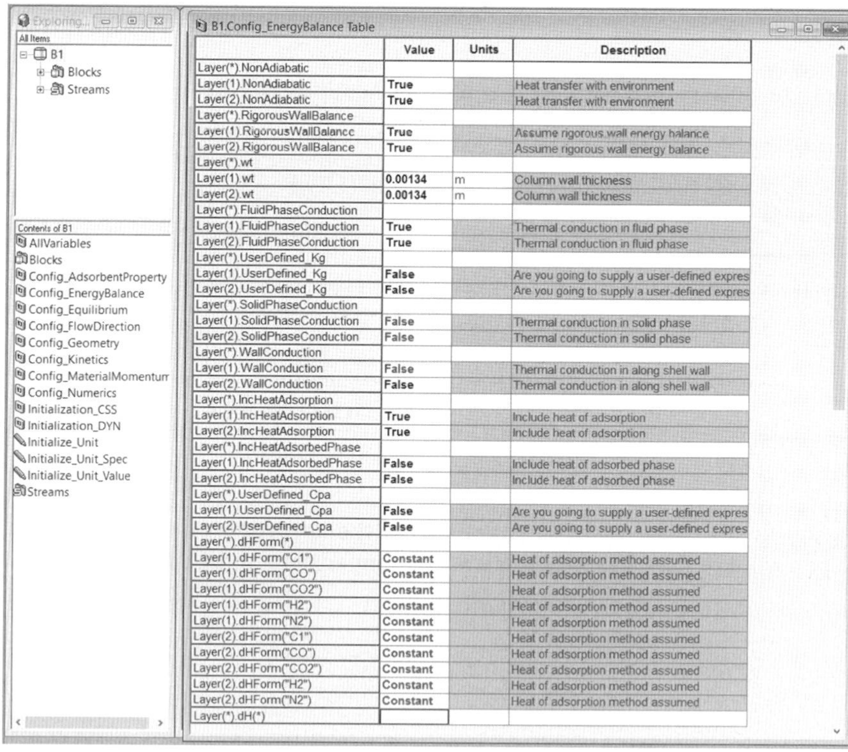

Figure 1.179 Using the explorer window to open the Energy Balance Properties form.

Step 13: Define the equilibrium behavior of the system

The next configuration form is the equilibrium behavior of the adsorption. We specify what form of the isotherm we use first, under "Equilibrium Model." There are a large number of options, and we can find their details in the help

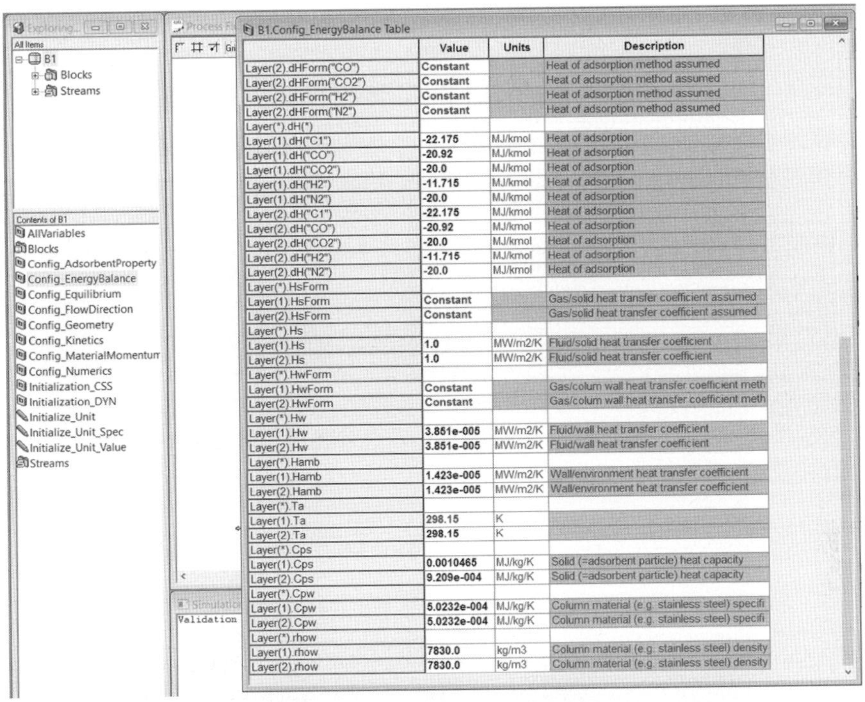

Figure 1.180 Specifying the rest of the Energy Balance Properties.

files. Once the isotherms form are selected, you must supply all the equilibrium property parameters (Figures 1.181 and 1.182).

To find out what isotherm form is assumed by "Loading Ratio Correlation 3" seen in Figure 1.183, we can look it up in the help files. Open "Aspen Adsorption Contents" under the help menu, then search for "loading ratio correlation" (Figures 1.183 and 1.184).

By clicking on the correct entry, we can find the full equilibrium equation assumed (Figure 1.184).

Step 14: Specify the flow direction options

If gas will only flow in one direction through a bed, calculations can be slightly more efficient. However, we let the pressure gradients determine the flow direction for us (Figure 1.185).

Step 15: Choose the mass transfer models to use

These options are identical to those mentioned in workshop 1. The most common model is lumped LDF, Eq. (1.15), since other models can add significant computational overhead, with very little improvement in accuracy. See Figure 1.186.

Step 16: Specify the momentum balance assumptions

In the momentum balance table, we may leave all values at their defaults. These options allow us to adjust how the adsorption bed calculates gas pressures drops and velocities. The well-known Ergun equation, Eq. (1.11), is suitable for most conditions (Figure 1.187), since it combines the equations useful at

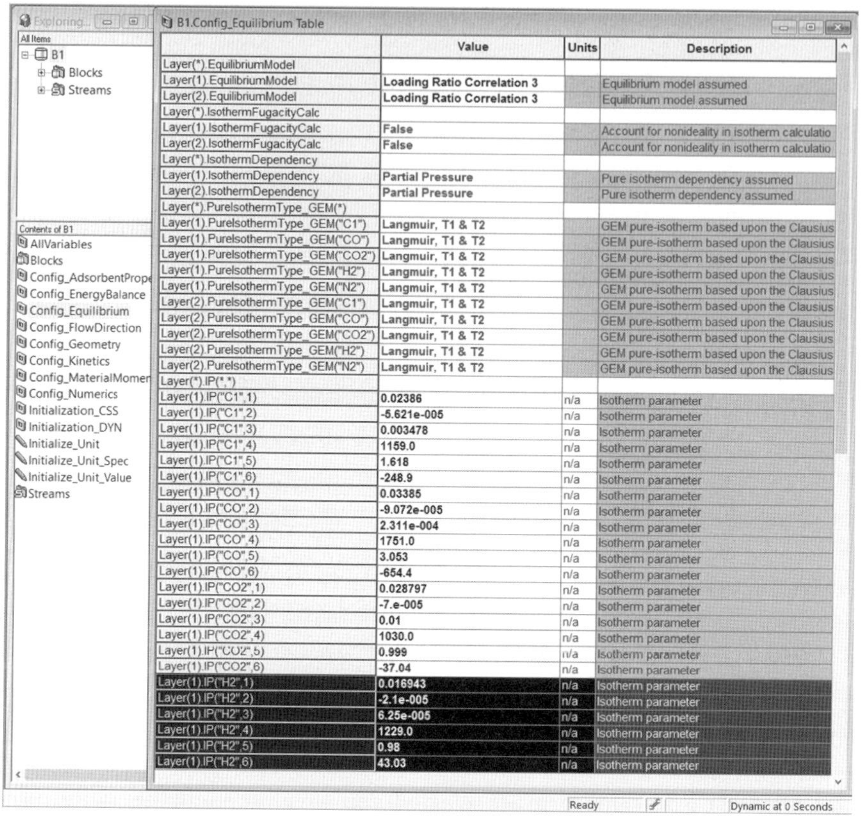

	Value	Units	Description
Layer(*).EquilibriumModel			
Layer(1).EquilibriumModel	Loading Ratio Correlation 3		Equilibrium model assumed
Layer(2).EquilibriumModel	Loading Ratio Correlation 3		Equilibrium model assumed
Layer(*).IsothermFugacityCalc			
Layer(1).IsothermFugacityCalc	False		Account for nonideality in isotherm calculatio
Layer(2).IsothermFugacityCalc	False		Account for nonideality in isotherm calculatio
Layer(*).IsothermDependency			
Layer(1).IsothermDependency	Partial Pressure		Pure isotherm dependency assumed
Layer(2).IsothermDependency	Partial Pressure		Pure isotherm dependency assumed
Layer(*).PureIsothermType_GEM(*)			
Layer(1).PureIsothermType_GEM("C1")	Langmuir, T1 & T2		GEM pure-isotherm based upon the Clausius
Layer(1).PureIsothermType_GEM("CO")	Langmuir, T1 & T2		GEM pure-isotherm based upon the Clausius
Layer(1).PureIsothermType_GEM("CO2")	Langmuir, T1 & T2		GEM pure-isotherm based upon the Clausius
Layer(1).PureIsothermType_GEM("H2")	Langmuir, T1 & T2		GEM pure-isotherm based upon the Clausius
Layer(1).PureIsothermType_GEM("N2")	Langmuir, T1 & T2		GEM pure-isotherm based upon the Clausius
Layer(2).PureIsothermType_GEM("C1")	Langmuir, T1 & T2		GEM pure-isotherm based upon the Clausius
Layer(2).PureIsothermType_GEM("CO")	Langmuir, T1 & T2		GEM pure-isotherm based upon the Clausius
Layer(2).PureIsothermType_GEM("CO2")	Langmuir, T1 & T2		GEM pure-isotherm based upon the Clausius
Layer(2).PureIsothermType_GEM("H2")	Langmuir, T1 & T2		GEM pure-isotherm based upon the Clausius
Layer(2).PureIsothermType_GEM("N2")	Langmuir, T1 & T2		GEM pure-isotherm based upon the Clausius
Layer(*).IP(*,*)			
Layer(1).IP("C1",1)	0.02386	n/a	Isotherm parameter
Layer(1).IP("C1",2)	-5.621e-005	n/a	Isotherm parameter
Layer(1).IP("C1",3)	0.003478	n/a	Isotherm parameter
Layer(1).IP("C1",4)	1159.0	n/a	Isotherm parameter
Layer(1).IP("C1",5)	1.618	n/a	Isotherm parameter
Layer(1).IP("C1",6)	-248.9	n/a	Isotherm parameter
Layer(1).IP("CO",1)	0.03385	n/a	Isotherm parameter
Layer(1).IP("CO",2)	-9.072e-005	n/a	Isotherm parameter
Layer(1).IP("CO",3)	2.311e-004	n/a	Isotherm parameter
Layer(1).IP("CO",4)	1751.0	n/a	Isotherm parameter
Layer(1).IP("CO",5)	3.053	n/a	Isotherm parameter
Layer(1).IP("CO",6)	-654.4	n/a	Isotherm parameter
Layer(1).IP("CO2",1)	0.028797	n/a	Isotherm parameter
Layer(1).IP("CO2",2)	-7.e-005	n/a	Isotherm parameter
Layer(1).IP("CO2",3)	0.01	n/a	Isotherm parameter
Layer(1).IP("CO2",4)	1030.0	n/a	Isotherm parameter
Layer(1).IP("CO2",5)	0.999	n/a	Isotherm parameter
Layer(1).IP("CO2",6)	-37.04	n/a	Isotherm parameter
Layer(1).IP("H2",1)	0.016943	n/a	Isotherm parameter
Layer(1).IP("H2",2)	-2.1e-005	n/a	Isotherm parameter
Layer(1).IP("H2",3)	6.25e-005	n/a	Isotherm parameter
Layer(1).IP("H2",4)	1229.0	n/a	Isotherm parameter
Layer(1).IP("H2",5)	0.98	n/a	Isotherm parameter
Layer(1).IP("H2",6)	43.03	n/a	Isotherm parameter

Figure 1.181 Using the explorer window to open the Equilibrium Properties form.

laminar conditions (Carman–Kozeny) with the equations for turbulent flow (Burke–Plummer).

Step 17: Choose the numerical discretization options

In this form, we may select the methods used for finite element discretization and estimation. The options are identical to those discussed in the first workshop. In this case, we use a second-order Central Finite Difference (CFD2) approximation with 16 nodes in each bed (Figure 1.188). This is adequate, and has good computational performance; for brand new simulations, it is generally a good idea to begin with a simulation using an Orthogonal Collocation Finite Element (OCFE) method. Once a solution is found with OCFE, the finite difference method can be compared to the more accurate result in terms of computation speed and accuracy.

With this, the adsorbent beds have been completely specified.

Step 18: Perform gCSS initializations

After completing all these steps, the simulation will still not be runnable. The reason is found at the very bottom of the Aspen window. The green box with red triangle indicates that our flowsheet is overspecified: there are negative degrees of freedom (Figure 1.189). The reason for this is that several variables

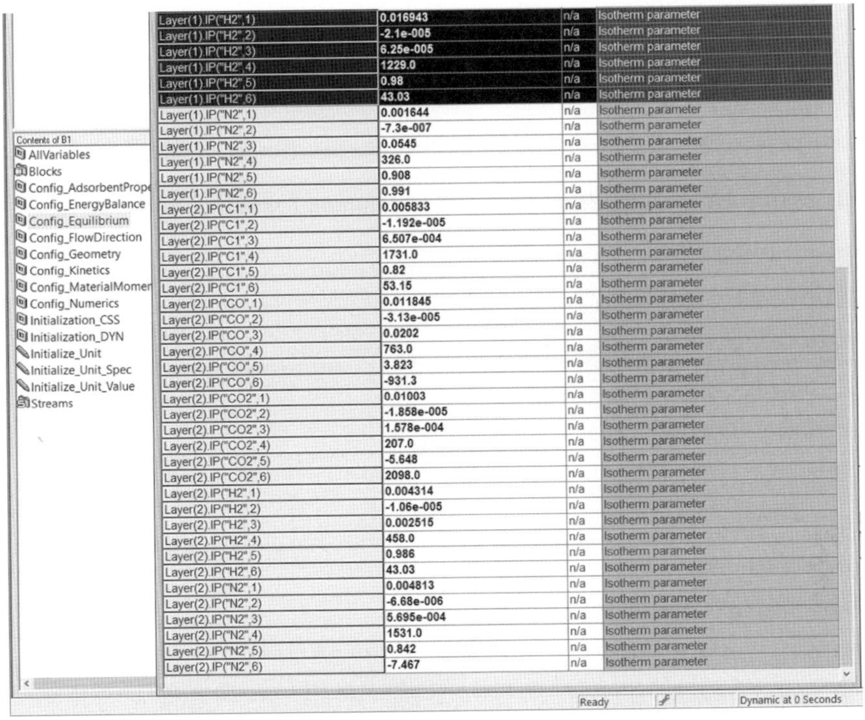

Figure 1.182 Specifying the rest of the Equilibrium Properties.

in the gCSS blocks (specifically, tanks and beds) are set to be fixed, when in a dynamic run, they should only represent initial conditions.

Fortunately, the gCSS blocks contain initialization scripts which fix this for us. To run the initialization script on a block, right-click the block, mouse over script, then click on Initialize_Unit (Figure 1.190). Save the file as **Workshop3d.ada**.

Once you do this, you should notice that the simulation will only be over-specified by six variables. You can manually run each initialization script, or we can once again run the flowsheet > "Check and initial."

1.9.5 Defining the Cycle Organizer

Step 19: Add the Cycle Organizer

Use the "Tools" menu to add the cycle organizer to the flowsheet.

Step 20: Define the cycle steps

As we saw previously, the gCSS valves use different variable names than the dynamic valves. To toggle the valves open and closed, we need to assign values to the "Use_Spec(1)" variable. Add the Use_Spec variables for the seven valves that change over time to step 1, then create six cycle steps containing a copy of those variables. Following this, assign the values according to Table 1.12. Refer to Table 1.5 in connection with Figure 1.7 for the meanings of valve ASs of 0–2.

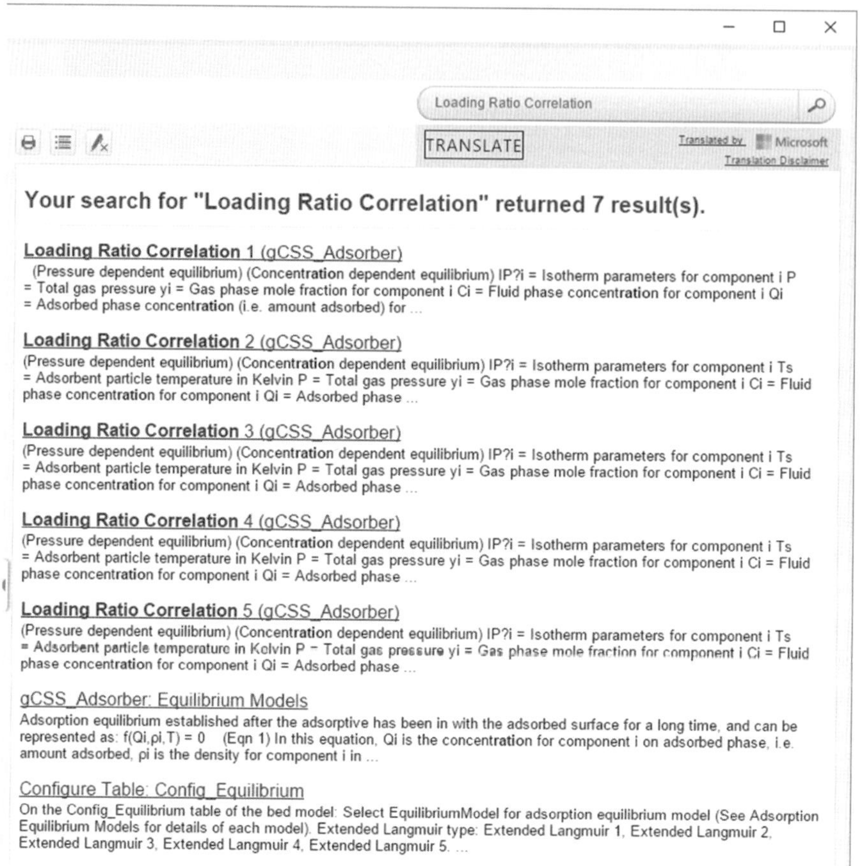

Figure 1.183 Searching for the isotherm equation.

Step 21: Wait! Do not generate the Cycle Task!
This simulation is more complex than our previous air separation example. Not only are there more components, but our adsorption beds have two layers, and we are using a more complex isotherm. All these factors can make it difficult for Aspen to "get things started." That is to say, the first few seconds of a dynamic simulation can be the hardest on the Aspen solver. To avoid having to repeatedly wait for this difficult solver step, we will perform a trick.

Step 22: Set the simulation to pause at one second
Use the "Run Control" button shown in Figure 1.191.

Step 23: Set the Solver Integration options
The Solver Options provide advanced control over the mathematical solution methods used in Aspen. For the vast majority of cases, the default values work very well. In this case, we will change how the numerical integration takes place. Specifically, we are going to set the initial time-integration step to be smaller, so that the changes in the various process variables will change by smaller amounts and be less likely to take on extreme values. Click on the "Solver" options button (Figure 1.191), and then select the Integrator tab and

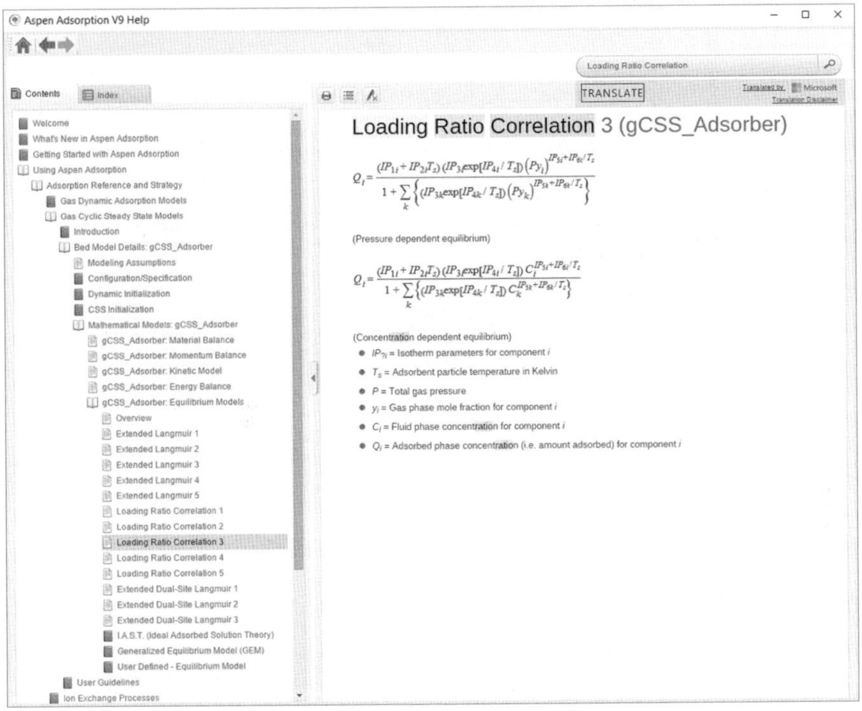

Figure 1.184 Viewing the full form of the Loading Ratio Correlation 3.

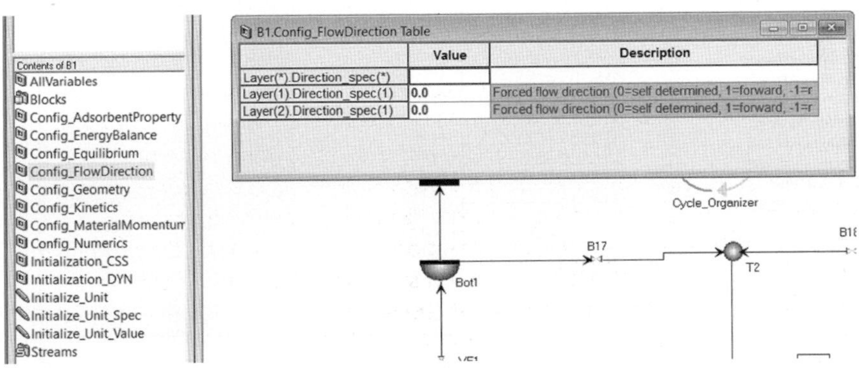

Figure 1.185 Set Aspen to automatically determine the gas flow direction in the adsorbent beds.

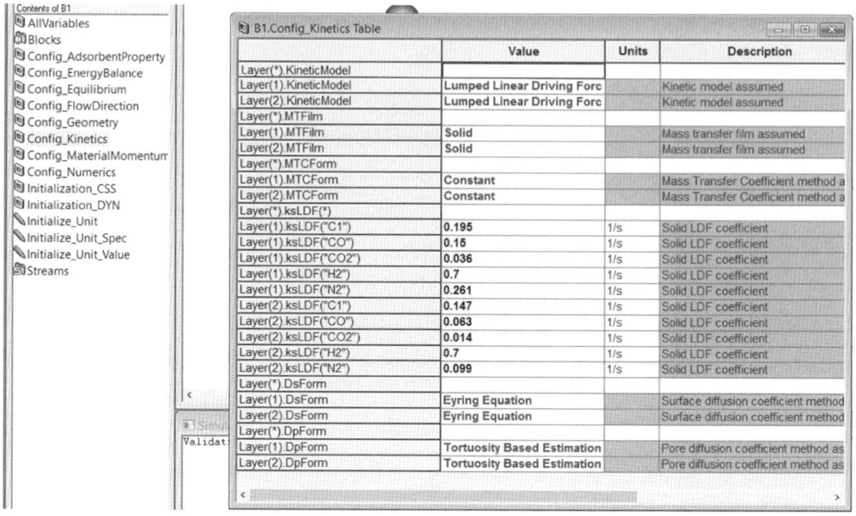

Figure 1.186 Specifying the kinetic properties of our adsorbent bed.

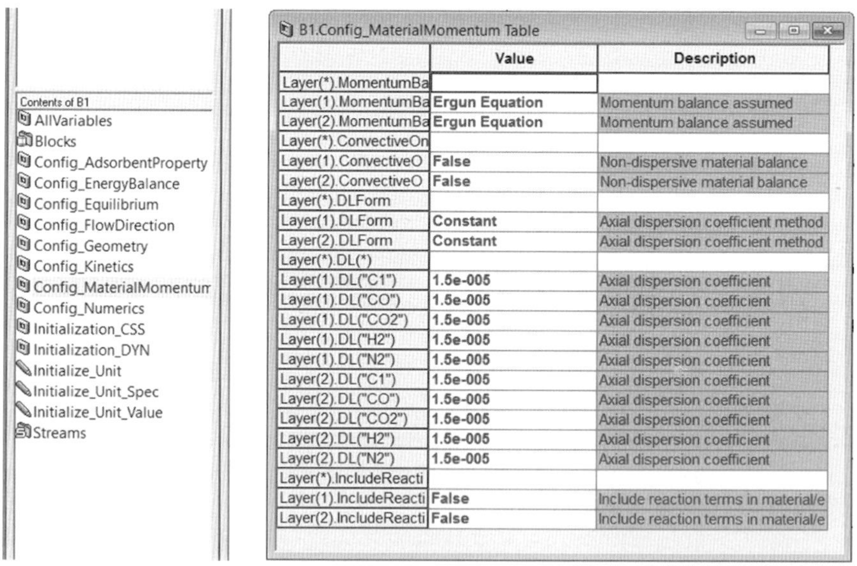

Figure 1.187 Choose the coefficients for the Ergun model in our momentum balance.

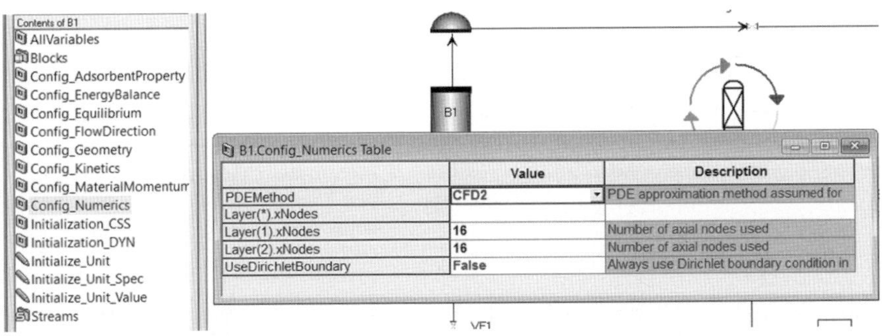

Figure 1.188 Choose the numerical PDE solution options.

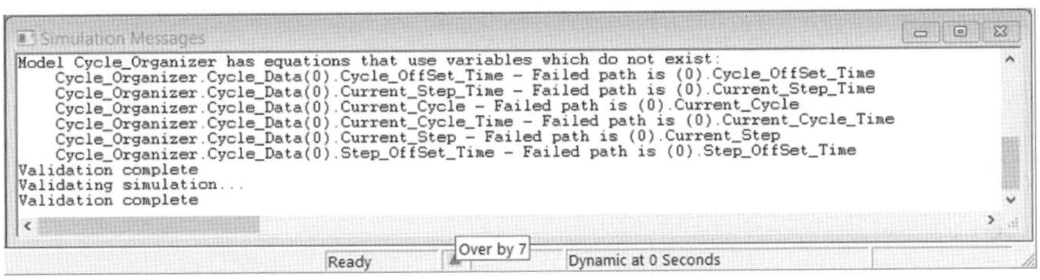

Figure 1.189 An overspecified flowsheet.

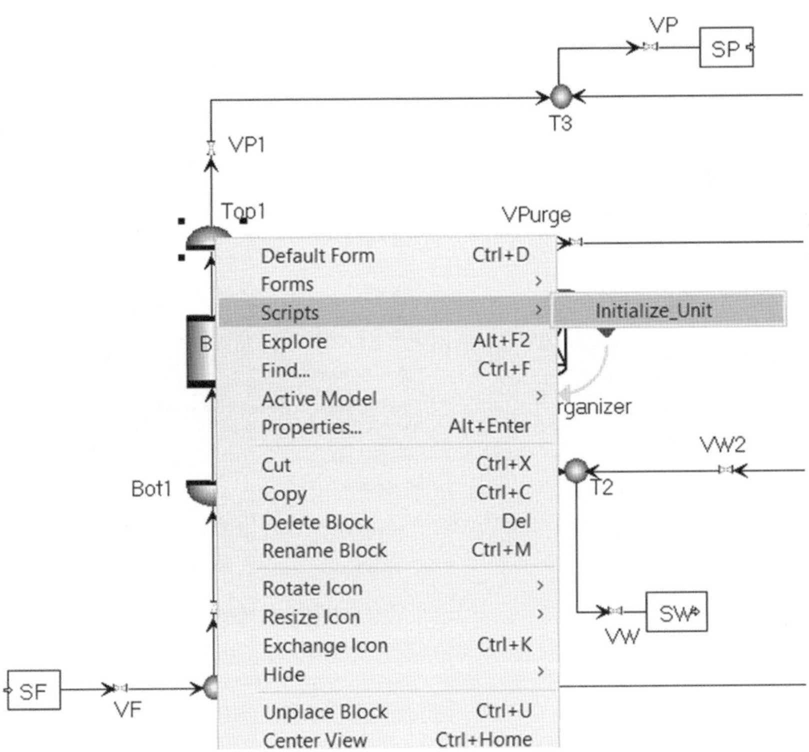

Figure 1.190 Initializing a gCSS block with an initialization script.

Table 1.12 Hydrogen PSA cycle and valve "Use_Specs."

Cycle	Step 1	Step 2	Step 3	Step 4	Step 5	Step 6
Role	Pressurize bed 1, depressurize bed 2	Adsorption bed 1, purge bed 2	Pressure equilibration	Pressurize bed 2, depressurize bed 1	Adsorption bed 2, purge bed 1	Pressure equilibration
Duration (s)	30	40	140	30	40	140
VF1	1	1	0	0	0	0
VF2	0	0	0	1	1	0
VP1	0	1	0	0	0	0
VP2	0	0	0	0	1	0
VPurge	0	2	2	0	2	2
VW1	0	0	0	1	1	0
VW2	1	1	0	0	0	0

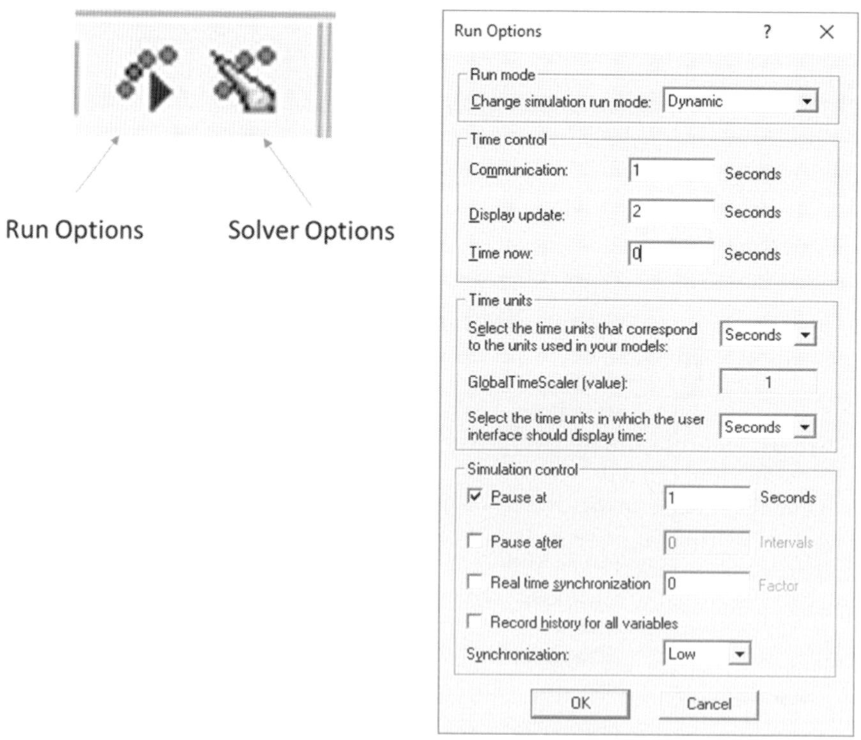

Figure 1.191 The run control button, and the run control panel.

change the “Initial step size” and “Minimum step size” to 0.1 (Figure 1.192). Save the file as **Workshop3e.ada**.

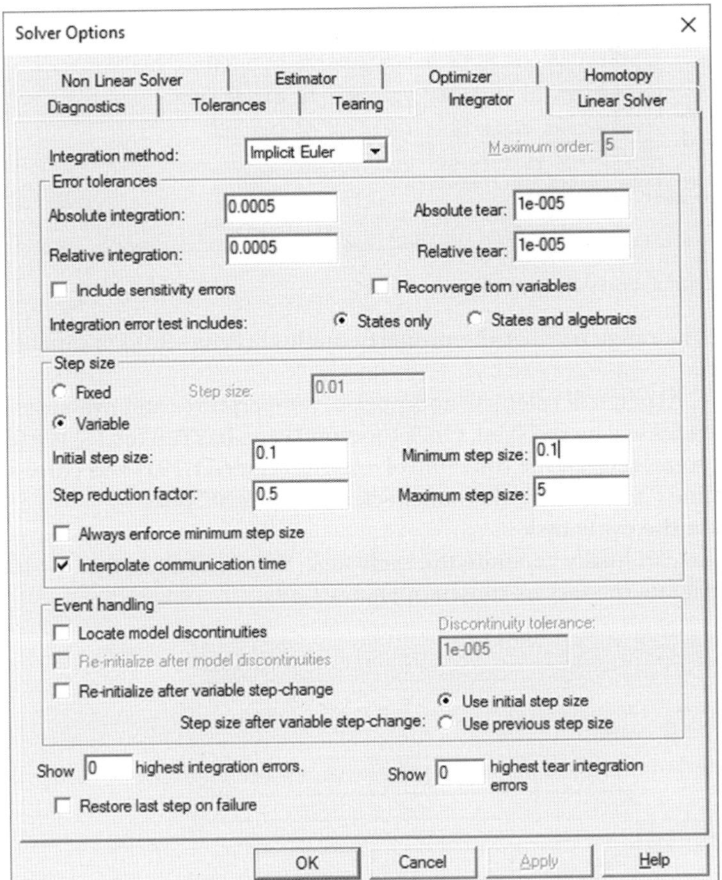

Figure 1.192 The Solver Options – Integrator tab.

Step 24: Run simulation

After confirming that the cycle *has not been generated*, and that the run mode is “Dynamic,” click on the “Run” button to begin the simulation. This will begin the dynamic run (Figure 1.193), and it will automatically end after 1 s. This may take some time, and there may be errors during the initial integration steps.

Figure 1.193 Run the dynamic simulation.

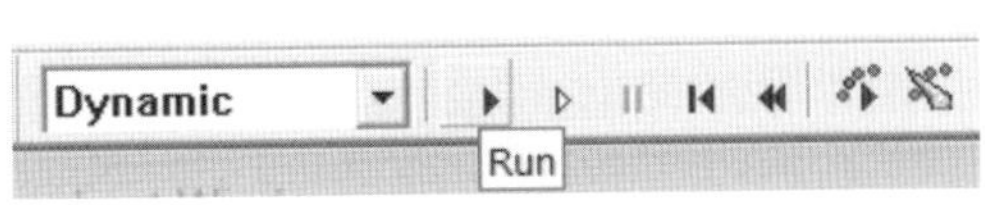

Examples of the types of errors that arise during the first few integration steps are shown in Figure 1.194. In this case, the solver has chosen a temperature

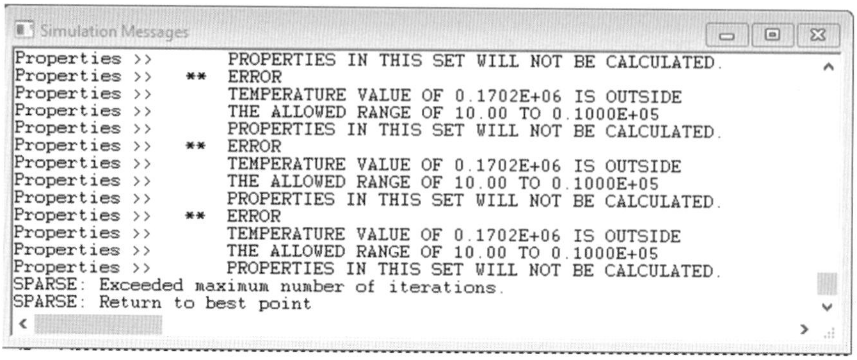

Figure 1.194 Initial simulation step example.

that is outside the range of what the property method (Peng–Rob) is capable of dealing with.

After a while, Aspen will calculate appropriate temperatures and pressures, and these messages will be resolved. Once the simulation has run for 1 s, it will pause. At this point, reopen the run control options (Figure 1.191) and enter a 0 for current time ("time now"), then uncheck the "pause at" box.

Step 25: Generate the cycle task

At this point, we can finally generate the cycle task. Set the cycle to run for 10 cycles, then generate the task as shown in Figure 1.195.

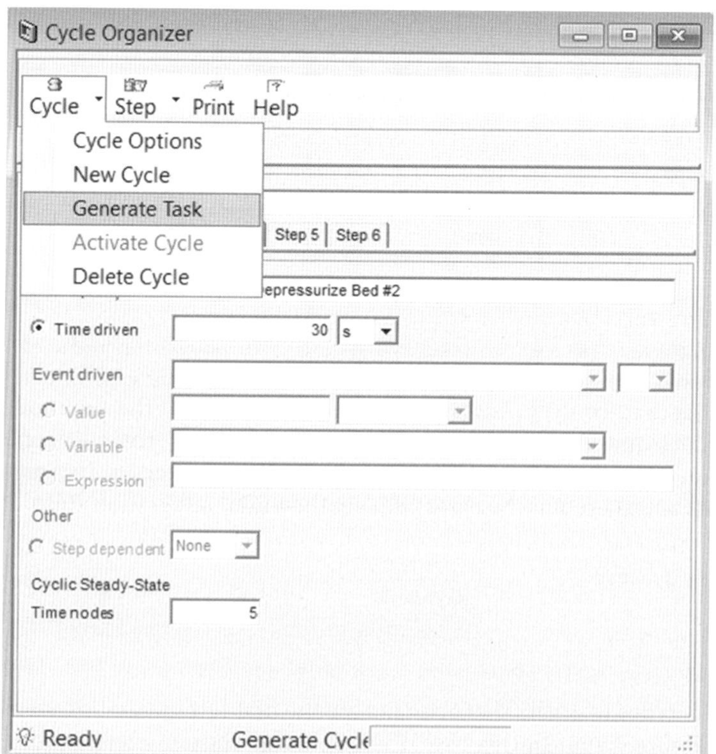

Figure 1.195 Generating the Cycle Organizer task.

Step 26: Run simulation

After confirming that the flowsheet is correct, and that the run mode is "Dynamic," click on the "Run" button to begin the simulation. This will begin the dynamic run, and it will automatically end after 10 cycles. See Figure 1.196.

Figure 1.196 Run the dynamic simulation.

1.9.6 Viewing Results

Step 27: Plot results

Once the simulation completes successfully, we would want to view the separation results in plot form. To do this, click on the "New form" button in the Explorer window, and add a new plot (Figure 1.197).

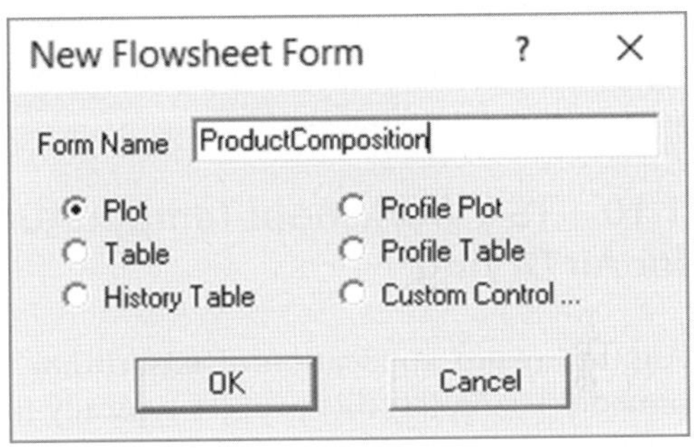

Figure 1.197 Creating a new plot.

To retrieve the variables we want to plot, right-click on the product tank, T3. By mousing over "Forms," we can see several predefined forms for the Tank blocks. For example, we could look at the Result_Plot_DYN to see one representation of the results (Figure 1.198). However, we plot the mole fraction of the various components, so we need to open the "AllVariables" form.

Once the AllVariables table is open, select all the mole fraction variables and drag them onto the *y*-axis of the plot (Figure 1.199). You can also drag them one at a time.

After adding the variables, right-click on the plot and "Zoom Full" to see the entire history (Figure 1.200).

When the full history is visible, the result should look like Figure 1.201. The product should be >99.7% pure hydrogen.

If your result matches, you have completed the workshop! Save the file as **Workshop3e_dyn.ada**.

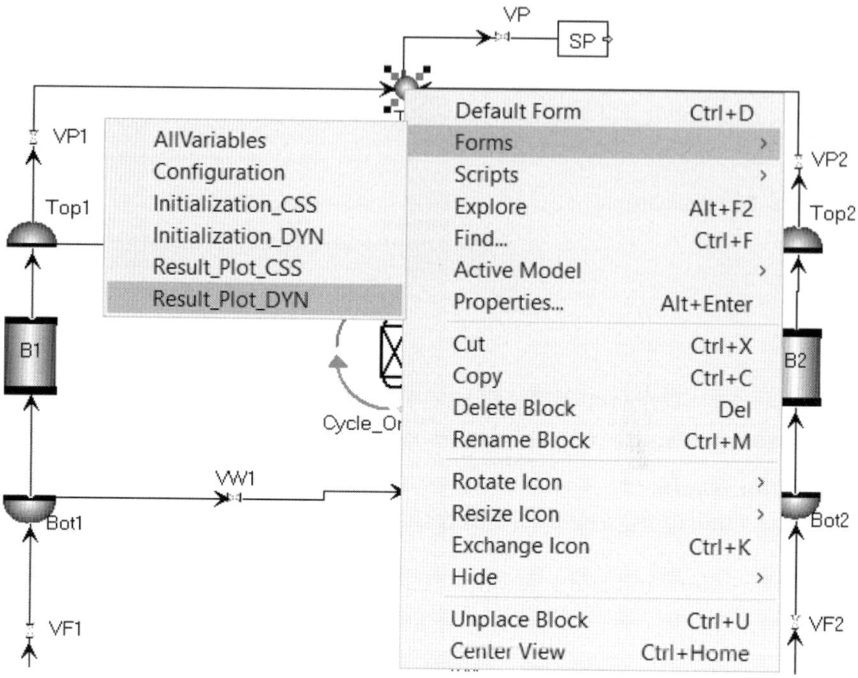

Figure 1.198 Locating variables to plot. Result_Plot_DYN contains a premade graph, but we want variables from AllVariables.

1.10 TSA Workshop: Temperature Swing Adsorption for Air Drying

Another major application of adsorption technology is air drying. Unlike the previous examples, this process typically involves TSA, as opposed to pressure swing. Let us consider a small process designed to dry air by passing moist air over an alumina adsorbent.

Step 1: Draw the flowsheet

Using the techniques learned in the first three workshops, create the flowsheet shown in Figure 1.202. Start with a new Aspen file called `Workshop4a.ada`. The components for the system are "Air" and "H_2O," both of which can be found in the Aspen Properties Databanks. In this process, we use the models in the gas_dynamic folder, rather than the gCSS folder.

The flowsheet shown in Figure 1.202 allows us to implement the cyclic procedure described in Figure 1.203. Not that this is very similar to the PSA case, but temperature changes as well.

Step 2: Enter the process data

Add the following information to the flowsheet as Figures 1.204 and 1.205 using the same techniques as in the PSA Workshop, Figures 1.45 and 1.46.

We also need to specify the pressure at the outlets to this process (Figures 1.206 and 1.207).

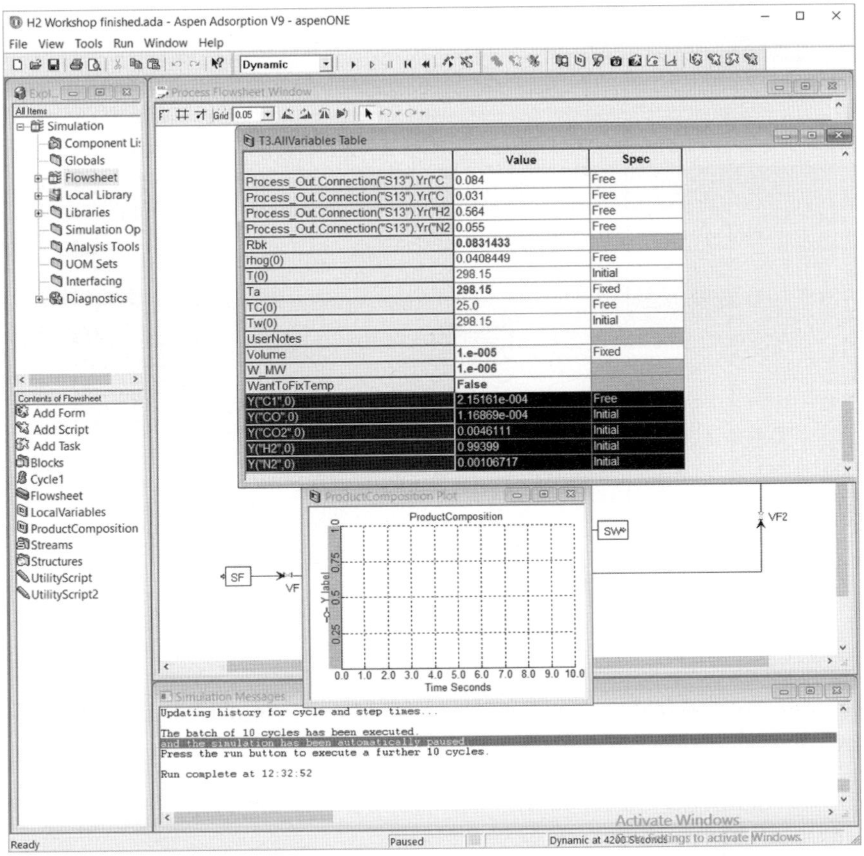

Figure 1.199 Dragging the mole fraction variables from the AllVariables table to the plot.

Step 3: Specify the tank volumes and options

For this problem, if we leave the tank volumes at their very small initial value, and try to calculate compressive terms, we will get occasional unreasonable spikes in temperature due to numerical issues at step changes. To mitigate this, we increase the tank volumes and ignore the compressibility of the gas in the tanks. See Figure 1.208.

Step 4: Determine the Isotherm parameters

Let us suppose that the supplier of your alumina desiccant was "creative" in choosing an isotherm to describe their desiccant's performance. They claim that air is never adsorbed, and provide you with the procedure described in Eq. (1.19) to calculate the equilibrium loading of water:

$$P_{\text{sat}} = \text{IP}_1 e^{\left(\text{IP}_2 - \frac{\text{IP}_3}{T}\right)}$$

$$P_{\text{rel}} = Y_{\text{H}_2\text{O}} \frac{P_{\text{Total}}}{P_{\text{sat}}}$$

$$W_{\text{H}_2\text{O}} = \text{IP}_4{}^* P_{\text{rel}} \qquad (1.19)$$

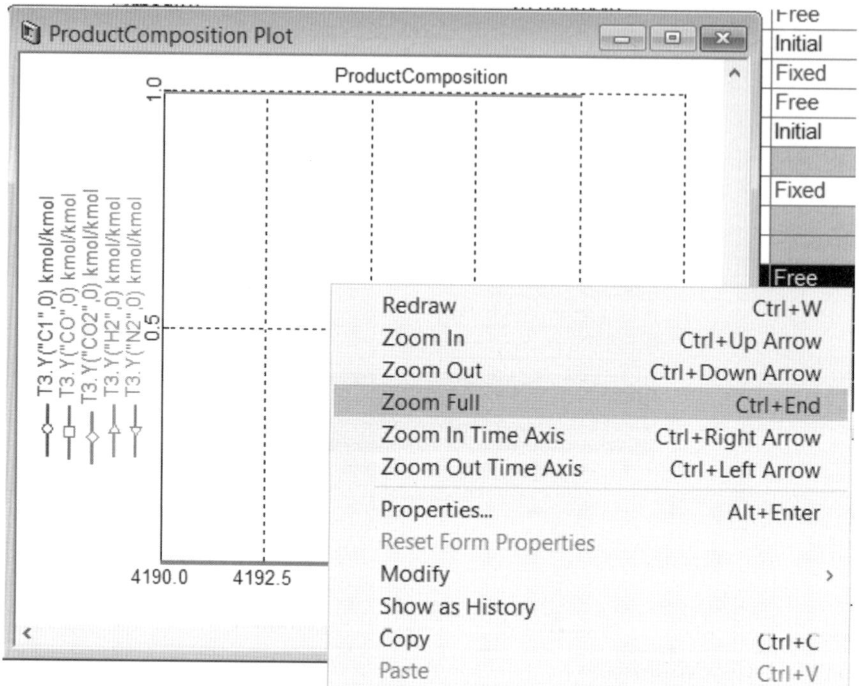

Figure 1.200 Zooming out on the Product Composition plot.

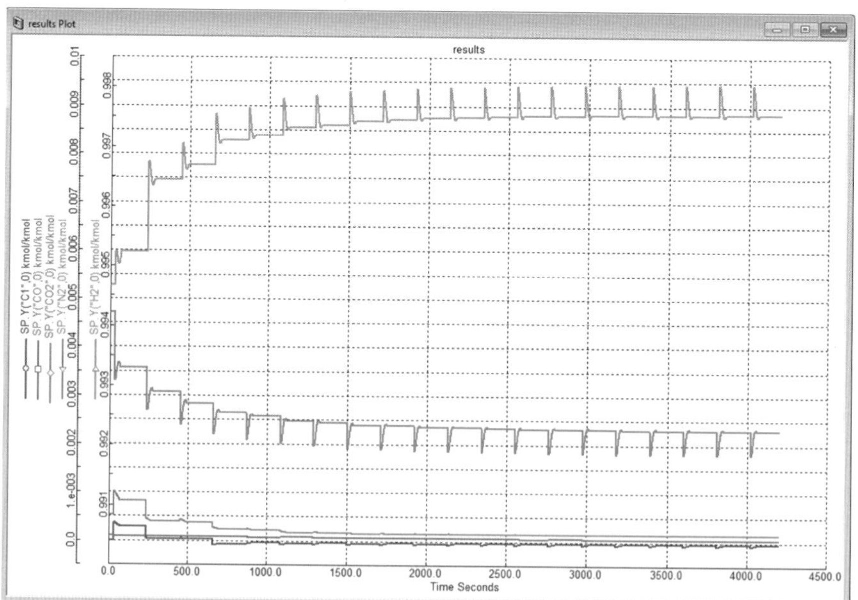

Figure 1.201 The final product composition result.

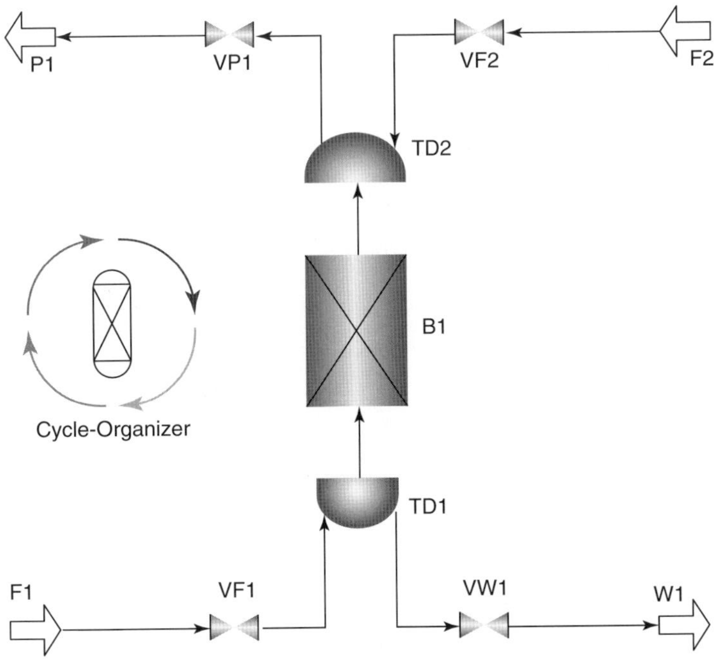

Figure 1.202 The TSA flowsheet. F1 represents the moist air to be dried, while F2 represents a partial recycle of the heated product air that is used to regenerate the adsorbent.

Can we input this multistep procedure into Aspen? What are our options?

- We could use *Flowsheet Constraints* to define the loading in exactly the form provided by the manufacturer.
- We could try to rearrange the equations into something Aspen can handle.

In this case, with some fairly simple observations, we can convert the isotherm into a form Aspen already has implemented. Specifically, let us start by combining all of those steps into one big equation

$$W_{H_2O} = IP_4 \frac{Y_{H_2O} P_{Total}}{IP_1 e^{\left(IP_2 - \frac{IP_3}{T}\right)}} \tag{1.20}$$

Right away, we should notice that $Y_{H_2O}P_{Total}$ is simply partial pressure (p_{H_2O}) and there is no reason to have IP_4/IP_2 as two separate isotherm parameters. We also know that we can simply make the exponent negative to move it into the numerator.

$$W_{H_2O} = IP_5 p_{H_2O} e^{\left(-IP_2 + \frac{IP_3}{T}\right)} \tag{1.21}$$

Now this looks much simpler. However, we still have the $e^{\left(-IP_2 + \frac{IP_3}{T}\right)}$ term, which does not appear in any comparable built-in isotherm. Fortunately, we know that it is possible to write $e^{\left(-IP_2 + \frac{IP_3}{T}\right)}$ as $e^{-IP_2} e^{\left(\frac{IP_3}{T}\right)}$ and that e^{-IP_2} is itself a constant. So it is finally possible to write the equation as

$$W_{H_2O} = IP_6 p_{H_2O} e^{\left(\frac{IP_3}{T}\right)} \tag{1.22}$$

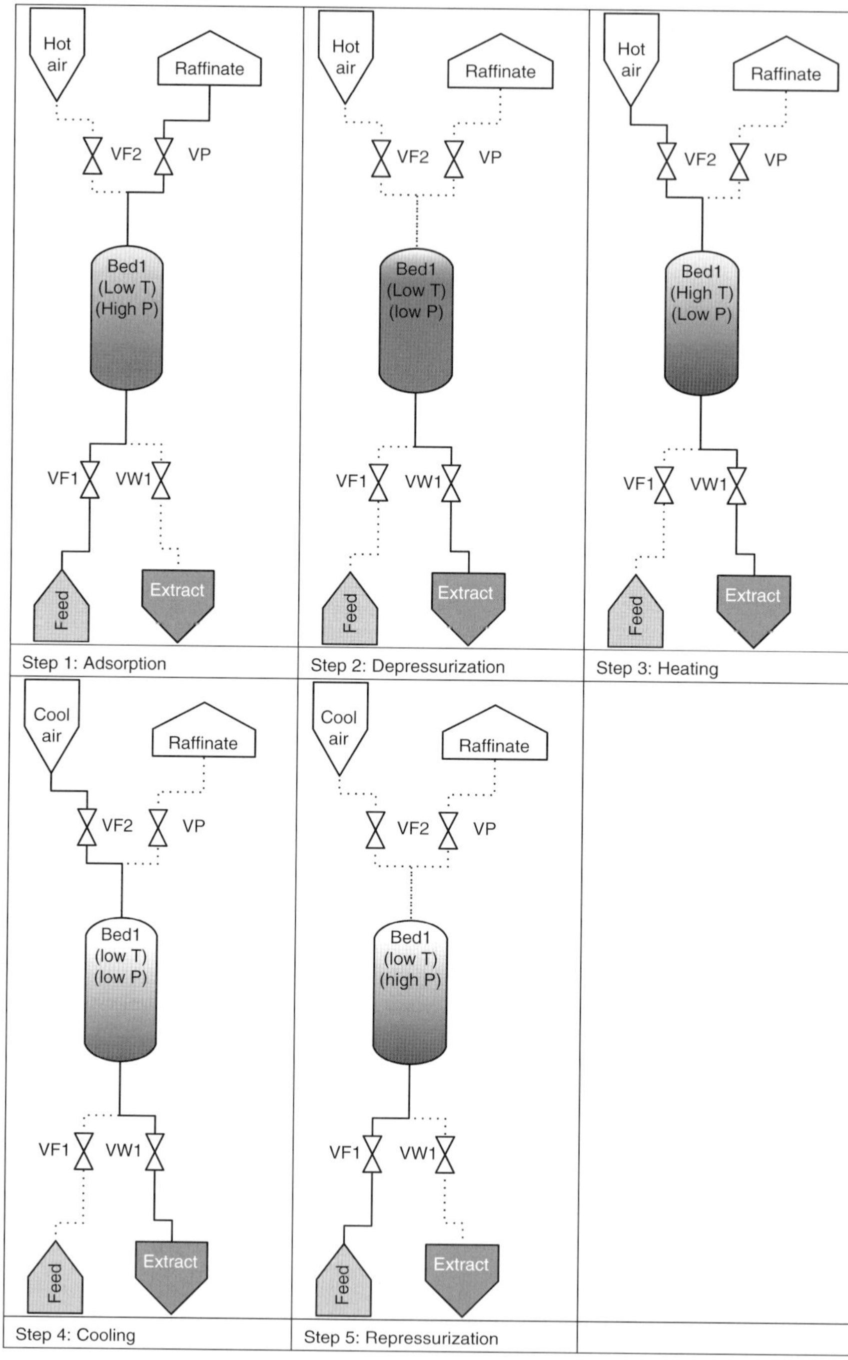

Figure 1.203 The one-bed TSA cycle.

F1.Specify Table

	Value	Units	Spec	Description
F	5.e-006	kmol/s	Free	Flowrate
Y_Fwd(*)				
Y_Fwd("Air")	0.99965	kmol/kmol	Fixed	Composition in forward
Y_Fwd("H2O")	3.5e-004	kmol/kmol	Fixed	Composition in forward
T_Fwd	289.15	K	Fixed	Temperature in forward
P	10.0	bar	Fixed	Boundary pressure

Figure 1.204 Moist air feed conditions.

F2.Specify Table

	Value	Units	Spec	Description
F	5.e-006	kmol/s	Free	Flowrate
Y_Fwd(*)				
Y_Fwd("Air")	1.0	kmol/kmol	Fixed	Composition in forward
Y_Fwd("H2O")	0.0	kmol/kmol	Fixed	Composition in forward
T_Fwd	413.15	K	Fixed	Temperature in forward
P	1.2	bar	Fixed	Boundary pressure

Figure 1.205 Hot regeneration air conditions.

P1.Specify Table

	Value	Units	Spec	Description
F	-10.2299	kmol/s	Free	Flowrate
Y_Rev(*)				
Y_Rev("AIR")	1.0	kmol/kmol	Fixed	Composition in reverse d
Y_Rev("H2O")	0.0	kmol/kmol	Fixed	Composition in reverse d
T_Rev	298.15	K	Fixed	Temperature in reverse d
P	9.99	bar	Fixed	Boundary pressure

Figure 1.206 Dry air collection conditions.

W1.Specify Table

	Value	Units	Spec	Description
F	0.0	kmol/s	Free	Flowrate
Y_Rev(*)				
Y_Rev("AIR")	0.5	kmol/kmol	Fixed	Composition in reverse d
Y_Rev("H2O")	0.5	kmol/kmol	Fixed	Composition in reverse d
T_Rev	298.15	K	Fixed	Temperature in reverse d
P	1.013	bar	Fixed	Boundary pressure

Figure 1.207 Wet product collection conditions.

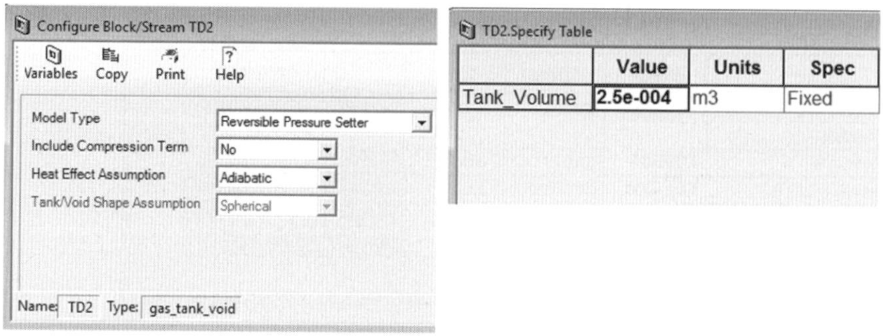

Figure 1.208 Specifying the tank options.

which is clearly a simple Henry's Law style of isotherm. In this case, IP_6 can be expressed as a function of the manufacturer's parameters as

$$IP_6 = \frac{IP_4}{IP_1} e^{-IP_2} \tag{1.23}$$

Step 5: Specify thecolumn behavior

Apply the following assumptions to the packed bed according to Table 1.13.

Table 1.13 Specifications of the adsorption column.

Discretization method	QDS (quadratic differencing scheme)	*Nodes*	20
Material balance	Convection only	*Pressure drop assumption*	Ergun
Kinetic film model	Linear lumped resistance	*Based on*	Solid phase concentration
Mass transfer coefficient	Constant		
Isotherm	Henry 2 (based on partial pressure)		
Energy balance	Nonisothermal	*Conduction*	No
Consider solid phase	No	*Heat of adsorption*	Constant
Heat transfer to environment	Adiabatic	*Heat capacity*	Constant

Once these options have been entered, specify the values for the column shown in Figure 1.209.

Step 6: Specify the initial conditions

We would like this separation to start with fully dry air in the column and tanks. To do this, we need to modify their initial conditions. On the column configuration sheet, click on the preset/initials button. The tanks and bed should be set to be completely dry at $t = 0$. See Figures 1.210 and 1.211.

B1.Layer(1).Specify Table

	Value	Units	Description
Hb	0.45	m	Height of adsorbent layer
Db	0.27	m	Internal diameter of adsorbent layer
Ei	0.35	m3 void/m3 bed	Inter-particle voidage
Ep	1.e-010	m3 void/m3 bea	Intra-particle voidage
RHOs	770.0	kg/m3	Bulk solid density of adsorbent
Rp	0.0045	m	Adsorbent particle radius
SFac	1.0	n/a	Adsorbent shape factor
MTC(*)			
MTC("AIR")	5.e-008	1/s	Constant mass transfer coefficients
MTC("H2O")	2.468e-004	1/s	Constant mass transfer coefficients
IP(*)			
IP(1,"AIR")	0.0	n/a	Isotherm parameter
IP(1,"H2O")	1.23e-008	n/a	Isotherm parameter
IP(2,"AIR")	0.0	n/a	Isotherm parameter
IP(2,"H2O")	5293.0	n/a	Isotherm parameter
Direction	0.0	n/a	Specified flow direction (self determined: 0, forwar
Cps	1.e-003	MJ/kg/K	Adsorbent specific heat capacity
DH(*)			
DH("AIR")	0.0	MJ/kmol	Constant for heat of adsorption
DH("H2O")	-64.0	MJ/kmol	Constant for heat of adsorption
HTC	1.0	MW/m2/K	Constant for the heat transfer coefficient
ap	500.0	1/m	Specific surface area of adsorbent

Figure 1.209 Column parameters for air drying.

TD2.Initials Table

	Value	Units	Spec
Y(*)			
Y("AIR")	1.0	kmol/kmol	Free
Y("H2O")	0.0	kmol/kmol	Initial
T	298.15	K	Initial
P	9.995	bar	Initial

Figure 1.210 Set the tanks to contain dry air.

B1.Layer(1).Initials_YVWCT Table

	Value	Units	Spec	Derivative	Description
ProfileType	Consta				Is the bed initially specified
Y_First_Node(*)					
Y_First_Node("AIR")	1.0	kmol/kmol	Initial		Mole fraction within first elem
Y_First_Node("H2O")	0.0	kmol/kmol	Initial		Mole fraction within first elem
Vg_First_Node	3.55e-004	m/s	Initial		Gas velocity within first elem
W_First_Node(*)					
W_First_Node("AIR")	0.0	kmol/kg	RateInitial	0.0	Solid loading within first ele
W_First_Node("H2O")	0.0	kmol/kg	RateInitial	0.0	Solid loading within first ele
Tg_First_Node	289.15	K	Initial		Gas temperature within first
Ts_First_Node	289.15	K	Initial		Solid temperature within first

Figure 1.211 Set the column to contain dry air.

Step 7: Create the Cycle Organizer steps

Create five steps according to Table 1.14. Figures 1.212–1.216 show the manipulated variables of steps 1–5. The steps will use different manipulated variables, so it is best to not copy the values from one step to the next.

Step 8: Specify the manipulated variables

Refer to Figure 1.7 and Table 1.5 for the four ASs for valves within Aspen Adsorption: (1) AS = 0, valve is fully closed; (2) AS = 1, valve is fully open; and (3) AS = 2, the gas flow rate through the valve is linearly dependent on the pressure drop across the valve according to Eq. (1.12); and (4) AS = 3, the valve has a fixed flow rate.

Step 9: Set the cycle to run twice

Table 1.14 A five-step TSA cycle.

	Description	Duration (min)
Step 1	Adsorption	151
Step 2	Depressurization	1
Step 3	Heating	47
Step 4	Cooling	72
Step 5	Repressurization	Dependent on step 2

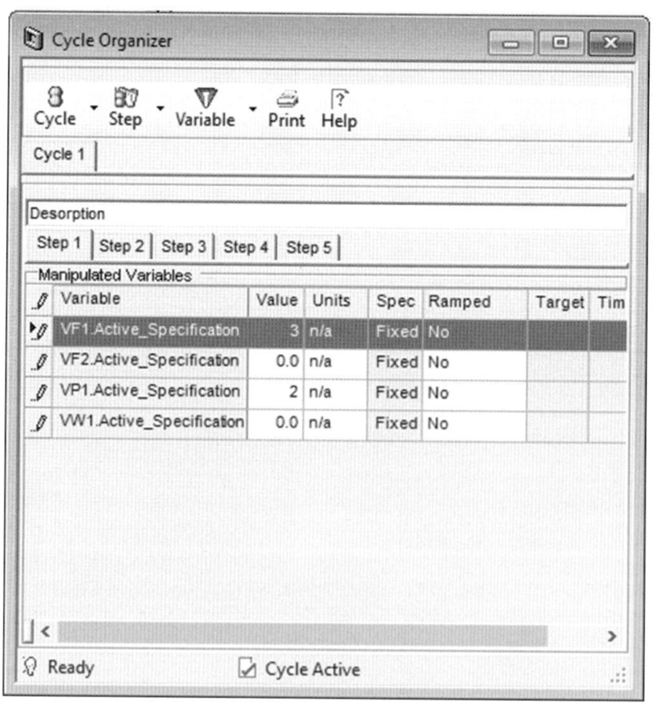

Figure 1.212 Step 1 (adsorption) manipulated variables.

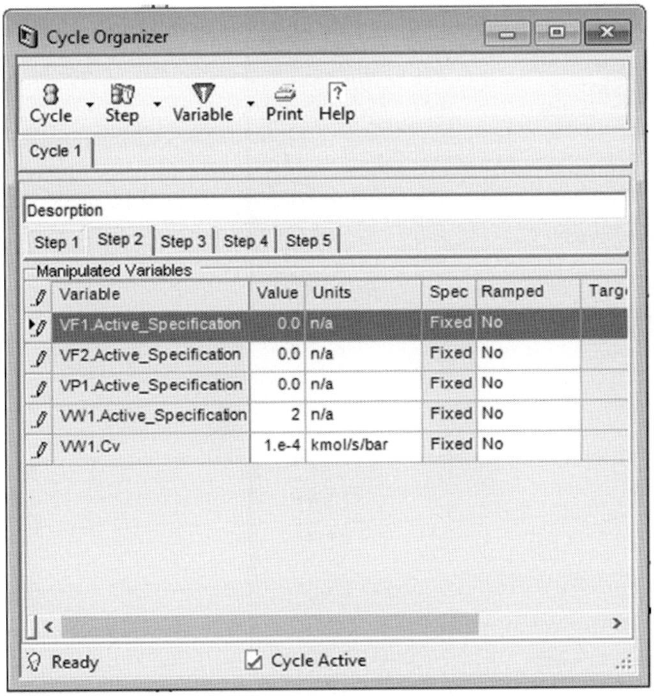

Figure 1.213 Step 2 (depressurizatin) manipulated variables.

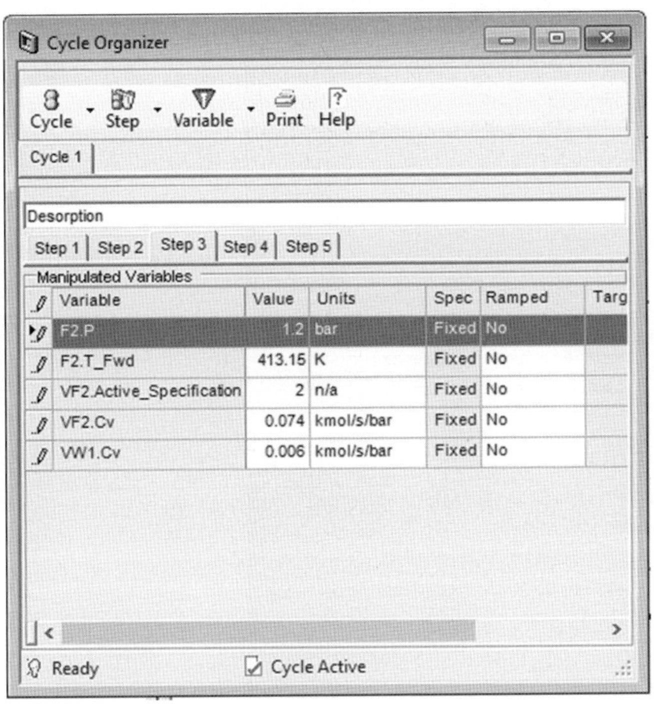

Figure 1.214 Step 3 (heating) manipulated variables.

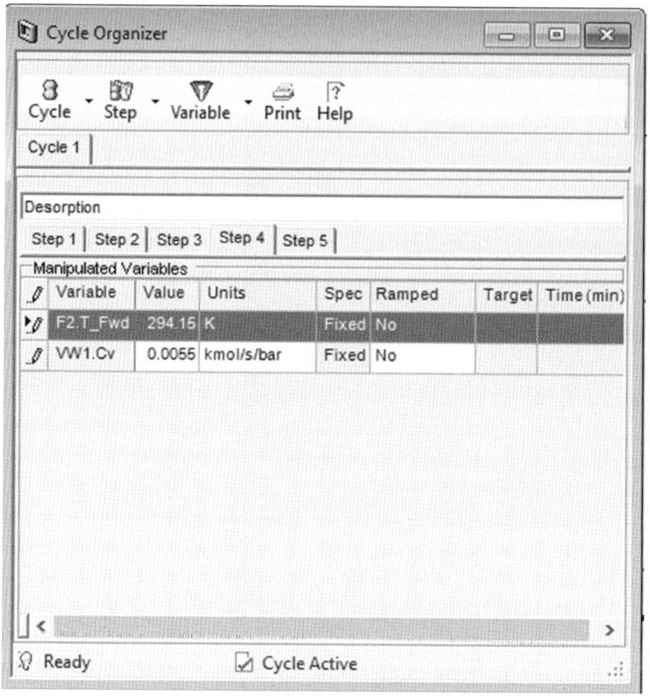

Figure 1.215 Step 4 (cooling) manipulated variables.

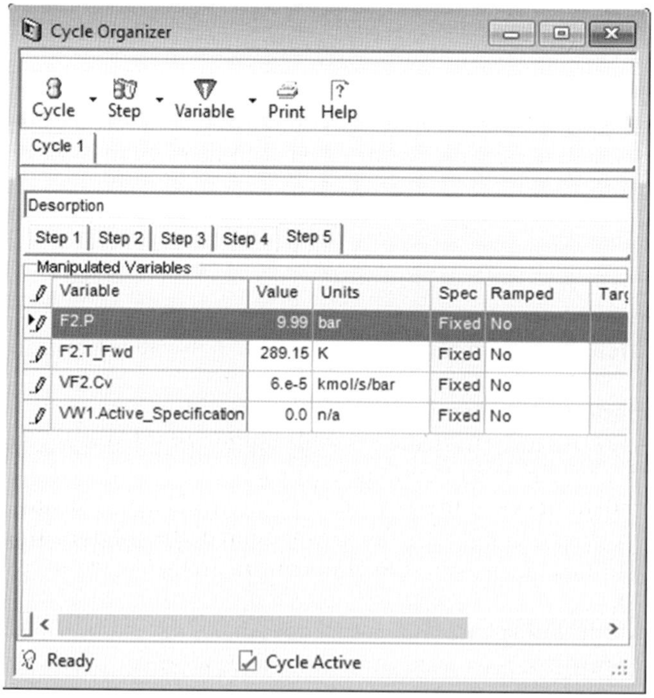

Figure 1.216 Step 5 (repressurization) manipulated variables.

This process reaches the steady state quite quickly, and therefore, we only need to run very few cycles to see the trends (Figure 1.217).

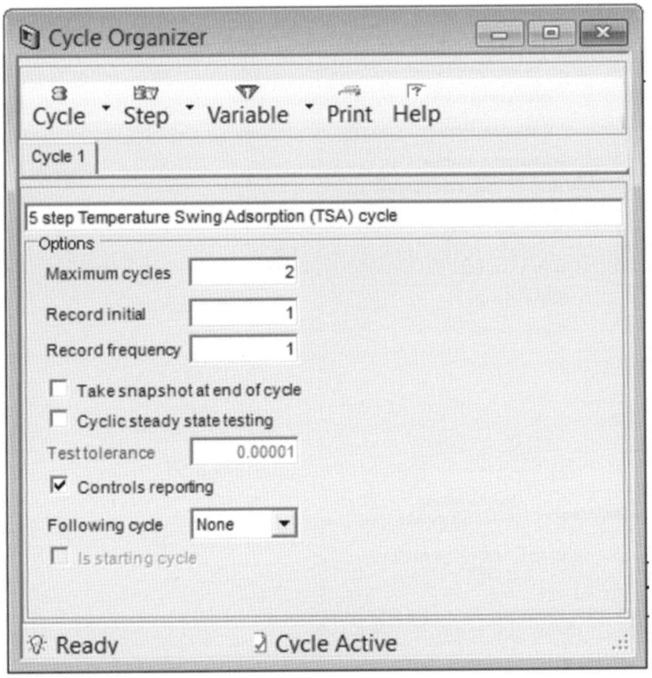

Figure 1.217 Limiting the cycle to two iterations.

Step 10: Change the integration options

Use the Solver Options button  to edit the integration options. Specifically, to improve the speed of the simulation, we want to increase the "Maximum step size" to 10 s (Figure 2.218).

Next, change the run options  so that the communication interval is also 10 s (Figure 2.219).

Step 11: Initialize the flowsheet

Use the "Check and Initial" entry in the flowsheet menu to initialize the flowsheet (Figure 1.220).

Step 12: Run the simulation and view the results

Create a plot to display the axial temperature profile in the bed (Figure 1.221), and leave this plot open while the simulation executes.

To see how the temperatures at either end of the column change over time, create a plot of the temperatures of the tanks at either end of the column. This will give us some additional insight into the temperature gradients across the column (Figure 1.222).

Finally, we will want to know what the actual product composition is. To do this, create a plot of the mole fraction of water and air in the P1 block. As we can see, we have successfully dried the air from an initial mole fraction of 3.5e−4 to below 1e−5 (Figure 1.223).

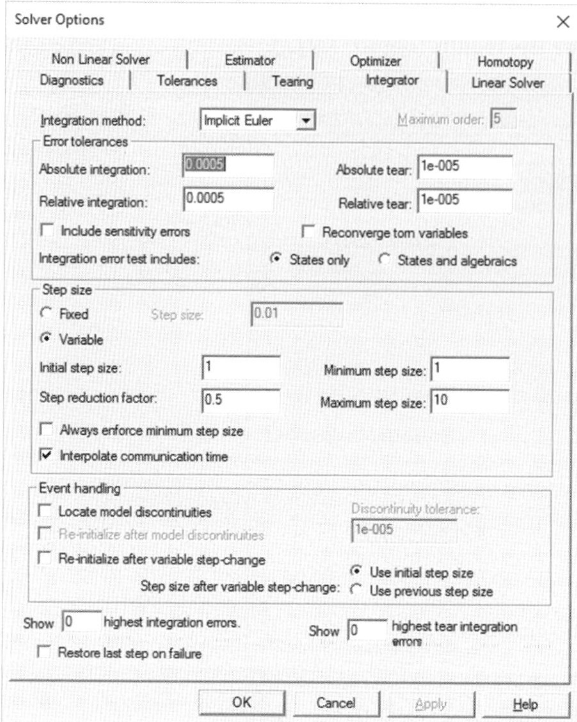

Figure 1.218 Increasing the maximum step size.

1.11 Conclusions

This chapter has covered a number of crucial modeling considerations involved in building simulations of gas adsorption processes. Specifically, the key requirements in each model are as follows:

1) Knowledge of which chemical species will be present in the simulation, as well as their properties. We discuss in Section 1.4 several possible methods for modeling the behavior of both the gas and solid phases and offered guidance on selecting the methods.
2) Knowledge of the solid–gas interaction equilibrium and kinetics. We cover several major equilibrium models in Section 1.3, and kinetic assumptions in Section 1.4. We show how to employ these equations in Aspen Adsorption.
3) Detailed knowledge of the model libraries available in Aspen Adsorption, and how they can combine to approximate real-world unit operations. Each workshop discussed covers a different set of the models and options available. At the conclusion of these three workshops, the reader should have a command of the majority of the important gas process models, and be well positioned to model a wide variety of industrial processes.
4) A detailed schedule of the valve operation required to implement the "swing" in temperature or PSA. Each workshop uses a slightly different switching scheme, and we discuss the methods for implementing both time-based and

Figure 1.219 Changing the Run options.

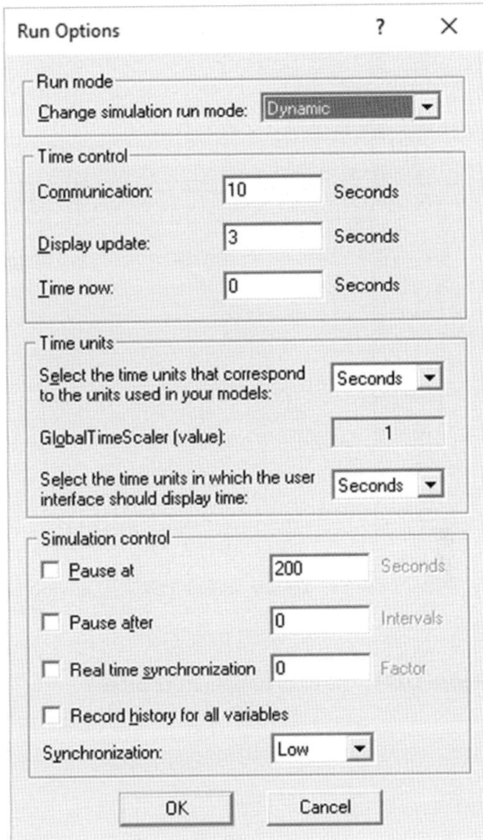

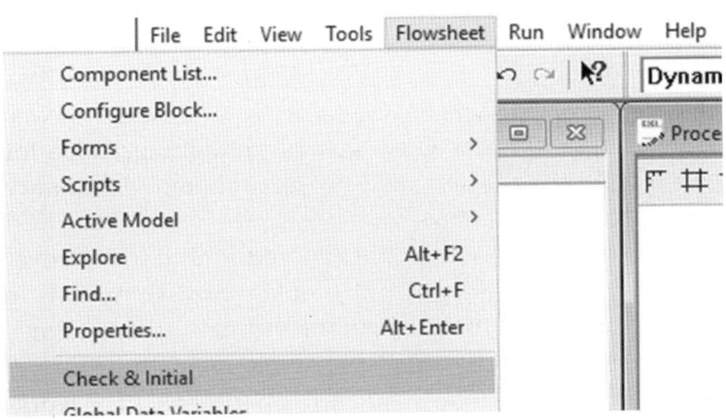

Figure 1.220 Initializing the flowsheet.

event-based scheduling. We also cover how to use a gas_interaction model to speed up computation, and how to schedule around that model.

5) Collecting data from the completed simulation. We discuss strategies for structuring the data collection in the H2 workshop. Specifically, we want to

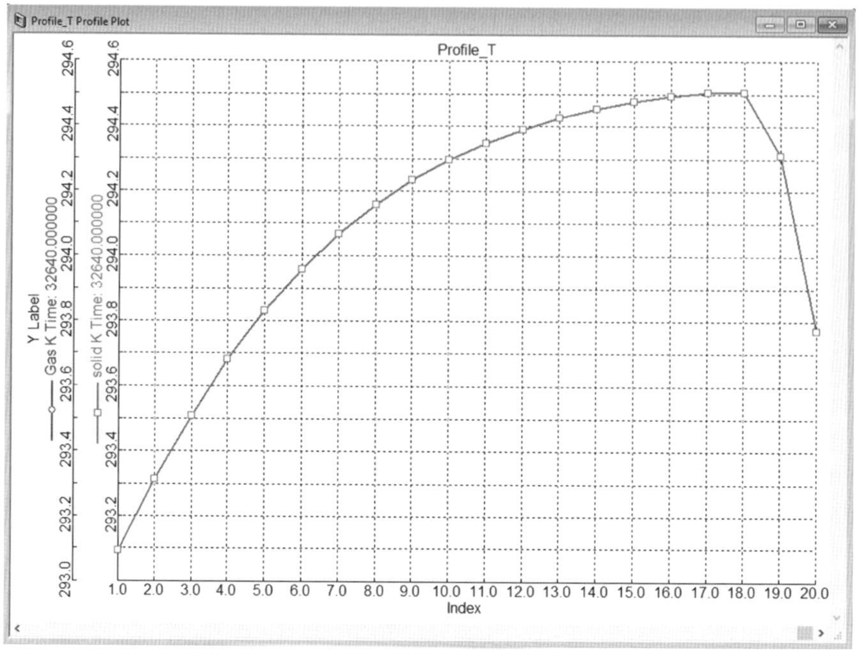

Figure 1.221 The axial temperature profile.

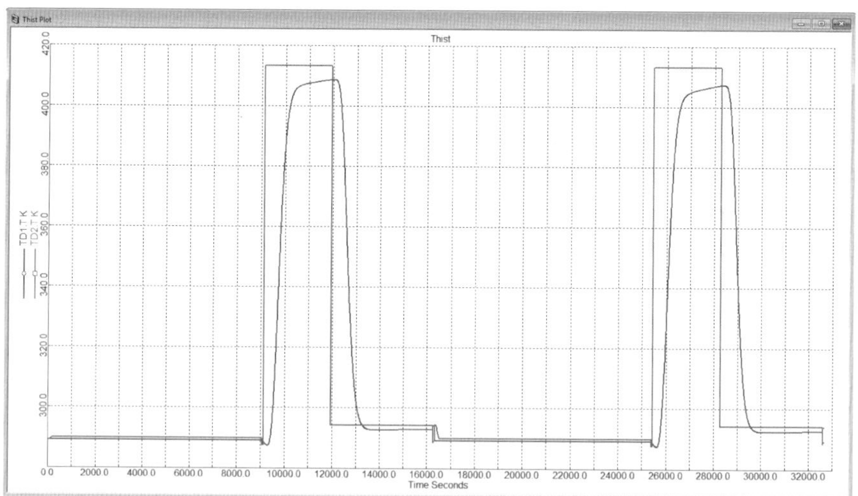

Figure 1.222 Temperature of the top and bottom tanks over time.

avoid errors due to division by trace quantities, and avoid making the results too dependent on the model's time discretization settings.

The techniques described in this chapter will allow the reader to quickly implement detailed, predictive models of gas adsorption processes. The lessons learned

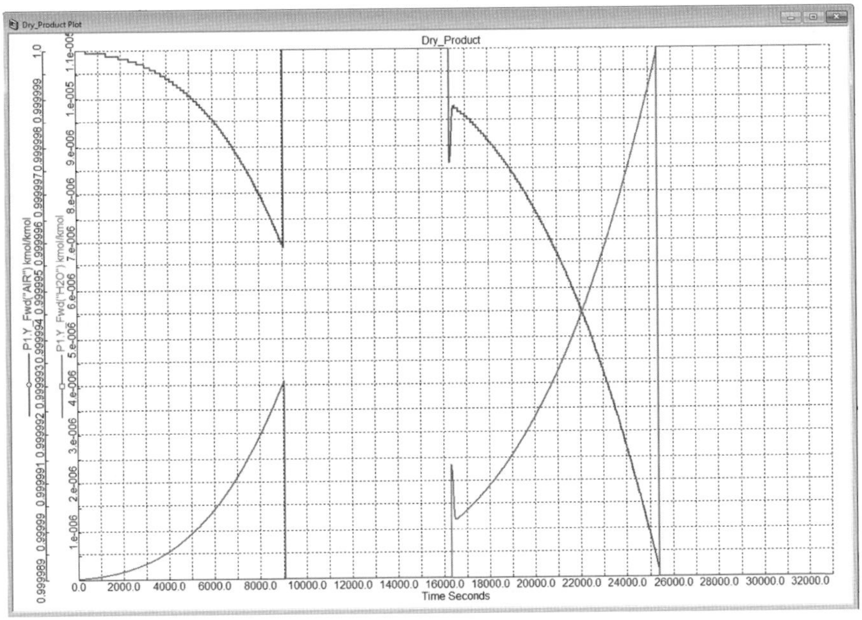

Figure 1.223 Product Composition. Notice that the mole fraction of water stays below 1.1e−5.

here will reappear in Chapter 3, when we employ them to model chromatographic processes.

1.12 Practice Problems

1.12.1 Introducing a gas_interaction Unit into Workshop 1

Using what you learned about the gas_interaction block in workshop 2, re-do workshop 1 using a single bed and gas_interaction unit (Figure 1.224). Consider the differences between the two flowsheets, especially about how the blowdown is implemented.

Here, the buffer tank should actually have a significant volume (e.g. 3m^3). Also, set the valves VP1 to have a Cv value of 0.00162 and the VB1 valve to have a Cv value of 1.5e−5.

Task 1: Re-create this plot of the LP stream using one gas_bed and one gas_interaction block (Figure 1.225). Once you have completed the process, save the file as **`AirSepSingleBed.ada`**.

Note that in this case, the simulation achieves a greater oxygen purity than the Air Separation workshop (nearly 99% purity in this case vs 95% purity in the workshop.) Why is that? Can you modify the Air Separation workshop to achieve similar purity? (Hint, change the VP1 and VP2 valves to use Cv instead of simply being on/off).

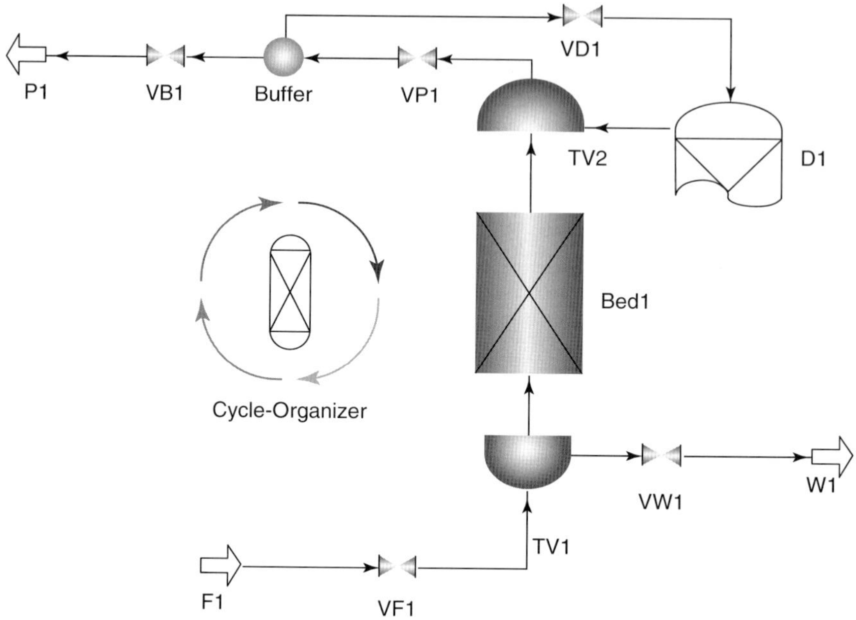

Figure 1.224 One possible flowsheet solution.

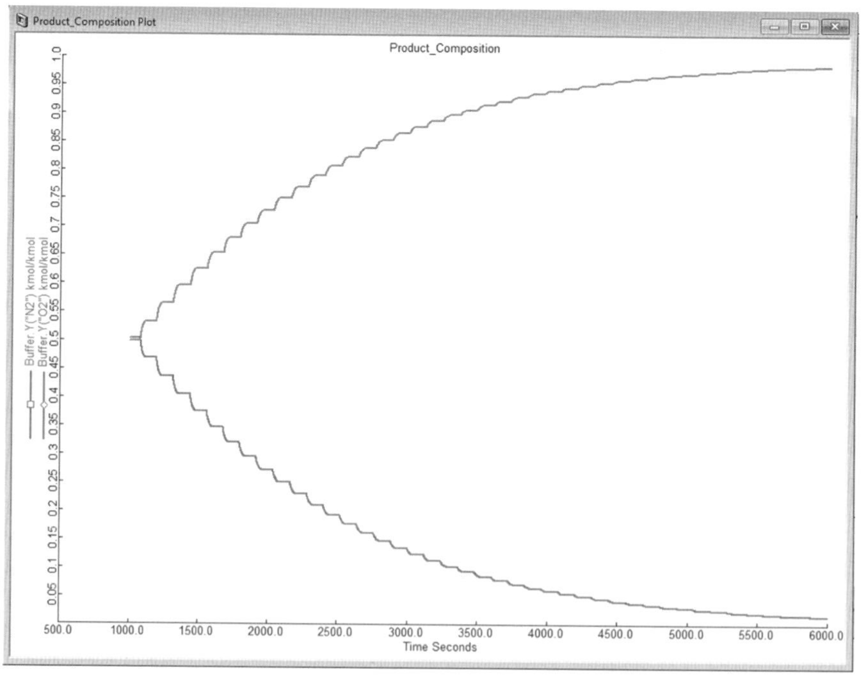

Figure 1.225 The product composition of a single bed air separator.

1.12.2 Naphtha Upgrading Using Adsorption

Bárcia et al. [25] discovered a new adsorbent (zeolite beta) that has properties very suitable for upgrading naphtha. Create a very simple flowsheet with only a feed, bed, and product block.

Make the following assumptions:

1) The gas phase is IDEAL, and consists of
   a) *n*-pentane (nP).
   b) *n*-hexane (nH).
   c) Iso-pentane (iP).
   d) Helium.
   e) 3-methylpentane (3MP).
   f) 2,3-dimethylbutane (23dmb).
   g) 2,2-dimethylbutane (22dmb).

   Task 1: Use the "Find" button to search for all of these components in the Aspen property databanks (Figure 1.226).
2) The dual-site Langmuir model is appropriate (partial pressure based), and Table 1.15 gives the isotherm parameters.
3) The bed has the following properties (Table 1.16).The mass transfer coefficients of the components are given in Table 1.17.Inside the bed, the Ergun equation, Eq. (1.11) for pressure drop applies, and mass transfer resistances are handled with the solid-phase linear lumped resistance model Eq. (1.15).

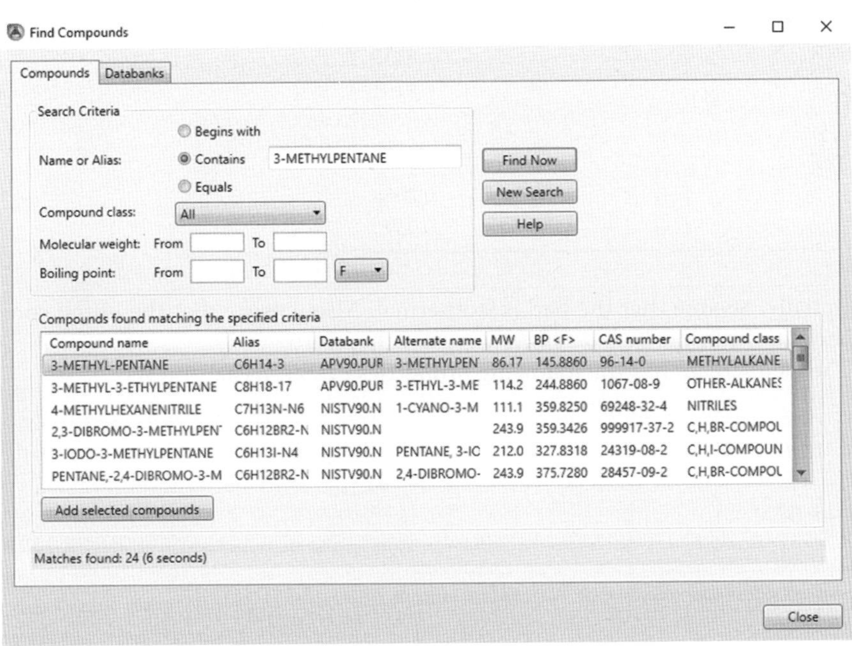

Figure 1.226 The Aspen tool for compound searches.

Table 1.15 Isotherm parameters.

	IP1	IP2	IP3	IP4	IP5	IP6	IP7	IP8
3MP	0.37	−1.50	0.80	−1.50	1.80	−5.25	3.62	−5.25
22DMB	0.42	−14.57	0.83	−14.57	0.23	−8.21	3.78	−8.21
23DMB	0.45	−12.17	0.82	−12.17	0.76	−7.63	3.76	−7.63
HE	0.00	0.00	0.00	0.00	0.00	0.00	0.00	0.00
IPEN	0.58	−11.04	0.82	−11.04	0.26	−4.99	3.60	−4.99
NHEX	0.31	20.44	0.75	20.44	4.09	−1.46	3.43	−1.46
NPEN	0.63	−8.42	0.82	−8.42	0.73	−6.70	3.71	−6.70

Table 1.16 Bed properties.

Diameter (cm)	0.46
Length (cm)	10
Bulk porosity (ε_b)	0.49
Adsorbent particle diameter (cm)	0.159
Bulk density (kg m^{-3})	1130

Table 1.17 Mass transfer coefficients.

3M	0.5549
22DMB	1.8355
23DMB	0.8165
He	5.0
IP	1.5698
NH	0.2668
NP	0.959

Initially, assume that the bed is isothermal. Also, ensure that the initial concentrations in the bed are *slightly* greater than zero.

4) The feed stream has the following properties (Table 1.18).

Table 1.18 Stream properties.

Pressure (kPa)	8
T (K)	523

There is an equimolar mixture of the components in the feed.

The product stream is collected at a pressure of 5 kPa.

Task 2: Ensure that your flowsheet is correctly constructed (Figure 1.227).

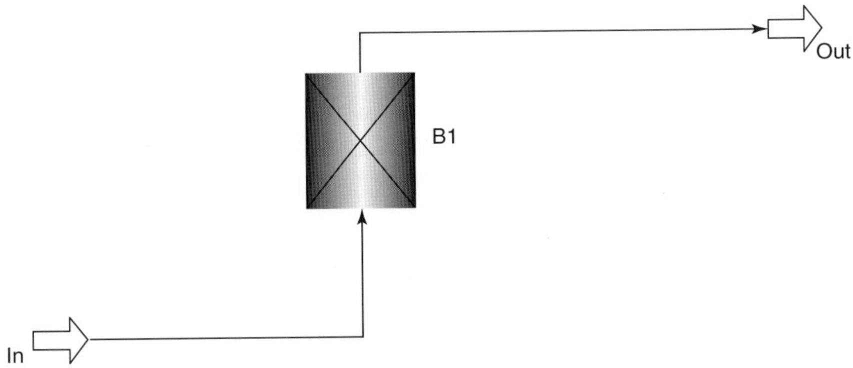

Figure 1.227 The correct flowsheet layout.

Task 3: Enter the following flowsheet constraints:

```
Fip as RealVariable;
Fnp as RealVariable;
F22 as RealVariable;
F23 as RealVariable;
F3mp as RealVariable;
FnHex as RealVariable;

Fip = (B3.F * B3.Y_Fwd("IP")) / (B2.F*B2.Y_Fwd("IP"));
Fnp = (B3.F * B3.Y_Fwd("NP")) / (B2.F*B2.Y_Fwd("NP"));
F22 = (B3.F * B3.Y_Fwd("22DMB")) / (B2.F*B2.Y_Fwd("22DMB"));
F23 = (B3.F * B3.Y_Fwd("23DMB")) / (B2.F*B2.Y_Fwd("23DMB"));
F3mp = (B3.F * B3.Y_Fwd("3M")) / (B2.F*B2.Y_Fwd("3M"));
FnHex = (B3.F * B3.Y_Fwd("NH")) / (B2.F*B2.Y_Fwd("NH"));
```

Ensure that these constrains are compiled.

Task 4: Run the simulation, and create a plot of the ratios defined in the flowsheet constraints.

Confirm that your enriched product stream achieves the correct concentrations of components (Figures 1.228 and 1.229).

Some notes on the discrepancy between the simulation and literature data:

- Our simulation is isothermal, while the actual adsorption unit is closer to adiabatic with a heat of adsorption.
- The literature uses a tri-site Langmuir model, while we restrict ourselves to a dual-site model. The isotherm parameters for this problem were generated by regressing a dual-site model against data generated by the tri-site model.
- Given these discrepancies, the qualitative agreement we have here is remarkably good.

Once you have completed this workshop, save the file as **NapthaIsothermal.ada**.

As an additional challenge, can you build that bed into a PSA unit?

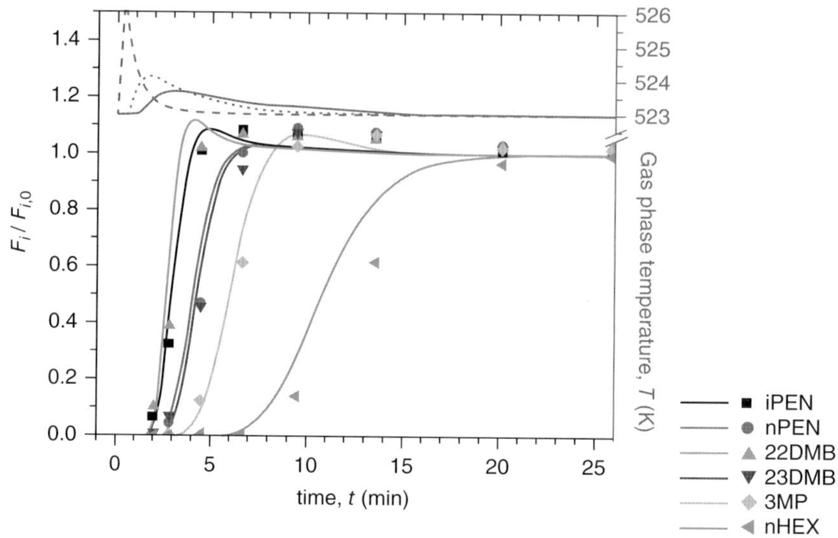

Figure 1.228 The product concentrations from literature. Kneabel and Hill 1985 [26]. Reproduced with permission of Elsevier.

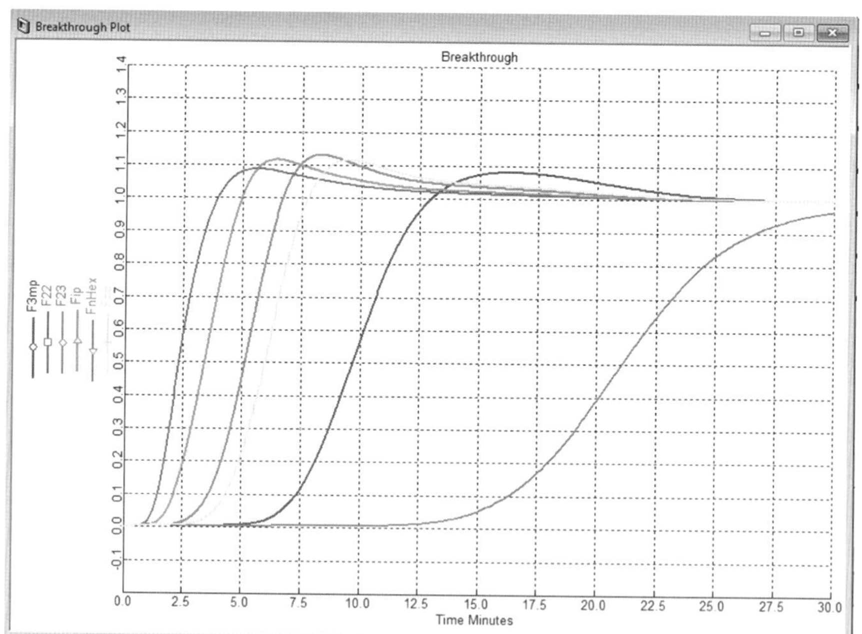

Figure 1.229 The actual simulation results.

1.13 Nomenclature

Variables	Units	Description
$a_{i,1}$	[—]	bi-Langmuir isotherm parameter a_1 for component i
$a_{i,2}$	[—]	bi-Langmuir isotherm parameter a_2 for component i
$b_{i,1}$	$[\mathrm{l\,g^{-1}}]$	bi-Langmuir isotherm parameter b_1 for component i
$b_{i,2}$	$[\mathrm{l\,g^{-1}}]$	bi-Langmuir isotherm parameter b_2 for component i
D_L	$[\mathrm{cm^2\,min^{-1}}]$	axial dispersion coefficient
Dia	[cm]	column diameter
F	N/A	Feed stream
IP1	[—]	isotherm parameter 1 in Aspen
IP2	[—]	isotherm parameter 2 in Aspen
k_i	$[\mathrm{1\,s^{-1}}]$	mass transfer coefficient for component i
K_i	[—]	equilibrium constant for component i
L	[m]	column length
P	[bar]	bed pressure drop
P_i	[bar]	partial pressure drop for component i
Pe	[—]	particle Peclet number
q_i	$[\mathrm{g\,g^{-1}}]$	solid loading for component i
q_i^*	$[\mathrm{g\,g^{-1}}]$	equilibrium solid loading for component i
r_p	[μm]	particle radius
S	$[\mathrm{cm^2}]$	cross-sectional area of the bed
U	$[\mathrm{m\,s^{-1}}]$	the superficial velocity of the gas flow
V_{bed}	$[\mathrm{cm^3}]$	bed volume per column
V_j^D	$[\mathrm{cm^3}]$	dead volume through the column
XFac	[—]	volume correction factor
Z	[m]	axial distance through the column
ΔP_j	[bar]	pressure drop in zone j
ΔP_{max}	[bar]	maximum pressure drop through the column
ϕ	$\left[\frac{\mathrm{m^3\,void}}{\mathrm{m^3}}\right]$	overall bed voidage
ε_i	$\left[\frac{\mathrm{m^3\,void}}{\mathrm{m^3}}\right]$	interparticle voidage
ε_p	$\left[\frac{\mathrm{m^3\,void}}{\mathrm{m^3}}\right]$	intraparticle voidage
μ	[cp]	liquid viscosity
ψ	[—]	particle shape factor
φ	$\left[\frac{\mathrm{bar^*min}}{\mathrm{cm^2}}\right]$	pressure drop coefficient

Bibliography

Books

Ruthven, D.M. (1984). *Principles of Adsorption and Adsorption Processes.* New York: Wiley.

Ruthven, D.M., Farooq, S., and Knaebel, K.S. (1994). *Pressure Swing Adsorption.* Hoboken, NJ: Wiley-VCH.

Yang, R.T. (1987). *Gas Separation by Adsorption Processes.* Waltham, MA: Butterworth-Heinemann.

Journal Articles Specifically Utilizing Aspen Adsorption

Bárcia, P.S., Silva, J.A.C., and Rodrigues, A.E. (2010a). Adsorption dynamics of C_5–C_6 isomerate fractions in zeolite beta for the octane improvement of gasoline. *Energy and Fuels* **24** (3): 1931.

Bárcia, P.S., Silva, J.A.C., and Rodrigues, A.E. (2010b). Octane upgrading of C_5/C_6 light naphtha by layered pressure swing adsorption. *Energy and Fuels* **24** (9): 5116.

Bhatt, T.S., Sliepcevich, A., Storti, G., and Rota, R. (2014). Experimental and modeling analysis of dual-reflux pressure swing adsorption process. *Industrial and Engineering Chemistry Research*, 140815153241007.

da Silva Bárcia, P. (2010). *Separation of Light Naphtha for the Octane Upgrading of Gasoline: Adsorption and Membrane Technologies and New Adsorbents.* http://hdl.handle.net/10198/3409.

da Silva Bárcia, P., Rodrigues, A.E., and Silva, J.A. (2012). Separation of light naptha by adsorption processes. *Conference HANQUEICCE2012*, ANQUE. http://hdl.handle.net/10198/7123.

Hu, Z., Zhang, D., and Wang, J. (2011). A simulation of citrus oil adsorption separation in the supercritical carbon dioxide. *Chemical Industry and Engineering Progress* **S2**.

Jang, S.-C., Yang, S.-I., Oh, S.-G., and Choi, D.-K. (2011). Adsorption dynamics and effects of carbon to zeolite ratio of layered beds for multicomponent gas adsorption. *Korean Journal of Chemical Engineering* **28** (2): 583.

Kostroski, K.P. and Wankat, P.C. (2006). High recovery cycles for gas separations by pressure-swing adsorption. *Industrial and Engineering Chemistry Research* **45** (24): 8117.

Li, S., Yang, H., and Zhang, D. (2013). Enrichment of CO from syngas with Cu(I)Y adsorbent by five-bed VPSA. *Frontiers of Chemical Science and Engineering* **7** (4): 472.

Liu, B.X. and Song, B.D. (2012). Simulation of hydrogen chloride dehydration by aspen adsorption. *Chemical Industry and Engineering* **29** (2): 58.

Liu, Z., Grande, C.A., Li, P. et al. (2011). Multi-bed vacuum pressure swing adsorption for carbon dioxide capture from flue gas. *Separation and Purification Technology* **81** (3): 307.

Moroi, K. and Sato, T. (1975). Comparison between procaine and Isocarboxazid metabolism in vitro by a liver microsomal Amidase-esterase. *Biochemical Pharmacology* **24** (16): 1517.
Natarajan, G. and Wankat, P.C. (2003). Thermal-adsorptive concentration. *Adsorption* **9** (1): 67.
Olson, D.H., Yang, X., and Camblor, M.A. (2004). ITQ-12: a zeolite having temperature dependent adsorption selectivity and potential for propene separation. *The Journal of Physical Chemistry B* **108** (30): 11044.
Ritter, J.A., Bhadra, S.J., and Ebner, A.D. (2011). On the use of the dual-process Langmuir model for correlating unary equilibria and predicting mixed-gas adsorption equilibria. *Langmuir* **27** (8): 4700.
Sharma, P.K. and Wankat, P.C. (2009). Hybrid cycles to purify concentrated feeds containing a strongly adsorbed impurity with a nonlinear isotherm: the PSA–TSA Supercycle. *Industrial and Engineering Chemistry Research* **48** (13): 6405.
Sharma, P.K. and Wankat, P.C. (2010). Solvent recovery by Steamless temperature swing carbon adsorption processes. *Industrial and Engineering Chemistry Research* **49** (22): 11602.
Silva, J.A., Rodrigues, A.E., and da Silva Bárcia, P. (2011). Octane upgrading of TIP processes by recycling in a layered zeolite 5A/BETA PSA. Chempor11. http://hdl.handle.net/10198/6371.
Stamm, O., Latscha, U., Janecek, P., and Campana, A. (1976). Development of a special electrode for continuous subcutaneous pH measurement in the infant scalp. *American Journal of Obstetrics and Gynecology* **124** (2): 193.
Thess, A., Spirn, D., and Jüttner, B. (1995). Viscous flow at infinite Marangoni Number. *Physical Review Letters* **75** (25): 4614.
Wu, F. (2012). New approach for modeling hybrid pressure swing adsorption-distillation processes. *Industrial and Engineering Chemistry Research* **51**: 9343–9355.
Yamaguchi, T., Aoki, K., Sakurai, M., and Kameyama, H. (2013). Development of new hybrid VOCs treatment process using activated carbon and electrically heated alumite catalyst. *Journal of Chemical Engineering of Japan* **46** (12): 802.
Yang, S., Zhang, J., Yang, Q., and Qian, Y. (2014). Development of an integrated oil shale refinery with retorting gas steam reforming for hydrogen production. *Energy and Fuels* **28** (8): 5557.
Yang, S.-I., Choi, D.-Y., Jang, S.-C. et al. (2008). Hydrogen separation by multi-bed pressure swing adsorption of synthesis gas. *Adsorption* **14** (4–5): 583.
Yangang, Y., Yanbing, D., Zhengfei, M., and Xiaoqin, L. (2012). Removal of CO and CO2 from ammonia synthesis gas by pressure swing adsorption with layered bed. *Chemical Engineering* **8** (40): 8.
Zhao, C., Ding, Y., Ma, Z. et al. (2011). Simulation and design for purification of CO from N2/CO gases by pressure swing adsorption. *Journal of Nanjing University of Technology (Natural Science Edition)* **4**: 019.
Zhou, L., Doong, S.-J., Russell, B.P., and Rastelli, H. (2014). Processes and apparatuses for preparing liquified natural gas. Google Patents.
Zhou, Y., Wei, X., and Zhang, D. (2011). Simulation of three-bed VPSA oxygen enrichment process. *Chemical Industry and Engineering Progress* **S2**.

Zhou, Y.Y., Yang, H.W., and Zhang, D.H. (2013). Modeling of energy-saving VPSA cycle for oxygen enrichment. *Chemical Industry and Engineering* **30** (2): 53.

Journal Articles

Bonjour, J., Chalfen, J.B., and Meunier, F. (2002). Temperature swing adsorption process with indirect cooling and heating. *Industrial and Engineering Chemistry Research* **41**: 5802–5811.

Chihara, K. and Suzuki, M. (1983). Simulation of nonisothermal pressure swing adsorption. *Journal of Chemical Engineering of Japan* **16**: 53–61.

Chue, K.T., Kim, J.N., Yoo, Y.J. et al. (1995). Comparison of activated carbon and zeolite 13X for CO2 recovery from flue gas by pressure swing adsorption. *Industrial and Engineering Chemistry Research* **34**: 591–598.

Clausse, M., Bonjour, J., and Meunier, F. (2004). Adsorption of gas mixtures in TSA adsorbers under various heat removal conditions. *Chemical Engineering Science* **59**: 3657.

Doong, S.J. and Yang, R.T. (1987). Hydrogen purification by the multibed pressure swing adsorption process. *Reactive Polymers* **6**: 7–13.

Farooq, S. and Ruthven, D.M. (1991). Numerical simulation of a kinetically controlled pressure swing adsorption bulk separation process based on a diffusion model. *Chemical Engineering Science* **46**: 2213–2224.

Farooq, S., Ruthven, D.M., and Boniface, H.A. (1990). Air separation by pressure swing adsorption. *Chemical Engineering Science* **44**: 2809–2816.

Fernandez, G.F. and Kenney, C.N. (1983). Modeling of the pressure swing air separation process. *Chemical Engineering Science* **38**: 827.

Gomes, V.G. and Yee, K.W.K. (2002). Pressure swing adsorption for carbon dioxide sequestration from exhaust gases. *Separation and Purification Technology* **28**: 161.

Grande, C.; Cavenati, S.; Rodrigues, A.E. (2015). Pressure swing adsorption for carbon dioxide sequestration. Second Mercosur Congress on Chemical Engineering and Fourth Mercosur Congress on Process System Engineering. https://www.researchgate.net/publication/238681749.

Hassan, M.M., Raghavan, N.S., and Ruthven, D.M. (1987). Pressure swing air separation on a carbon molecular sieve-II investigation of a modified cycle with pressure equalization and no purge. *Chemical Engineering Science* **42**: 2037–2043.

Hassan, M.M., Ruthven, D.M., and Raghavan, N.S. (1986). Air separation by pressure swing adsorption on a carbon molecular sieve. *Chemical Engineering Science* **41**: 1333–1343.

Jee, J.-G., Kim, M.-B., and Lee, C.-H. (2005). Pressure swing adsorption to purify oxygen using a carbon molecular sieve. *Chemical Engineering Science* **60**: 869.

Jee, J.G., Lee, S.J., and Lee, C.H. (2004). Comparison of the adsorption dynamics of air on zeolite 5A and carbon molecular sieve beds. *Korean Journal of Chemical Engineering* **21**: 1183.

Kalbassi, M.A., Allam, R.J., and Golden, T.C. (1997). Temperature swing adsorption. US Patent 5,846,295.

Leavitt, F.W. (1990). Air separation pressure swing adsorption process. US Patent 5,074,892 A.

Ling, J., Ntiamoah, A., Xiao, P. et al. (2014). Overview of CO2 capture from flue gas streams by vacuum pressure swing adsorption technology. *Austin Journal of Chemical Engineering* **1**: 1009.

Mendes, A.M.M., Costa, C.A.V., and Rodrigues, A.E. (2001). Oxygen separation from air by PSA: modelling and experimental results part I: isothermal operation. *Separation and Purification Technology* **24**: 173–188.

Merel, J., Clausse, M., and Meunier, F. (2006). Carbon dioxide capture by indirect thermal swing adsorption using 13X zeolite. *Environmental Progress* **25**: 327–333.

Merel, J., Clausse, M., and Meunier, F. (2008). Experimental investigation on CO2 post-combustion capture by indirect thermal swing adsorption using 13X and 5A zeolites. *Industrial and Engineering Chemistry Research* **47**: 209.

Mulgundmath, V. and Tezel, F.H. (2010). Optimization of carbon dioxide recovery from flue gas in a TPSA system. *Adsorption* **16**: 587.

Raghavan, N.S., Hassan, M.M., and Ruthven, D.M. (1986). Numerical simulation of a PSA system using a pore diffusion model. *Chemical Engineering Science* **41**: 2787–2793.

Raghavan, N.S. and Ruthven, D.M. (1985). Pressure swing adsorption part II: numerical simulation of a kinetically controlled bulk gas separation. *AIChE Journal* **31**: 2017–2025.

Ritter, J.A. and Yang, R.T. (1991). Pressure swing adsorption: experimental and theoretical study on air purification and vapor recovery. *Industrial and Engineering Chemistry Research* **30**: 1023–1032.

Shin, H.S. and Knaebel, K.S. (1988). Pressure swing adsorption: an experimental study of diffusion-induced separation. *AIChE Journal* **34**: 1409–1416.

Tlili, N., Grevillot, G., and Vallieres, C. (2009). Carbon dioxide capture and recovery by means of TSA and/or VSA. *International Journal of Greenhouse Gas Control* **3**: 519–527.

Yang, M., Chen, N., Huang, C. et al. (2014). Temperature swing adsorption process for CO2 capture using polyaniline solid sorbent. *Energy Procedia* **63**: 2351.

Yu, C.H., Huang, C.H., and Tan, C.S.A. (2012). Review of CO2 capture by absorption and adsorption. *Aerosol and Air Quality Research* **12**: 745.

Zaman, M. and Lee, J.H. (2013). Carbon capture from stationary power generation sources: a review of the current status of the technologies. *Korean Journal of Chemical Engineering* **30**: 1794.

2

Simulation of SMB Chromatographic Processes

This chapter covers the major features of chromatographic separations, including several critical differences between chromatographic and adsorptive separations before showing how these concepts are applied in rigorous simulations. Section 2.1 gives a high-level overview of chromatography, emphasizing areas where chromatography differs from the adsorptive separations described in Chapter 1. Sections 2.2 and 2.3 discuss simulated moving bed (SMB) chromatography, covering specifically the overall concept, and implementation details, respectively. Sections 2.4 and 2.5 describe how to implement SMB separations in Aspen chromatography, specifically looking at a binary chiral separation and a ternary separation of insulin. Section 2.6 presents practice problems for the readers to apply the concepts described in this chapter, and Section 2.7 gives the bibliography.

2.1 Introduction to Chromatography

Chromatography deals with the separations that occur when a liquid (called the **mobile phase**) interacts with a fixed scaffold of solid material referred to as the **stationary phase**. When that interaction is adsorptive, chromatographic separations are conceptually very similar to gas adsorption. As such, all of the general concepts introduced in Chapter 1 regarding the adsorption process and physical properties of the system still apply in many forms of chromatography. Chromatography, however, encompasses a wider variety of solid–fluid interactions. Fortunately, liquid systems afford several simplifying assumptions that we cannot make in gas systems.

2.1.1 Mathematical Differences from Gas Adsorption

2.1.1.1 The Trace Liquid Assumption

The most immediate and important assumption of a liquid system is that the fluid phase is incompressible. More than that, for liquid systems that are predominantly solvent (i.e. the components being separated are only present in *trace* quantities) we may assume that the density of the fluid phase is not a function of the mobile-phase composition. This trace liquid assumption

Design, Simulation, and Optimization of Adsorptive and Chromatographic Separations: A Hands-On Approach, First Edition. Kevin R. Wood, Y. A. Liu, and Yueying Yu.

allows for dramatic simplifications in the differential equations that govern chromatographic separations.

2.1.1.2 Concentration Versus Partial Pressure

It is very common in gas separations to express the composition of the gas phase as either mole fraction or partial pressure. In liquid chromatography, the convention is to use concentration (mass/volume) for virtually all equations.

2.1.2 Thermodynamic Differences from Gas Adsorption

2.1.2.1 Isotherms

The isotherms for liquid systems generally derive from research on gas systems. Indeed, the very popular Langmuir isotherm is derived by explicitly assuming that the fluid behaves as an ideal gas, despite the fact that the liquid phase does not behave as an ideal gas. The development of new isotherms that are thermodynamically consistent with liquid conditions is still an area of active research; empirical isotherms are often used.

2.1.2.2 Physical Property Models

There are entirely different methods available to calculate the physical properties of liquid systems. Unlike gas systems that emphasize equation-of-state models, liquid systems frequently invoke activity coefficient methods, such as the nonrandom two liquid (NRTL) or UNIversal QUAsi-Chemical (UNIQUAC) methods.

2.2 Introduction to SMB Chromatography

Chromatographic separations are a powerful technology and have been in use for a long time. However, such separations were traditionally carried out in batches, which limited their efficiency. The search for a solid–liquid separation method that would eliminate the inefficiencies of batch operation by working as an analog of an absorber/stripper came to an end in 1961, when Broughton and Gerhold [27] developed the concept of an SMB and found applications in petrochemical and sugar separations. The SMB concept allows a column to approximate the countercurrent solid/liquid separation, without the need for solid-handling equipment.

The complexity and novelty of the SMB approach restricted its use to high-value applications for a long time, but this has changed as our conceptual understanding has improved. There has been a surge in interest in applying SMB to much smaller-scale problems in pharmaceutical, biochemical, and chiral separations. Depending on the economics of the separations, SMB can offer a number of significant benefits over a more conventional batch chromatographic separation, from reduced desorbent consumption and lower adsorbent requirements to higher productivity or recovery.

Zone 1: To understand how SMB chromatography works, let us first consider an ideal system illustrated in Figure 2.1. In an ideal binary chromatographic

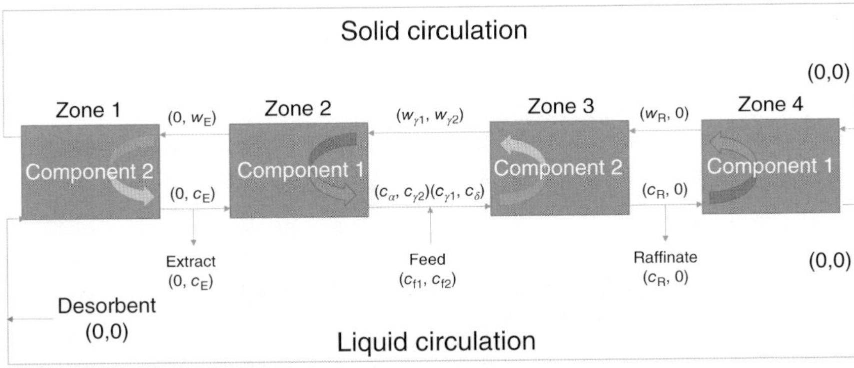

Figure 2.1 SMB unit under ideal separation conditions.

system, the two components introduced in the feed stream would be perfectly separated into the extract and raffinate streams. Mass-transfer resistance would be absent, and the liquid phase would instantly reach equilibrium with the solid adsorbent. Under such conditions, we would have the situation depicted in Figure 2.1. The net effect is that component 1 will travel through zone 3 in the liquid phase to be removed in the raffinate stream. Component 2, on the other hand, will move through zone 2 to be removed in the extract stream. Component 2 is desorbed from the solid adsorbent into the liquid phase. This ensures that all of component 2 is present in the extract by preventing the solid phase from carrying it through to zones 4 and 3.

Zone 2: Component 1 is desorbed from the solid adsorbent into the liquid phase. This ensures that there will not be any of this component in zone 1 and therefore the extract will be uncontaminated.

Zone 3: Component 2 is adsorbed into the solid phase. This means the liquid will only contain component 1 so the raffinate will remain uncontaminated.

Zone 4: The remainder of component 1 is adsorbed onto the solid. This ensures that component 1 will not be present in the liquid recycle, and will therefore not contaminate zone 1 and the extract.

2.3 SMB Implementation Details

2.3.1 Common Process Configurations

The most common SMB configuration is the basic four-zone separator used for binary separation. As we mentioned previously, the countercurrent separator to perform this separation would require four ports and four zones. Figure 2.2 shows a more concrete view of what such a process would look like.

In a true moving bed (TMB) unit, the liquid and solid phases continuously flow in the opposite direction. Figure 2.2 shows the two required inlet streams (feed and desorbent), and two outlet streams (extract and raffinate). This configuration would allow for the attainment for high-purity products even when using a low-selectivity adsorbent. However, from an engineering point of view, the use of

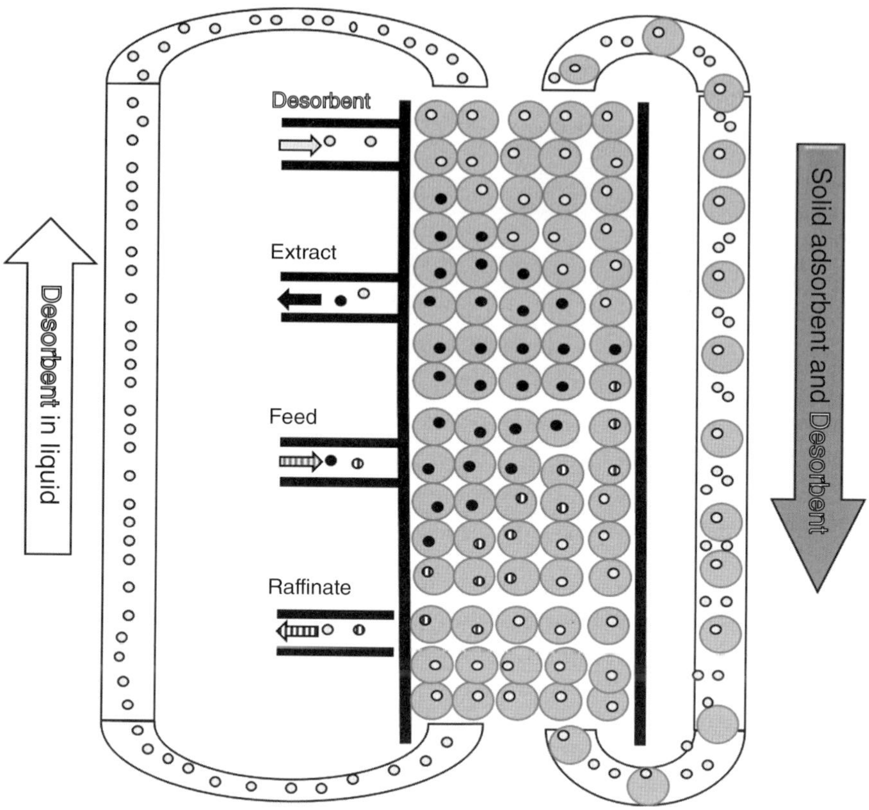

Figure 2.2 Diagram of a basic 4-zone true moving bed (TMB).

this type of equipment with the actual movement of the stationary phase is not practical for industrial implementation. This complex *solid handling may lead to some technical problems, namely, equipment abrasion, mechanical erosion of* adsorbent, and difficulties in maintaining plug flow for the solid (especially in beds with large diameters).

The key insight that allows engineers to sidestep these drawbacks is to realize that this separation is possible because of *the relative motion of the solid and the ports.* However, we do not actually need to move the solids to create this relative motion; we can move the ports instead, as shown in Figure 2.3. Ideally, we could move the ports continuously, and perfectly replicate a TMB. If we only switch the port locations at discrete times and positions, we can *simulate* TMB performance. The finer in space and time we make these steps, the closer we approximate TMB performance. In practice, this means that we can more closely approximate TMB performance by creating more possible locations for the ports to switch to.

The four streams in Figures 2.2 and 2.3 divide the column into four regions that have different roles in the separation. If we assume that ***component A is attracted to the solid phase more strongly than component B***, our separator can achieve the state illustrated in Figure 2.4. In the figure, Q_1, Q_2, Q_3, Q_4 are the internal flow rates in zone 1, 2, 3, and 4, respectively; Q_F, Q_R, Q_D, Q_E are the feed,

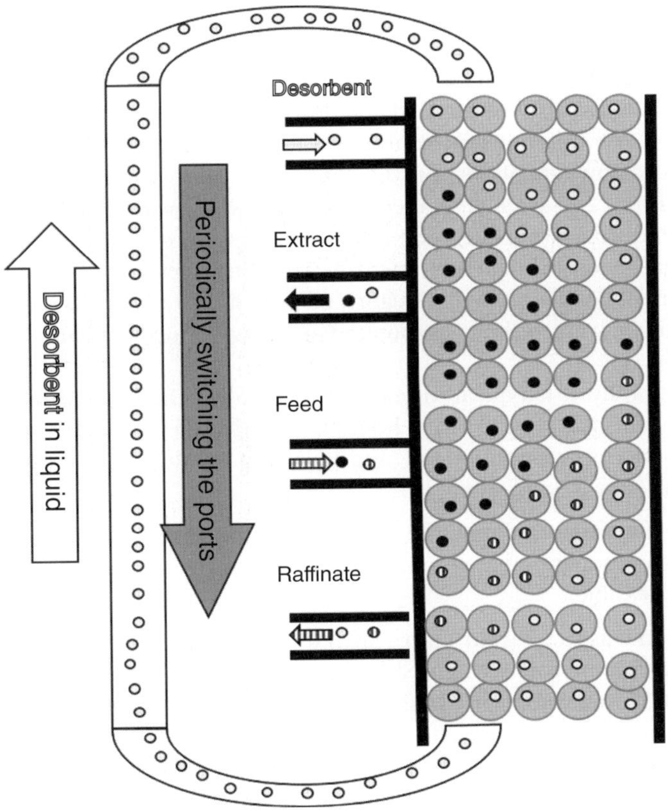

Figure 2.3 Diagram of 4-zone simulated moving bed operation.

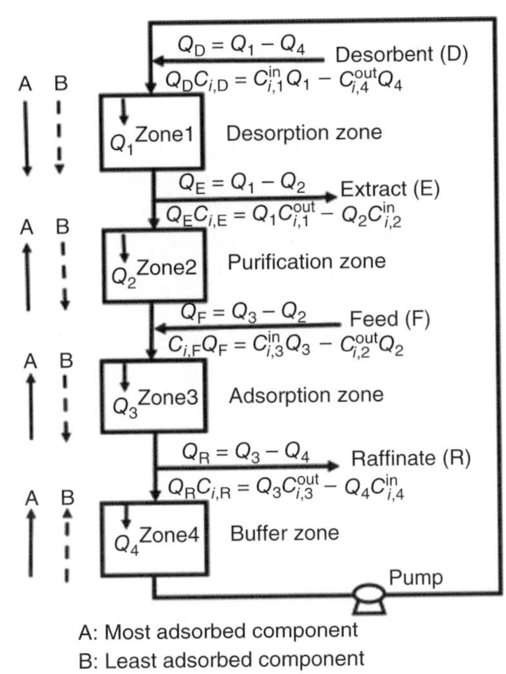

Figure 2.4 Net motion of the components in a 4-zone system.

raffinate, desorbent, and extract flow rates, respectively; and $C_{i,j}^{\text{in}}$ and $C_{i,j}^{\text{out}}$ are the concentrations of component i at the inlet and outlet of zone j, respectively. In this four-zone separation system, we fix the external flow rates Q_F, Q_D, and Q_R, the internal recycle flow rate Q_4, and the feed and desorbent concentrations. The figure also shows that the difference in volumetric flow rate between zones is equal to the volumetric flow rate of the port (Q_D, Q_E, Q_F, Q_R) between the zones.

The conditions in zones 2–4 are such that component A will get adsorbed onto the solid phase, and moved upwards. In zone 1, the desorbent is capable of regenerating the solid phase by desorbing component A. The conditions in zones 1–3 are such that component B will not be significantly adsorbed, and move down the column. In zone 4, the solvent will be regenerated and B will move back up the column with the solid.

In SMB chromatography, it is typical to make what is called **the trace liquid assumption,** which states the density of the liquid is not affected by the concentrations of the components being separated, and therefore the fluid flow rate in each zone is constant. This is valid when

- the components being separated have a similar density to the solvent;
- the components being separated are present in low (i.e. trace) concentrations;
- the fluid is incompressible.

These properties allow us to treat the entire SMB unit as though the volume is conserved; we can do volume balances instead of mass balances if we so choose.

Designing the system on a volumetric basis is handy, but what would really be nice is a scale-independent parameter that can characterize the operation of each zone. If we were able to create that kind of dimensionless quantity, we would be able to easily design and scale up SMB processes.

2.3.2 M-Values

In each zone, it turns out that the key parameters to characterize the relative movement of the solid and liquid are the flow rate ratios. These are defined differently, depending on whether the system is a TMB or an SMB. The specific equations are given in Eqs. (2.1) and (2.2):

$$\text{TMB} \quad m_j = \frac{(Q_j - \varepsilon_p Q_s)}{Q_s(1-\varepsilon_p)} \tag{2.1}$$

$$\text{SMB} \quad m_j = \frac{(Q_j t_{sw} - \varepsilon^* V_{bed})}{V_{bed}(1-\varepsilon^*)} \tag{2.2}$$

In these equations, Q refers to volumetric flow rates, the j subscript denotes the zone, while the s subscript denotes the solid phase. The solid phase flow rate Q_s does not change between the zones, so it does not need a zone subscript. Since the SMB column does not have an actual solid flow rate, the ratio $\varepsilon^* V_{bed}/t_{sw}$ is used instead, where ε^* is the total porosity, V_{bed} is the volume of the bed, and t_{sw} is the time between port switching events in the SMB column.

There are two key benefits of using these ratios:

First, they allow us to flip back and forth between TMB and SMB models. Because the equations are different, the exact flow rates will be different, but they will have the same theoretical performance. That is to say, if an SMB unit has the same m-values as a TMB unit, the SMB unit will be *simulating* the TMB unit; and the TMB unit represents, in some sense, a bound on the SMB's performance. If the SMB unit is underperforming its TMB counterpart, adding more beds to each zone will cause the SMB unit's performance to get closer to the theoretical TMB performance. Note that this is only strictly true in the case of basic SMB systems; units that employ the advanced operational modes described in Chapter 4 no longer strictly represent approximations of TMB units.

Second, these ratios are scale-independent and are not enough to completely satisfy all the degrees of freedom in the system. This is shown in Figure 2.5, with the final degree of freedom from a pressure drop constraint. In practice, the pressure drop constraint is generally a nuanced parameter to pin down. It is most common to assume a whole-column pressure drop, but there are cases where the pressure drop may apply on a per-zone basis. Regardless, after applying the pressure drop constraint, we can calculate the required flow rates for any scale of column.

2.3.3 Scale-Up Concerns

In the previous section, we made the bold claim that the flow rate ratios can calculate the required flow rates for any scale of column. This is true so long as the columns behave in a way that is consistent with the TMB approximation. There are two main ways that physical columns break this approximation such that the scale is important.

The first, and most important way that columns may behave in a non-TMB way is if they contain large dead volumes. These are regions of the SMB unit that do not perform any separation. We need to consider these volumes because they introduce lag times in the movement of material through the systems. The most common sources of dead volumes in both gas and liquid systems are the *bed head* and the *bed lines*, as shown in Figure 2.6. Each of the multiple beds in an SMB, temperature swing adsorption (TSA), or pressure swing adsorption (PSA) system will have these voids. In Chapter 1, we managed these dead volumes by adding explicit tanks to our flowsheet. However, as we see in our first SMB workshop (Section 2.4), things are not quite so straightforward with SMB systems.

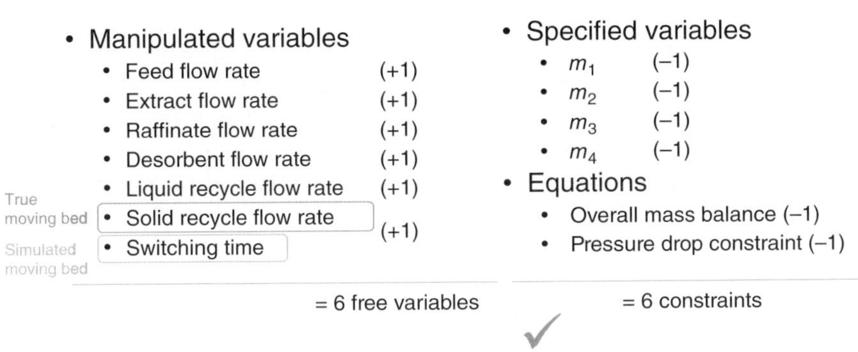

Figure 2.5 Simplified degrees of freedom in an SMB system.

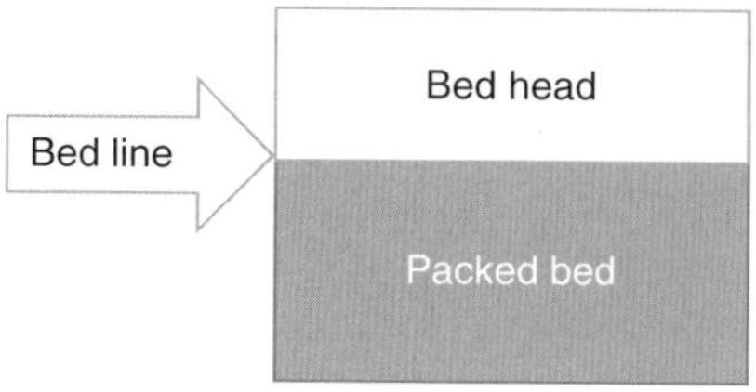

Figure 2.6 The locations of dead volumes in a packed bed.

Also, it is possible to somewhat accommodate the dead volume into the m-values. By modifying Eq. (2.2) as shown in Eq. (2.3), we can, to some extent, account for lag times introduced by dead volumes (V_{dead}). Of course, this requires some knowledge of the expected size of the dead volumes during the design process. Typically, the dead volume is assumed to be around 5% of the total bed volume, but this number may vary depending on the scale of the process.

$$m_j = \frac{(Q_j t_{\text{sw}} - \varepsilon^* V_{\text{bed}} - V_{\text{dead}})}{V_{\text{bed}}(1 - \varepsilon^*)} \tag{2.3}$$

2.3.4 Pressure Drop Limitations

The final piece to the SMB design puzzle is the pressure drop. Pressure drop constraints generally come from the need to prevent the solid adsorbent from shifting and breaking during column operation. The overall strength of the chromatographic column may also impose constraints. If we assume that the overall pressure drop is given by the Carman–Kozeny equation (2.4), then we can lump the Carman–Kozeny parameters into a single variable ϕ. This makes it easy to show through simple algebra that in order for the pressure drop inside the column to be less than the maximum pressure drop, the operating conditions must satisfy the inequality given in Eq. (2.5).

$$\frac{\Delta P_j}{L} = \frac{-150\mu(1-\varepsilon)^2}{(2r_{\text{p}}\psi)^2\varepsilon^3}\frac{Q_j}{S} = \phi\frac{Q_j}{S} \tag{2.4}$$

$$t_{\text{sw}} \geq \frac{\phi L^2}{\Delta P_{\text{max}}}\sum_{j=1}^{j=4} n_j(m_j(1-\varepsilon)+\varepsilon) \tag{2.5}$$

In these equations, Q once again refers to volumetric flow rates. S is the cross-sectional area, L the column length, and r_{p} the particle radius. n_j is the number of beds in each zone, and m_j the flow rate ratio. ψ is a *shape factor* that is 1 for perfect spheres, and ϕ is a linear pressure drop coefficient.

We can perform a similar analysis should our pressure drop constraint be phrased in terms of a single zone, but only the zone with the largest m-value will actually limit the pressure; so the constraint in that case would require us to know which zone will have the greatest flow rate. This will always be either zone 1 or zone 3, and is much more likely to be zone 1. In that case, the constraint can simply be expressed as Eq. (2.6).

$$t_{\text{sw}} \geq \frac{\phi L^2}{\Delta P_{\text{max,zone}}}(n_1(m_1(1-\varepsilon)+\varepsilon)) \tag{2.6}$$

2.3.5 Introduction to Operational Modes

The success of SMB applications has led to the development of new methods for operating SMB units to further improve the separation efficiency. Specifically, the goal is to either exceed the separation efficiency of the TMB approximation, or improve efficiency in columns that do not approximate the TMB well. The literature contains a number of proposed changes to SMB operations, which we refer to as operational modes. The latest review [28] details the most recent SMB operational modes, applications, and developments. In Chapter 4, we focus particularly from a practical point of view on several operational modes applied to the existing SMB units by switching ports or adjusting control valves. Specifically, we provide heuristics for selecting from the following operational modes:

1) Varicol – switching the inlet and outlet ports asynchronously [28–33]
2) PowerFeed or Partial Feeding – varying feed flow rates [34–36]
3) ModiCon – varying feed concentrations [29, 37]
4) Combinations of the previous three operational modes [38]

2.4 SMB Workshop: Simulate a Four-Zone SMB in Aspen Chromatography for the Separation of Tröger's Base Enantiomers

This workshop provides a step-by-step illustration to build a four-zone basic SMB model for the binary chiral separation of the Tröger's base.

2.4.1 Creating a Flowsheet in Aspen Chromatography

Step 1: Start a new case in Aspen Chromatography, and save it as "4-zone basic SMB model.cra" (Figure 2.7**).**

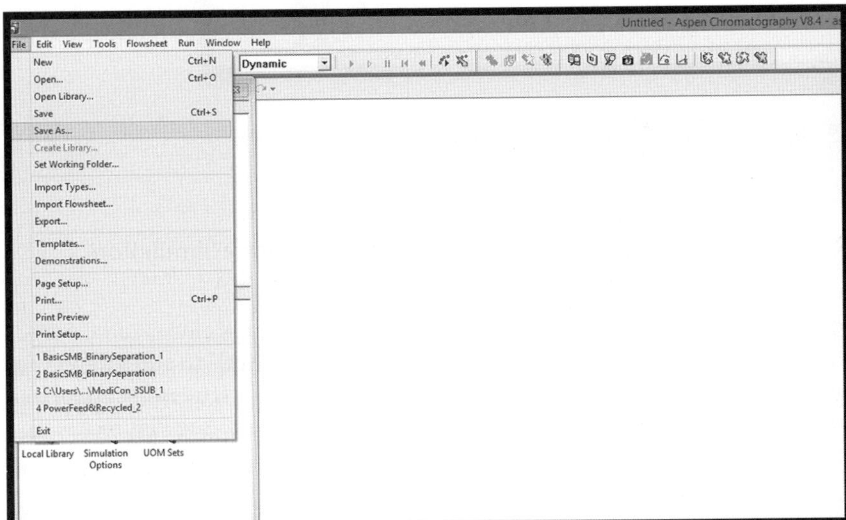

Figure 2.7 A new file named **4-zone basic SMB model.cra**.

Step 2: Lay out the flowsheet blocks

Go to "Model Library," then choose the blocks in Figure 2.8, and create the flowsheet for the basic SMB model. Select "Chromatography: Non-Reversible" → "chrom_ccc_separator2" for the SMB column; select "Chromatography: Non-Reversible" →> "chrom_feed" for the DESORBENT and FEED blocks; select "Chromatography: Non-Reversible" →> "chrom_product" for the EXTRACT and RAFFINATE blocks. Name the desorbent-in, extract-out, feed-in, and raffinate-out streams as D, E, F, and R.

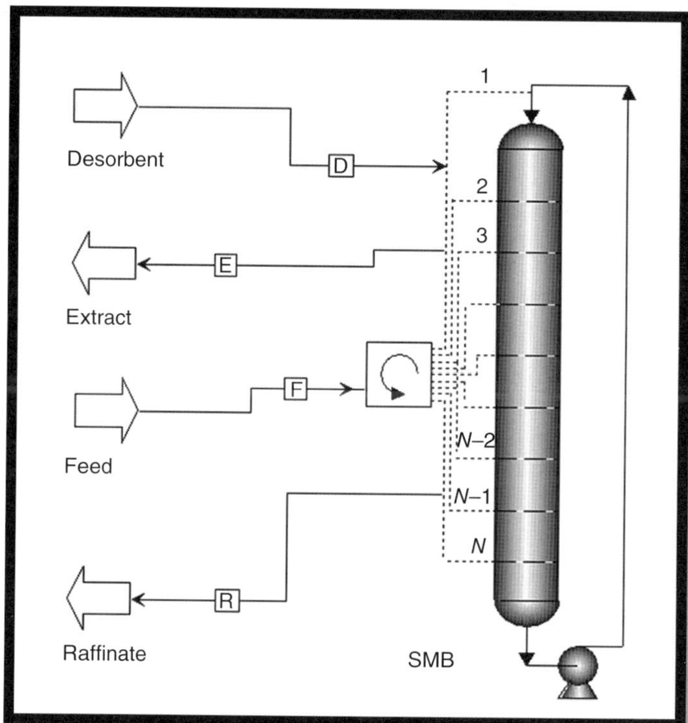

Figure 2.8 Flowsheet of the basic SMB model in Aspen Chromatography.

2.4.2 Adding Components to an Aspen Chromatography Simulation

Step 3: Convert the Component List to a Set.

Go to "Component Lists" in the Exploring window, double click "Add Component List," and choose "Do not use Physical Properties (Set)" to add a new component set into the model. Name the components as **TB+ (the most adsorbed component) and TB− (the least adsorbed component)**. See Figure 2.9.

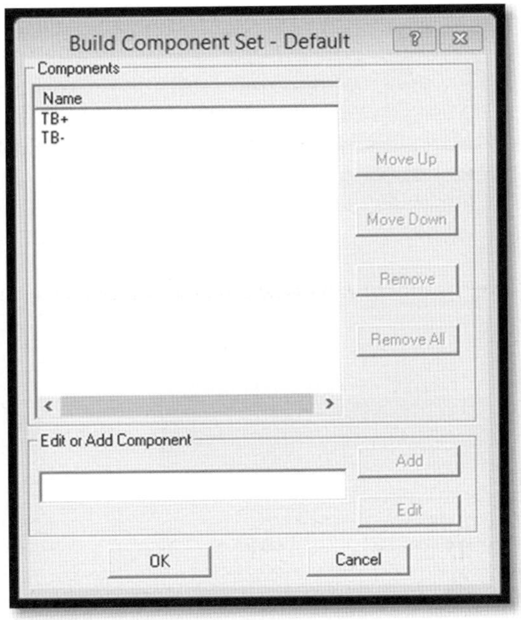

Figure 2.9 Build a new component set as default.

Because we did not specify physical properties, Aspen will be unable to automatically calculate liquid density, viscosity, or other properties. However, since we are making the trace liquid assumption, we assume that the density is constant. Moreover, *we do not need to specify the solvent as a component*. This is only possible because we are manually providing the density and viscosity as constants. This allows the liquid flow rates to be correctly calculated without specifically considering the solvent. Most importantly, however, the isotherm is *only* a function of TB+ and TB−. The solvent concentration does not actually factor into the adsorption behavior.

2.4.3 The Chrom_CCC_separator2 Block

In this section, we investigate the "Chrom_CCC_Separator2" model that we added in the second step. This model was built by Aspen Tech to allow an incredibly simple and efficient simulation of SMB columns. It is capable of simulating a wide variety of SMB configurations, including operational modes. Moreover, it does not require a complex cycle organizer to define the port switching.

Step 4: Open the default configuration form for the SMB column.

Figure 2.10 shows the final state of the column configuration form. We have made the following changes:

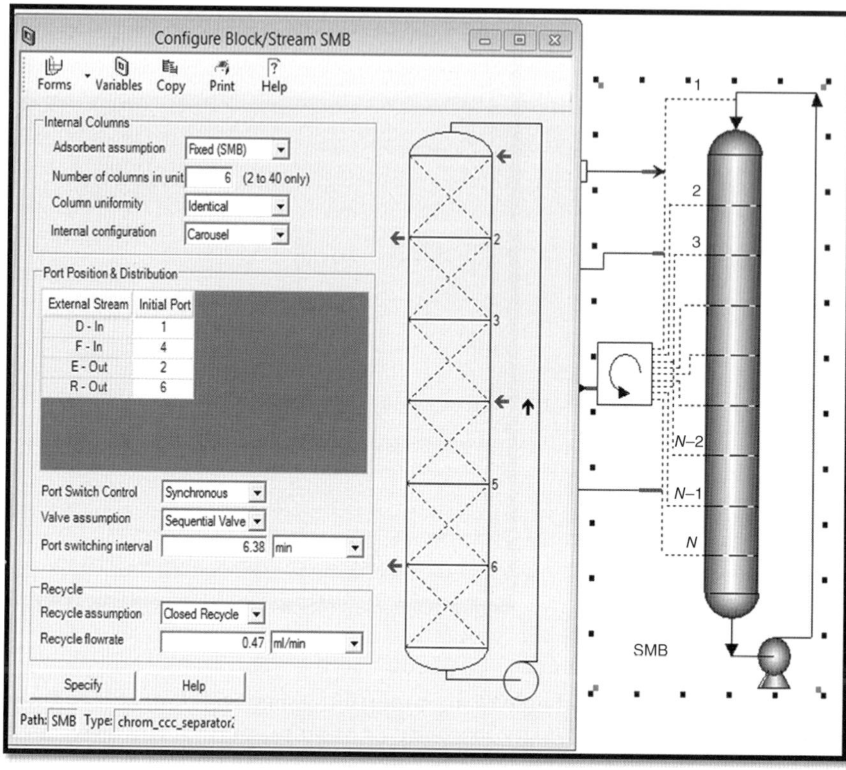

Figure 2.10 The SMB configuration form.

1) We use an SMB unit with six beds in the column, which considers the solid adsorbent as a stationary phase packed in a fixed column.
2) All of the beds in this column have an identical cross-section with the same bed diameter and the same bed length.
3) We assume this column as a carousel internal configuration.
4) Since the column configuration for this four-zone system is (1-2-2-1), the initial ports for D-In, F-In, E-Out, and R-Out, respectively, are 1, 4, 2, and 6.
5) We apply the synchronous port switching control in the base case.
6) We use the sequential valves to deal with the port switching.
7) All the inlet and outlet ports would move to the next bed every 6.38 min as a switching time.
8) The closed recycle flow rate in this system is 0.47 ml/min.

Step 5: Enter the column specifications using the "specify" button, starting with the discretization method.

We use the quadratic upwind differencing scheme (QUDS; QDS in the software) with 40 element nodes as a discretization method (Figure 2.11). QUDS is a linear finite-element scheme that has good computational performance. For this binary separation, QUDS method has very good accuracy and little numerical dispersion, and is well suited to time integration. It also produces results that match experimental results well.

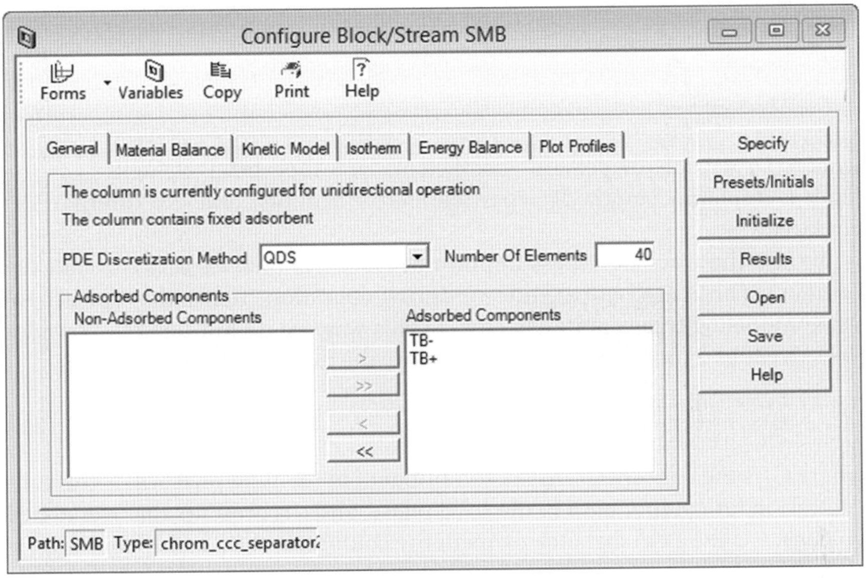

Figure 2.11 Partial differential equation discretization method in the basic SMB model.

Step 6: Specify the material balance assumptions.

Click on the "Material Balance" tab, and set the material balance to Convection with Dispersion Based on Peclet number (*Pe*). See Figure 2.12. We use the convection with dispersion based on Peclet number as a material balance assumption. This assumption evaluates the axial

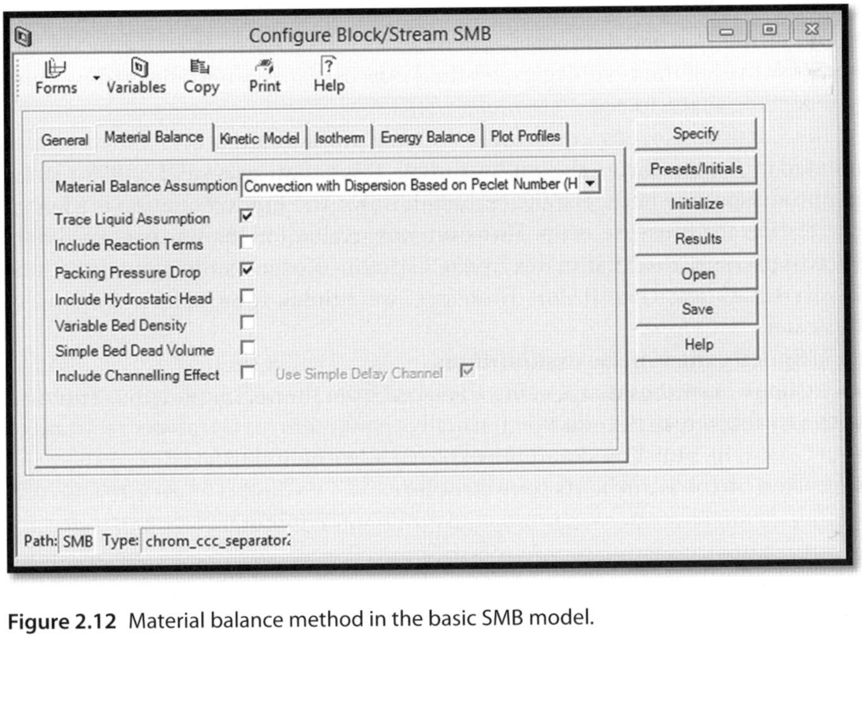

Figure 2.12 Material balance method in the basic SMB model.

dispersion coefficient as a function of the Peclet number based on the column length:

$$D_L = \frac{uL}{Pe\, \varepsilon_i} \tag{2.7}$$

where D_L = axial dispersion coefficient, u = superficial velocity of the fluid, L = column length, Pe = particle Peclet number, and ε_i = interparticle voidage.

We use the trace liquid assumption because the bulk density through the column is considered not significant during operation. This is the typical case when modeling trace solute systems. In this way, we do not need to consider the solvent as a modeled component. Then, the mass balance equation is

$$\frac{\partial C_i}{\partial t} + \frac{(1-\varepsilon)}{\varepsilon}\frac{\partial q_i}{\partial t} + u\frac{\partial C_i}{\partial z} = D_L \frac{\partial^2 C_i}{\partial z^2} \tag{2.8}$$

where C_i is the concentration for component i in the fluid phase, q_i is the concentration for component i in the solid phase, and ε represents the overall bed voidage, z is the axial distance through the column, and D_L is the axial dispersion coefficient.

For this four-zone SMB model, we simply choose the assumption of packing pressure drop in the column. It applies the Carman–Kozeny equation, which relates the pressure drop inside the chromatographic column to the superficial velocity:

$$\frac{\Delta P_j}{n_j{}^* l} = \frac{-150\mu(1-\varepsilon)^2}{(2r_p\psi)^2\varepsilon^3}\frac{Q_j}{S} \tag{2.9}$$

In the equations, ΔP_j = pressure drop in each zone; n_j = bed number in each zone; l = length of each bed; ΔP_{max} = maximum pressure drop through the column; ϕ = pressure drop coefficient; S = cross-sectional area of the bed; ε = overall bed voidage; μ = fluid viscosity; r_p = particle radius; and ψ = particle shape factor.

The Carman–Kozeny equation is the only pressure drop correlation implemented in Aspen Chromatography, though this equation is valid only for laminar flow in packed beds. For the turbulent flows, the Ergun equation is a better correlation for pressure drop. However, our evaluation shows that the difference in pressure drop estimates by the Carman–Kozeny and Ergun equations is very small (less than 0.5%). Therefore, we employ the Carman–Kozeny in this case.

Step 7: Specify the kinetic assumptions.

As we know, a species that is being adsorbed from the bulk liquid phase onto an active surface site of the adsorbent usually encounters mass transfer resistances

- between the bulk liquid and the external adsorbents surface;
- exerted by the adsorbents pore structure.

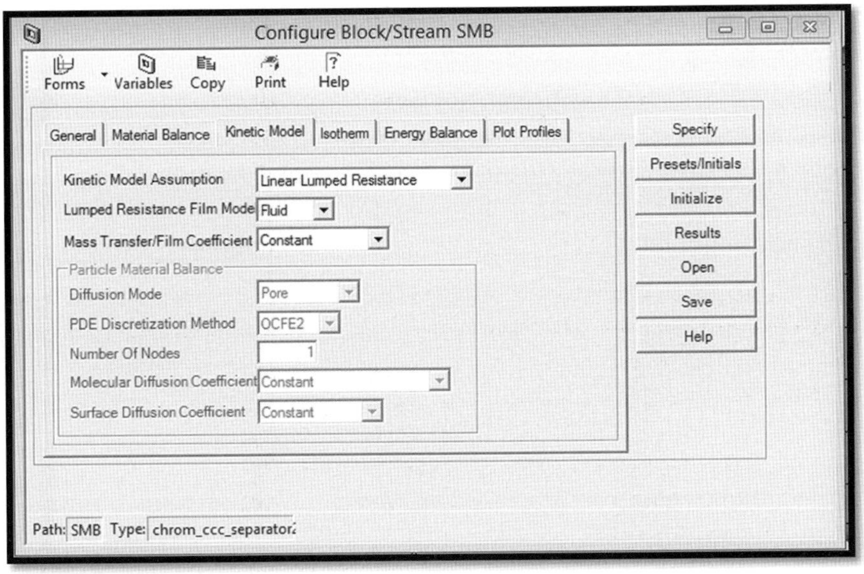

Figure 2.13 Kinetic method in the basic SMB model.

We select the linear lumped resistance in this model (Figure 2.13), where we lump together these resistances. The mass-transfer driving force for component i is a linear function of the solid phase loading:

$$\frac{\partial q_i}{\partial t} = k_i(q_i^* - q_i) \tag{2.10}$$

where k_i is the linear lumped mass transfer coefficient, q_i is solute concentrations of the solid phase, and q_i^* represents the adsorbed phase concentration at equilibrium with the liquid phase.

Step 8: Specify the adsorption assumptions

Use the Isotherm tab to select the Dual-Site Langmuir Model.

Adsorption isotherms are important data in the design of chromatographic systems, because the driving force behind all adsorptive liquid separation processes lies in the departure from adsorption equilibrium.

We assume the bi-Langmuir adsorption isotherm (Figure 2.14) based on volume loading to predict the performance of the chromatographic column for the specified operating conditions. The bi-Langmuir adsorption isotherm equation is

$$C_i^* = \frac{a_{i,1}C_i}{1+\sum b_{i,1}C_i} + \frac{a_{i,2}C_i}{1+\sum b_{i,2}C_i} \ (i = A, B) \tag{2.11}$$

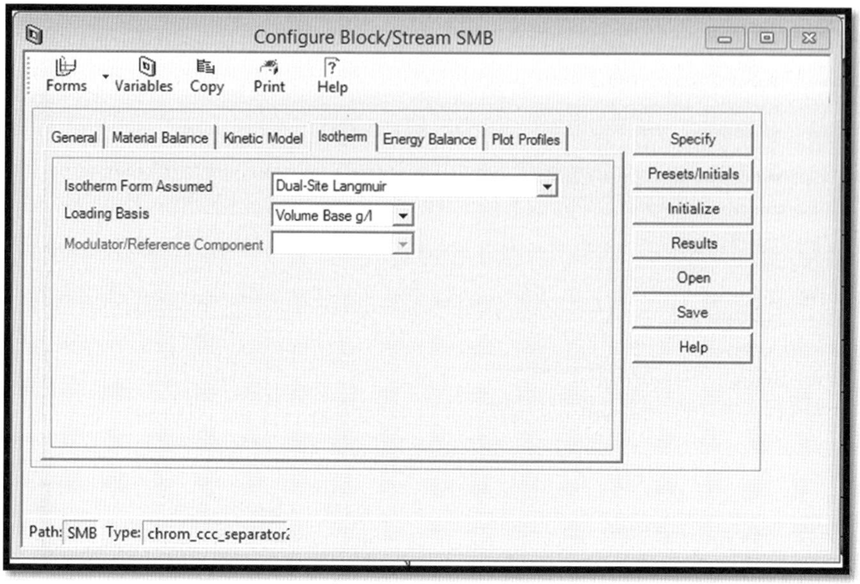

Figure 2.14 Isotherm method in the basic SMB model.

In the equation, C_i^* is the adsorbed phase concentration of component i in equilibrium with the mobile phase. $a_{i,1}$, $a_{i,2}$, $b_{i,1}$, $b_{i,2}$ are the bi-Langmuir isotherm parameters for component i.

Step 9: Energy Balance → Isothermal Operation at 23 °C.

We select the isotherm operation at 23 °C (Figure 2.15), where the variables for fluid, solid, and wall temperature would become inactive in this mode.

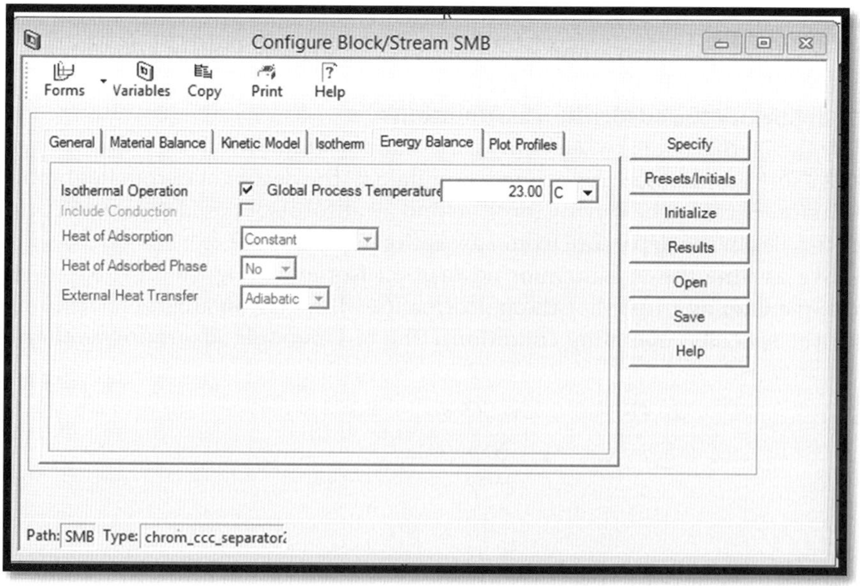

Figure 2.15 Energy balance method in the basic SMB model.

Step 10: Plot Profiles → Available Plot Profiles.

The Plot Profiles tab in Figure 2.16 contains the most commonly used profile variables for the generation of profiles plots. We can simply copy and paste the variables when generating a profile plot.

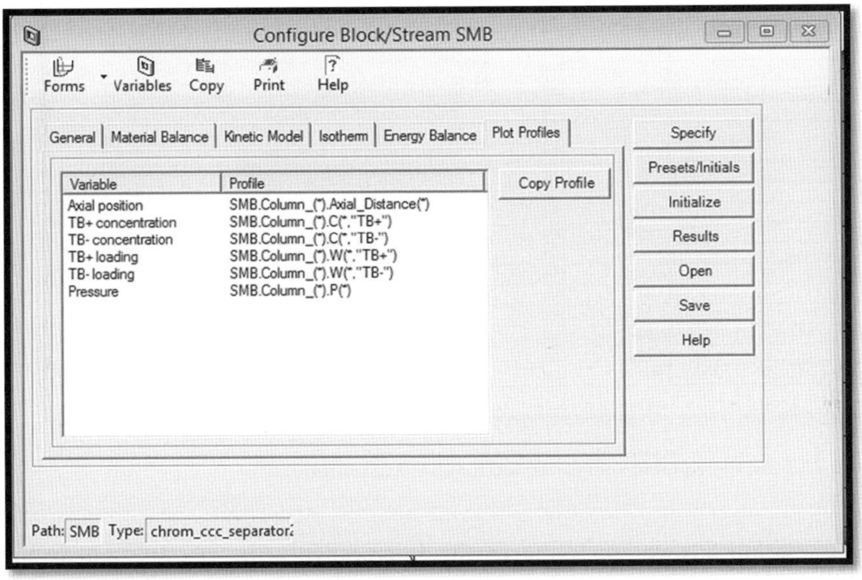

Figure 2.16 Available plot profiles in the basic SMB model.

Step 11: SMB Column → Specify.

Table 2.1 summarizes all the variable values assumed in steps 5–11 for the basic SMB column. Figure 2.17 shows the implementation. Since Eq. (2.12) is the

Table 2.1 List of column parameter definitions and values.

Column length	Hb	15 cm
Column diameter	Db	0.46 cm
Interparticle voidage	ε	0.68 m^3 void/m^3 bed
Adsorbent particle radius	r_p	10 μm
Particle shape factor	ψ	0.1475
Liquid (ethanol) viscosity	μ	1.07 cP
Recycle pressure	$P_{Recycle}$	15 bar
Dual-site Bi-Langmuir isotherm	A (TB+)	B (TB−)
$a_{i,1}$	3.99	1.56
$b_{i,1}$ (l/g)	0.0107	0.0132
$a_{i,2}$	0.986	0.304
$b_{i,2}$ (l/g)	0.601	0.136
Mass transfer coefficient k_i (1/s)	1.81	2.96
Peclet number	500	500

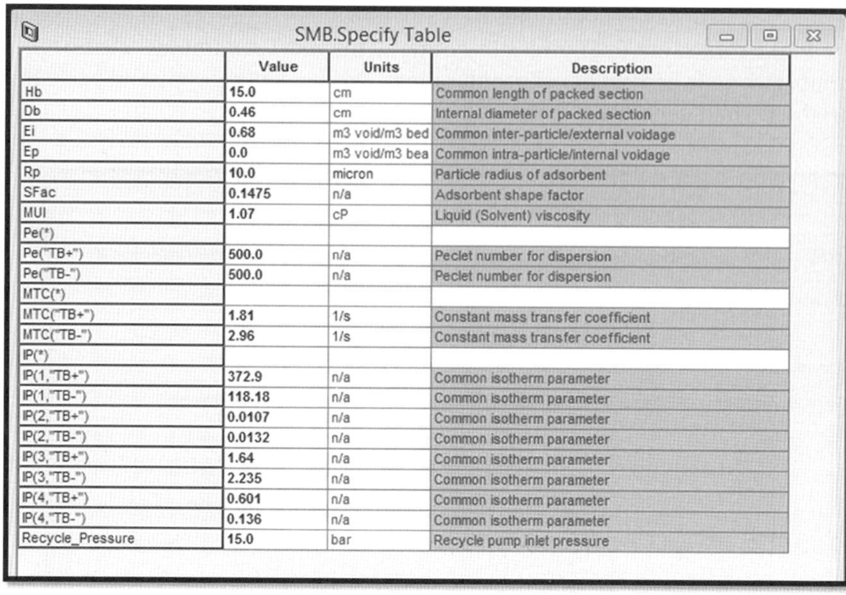

SMB.Specify Table

	Value	Units	Description
Hb	15.0	cm	Common length of packed section
Db	0.46	cm	Internal diameter of packed section
Ei	0.68	m3 void/m3 bed	Common inter-particle/external voidage
Ep	0.0	m3 void/m3 bea	Common intra-particle/internal voidage
Rp	10.0	micron	Particle radius of adsorbent
SFac	0.1475	n/a	Adsorbent shape factor
MUI	1.07	cP	Liquid (Solvent) viscosity
Pe(*)			
Pe("TB+")	500.0	n/a	Peclet number for dispersion
Pe("TB-")	500.0	n/a	Peclet number for dispersion
MTC(*)			
MTC("TB+")	1.81	1/s	Constant mass transfer coefficient
MTC("TB-")	2.96	1/s	Constant mass transfer coefficient
IP(*)			
IP(1,"TB+")	372.9	n/a	Common isotherm parameter
IP(1,"TB-")	118.18	n/a	Common isotherm parameter
IP(2,"TB+")	0.0107	n/a	Common isotherm parameter
IP(2,"TB-")	0.0132	n/a	Common isotherm parameter
IP(3,"TB+")	1.64	n/a	Common isotherm parameter
IP(3,"TB-")	2.235	n/a	Common isotherm parameter
IP(4,"TB+")	0.601	n/a	Common isotherm parameter
IP(4,"TB-")	0.136	n/a	Common isotherm parameter
Recycle_Pressure	15.0	bar	Recycle pump inlet pressure

Figure 2.17 All the variable values needed for all the assumptions.

bi-Langmuir adsorption isotherm, the equation with IP_1, IP_2, IP_3, and IP_4 in Aspen Chromatography is given by

$$C_i^* = \frac{IP_{1i}IP_{2i}C_i}{1+\sum IP_{2i}C_i} + \frac{IP_{3i}IP_{4i}C_i}{1+\sum IP_{4i}C_i} \quad (i = A, B) \tag{2.12}$$

By comparing Eq. (2.12) and Table 2.1, we get

$$IP_{2A} = b_{A,1} = 0.0107;$$

$$IP_{2B} = b_{B,1} = 0.0132;$$

$$IP_{4A} = b_{A,2} = 0.601;$$

$$IP_{4B} = b_{B,2} = 0.136;$$

$$IP_{1A} = \frac{a_{A,1}}{IP_{2A}} = \frac{3.99}{0.0107} = 372.9;$$

$$IP_{1B} = \frac{a_{B,1}}{IP_{2B}} = \frac{1.56}{0.0132} = 118.18;$$

$$IP_{3A} = \frac{a_{A,2}}{IP_{4A}} = \frac{0.986}{0.601} = 1.64;$$

$$IP_{3B} = \frac{a_{B,2}}{IP_{4B}} = \frac{0.304}{0.136} = 2.235.$$

Step 12: SMB Column → Presets/Initials.

We set all the initial variable values as zero for the model, and then press the "Initialize" button to initialize the column, as illustrated in Figure 2.18.

SMB.Initials Table

	Value	Units	Spec	Derivative	Description
Column_(1).C(1,*)					
Column_(1).C(1,"TB+")	0.0	g/l	Initial	0.0	Bulk concentration
Column_(1).C(1,"TB-")	0.0	g/l	Initial	0.0	Bulk concentration
Column_(1).W(1,*)					
Column_(1).W(1,"TB+")	0.0	n/a	Initial	0.0	Solid loading, g/g(MassBase
Column_(1).W(1,"TB-")	0.0	n/a	Initial	0.0	Solid loading, g/g(MassBase

Specify | Presets/Initials | Initialize | Results | Open | Save | Help

Figure 2.18 All the initial variable values needed for the model.

Step 13: Feed block → Feed Specification.

As illustrated in Figure 2.19, the feed-in flow rate is 0.38 ml/min. We use a feed concentration of 0.5 g/l for each component, and the feed contains 50% component A and 50% component B by weight. We set the feed pressure as a free variable.

FEED.Specify Table

	Value	Units	Spec	Description
Flowrate	0.38	ml/min	Fixed	Feed flowrate
Overall_Concentration	1.0	g/l	Fixed	Total concentration of the feed
Material_Fraction(*)				
Material_Fraction("TB+")	50.0	%	Fixed	Mass fraction of the feed
Material_Fraction("TB-")	50.0	%	Fixed	Mass fraction of the feed
Pressure	35.2094	bar	Free	Feed pressure
Cref(*)				
Cref("TB+")	1.0	g/l	Fixed	Reference concentration
Cref("TB-")	1.0	g/l	Fixed	Reference concentration

Figure 2.19 Feed specification for the model.

Step 14: Desorbent block → Desorbent Specification.

In this process, the desorbent-in flow rate is 0.65 ml/min. Since the liquid desorbent does not contain TB+ and TB–, we set both concentrations as zero in Figure 2.20. We set the pressure as a free variable.

DESORBENT.Specify Table

	Value	Units	Spec	Description
Flowrate	0.65	ml/min	Fixed	Feed flowrate
Component_Concentration(*)				
Component_Concentration("TB+")	0.0	g/l	Fixed	Component concentration of the
Component_Concentration("TB-")	0.0	g/l	Fixed	Component concentration of the
Pressure	54.4394	bar	Free	Feed pressure
Cref(*)				
Cref("TB+")	1.0	g/l	Fixed	Reference concentration
Cref("TB-")	1.0	g/l	Fixed	Reference concentration

Figure 2.20 Desorbent specification for the model.

Step 15: Raffinate block → Raffinate Specification.

The raffinate-out flow rate is 0.43 ml/min. We set the pressure as a free variable (Figure 2.21).

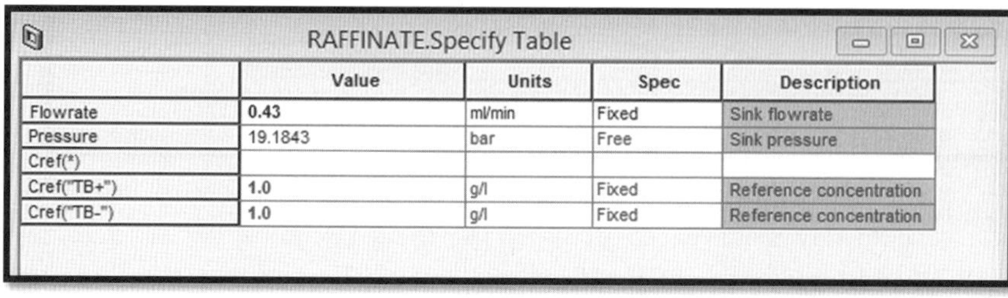

RAFFINATE.Specify Table

	Value	Units	Spec	Description
Flowrate	0.43	ml/min	Fixed	Sink flowrate
Pressure	19.1843	bar	Free	Sink pressure
Cref(*)				
Cref("TB+")	1.0	g/l	Fixed	Reference concentration
Cref("TB-")	1.0	g/l	Fixed	Reference concentration

Figure 2.21 Raffinate specification for the model.

Step 16: Extract block → Extract Specification.

In the Extract block, we set both extract flow rate and the pressure as free in Figure 2.22. Since all the other three (feed, desorbent, raffinate) flow rates are fixed, we can calculate the extract flow rate through the mass balance in the model.

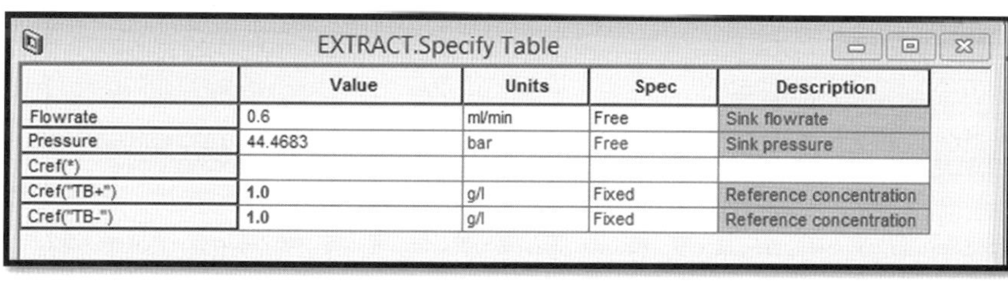

EXTRACT.Specify Table

	Value	Units	Spec	Description
Flowrate	0.6	ml/min	Free	Sink flowrate
Pressure	44.4683	bar	Free	Sink pressure
Cref(*)				
Cref("TB+")	1.0	g/l	Fixed	Reference concentration
Cref("TB-")	1.0	g/l	Fixed	Reference concentration

Figure 2.22 Extract specification for the model.

Step 17: Dead Volume Specification → Pipe Volume in the Inlet and Outlet Streams.

For all the inlet and outlet streams Desorbent (D), Extract (E), Feed (F), and Raffinate (R), we set the same dead volume with the pipe length of 18 cm and pipe diameter as 0.075 cm (Figure 2.23).

Step 18: Dead Volume Specification in the Column.

Go to the "Forms" in the "Configure Block/Stream SMB," select the streams, and press "Open" (Figure 2.24).

We make sure that the total dead volume equals 0.23 ml, and assume an effective dead volume with fixed pipe diameter and pipe length. For all the dead volumes in the column, we define the same dead volume with the pipe length of 20 cm and pipe diameter as 0.075 cm for each inlet and outlet void (Figure 2.25).

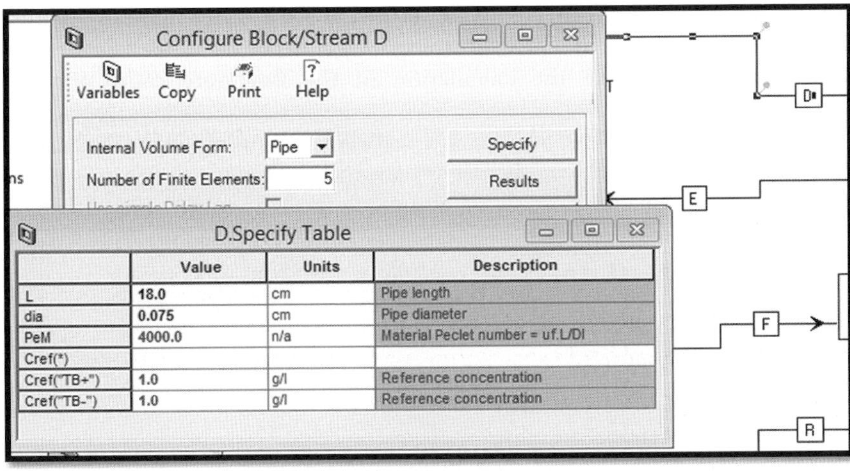

Figure 2.23 Dead volume specification in all the inlet and outlet steams.

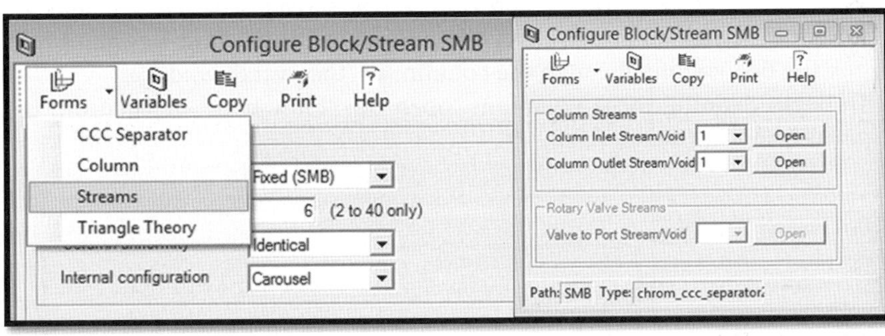

Figure 2.24 Open the inlet/outlet streams in the column.

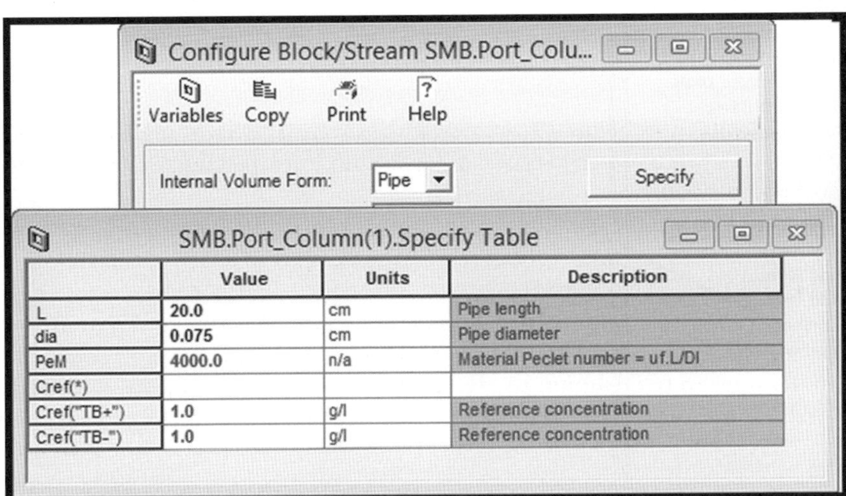

Figure 2.25 Dead volume specification in the column.

Step 19: Flowsheet Constraints → Calculation of Flow Rates form m-Values. This step shows the detailed programming we put in the flowsheet constraints of the model to define the calculation of the flow rate ratios in Eqs. (2.13)–(2.16).

$$m_1 = \frac{Q_1 t_{\text{sw}} - V_{\text{bed}}\varepsilon - V_1^{\text{D}}}{V_{\text{bed}}(1-\varepsilon)} \tag{2.13}$$

$$m_2 = \frac{Q_2 t_{\text{sw}} - V_{\text{bed}}\varepsilon - V_2^{\text{D}}}{V_{\text{bed}}(1-\varepsilon)} \tag{2.14}$$

$$m_3 = \frac{Q_3 t_{\text{sw}} - V_{\text{bed}}\varepsilon - V_3^{\text{D}}}{V_{\text{bed}}(1-\varepsilon)} \tag{2.15}$$

$$m_4 = \frac{Q_4 t_{\text{sw}} - V_{\text{bed}}\varepsilon - V_4^{\text{D}}}{V_{\text{bed}}(1-\varepsilon)} \tag{2.16}$$

In the equations, m_1, m_2, m_3, and m_4 represent the flow rate ratios in zone 1, zone 2, zone 3, and zone 4, respectively; Q_1, Q_2, Q_3, and Q_4 represent the flow rates in zone 1, zone 2, zone 3, and zone 4, respectively; V_1^{D}, V_2^{D}, V_3^{D}, and V_4^{D} represent the dead volume in zone 1, zone 2, zone 3, and zone 4, respectively; V_{bed} = the bed volume through the column; ε = the overall voidage.

To begin defining these equations, open the "Flowsheet" object from the Explorer window (Figure 2.26).

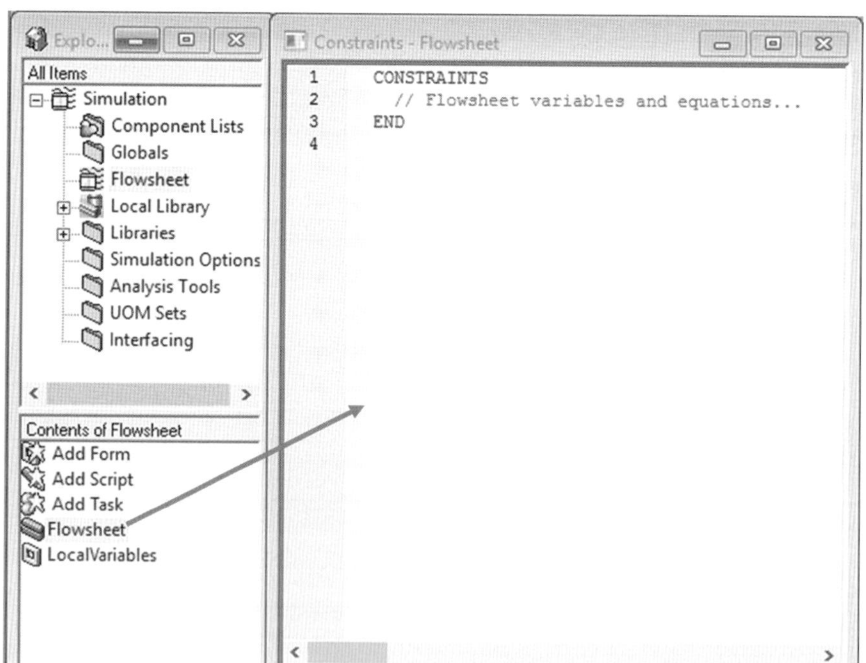

Figure 2.26 Accessing the flowsheet constraints text editor.

First, we define m values as real variables at fixed values. To do this, enter the following lines of code between the "CONSTRAINTS" and "END" text.

```
m1 as RealVariable(fixed);
m2 as RealVariable(fixed);
m3 as RealVariable(fixed);
m4 as RealVariable(fixed);
```

Second, we include all the definition of the volume we use in the equations, including the whole column volume (ColVol) and the dead volume in each zone (DeadVol):

```
ColVol as c_Volume(free);
DeadVol as c_Volume(free);
DeadVol = 0.23;
ColVol = SMB.Hb * 3.1415926 * (SMB.Db/2)^2;
```

Third, we follow Eqs. (2.13)–(2.16) to define the flow rate ratio of each zone:

```
m1 * ColVol * (SMB.Ei - 1) = ColVol * SMB.Ei - SMB.Switch_
  Time * 1000 * (DESORBENT.Flowrate + SMB.Recycle_Flowrate)
  + DeadVol;
m2 * ColVol * (SMB.Ei - 1) = ColVol * SMB.Ei - SMB.Switch_
  Time * 1000 * (RAFFINATE.Flowrate - FEED.Flowrate +
  SMB.Recycle_Flowrate) + DeadVol;
m3 * ColVol * (SMB.Ei - 1) = ColVol * SMB.Ei - SMB.Switch_
  Time * 1000 * (RAFFINATE.Flowrate + SMB.Recycle_Flowrate)
  + DeadVol;
m4 * ColVol * (SMB.Ei - 1) = ColVol * SMB.Ei - SMB.Switch_
  Time * 1000 *SMB.Recycle_Flowrate + DeadVol;
```

After entering these codes, we find the constraints as in Figure 2.27.

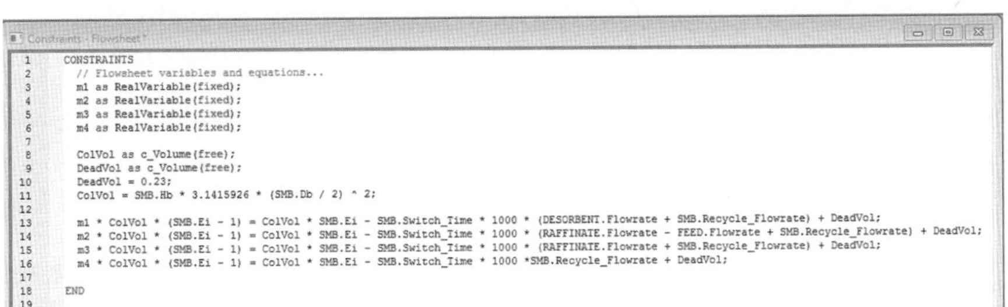

Figure 2.27 Flowsheet constraints from Step 19.

Step 20: Flowsheet Constraints → Calculation of Performance Indices. This step shows the detailed programming in the flowsheet constraints of the model to define the calculation of performance indices in Eqs. (2.17)–(2.20).

First, we set PurityA, PurityB, RecoveryA, and RecoveryB as fraction variables in Eqs. (2.17)–(2.20).

$$\text{PurityA} = \overline{\text{Pur}_\text{A}} = \frac{\overline{C}_\text{A}^\text{E}}{\overline{C}_\text{A}^\text{E} + \overline{C}_\text{B}^\text{E}} \tag{2.17}$$

$$\text{PurityB} = \overline{\text{Pur}_\text{B}} = \frac{\overline{C}_\text{B}^\text{R}}{\overline{C}_\text{A}^\text{R} + \overline{C}_\text{B}^\text{R}} \tag{2.18}$$

$$\text{RecoveryA} = \overline{\text{Rec}_\text{A}} = \frac{Q_\text{E}\overline{C}_\text{A}^\text{E}}{Q_\text{F}\overline{C}_\text{A}^\text{F}} \tag{2.19}$$

$$\text{RecoveryB} = \overline{\text{Rec}_\text{B}} = \frac{Q_\text{R}\overline{C}_\text{B}^\text{R}}{Q_\text{F}\overline{C}_\text{B}^\text{F}} \tag{2.20}$$

```
PurityA as Fraction;
PurityA = Min(E.Total_Component("TB+") / Max(E.Total_Mass,
 0.0000001),1);

PurityB as Fraction;
PurityB = Min(R.Total_Component("TB-") / Max(R.Total_Mass,
 0.0000001),1);

RecoveryA as Fraction;
RecoveryA=Min(1-R.Total_Component("TB+")/Max(F.Total_
 Component("TB+"),0.0000001),1);

RecoveryB as Fraction;
RecoveryB = Min(1-E.Total_Component("TB-")/Max(F.Total_
 Component("TB"), 0.0000001), 1);
```

Second, we set ProductivityA, ProductivityB, and ProductivityTotal as mass flow variables in Eqs. (2.21)–(2.23). The unit of the productivity is gram per liter per hour.

$$\text{ProductivityA} = \frac{Q_\text{E}\overline{C}_\text{A}^\text{E}}{nV_\text{bed}} \tag{2.21}$$

$$\text{ProductivityB} = \frac{Q_\text{R}\overline{C}_\text{B}^\text{R}}{nV_\text{bed}} \tag{2.22}$$

$$\text{ProductivityTotal} = \text{ProductivityA} + \text{ProductivityB} = \frac{Q_\text{E}\overline{C}_\text{A}^\text{E} + Q_\text{R}\overline{C}_\text{B}^\text{R}}{nV_\text{bed}} \tag{2.23}$$

```
ProductivityA as c_Flow_Mass;
ProductivityB as c_Flow_Mass;
ProductivityTotal as c_Flow_Mass;

ProductivityA = (E.Total_Component("TB+") / SMB.Switch_
 Time)*60*1000/ColVol;
ProductivityB = (R.Total_Component("TB-") / SMB.Switch_
 Time)*60*1000/ColVol;
ProductivityTotal = ProductivityA + ProductivityB;
```

Step 21: Compile the flowsheet.

Once all the flowsheet constraints are entered, we need to compile the flowsheet in order to activate them. To do this, right-click on the flowsheet text editor and choose "Compile" (Figure 2.28).

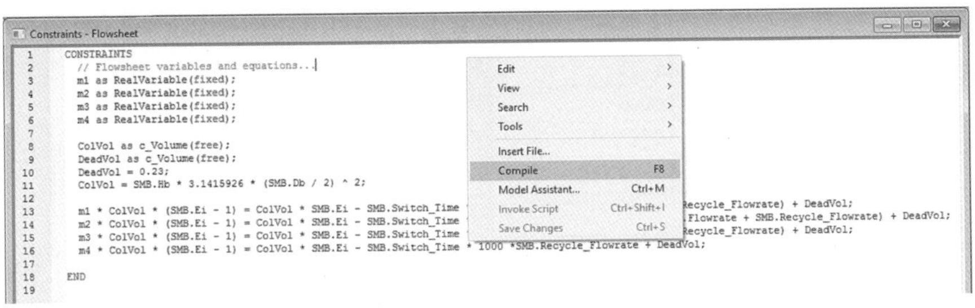

Figure 2.28 Compiling the flowsheet.

Step 22: Add Form → Table → FlowSpecification.

Figure 2.29 shows the flow specifications based on the flowsheet constraints in step 19.

We show the following calculations on how to get the flow rate ratios m values in Figure 2.29 from the zone flow rate values in terms of Eqs. (2.24)–(2.27).

$$m_1 = \frac{Q_1 t_{\text{sw}} - V_{\text{c}}\varepsilon - V_1^{\text{D}}}{V_{\text{bed}}(1-\varepsilon)} = \frac{(1.12\,\text{ml/min}) * 6.38\ \text{min} - 2.493\,\text{cm}^3 * 0.68 - 0.23\,\text{ml}}{2.493\,\text{cm}^3 * (1-0.68)} = 6.54 \tag{2.24}$$

$$m_2 = \frac{Q_2 t_{\text{sw}} - V_{\text{c}}\varepsilon - V_2^{\text{D}}}{V_{\text{bed}}(1-\varepsilon)} = \frac{(0.52\,\text{ml/min}) * 6.38\,\text{min} - 2.493\,\text{cm}^3 * 0.68 - 0.23\,\text{ml}}{2.493\,\text{cm}^3 * (1-0.68)} = 1.75 \tag{2.25}$$

$$m_3 = \frac{Q_3 t_{\text{sw}} - V_{\text{c}}\varepsilon - V_3^{\text{D}}}{V_{\text{bed}}(1-\varepsilon)} = \frac{(0.9\,\text{ml/min}) * 6.38\,\text{min} - 2.493\,\text{cm}^3 * 0.68 - 0.23\,\text{ml}}{2.493\,\text{cm}^3 * (1-0.68)} = 4.785 \tag{2.26}$$

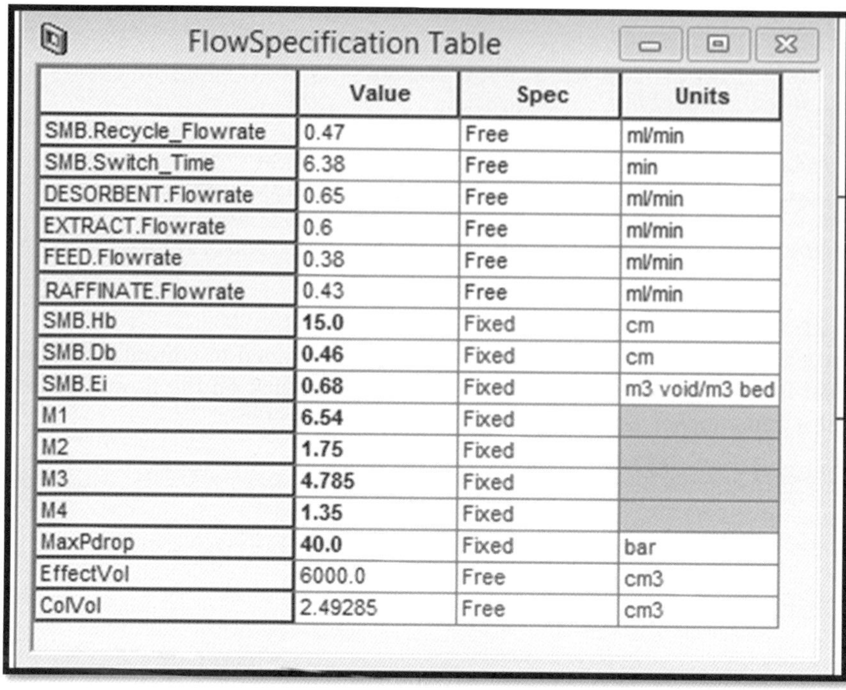

FlowSpecification Table

	Value	Spec	Units
SMB.Recycle_Flowrate	0.47	Free	ml/min
SMB.Switch_Time	6.38	Free	min
DESORBENT.Flowrate	0.65	Free	ml/min
EXTRACT.Flowrate	0.6	Free	ml/min
FEED.Flowrate	0.38	Free	ml/min
RAFFINATE.Flowrate	0.43	Free	ml/min
SMB.Hb	15.0	Fixed	cm
SMB.Db	0.46	Fixed	cm
SMB.Ei	0.68	Fixed	m3 void/m3 bed
M1	6.54	Fixed	
M2	1.75	Fixed	
M3	4.785	Fixed	
M4	1.35	Fixed	
MaxPdrop	40.0	Fixed	bar
EffectVol	6000.0	Free	cm3
ColVol	2.49285	Free	cm3

Figure 2.29 The flow specification defined in the flowsheet constraints.

$$m_4 = \frac{Q_4 t_{\mathrm{sw}} - V_{\mathrm{c}}\varepsilon - V_4^{\mathrm{D}}}{V_{\mathrm{bed}}(1-\varepsilon)} = \frac{(0.47\,\mathrm{ml/min}) * 6.38\ \mathrm{min} - 2.493\,\mathrm{cm}^3 * 0.68 - 0.23\,\mathrm{ml}}{2.493\,\mathrm{cm}^3 * (1-0.68)} = 1.35 \tag{2.27}$$

The criterion of the switching time is

$$t_{\mathrm{sw}} \geq \frac{\phi L^2}{\Delta P_{\mathrm{max}}} \sum_{j=1}^{j=4} n_j \left[(m_j(1-\varepsilon)+\varepsilon) + \frac{V_j^{\mathrm{D}}}{V} \right] \tag{2.28}$$

$$\begin{aligned}\sum_{j=1}^{j=4} n_j(m_j(1-\varepsilon)+\varepsilon) &= 1^*(6.54^*(1-0.68)+0.68+0.23/2.493)\\ &\quad + 2^*(1.75^*(1-0.68)+0.68+0.23/2.493)\\ &\quad + 2^*(4.785^*(1-0.68)+0.68+0.23/2.493)\\ &\quad + 1^*(1.35^*(1-0.68)+0.68+0.23/2.493)\\ &= 2.8551 + 2.66452 + 4.60692 + 1.20426\\ &= 11.3308\end{aligned}$$

$$t_{\mathrm{sw}} \geq \frac{0.1 * 15^2}{40} * 11.3308 = 6.38\,\mathrm{min} \tag{2.29}$$

Step 22: Add Form → Table → Results.

Figure 2.30 shows the results table that displays all the performance indices defined in step 20. It shows all the results as zero before running, and will display the performance results after running.

Results Table

	Value	Spec	Units
PurityA	0.0	Free	%
PurityB	0.0	Free	%
RecoveryA	0.0	Free	%
RecoveryB	0.0	Free	%
ProductivityA	0.0	Free	g/min
ProductivityB	0.0	Free	g/min
ProductivityTotal	0.0	Free	g/min

Figure 2.30 The performance indices defined in the flowsheet constraints.

Step 23: Add Form → Plot → Purity_Recovery.

In order to clearly see the changing of both product purity and recovery over the simulation time, we make a performance plot of the product purity and recovery results named as Purity_Recovery (Figure 2.31).

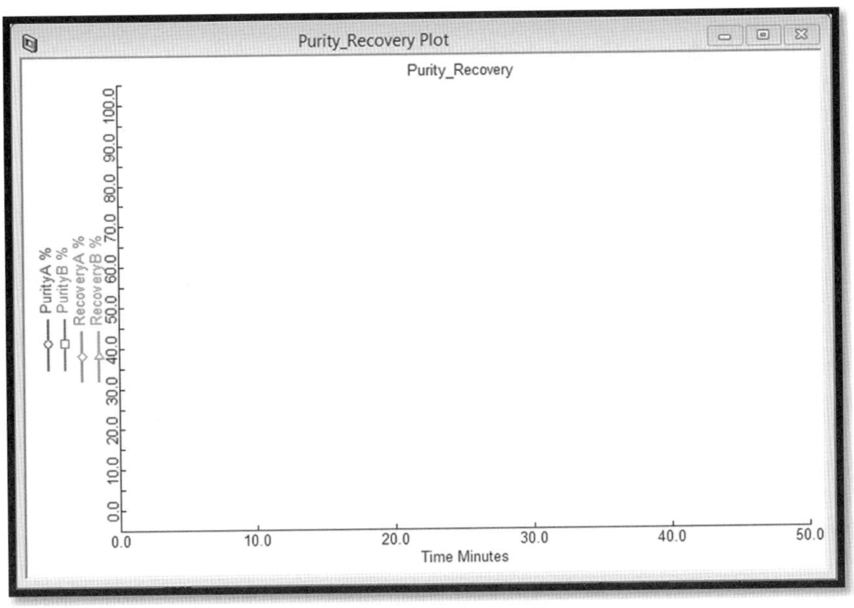

Figure 2.31 The performance plot of the product purity and recovery results.

Step 24: Add Form → Plot → Extract Plot.

The extract stream receives most of the component A, so we make a plot "Extract" to see the concentration and mole fraction for both components in the extract-out stream (Figure 2.32).

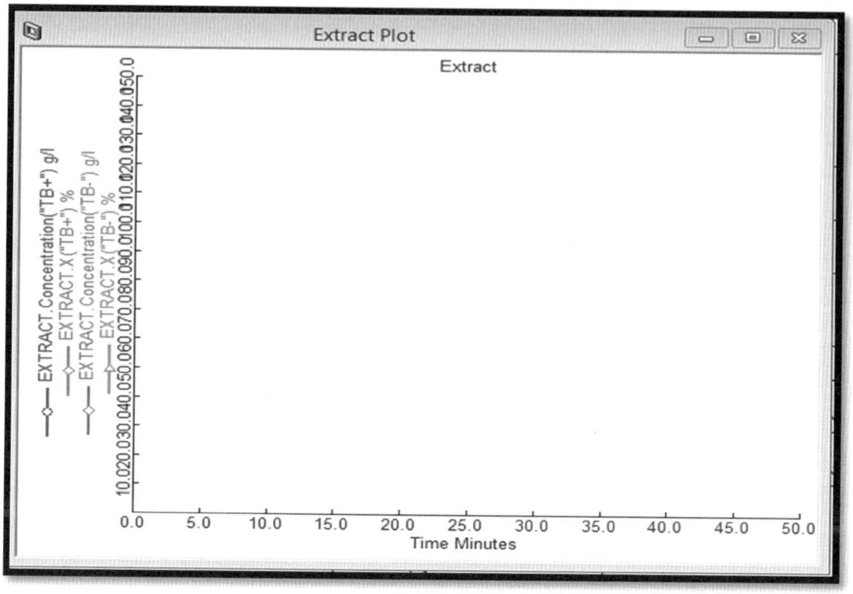

Figure 2.32 The extract concentration and mole fraction results for both components.

Step 25: Add Form → Plot → Raffinate Plot.

The raffinate stream receives most of component B, so we make a plot "Raffinate" to see the concentration and mole fraction for both components in the raffinate-out stream (Figure 2.33).

Step 26: Add Form → Profile Plot → Concentration Profile Plot.

Our next goal is to make a concentration profile, which gives physical insights within the SMB column of their separation performance for both components. It reveals to us how the concentrations change through the axial distance of the column over time.

We define the axial position of the column by "SMB. Column_(*).Axial_Distance(*)"; TB+ Concentration by "SMB. Column_(*).C (*,"TB+")"; and TB- Concentration by "SMB. Column_(*). C (*,"TB-")," as shown in Figure 2.34.

After we press "Apply," it will show the resulting column concentration profile for both TB+ and TB− at zero simulation time, as shown in Figure 2.35.

Step 27: Run → Solver Options → Integrator.

After we finish making all the previous plots and profile plots, we define the solver options. The Solver Options dialog box in Aspen Chromatography enables us to change the default settings for numerical algorithms used

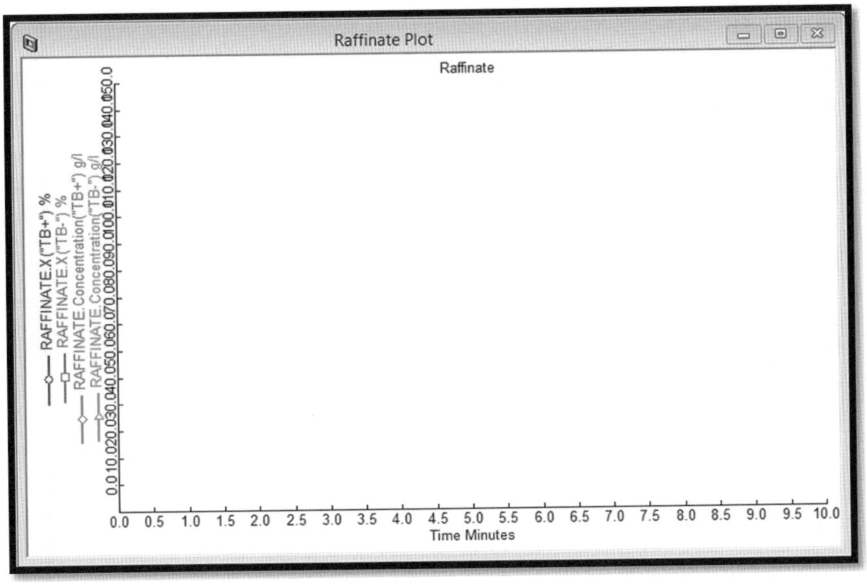

Figure 2.33 Setting up a plot of raffinate concentration and mole fraction results for both components.

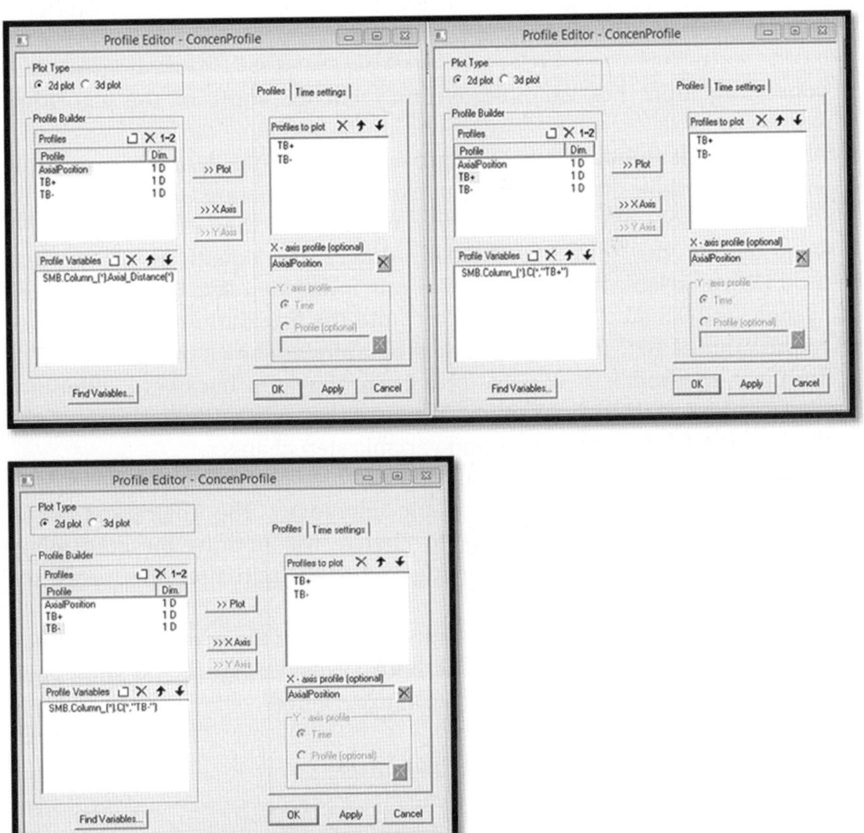

Figure 2.34 The variable definition of the concentration profile within the column.

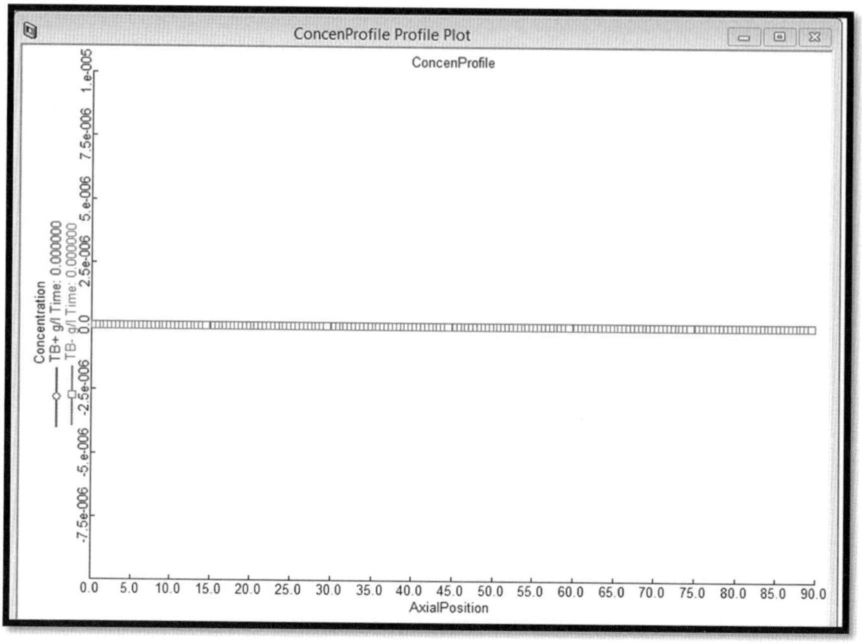

Figure 2.35 The resulting concentration profile for both TB+ and TB− at zero simulation time.

in your simulation. For example, we may change some of the algorithm properties to

- solve a difficult simulation or to improve the speed of a dynamic run;
- get comprehensive diagnostic output from our simulation to determine why a simulation does not converge or runs more slowly than expected.

We integrate the resulting ordinary differential equations using Gear method with a variable time step size (0.1 min initial step size, 0.1–0.5 min step size range, and 0.5 step reduction factor) and the integration error tolerance set to the value of 10^{-4} in Aspen Chromatography, as shown in Figure 2.36.

The Gear method is a variable-order, variable-step backward difference implicit method with high accuracy. The user can define the behavior of the method for model discontinuities and variable step changes.

Step 28: Run → Solver Options → Tolerances.

We set the absolute and relative variable tolerances. The Tolerances tab in Aspen Chromatography enables us to change the options for controlling tolerances, and to control solver scaling and the elimination of equivalence equations.

We can change the value of the tolerances to affect the speed and accuracy of the solution under most solution methods available. Figure 2.37 shows our tolerance settings based on our simulation experiences.

Step 29: Run Options → Initialization.

We are ready to initialize the model (Figure 2.38).

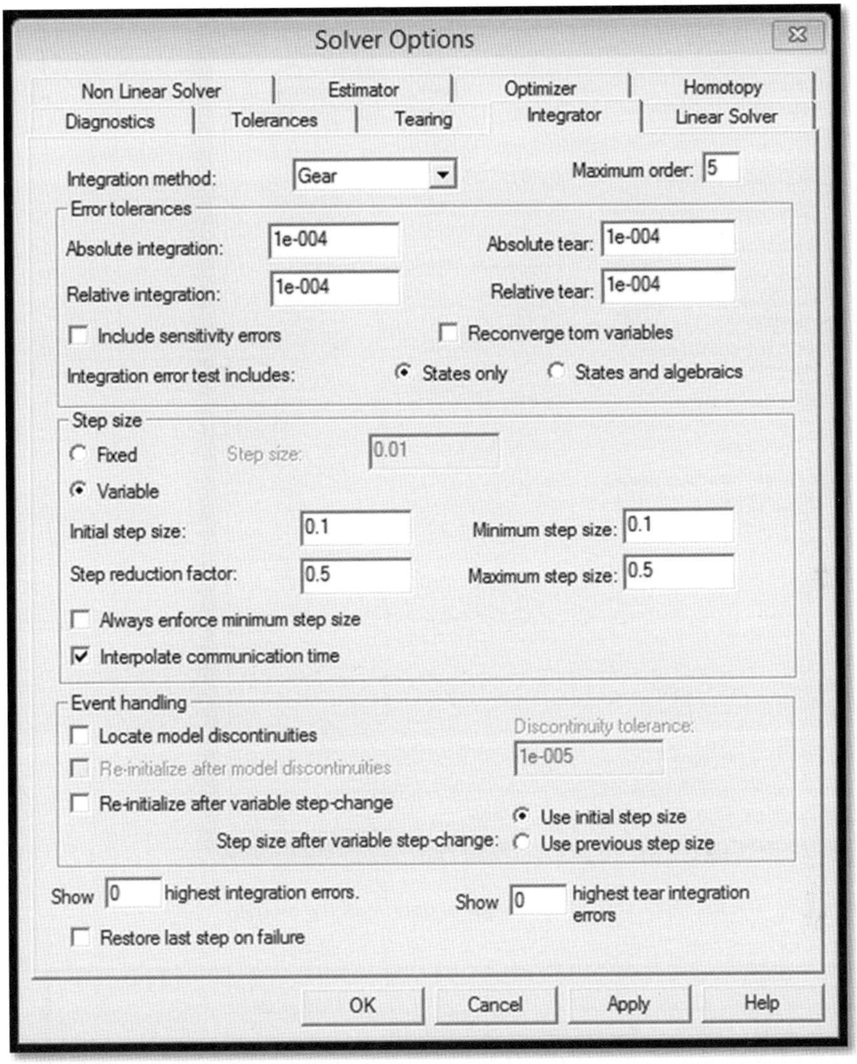

Figure 2.36 The integrator settings in the solver options of the model.

Step 30: Run Options → Pause at 382.8 s.

Since we consider the dynamic state in all the models, the results are recorded once the system reaches a cyclic steady state. We determine the steady state by finding the time at which the average of the performance indices over one step becomes constant. The steady state is attained after 10 cycles (each cycle contains six switching time rounds). Our simulation time is therefore 6.38*6*10 = 382.8 min (Figure 2.39).

Step 31: Run Options → Dynamic.

The model is ready for the dynamic run after reaching the pause time (382.8 min) we set (Figure 2.40).

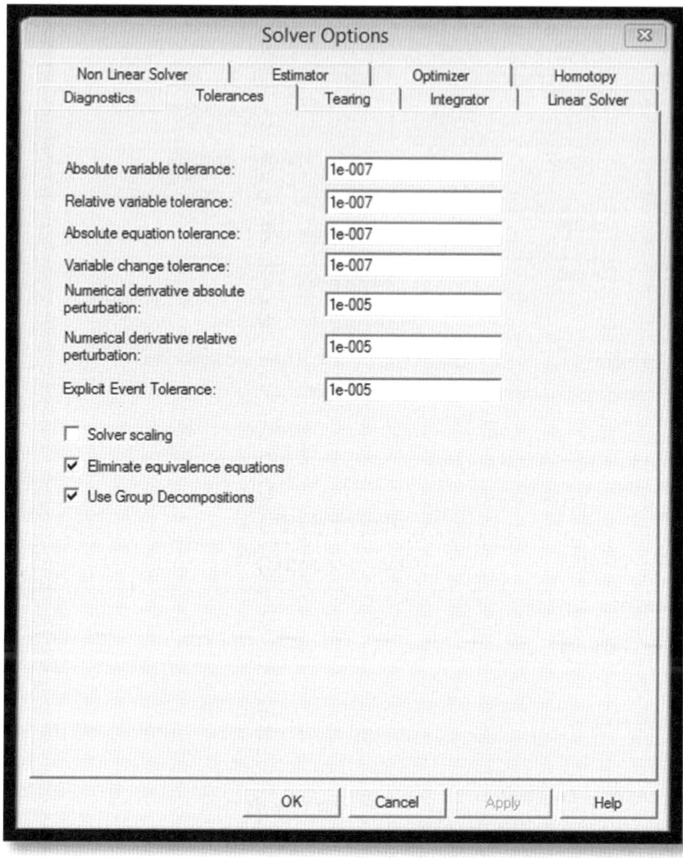

Figure 2.37 The tolerance settings in the solver options of the model.

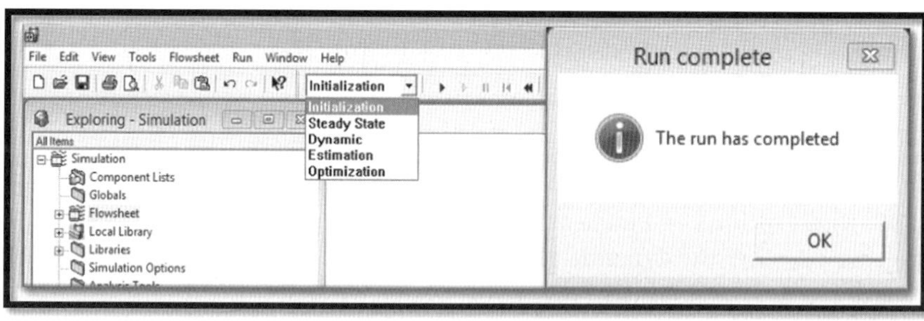

Figure 2.38 The initialization of the SMB model.

Figure 2.39 The simulation running time setting in the run options of the model.

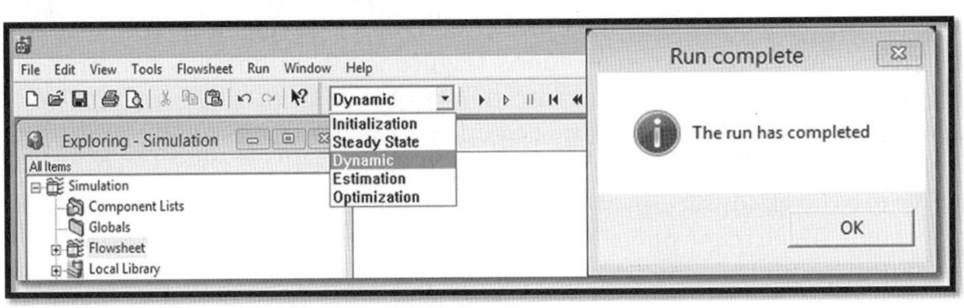

Figure 2.40 The dynamic run of the SMB model.

2.4.4 Viewing Results

Step 23: Results → Purity and Recovery Results.

After each simulation run, we would open the purity and recovery plot to either see the plot or to look at the detailed result history through the simulation times. Figure 2.41 illustrates the Purity_Recovery plot results for both product.

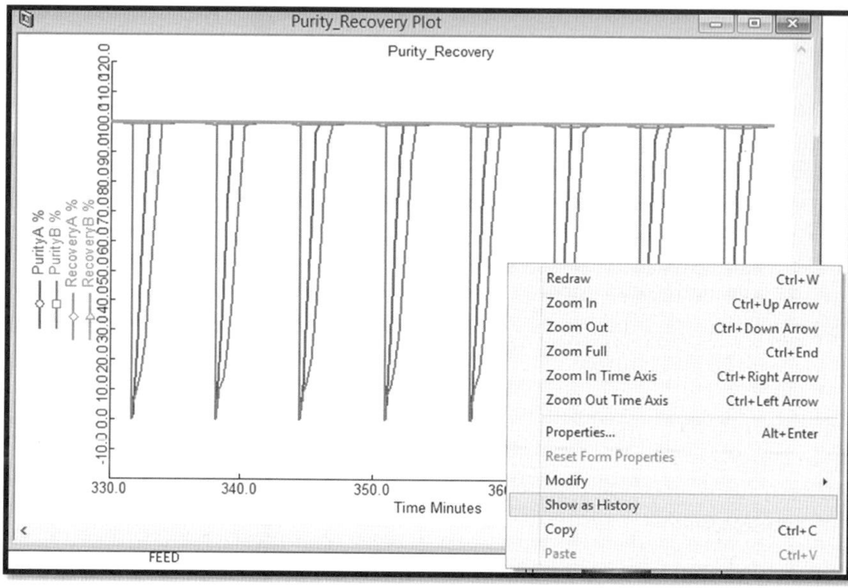

Figure 2.41 The Purity_Recovery plot results for both product.

If we want to look at the detailed values for both product purity and recovery, right-click the plot, and then choose the "Show as History" in Figure 2.41. We get the Purity_Recovery history table as shown in Figure 2.42. The blue highlighted region in the figure represents the purity and recovery results through one whole switching time. We can calculate the average purity and recovery based on these results.

Step 24: Results → Plots.

This step presents all the plots and profile plots we made in steps 23–26, including the extract concentrations (Figure 2.43), raffinate concentrations (Figure 2.44), the column concentration profile (Figure 2.45), and the concentration profile table (Figure 2.46). Save the file as **4-zone basic SMB model.cra** (Figure 2.47).

We use the quadratic upwind differencing scheme (QUDS; QDS in the software) with 40 element nodes as a discretization method (Figure 2.48). QUDS is a linear finite-element scheme that has good computational performance. For this binary separation, QUDS method has very good accuracy and little numerical dispersion, and is well suited to time integration. It also produces results that match experimental data well.

Purity_Recovery History Table

Time	PurityA	PurityB	RecoveryA	RecoveryB
Minutes	%	%	%	%
375.6	99.997	99.4167	99.4228	99.9966
375.9	99.997	99.2073	99.2088	99.9968
376.2	0.0	0.0	100.0	100.0
376.5	3.55675	7.26983	99.2555	99.9977
376.8	30.7655	11.2622	98.9861	99.9897
377.1	59.1493	16.8796	98.985	99.9895
377.4	87.5372	28.8567	98.9849	99.9893
377.7	99.9907	46.3191	98.9849	99.9892
378.0	99.9921	66.2538	98.9849	99.9891
378.3	99.9936	86.5835	99.1329	99.9903
378.6	99.9943	99.0588	99.2496	99.9915
378.9	99.995	99.2075	99.3387	99.9928
379.2	99.9956	99.3157	99.4088	99.9933
379.5	99.9959	99.3979	99.4655	99.9938
379.8	99.9961	99.4625	99.5123	99.9943
380.1	99.9963	99.5145	99.5515	99.9948
380.4	99.9966	99.5565	99.5842	99.9953
380.7	99.9968	99.5886	99.6093	99.9956
381.0	99.9969	99.6066	99.6223	99.9959
381.3	99.9969	99.6012	99.6133	99.9962
381.6	99.997	99.5565	99.566	99.9965
381.9	99.9971	99.452	99.459	99.9969
382.2	99.9971	99.2658	99.2688	99.9972
382.5	0.0	0.0	100.0	100.0
382.8	0.23365	3.2132	99.6733	99.9998
383.1	23.28	10.5545	98.9822	99.9904
[illegible]	[illegible]	[illegible]	[illegible]	[illegible]

Figure 2.42 The Purity_Recovery history table.

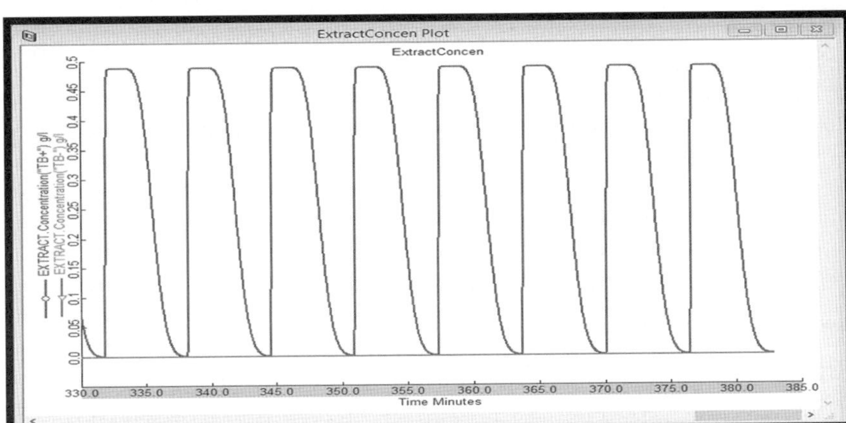

Figure 2.43 The resulting extract concentration results for both components.

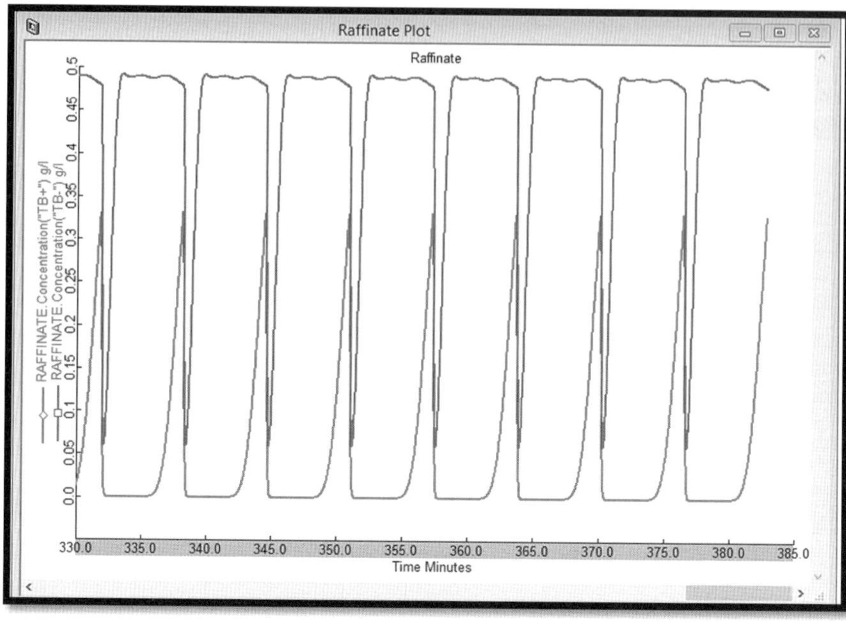

Figure 2.44 The resulting raffinate concentration results for both components.

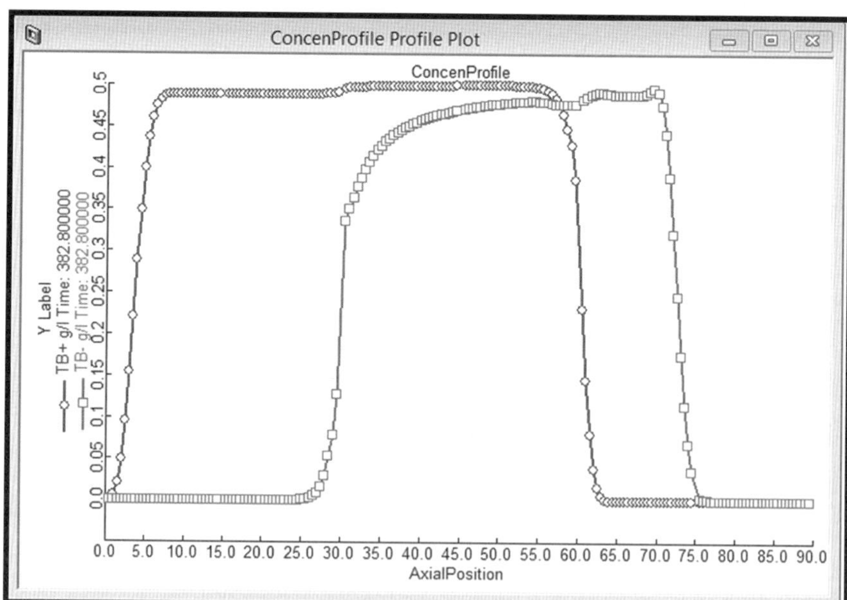

Figure 2.45 The resulting concentration profile within the SMB column.

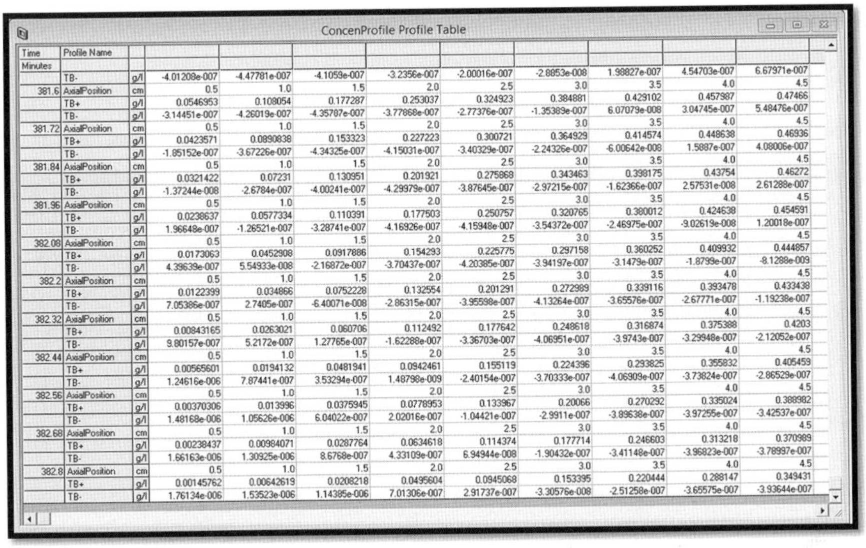

ConcenProfile Profile Table

Time	Profile Name										
Minutes											
	TB-	g/l	-4.01208e-007	-4.47781e-007	-4.1059e-007	-3.2356e-007	-2.00016e-007	-2.8853e-008	1.98827e-007	4.54703e-007	6.67971e-007
381.6	AxialPosition	cm	0.5	1.0	1.5	2.0	2.5	3.0	3.5	4.0	4.5
	TB+	g/l	0.0546953	0.108054	0.177287	0.253037	0.324923	0.384881	0.429102	0.457987	0.47466
	TB-	g/l	-3.14451e-007	-4.26019e-007	-4.35787e-007	-3.77868e-007	-2.77376e-007	-1.35389e-007	6.07079e-008	3.04745e-007	5.48476e-007
381.72	AxialPosition	cm	0.5	1.0	1.5	2.0	2.5	3.0	3.5	4.0	4.5
	TB+	g/l	0.0423571	0.0890838	0.153323	0.227223	0.300721	0.364929	0.414574	0.448638	0.46936
	TB-	g/l	-1.85152e-007	-3.67226e-007	-4.34325e-007	-4.15031e-007	-3.40329e-007	-2.24326e-007	-6.00642e-008	1.5887e-007	4.08006e-007
381.84	AxialPosition	cm	0.5	1.0	1.5	2.0	2.5	3.0	3.5	4.0	4.5
	TB+	g/l	0.0321422	0.07231	0.130951	0.201921	0.275868	0.343463	0.398175	0.43754	0.46272
	TB-	g/l	-1.37244e-008	-2.6784e-007	-4.00241e-007	-4.29979e-007	-3.87645e-007	-2.97215e-007	-1.62366e-007	2.57531e-008	2.61288e-007
381.96	AxialPosition	cm	0.5	1.0	1.5	2.0	2.5	3.0	3.5	4.0	4.5
	TB+	g/l	0.0238637	0.0577334	0.110391	0.177503	0.250757	0.320765	0.380012	0.424638	0.454591
	TB-	g/l	1.96648e-007	-1.26521e-007	-3.28741e-007	-4.16926e-007	-4.15948e-007	-3.54372e-007	-2.46975e-007	-9.02619e-008	1.20018e-007
382.08	AxialPosition	cm	0.5	1.0	1.5	2.0	2.5	3.0	3.5	4.0	4.5
	TB+	g/l	0.0173063	0.0452908	0.0917886	0.154293	0.225775	0.297158	0.360252	0.409932	0.444857
	TB-	g/l	4.39639e-007	5.54933e-008	-2.16872e-007	-3.70437e-007	-4.20385e-007	-3.94197e-007	-3.1479e-007	-1.8799e-007	-8.1288e-009
382.2	AxialPosition	cm	0.5	1.0	1.5	2.0	2.5	3.0	3.5	4.0	4.5
	TB+	g/l	0.0122399	0.034866	0.0752228	0.132554	0.201291	0.272989	0.339116	0.393478	0.433438
	TB-	g/l	7.05386e-007	2.7405e-007	-6.40071e-008	-2.86315e-007	-3.95598e-007	-4.13264e-007	-3.65576e-007	-2.67771e-007	-1.19238e-007
382.32	AxialPosition	cm	0.5	1.0	1.5	2.0	2.5	3.0	3.5	4.0	4.5
	TB+	g/l	0.00843165	0.0263021	0.060706	0.112492	0.177642	0.248618	0.316874	0.375388	0.4203
	TB-	g/l	9.80157e-007	5.2172e-007	1.27765e-007	-1.62288e-007	-3.36703e-007	-4.06951e-007	-3.9743e-007	-3.29948e-007	-2.12052e-007
382.44	AxialPosition	cm	0.5	1.0	1.5	2.0	2.5	3.0	3.5	4.0	4.5
	TB+	g/l	0.00565601	0.0194132	0.0481941	0.0942461	0.155119	0.224396	0.293825	0.355832	0.405459
	TB-	g/l	1.24616e-006	7.87441e-007	3.53294e-007	1.48798e-009	-2.40154e-007	-3.70333e-007	-4.06909e-007	-3.73824e-007	-2.86529e-007
382.56	AxialPosition	cm	0.5	1.0	1.5	2.0	2.5	3.0	3.5	4.0	4.5
	TB+	g/l	0.00370306	0.013996	0.0375945	0.0778953	0.133967	0.20066	0.270292	0.335024	0.388982
	TB-	g/l	1.48168e-006	1.05626e-006	6.04022e-007	2.02016e-007	-1.04421e-007	-2.9911e-007	-3.89638e-007	-3.97255e-007	-3.42537e-007
382.68	AxialPosition	cm	0.5	1.0	1.5	2.0	2.5	3.0	3.5	4.0	4.5
	TB+	g/l	0.00238437	0.00984071	0.0287764	0.0634618	0.114374	0.177714	0.246603	0.313218	0.370989
	TB-	g/l	1.66163e-006	1.30925e-006	8.6768e-007	4.33109e-007	6.94944e-008	-1.90432e-007	-3.41148e-007	-3.96823e-007	-3.78997e-007
382.8	AxialPosition	cm	0.5	1.0	1.5	2.0	2.5	3.0	3.5	4.0	4.5
	TB+	g/l	0.00145762	0.00642619	0.0208218	0.0495604	0.0945068	0.153395	0.220444	0.288147	0.349431
	TB-	g/l	1.76134e-006	1.53523e-006	1.14385e-006	7.01306e-007	2.91737e-007	-3.30576e-008	-2.51258e-007	-3.65575e-007	-3.93644e-007

Figure 2.46 The resulting concentration profile table within the SMB column.

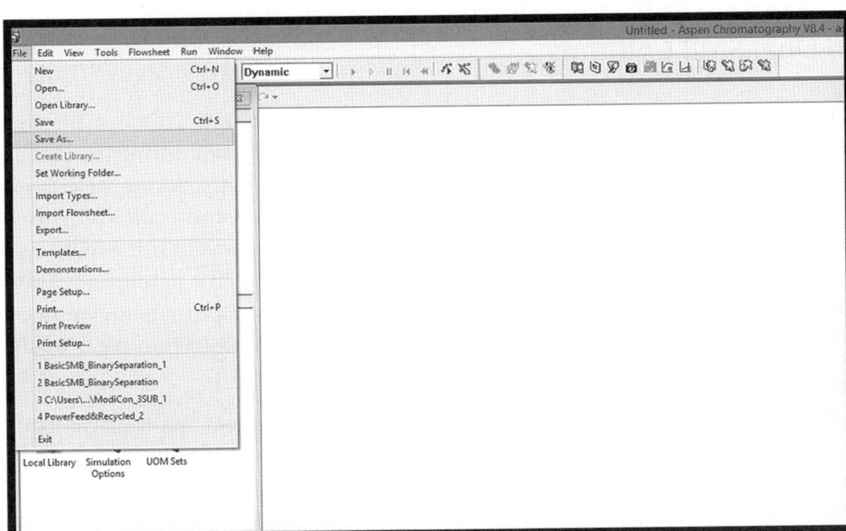

Figure 2.47 Save the results as a new file named **4-zone basic SMB model.cra.**

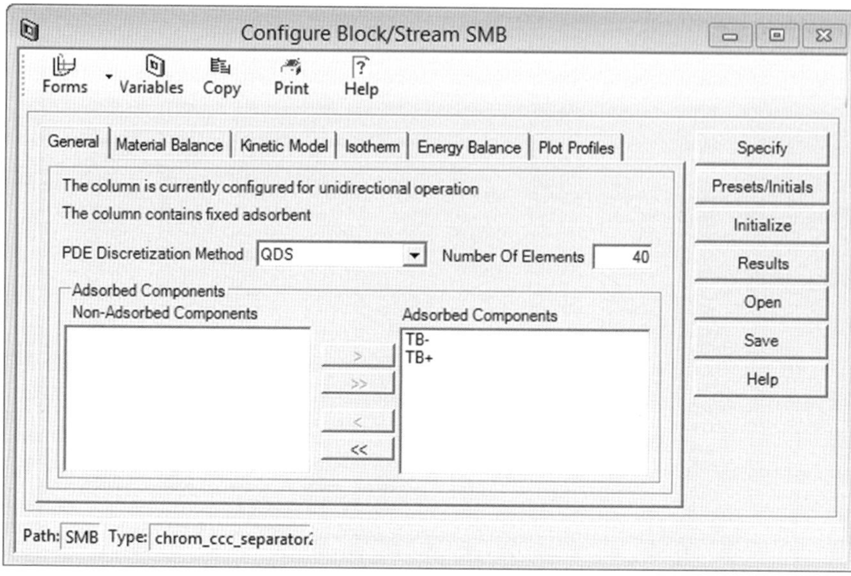

Figure 2.48 Partial differential equation discretization method in the basic SMB model.

2.5 Tandem SMB Workshop: Simulate a Separation with Dual SMB Columns

Insulin production is a major application of SMB technology. This is a ternary separation, where insulin needs to be separated from a salt impurity $ZnCl_2$ and a collection of proteins with very high molecular weights. As we discussed in Section 1.5, adsorbents frequently separate based on a molecule's effective diameter. Since the insulin molecule's size falls in between the small salt molecules and high-molecular-weight proteins (HMWPs), we cannot separate the small molecules from the large molecules in a single step. Instead, we need to first remove the HMWPs, and then use a second SMB unit to remove the salt.

Step 1: Add the Component Set.

Just as with the previous example, we use a component set without rigorous properties (Figure 2.49).

Step 2: Define the flowsheet.

Using the same feed/product blocks and Chrom _CCC _Separator from the first tutorial, set up the flowsheet as shown in Figure 2.50.

Notice that the F2 and E blocks are not connected to a feed or product block. This is because we want to connect the two columns to one another. To do this, place a chrom_r_pump model on the flowsheet, and connect the two streams (Figure 2.51).

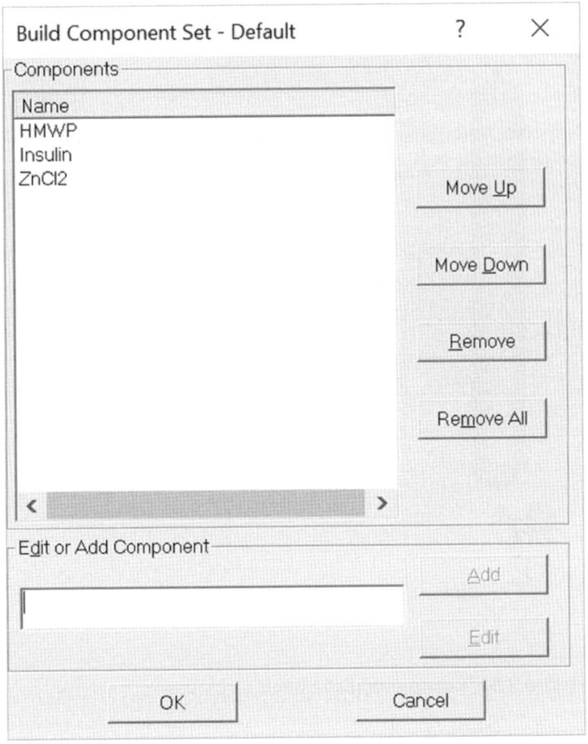

Figure 2.49 The three components of this separation.

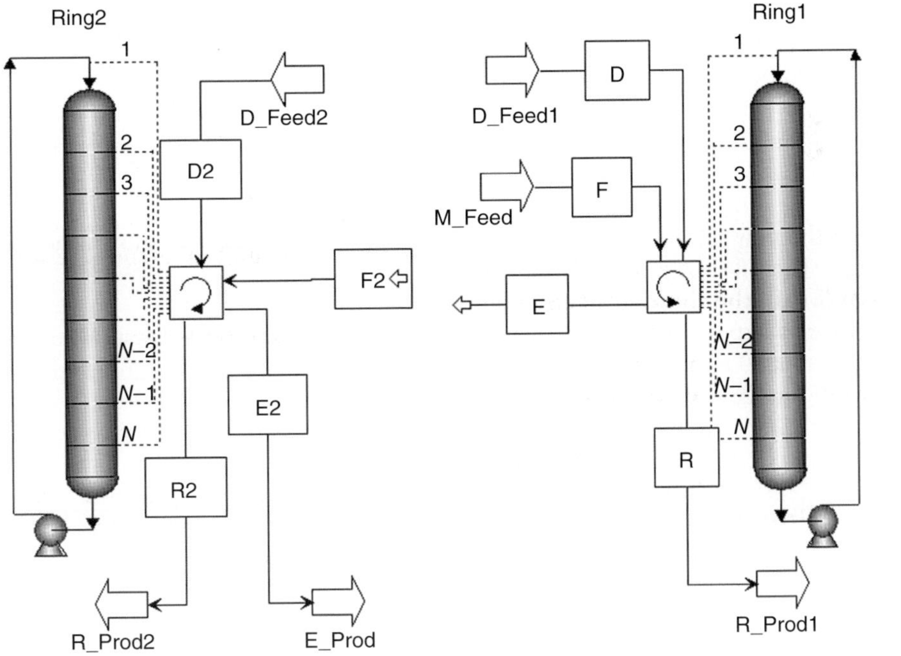

Figure 2.50 Initial flowsheet specification.

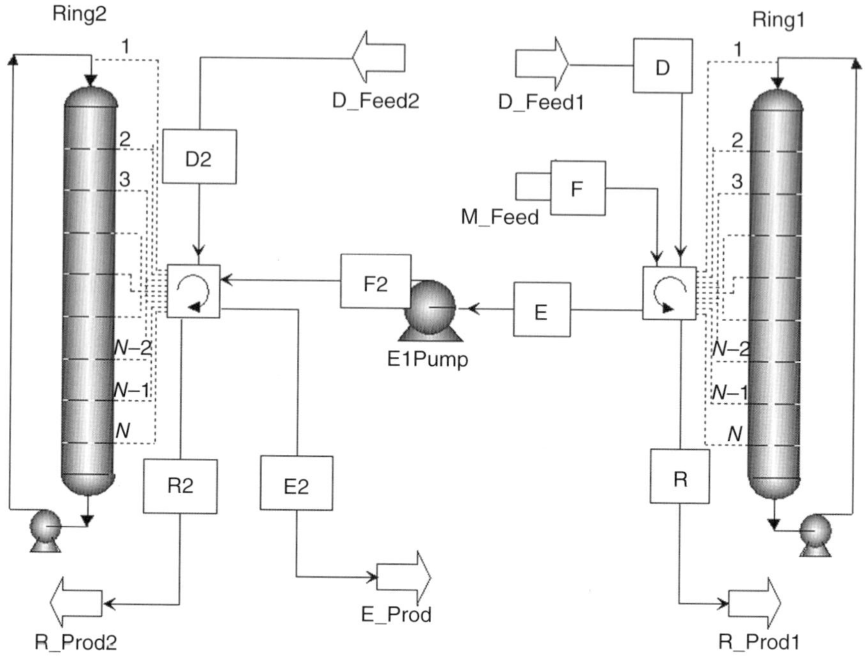

Figure 2.51 Completed flowsheet.

Step 3: Configure the Two SMB Units.
Now that the flowsheet is connected, let us specify the SMB configurations (Figures 2.52 and 2.53)

Step 4: Specify the column values.
Just as in the previous workshop, we open the "specify" form to define the physical dimensions of the system. Figures 2.54 and 2.55 show the values for ring 1 and ring 2, respectively.

Step 5: Specify the feed flow rates.
Set the flow rates according to Figure 2.56. Any feed blocks not shown should have their flow rate set to "free."

Step 6: Run the simulation and view results.
After specifying the flow rates, we run the simulation. Set the dynamic run to end at 6000 min. Once it has completed, create a plot to view the composition of R_Prod2, E_Prod and R_Prod1 (Figures 2.57–2.59).

2.6 Practice Problems

2.6.1 Run Workshop 2.4 as a Steady-State Simulation

Thus far, we have executed all of our Aspen runs as "dynamic" time-based simulations. To run the workshop as a "Steady-State" model, open the configure form for the SMB block, and change the "Adsorbent Assumption" to "Moving (TMB)."

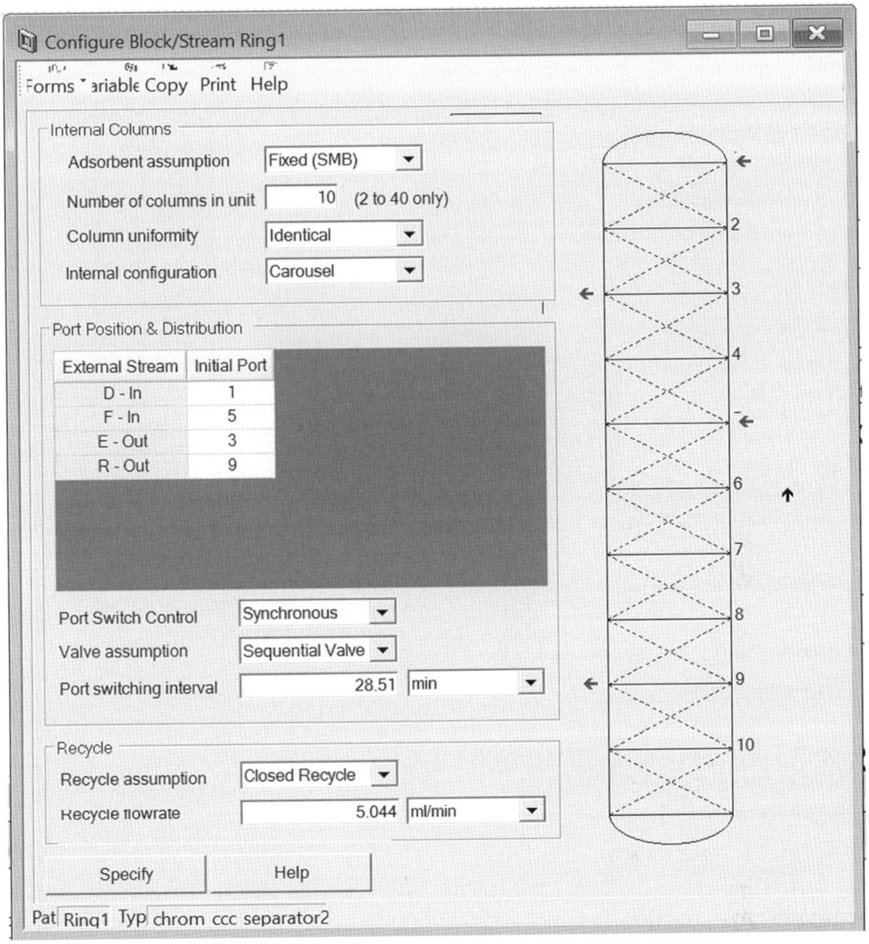

Figure 2.52 The Ring1 SMB unit.

This will automatically change the run mode to "Steady State"; you can verify this by looking at the drop-down box next to the "run simulation" button. **Before you run the simulation, be sure to save the file with a new name.**

Using the definitions of the m-values supplied in Eqs. (2.1) and (2.2), calculate the flow rates for the TMB case that will be theoretically equivalent to the SMB case, and enter them into the flow sheet. Remember that $Q_{\text{solid}} = \frac{\varepsilon_i V_{\text{bed}}}{t_{\text{sw}}}$ and $v_{\text{SMB}} = v_{\text{TMB}} + v_{\text{solid}}$ so we would expect $Q_{\text{recycle,TMB}} = Q_{\text{recycle,SMB}} - \frac{\varepsilon_i V_{\text{bed}}}{t_{\text{sw}}}$

Once you have run the simulation in Steady State mode, compare the axial concentration profiles between the steady state and dynamic simulations. What went wrong in the steady-state case (Figure 2.60)?

Use the "Help" button and navigate to "Recycle Control" (Figure 2.61). Notice that the configure form expects the SMB recycle flow rate, rather than the TMB

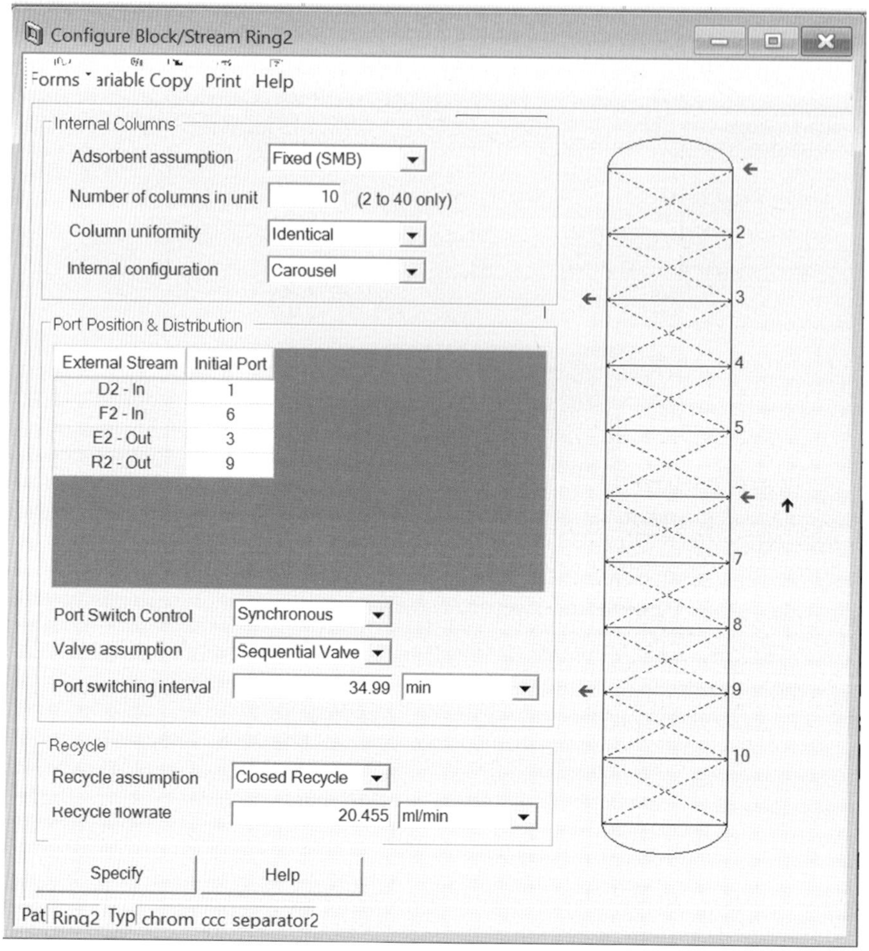

Figure 2.53 The Ring2 SMB unit.

Ring1.Specify Table

	Value	Units	Description
Hb	15.0	cm	Common length of packed section
Db	2.0	in	Internal diameter of packed section
Ei	0.35	m3 void/m3 bed	Common inter-particle/external voidage
Ep	0.89	m3 void/m3 bea	Common intra-particle/internal voidage
Rp	54.0	micron	Particle radius of adsorbent
SFac	1.0	n/a	Adsorbent shape factor
RHOl	996.0	kg/m3	Mass density of eluent
MUl	1.57	cP	Liquid (Solvent) viscosity
fex(*)			
fex("HMWP")	19.0	%	Size exclusion factor/accessible pore
fex("Insulin")	74.0	%	Size exclusion factor/accessible pore
fex("ZnCl2")	99.0	%	Size exclusion factor/accessible pore
Dm(*)			
Dm("HMWP")	4.8e-005	cm2/min	Liquid phase diffusion coefficient
Dm("Insulin")	5.49e-005	cm2/min	Liquid phase diffusion coefficient
Dm("ZnCl2")	3.96e-004	cm2/min	Liquid phase diffusion coefficient
Recycle_Pressure	12.0	bar	Recycle pump inlet pressure

Figure 2.54 Ring 1 column specification.

Ring2.Specify Table

	Value	Units	Description
Hb	15.0	cm	Common length of packed section
Db	3.464	in	Internal diameter of packed section
Ei	0.35	m3 void/m3 bed	Common inter-particle/external voidage
Ep	0.89	m3 void/m3 bea	Common intra-particle/internal voidage
Rp	54.0	micron	Particle radius of adsorbent
SFac	1.0	n/a	Adsorbent shape factor
RHOl	996.0	kg/m3	Mass density of eluent
MUl	1.57	cP	Liquid (Solvent) viscosity
fex(*)			
fex("HMWP")	19.0	%	Size exclusion factor/accessible pore
fex("Insulin")	74.0	%	Size exclusion factor/accessible pore
fex("ZnCl2")	99.0	%	Size exclusion factor/accessible pore
Dm(*)			
Dm("HMWP")	4.8e-005	cm2/min	Liquid phase diffusion coefficient
Dm("Insulin")	5.49e-005	cm2/min	Liquid phase diffusion coefficient
Dm("ZnCl2")	3.96e-004	cm2/min	Liquid phase diffusion coefficient
Recycle_Pressure	12.0	bar	Recycle pump inlet pressure

Figure 2.55 Ring 2 column specification.

flowrates Table

	Value	Spec
E1Pump.Flowrate	3.504	Fixed
E_Prod.Flowrate	3.692	Fixed
M_Feed.Flowrate	2.472	Fixed
R_Prod1.Flowrate	2.559	Fixed
R_Prod2.Flowrate	4.021	Fixed

Figure 2.56 The feed flow rates for the Tandem SMB unit.

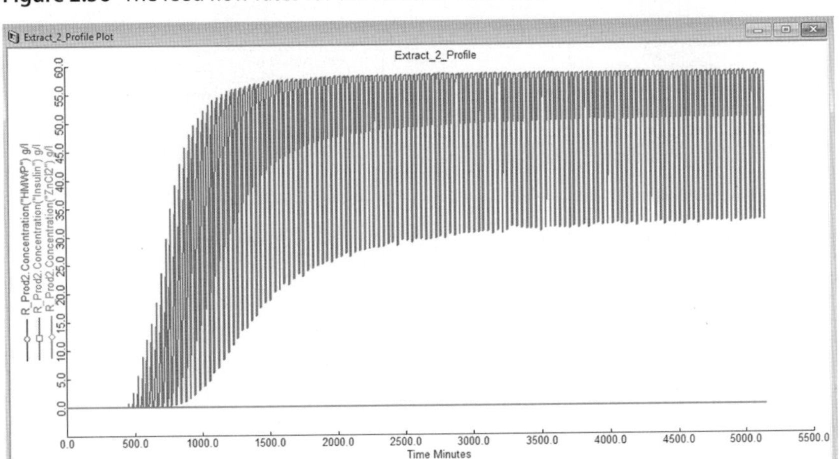

Figure 2.57 The Extract 2 profile.

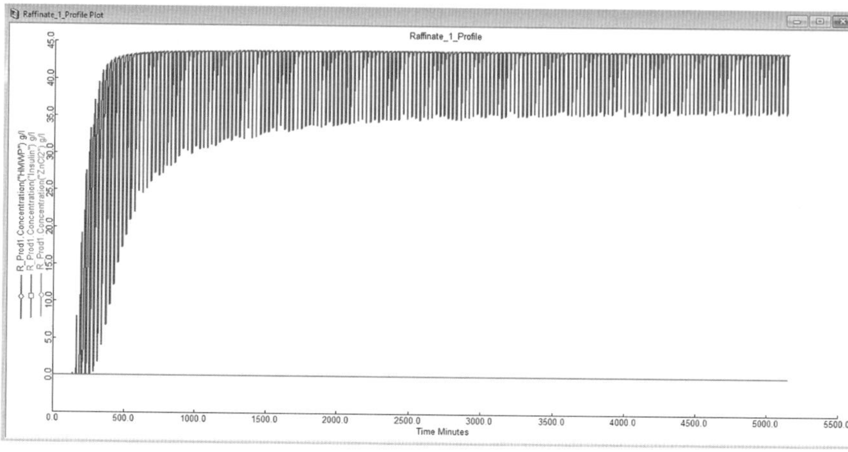

Figure 2.58 The Raffinate 1 profile.

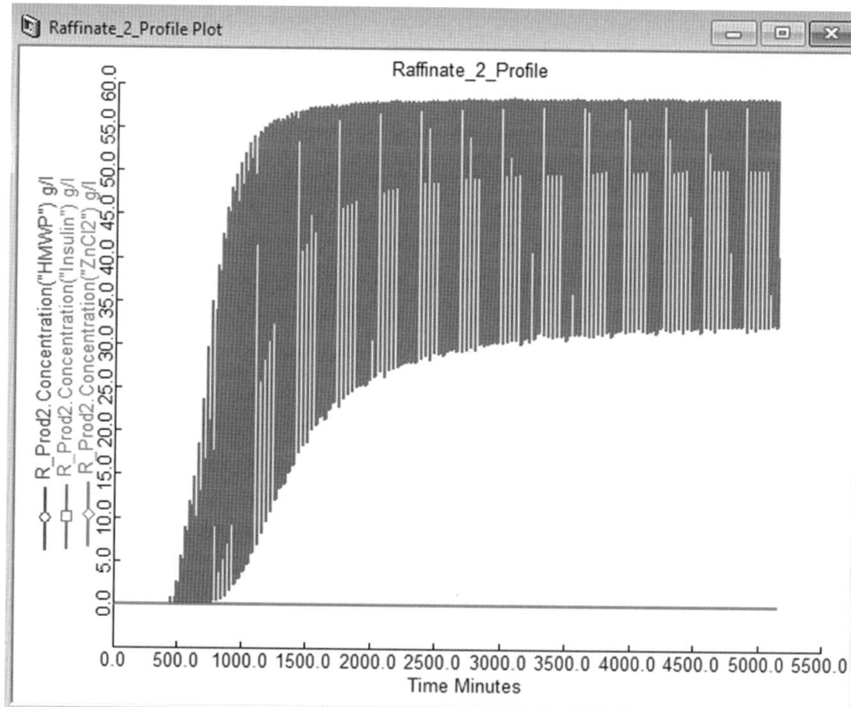

Figure 2.59 The insulin product (Raffinate 2) profile.

recycle flow rate. Update the unit with the correct recycle flow rate and rerun the simulation.

Once you enter the correct recycle flow rate, you should be able to recreate the concentration profile plot shown in Figure 2.62.

Save this file as **SSWorkshop2_4.cra.**

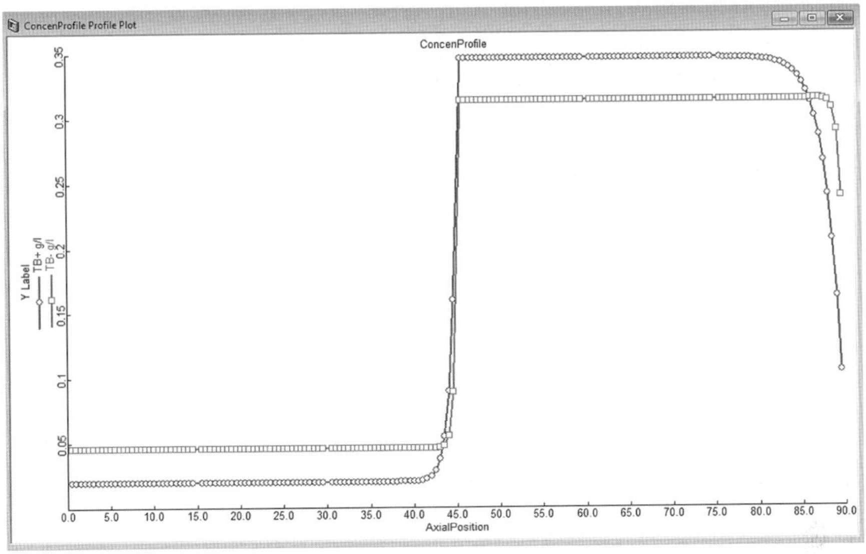

Figure 2.60 Invalid concentration profile.

Figure 2.61 The help button.

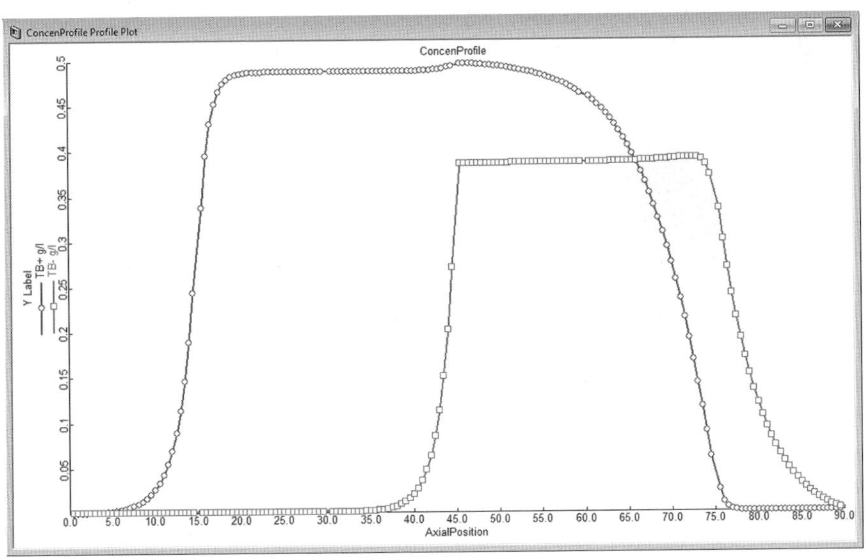

Figure 2.62 The steady state concentration profile.

2.6.2 Simulation of an Industrial-Scale Xylene Separation Using Literature Data

Professor Wankat has done a significant amount of work on optimizing xylene separation processes using Aspen Chromatography [39, 40]. Build your own 4-zone SMB unit for xylene separation using the same parameters.

1) Use the Peng-Robinson physical property model for the components.
   a) Ethyl-benzene (EB)
   b) M-xylene (MX)
   c) O-xylene (OX)
   d) P-xylene (PX)
   e) Para diethylbenzene (PDEB)
2) The Langmuir Isotherm is applicable.
3) OCFE 4 with four elements is the appropriate discretization method.
4) Dispersion is based on Peclet number, and the trace liquid assumption applies.
5) The mass transfer resistance is linear and based on the solid phase concentration.
6) The SMB unit is constructed as a twin stack with a rotary valve.
7) The data in Tables 2.2 and 2.3 apply.

Task 1: Build a Steady-State simulation and run it. Create two column profile plots, one of concentration, the other of liquid velocity. Compare them with Figures 2.63 and 2.64.

Save this file as **Xylene.cra**

As we can see in Figure 2.63, PX is successfully isolated from the rest of the components. Unlike the other chromatographic simulations we have made, this simulation explicitly models the "solvent" PDEB. This means that the extract product

Table 2.2 Column configuration and isotherm parameters for xylene separation.

Column configuration	6-9-6-3				
Column length, L (cm)	113.5				
Column diameter, Dia (cm)	411.7				
Bed volume per column, Vc (cm^3)	122.73				
Internal bed voidage (m^3 void/m^3 bed)	0.39				
External bed voidage (m^3 void/m^3 bed)	0.0				
Adsorbent particle diameter (cm)	0.046				
Stationary phase density (kg/m^3)	2253				
Peclet number (all components)	2000				
Amino acids	MX	OX	PX	PDEB	EB
Feed concentration (Mass %)	49.7	12.7	23.6	0	14
Langmuir isotherm parameter 1	0.1303	0.1303	0.1303	0.1077	0.1303
Langmuir isotherm parameter 2	0.2299	0.1884	1.0658	1.2935	0.3067
Mass transfer coefficient (cm/min)	0.031				

Source: Adapted from Wankat [41] and Venkatesan and Wankat [42].

Table 2.3 Operating conditions in the basic 5-zone SMB model for xylene separation.

Inlet and outlet flow rates (m^3/h)	Q_D	Free
	Q_E	99
	Q_F	87
	Q_R	161.4
Recycle flow rate (m^3/h)	$Q_{Recycle}$	323.4
Switching time (min)	t_{sw}	1.16

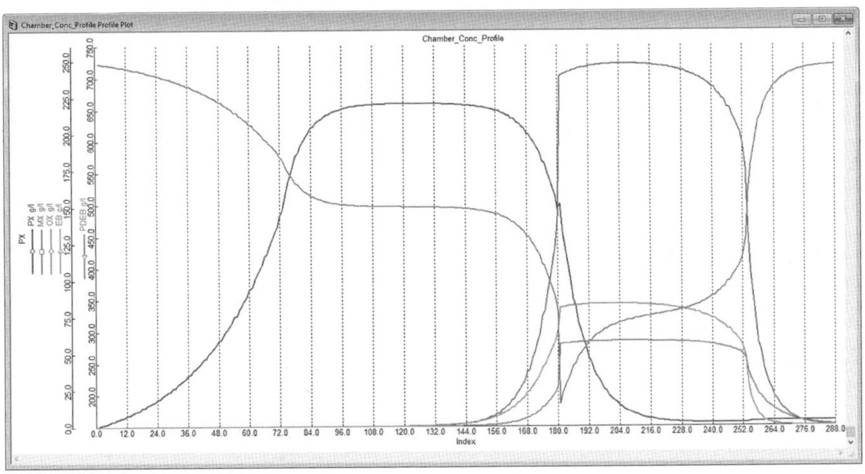

Figure 2.63 Concentration profile inside the SMB unit.

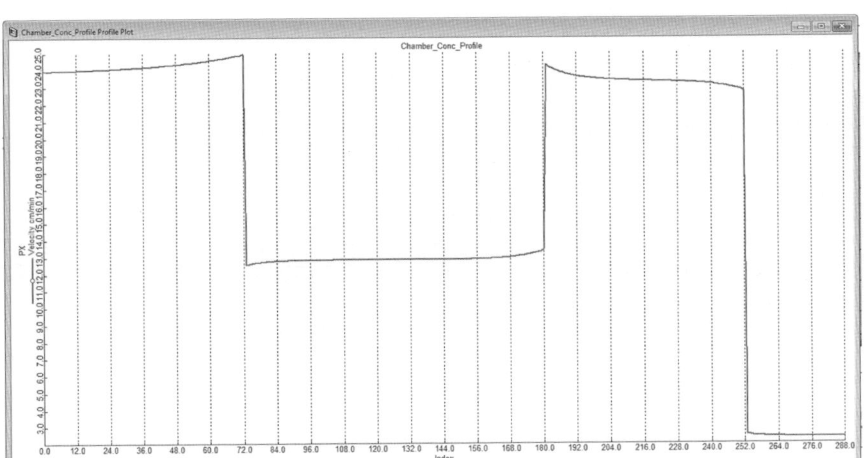

Figure 2.64 Liquid velocity profile.

has a density equal to the sum of the concentrations of PX and PDEB. In industrial practice, PX and PDEB will later be separated by distillation, while the other products will be recycled to an isomerization reactor [43].

2.6.3 Simulate a Five-Zone SMB System for Separating Phenylalanine, Tryptophan, and Methionine

A second method for performing a ternary separation is the use of extra zones within the SMB unit. The specific chemicals are phenylalanine, tryptophan, and methionine. Adding a second extract stream creates the additional zone by dividing zone2. Assume that the isotherm is linear, and that rigorous properties are not required because the trace liquid assumption is valid. Use your own judgment regarding the selection of partial differential equation (PDE) discretization method within the column. Table 2.4 lists the required physical property data for this simulation, while Table 2.5 provides the recommended operating conditions.

Save this file as **5ZonePractice.cra**

Run the simulation, and then create a column concentration profile for the SMB unit (Figure 2.65).

Reset the simulation, and then create some flowsheet constraints to define purity, recovery, and productivity. Specifically, you can add the following constraints:

```
PurityA as Fraction;
PurityA = Min(E1.Total_Component("Phenylalanine")/
Max(E1.Total_Mass,0.0000001),1);

PurityB as Fraction;
PurityB = Min(E2.Total_Component("Tryptophan")/
Max(E2.Total_Mass,0.0000001),1);

PurityC as Fraction;
PurityC = Min(R.Total_Component("Methionine")/
Max(R.Total_Mass,0.0000001),1);

RecoveryA as Fraction;
RecoveryA = Max(Min(1 - (E2.Total_Component
("Phenylalanine")+
R.Total_Component("Phenylalanine"))/
Max(F.Total_Component("Phenylalanine"),0.0000001),1), 0);

RecoveryB as Fraction;
RecoveryB = Max(Min(1 - (E1.Total_Component("Tryptophan")+
R.Total_Component("Tryptophan"))/
Max(F.Total_Component("Tryptophan"),0.0000001),1), 0);

RecoveryC as Fraction;
RecoveryC = Max(Min(1 - (E1.Total_Component("Methionine")+
E2.Total_Component("Methionine"))/
Max(F.Total_Component("Methionine"),0.0000001),1), 0);
```

Table 2.4 Column configuration and isotherm parameters for ternary separation.

Column configuration	1-2-1-2-2		
Column length, L (cm)	25		
Column diameter, Dia (cm)	2.5		
Bed volume per column, Vc (cm^3)	122.73		
Internal bed voidage (m^3 void/m^3 bed)	0.346		
External bed voidage (m^3 void/m^3 bed)	0.55		
Adsorbent particle diameter (cm)	0.0362		
Mobile phase density (g/cm^3)	0.996		
Mobile phase viscosity (cP)	0.89		
Amino acids	Methionine (B)	Phenylalanine (A)	Tryptophan (I)
Feed concentration (g/l)	0.5	0.5	0.5
Mass transfer coefficient (cm/min)	0.0518	0.0171	0.0157
Linear isotherm parameter (l/l)	0.3537	11.771	1.6015
Axial dispersion coefficient (cm^2/min)	Estimated		

Table 2.5 Operating conditions in the basic 5-zone SMB model for ternary separation.

Flow rate ratio	m_{I}	18.83
	m_{II}	2.56
	m_{III}	0.56
	m_{IV}	1.001
	m_{V}	0.22
Zone flow rates (cm^3/min)	Q_{I}	48.742
	Q_{II}	11.384
	Q_{III}	6.792
	Q_{IV}	7.804
	Q_{V}	6.011
Inlet and outlet flow rates (cm^3/min)	Q_{D}	42.731
	Q_{E1}	37.358
	Q_{E2}	4.592
	Q_{F}	1.013
	Q_{R}	1.793
Recycle flow rate (cm^3/min)	$Q_{Recycle}$	6.011
Recycle pressure (bar)	$P_{Recycle}$	15
Switching time (min)	t_{sw}	15.73

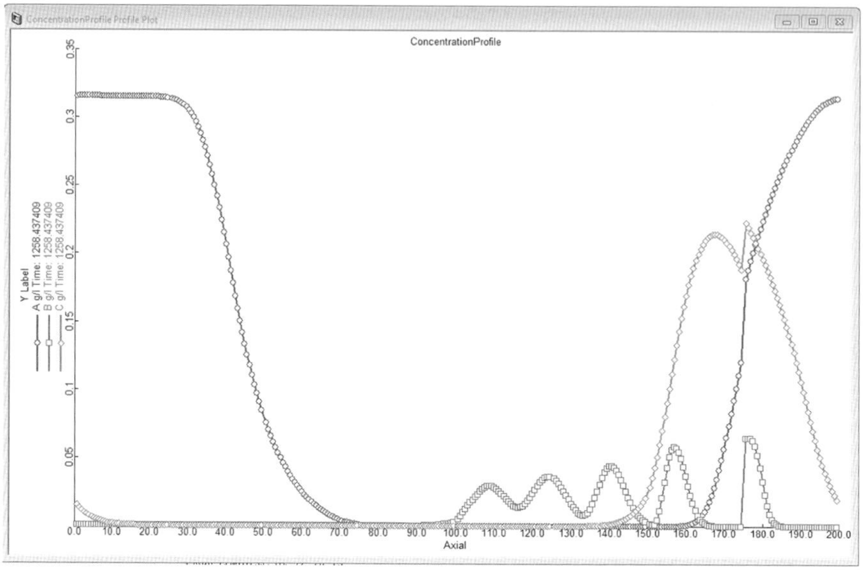

Figure 2.65 The column concentration profile from the practice problem.

Note the Max(Min()) construct. This is to ensure that we do not end up dividing by small negative numbers and calculating very large or very negative fractions. In Aspen, variables defined as Fraction must stay between −2 and 2 or the simulation will fail.

One final modification is to make a task that resets the Total_Component quantity each switching interval. This will allow us to get a "steady state" value for the purity and recoveries that discounts the losses during startup. Add the following task:

```
Task ResetTask runs always

  E1.Total_Component("Methionine"):0;
  E1.Total_Component("Tryptophan"):0;
  E1.Total_Component("Phenylalanine"):0;
  E1.Total_Mass:0;

  E2.Total_Component("Methionine"):0;
  E2.Total_Component("Tryptophan"):0;
  E2.Total_Component("Phenylalanine"):0;
  E2.Total_Mass:0;

  R.Total_Component("Methionine"):0;
  R.Total_Component("Tryptophan"):0;
  R.Total_Component("Phenylalanine"):0;
  R.Total_Mass:0;

  F.Total_Component("Methionine"):0;
  F.Total_Component("Tryptophan"):0;
```

```
  F.Total_Component("Phenylalanine"):0;
  F.Total_Mass:0;

  wait SMB.Switch_Time;
End
```

Once you have defined these (remember to compile the constraints, and activate the task,) rerun the simulation and create plots as shown in Figures 2.66 and 2.67 of purities and recoveries.

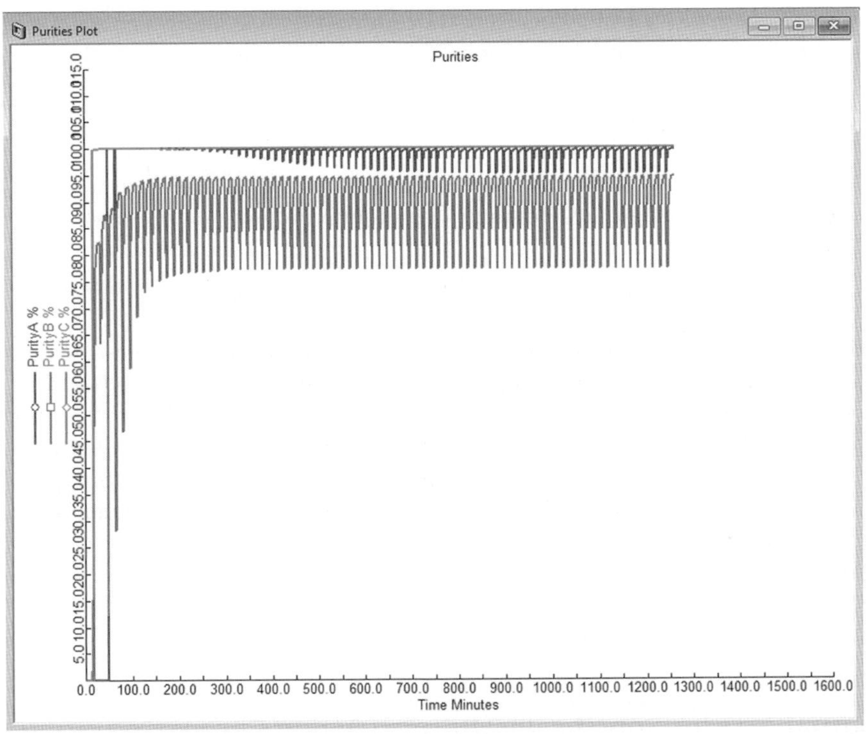

Figure 2.66 Product purities over time.

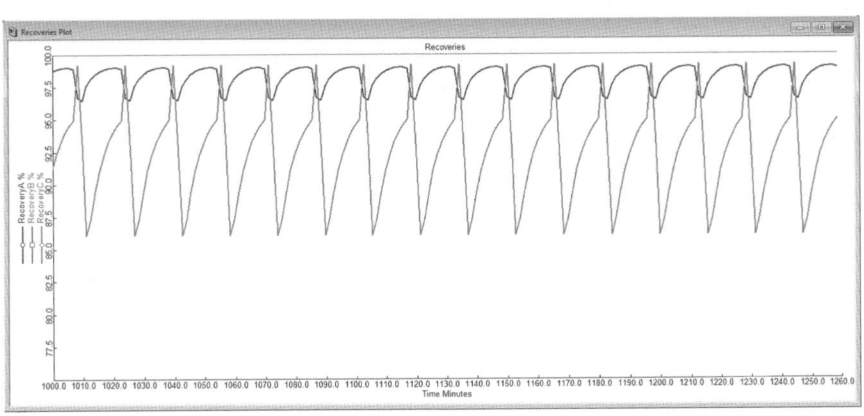

Figure 2.67 Product recoveries over time.

We can clearly see the relationship between the purity/recovery graphs and the column profile. The "B" component (tryptophan) has a lower purity with higher recovery while the C component (methionine) is the reverse.

Bibliography

Books

Belen'kii, B.G. and Vilenchik, L.Z. (1983). *Modern Liquid Chromatography of Macromolecules.* Amsterdam; New York: Elsevier.

Freitag, R. (2002). *Modern Advances in Chromatography.* Berlin; New York: Springer.

Gu, T. (1995). *Mathematical Modeling and Scale-Up of Liquid Chromatography.* Berlin ; New York: Springer.

Nicoud, R.-M. (2015). *Chromatographic Processes: Modeling, Simulation and Design.* Cambridge, UK: Cambridge University Press.

Rathore, A.S. and Velayudhan, A. (2003). *Scale-Up and Optimization in Preparative Chromatography Principles and Biopharmaceutical Applications.* New York: Marcel Dekker.

Rodrigues, A.E., Pereira, C., Minceva, M. et al. *Simulated Moving Bed Technology: Principles, Design and Process Applications.* Waltham, MA: Butterworth-Heinemann, Elsevier, Inc.

Schmidt-Traub, H., Schulte, M., and Seidel-Morgenstern, A. (2013). *Preparative Chromatography,* seconde. Weinheim: Wiley-VCH.

Subramanian, G. (1995). *Process Scale Liquid Chromatography.* Weinheim; New York: Wiley-VCH.

Turkov, J. (1993). *Bioaffinity Chromatography.* Burlington: Elsevier.

Wankat, P.C. (1986). *Large-Scale Adsorption and Chromatography,* vol. **1** and **2**. Boca Raton, FL: CRC Press.

Journal Articles Specifically Utilizing Aspen Chromatography

Abunasser, N. and Wankat, P.C. (2004). One-column chromatograph with recycle analogous to simulated moving bed adsorbers: analysis and applications. *Industrial and Engineering Chemistry Research* **43** (17): 5291–5299.

Abunasser, N., Wankat, P.C., Kim, Y.-S., and Koo, Y.M. (2003). One-column chromatograph with recycle analogous to a four-zone simulated moving bed. *Industrial and Engineering Chemistry Research* **42** (21): 5268–5279.

Altenhöner, U., Meurer, M., Strube, J., and Schmidt-Traub, H. (1997). Parameter estimation for the simulation of liquid chromatography. *Journal of Chromatography A* **769** (1): 59–69.

Hur, J.S. and Wankat, P.C. (2005). New design of simulated moving bed (SMB) for ternary separations. *Industrial and Engineering Chemistry Research* **44** (6): 1906–1913.

Hur, J.S. and Wankat, P.C. (2006a). Hybrid simulated moving bed and chromatography systems for center-cut separation from quaternary mixtures: linear isotherm systems. *Industrial and Engineering Chemistry Research* **45** (25): 8713–8722.

Hur, J.S. and Wankat, P.C. (2006b). Two-zone SMB/chromatography for center-cut separation from ternary mixtures: linear isotherm systems. *Industrial and Engineering Chemistry Research* **45** (4): 1426–1433.

Hur, J.S., Wankat, P.C., Kim, J. et al. (2007). Purification of L-phenylalanine from a ternary amino acid mixture using a two-zone SMB/chromatography hybrid system. *Separation Science and Technology* **42** (5): 911–930.

Jeon, Y.-J., Park, M.B., and Kim, I.-H. (2010). L-Ribose from L-arabinose by epimerization and its purification by 3-zone simulated moving bed chromatography. *Bioprocess and Biosystems Engineering* **33** (1): 87–95.

Jin, W. and Wankat, P.C. (2005). Two-zone SMB process for binary separation. *Industrial and Engineering Chemistry Research* **44** (5): 1565–1575.

Jin, W. and Wankat, P.C. (2006). Scaling rules and increasing feed rates in two-zone and four-zone simulated moving bed systems. *Industrial and Engineering Chemistry Research* **45** (8): 2793–2807.

Jin, W. and Wankat, P.C. (2007a). Hybrid simulated moving bed processes for the purification of *p* -xylene. *Separation Science and Technology* **42** (4): 669–700.

Jin, W. and Wankat, P.C. (2007b). Thermal operation of four-zone simulated moving beds. *Industrial and Engineering Chemistry Research* **46** (22): 7208–7220.

Kang, S.-H., Kim, J.-H., and Mun, S. (2010). Optimal design of a tandem simulated moving bed process for separation of paclitaxel, 13-dehydroxybaccatin III, and 10-deacetylpaclitaxel. *Process Biochemistry* **45** (9): 1468–1476.

Kim, J.K., Abunasser, N., and Wankat, P.C. (2005). Use of two feeds in simulated moving beds for binary separations. *Korean Journal of Chemical Engineering* **22** (4): 619–627.

Kim, J.K., Abunasser, N., Wankat, P.C. et al. (2005). Thermally assisted simulated moving bed systems. *Adsorption* **11**: 579–584.

Kim, J.K. and Wankat, P.C. (2004). Designs of simulated-moving-bed cascades for quaternary separations. *Industrial and Engineering Chemistry Research* **43** (4): 1071–1080.

Kim, J.K., Zang, Y., and Wankat, P.C. (2003). Single-cascade simulated moving bed systems for the separation of ternary mixtures. *Industrial and Engineering Chemistry Research* **42** (20): 4849–4860.

Kim, Y.S., Lee, C.H., Wankat, P.C., and Koo, Y.M. (2004). Comparing the performance of one-column process and four-zone simulated moving bed by computer simulation. *Biotechnology and Bioprocess Engineering* **9** (5): 362–368.

Kumar, V. and Prud'homme, R.K. (2009). Nanoparticle stability: processing pathways for solvent removal. *Chemical Engineering Science* **64** (6): 1358–1361.

Kurup, A.S., Hidajat, K., and Ray, A.K. (2006). Comparative Study of Modified Simulated Moving Bed Systems at Optimal Conditions for the Separation of Ternary Mixtures under Nonideal Conditions. *Industrial and Engineering Chemistry Research* **45** (11): 3902–3915.

Lee, A.-L. and Lim, Y.-I. (2009). Performance evaluation of the conservation element and solution element method in SMB process simulation. *Chemical Engineering and Processing: Process Intensification* **48** (4): 878–884.

Lee, J.W. and Wankat, P.C. (2008). Optimized design of recycle chromatography for separation of a single component from a ternary mixture. *Industrial and Engineering Chemistry Research* **47** (23): 9601–9610.

Lee, J.W. and Wankat, P.C. (2009). Comparison of recycle chromatography and simulated moving bed for pseudobinary separations. *Industrial and Engineering Chemistry Research* **48** (16): 7724–7732.

Lee, J.W. and Wankat, P.C. (2010). Design of pseudo-simulated moving bed process with multi-objective optimization for the separation of a ternary mixture: linear isotherms. *Journal of Chromatography A* **1217** (20): 3418–3426.

Lee, J.W. and Wankat, P.C. (2011). Thermal simulated moving bed concentrator. *Chemical Engineering Journal* **166** (2): 511–522.

Mai, N.L., Nguyen, N.T., Kim, J.-I. et al. (2012). Recovery of ionic liquid and sugars from hydrolyzed biomass using ion exclusion simulated moving bed chromatography. *Journal of Chromatography A* **1227**: 67–72.

Mun, S. (2010a). Enhanced separation performance of a five-zone simulated moving bed process by using partial collection strategy based on alternate opening and closing of a product port. *Industrial and Engineering Chemistry Research* **49** (19): 9258–9270.

Mun, S. (2010b). Partial port-closing strategy for obtaining high throughput or high purities in a four-zone simulated moving bed chromatography for binary separation. *Journal of Chromatography A* **1217** (42): 6522–6530.

Mun, S. (2011). Effect of a partial-feeding application on product purities and throughput of a five-zone simulated moving bed process for the separation of a ternary nucleoside mixture. *Process Biochemistry* **46** (4): 977–986.

Mun, S. (2012a). Development of a non-triangular separation region for improving the performance of a three-zone simulated moving bed chromatography for binary separation with linear isotherms. *Journal of Chromatography A* **1256**: 46–57.

Mun, S. (2012b). Strategy of rearranging the port locations in a three-zone simulated moving bed chromatography for binary separation with linear isotherms. *Journal of Chromatography A* **1230**: 100–109.

Nam, H.-G., Jo, S.-H., and Mun, S. (2011). Comparison of Amberchrom-CG161C and Dowex99 as the adsorbent of a four-zone simulated moving bed process for removal of acetic acid from biomass hydrolyzate. *Process Biochemistry* **46** (10): 2044–2053.

Nam, H.-G. and Mun, S. (2012). Optimal design and experimental validation of a three-zone simulated moving bed process based on the Amberchrom-CG161C adsorbent for continuous removal of acetic acid from biomass hydrolyzate. *Process Biochemistry* **47** (5): 725–734.

Song, S.-M., Park, M.-B., and Kim, I.H. (2012). Three-zone simulated moving-bed (SMB) for separation of cytosine and guanine. *Korean Journal of Chemical Engineering* **29** (7): 952–958.

Venkatesan, A. and Wankat, P.C. (2011). Simulation of ion exchange water softening pretreatment for reverse osmosis desalination of brackish water. *Desalination* **271** (1–3): 122–131.
Wankat, P.C. (2001). Simulated moving bed cascades for ternary separations. *Industrial and Engineering Chemistry Research* **40** (26): 6185–6193.
Yang, Y.J., Lee, C.H., and Koo, Y.M. (2004). Separation of amino acids by simulated moving bed using competitive Langmuir isotherm. *Biotechnology and Bioprocess Engineering* **9** (5): 331–338.
Yoon, T.H., Lee, E., Kim, J.M. et al. (2008). The enrichment of loxoprofen enantiomer by 6-columns simulated moving bed chromatography. *Korean Journal of Chemical Engineering* **25** (2): 285.
Zang, Y. and Wankat, P.C. (2002). SMB operation strategy–partial feed. *Industrial and Engineering Chemistry Research* **41** (10): 2504–2511.
Zang, Y. and Wankat, P.C. (2003). Variable flow rate operation for simulated moving bed separation systems: simulation and optimization. *Industrial and Engineering Chemistry Research* **42** (20): 4840–4848.

Journal Articles

Campo, M.C., Baptista, M.C., Ribeiro, A.M. et al. (2013). Gas phase SMB for propane/propylene separation using enhanced 13X zeolite beads. *Adsorption* **20** (1): 61–75.
Da Silva, E.A.B., De Souza, A.A.U., De Souza, S.G.U., and Rodrigues, A.E. (2005). Simulated moving bed technology in the reactive process of glucose isomerization. *Adsorption* **11** (1): 847.
Dünnebier, G., Fricke, J., and Klatt, K.-U. (2000). Optimal design and operation of simulated moving bed chromatographic reactors. *Industrial and Engineering Chemistry Research* **39** (7): 2290.
Fanali, S. (2009). Editorial on "Simulated moving bed chromatography for the separation of enantiomers" by A. Rajendran, G. Paredes and M. Mazzotti. *Journal of Chromatography A* **1216** (4): 708.
Fargues, C., Bailly, M., and Grevillot, G. (1998). Adsorption of BSA and hemoglobin on hydroxyapatite support: equilibria and multicomponent dynamic adsorption. *Adsorption* **4** (1): 5.
Faria, R.P.V. and Rodrigues, A.E. (2015). Industrial aspects of simulated moving bed chromatography. *Journal of Chromatography A* **1421**: 82–102.
Fricke, J. and Schmidt-Traub, H.A. (2003). New method supporting the design of simulated moving bed chromatographic reactors. *Chemical Engineering and Processing: Process Intensification* **42** (3): 237.
Graça, N.S., Pais, L.S., Silva, V.M.T.M., and Rodrigues, A.E. (2011). Analysis of the synthesis of 1,1-dibutoxyethane in a simulated moving-bed adsorptive reactor. *Chemical Engineering and Processing: Process Intensification* **50** (11–12): 1214.
Jarvelin, H. and Fair, J.R. (1993). Adsorptive separation of propylene-propane mixtures. *Industrial and Engineering Chemistry Research* **32** (10): 2201.

Jermann, S., Katsuo, S., and Mazzotti, M. (2012). Intermittent simulated moving bed processes for chromatographic three-fraction separation. *Organic Process Research and Development* **16** (2): 311.

Jungbauer, A. (2013). Continuous downstream processing of biopharmaceuticals. *Trends in Biotechnology* **31** (8): 479.

Katsuo, S. and Mazzotti, M. (2010). Intermittent simulated moving bed chromatography: 1. design criteria and cyclic steady-state. *Journal of Chromatography A* **1217** (8): 1354.

Kawase, M., Suzuki, T.B., Inoue, K. et al. (1996). Increased esterification conversion by application of the simulated moving-bed reactor. *Chemical Engineering Science* **51** (11): 2971.

Khalighi, M., Chen, Y.F., Farooq, S. et al. (2013). Propylene/propane separation using SiCHA. *Industrial and Engineering Chemistry Research* **52** (10): 3877–3892.

Kostroski, K.P. and Wankat, P.C. (2008). Separation of dilute binary gases by simulated-moving bed with pressure-swing assist: SMB/PSA processes. *Industrial and Engineering Chemistry Research* **47** (9): 3138.

Kostroski, K.P. and Wankat, P.C. (2009). Separation of concentrated binary gases by hybrid pressure-swing adsorption/simulated-moving bed processes. *Industrial and Engineering Chemistry Research* **48** (9): 4445.

Krishna, R. (2002). Reactive separations: more ways to skin a cat. *Chemical Engineering Science* **57** (9): 1491.

Kurup, A.S., Hidajat, K., and Ray, A.K. (2006). Optimal operation of a pseudo-SMB process for ternary separation under non-ideal conditions. *Separation and Purification Technology* **51** (3): 387.

Liang, M.-T. and Liang, R.-C. (2012). Fractionation of polyethylene glycol particles by simulated moving bed with size-exclusion chromatography. *Journal of Chromatography A* **1229**: 107.

Lode, F., Mazzotti, M., and Morbidelli, M.A. (2001). New reaction-separation unit: the simulated moving bed reactor. *Chimia International Journal for Chemistry* **55** (10): 883.

Martínez Cristancho, C.A., Peper, S., and Johannsen, M. (2012). Supercritical fluid simulated moving bed chromatography for the separation of ethyl linoleate and ethyl oleate. *The Journal of Supercritical Fluids* **66**: 129.

Mazzotti, M. (2006). Design of simulated moving bed separations: generalized Langmuir isotherm. *Industrial and Engineering Chemistry Research* **45** (18): 6311.

Minceva, M.; Rodrigues, A. E. UOP's Parex: Modeling, Simulation and Optimization, 2nd Mercosur Congress on Chemical Engineering.

Minceva, M., Silva, V.M.T., and Rodrigues, A.E. (2005). Analytical solution for reactive simulated moving bed in the presence of mass transfer resistance. *Industrial and Engineering Chemistry Research* **44** (14): 5246.

Pais, L.S., Loureiro, J., and Rodrigues, A.E. (1997). Separation of 1,1′-bi-2-naphthol enantiomers by continuous chromatography in simulated moving bed. *Chemical Engineering Science* **52** (2): 245–257.

Pais, L.S., Loureiro, J.M., and Rodrigues, A.E. (1998a). Modeling strategies for enantiomers separation by SMB chromatography. *AIChE Journal* **44** (3): 561.

Pais, L.S., Loureiro, J.M., and Rodrigues, A.E. (1998b). Separation of enantiomers of a chiral epoxide by simulated moving bed chromatography. *Journal of Chromatography A* **827** (2): 215.

Paredes, G., Abel, S., Mazzotti, M. et al. (2004). Analysis of a simulated moving bed operation for three-fraction separations (3F-SMB). *Industrial and Engineering Chemistry Research* **43** (19): 6157.

Pereira, C.S.M., Zabka, M., Silva, V.M.T.M., and Rodrigues, A.E.A. (2009). Novel process for the ethyl lactate synthesis in a simulated moving bed reactor (SMBR). *Chemical Engineering Science* **64** (14): 3301.

Seidel-Morgenstern, A., Keßler, L.C., and Kaspereit, M. (2008). New developments in simulated moving bed chromatography. *Chemical Engineering and Technology* **31** (6): 826.

Subramani, H.J., Hidajat, K., and Ray, A.K. (2003). Optimization of reactive SMB and varicol systems. *Computers and Chemical Engineering* **27** (12): 1883.

Thomas, W.J. and Crittenden, B.D. (1998). *Adsorption Technology and Design.* Oxford; Boston: Butterworth-Heinemann.

Wagner, N., Fuereder, M., Bosshart, A. et al. (2012). Practical aspects of integrated operation of biotransformation and SMB separation for fine chemical synthesis. *Organic Process Research and Development* **16** (2): 323.

Wooley, R., Ma, Z., and Wang, N.-H. (1998). A nine-zone simulating moving bed for the recovery of glucose and xylose from biomass hydrolyzate. *Industrial and Engineering Chemistry Research* **37** (9): 3699.

Xie, Y., Farrenburg, C.A., Chin, C.Y. et al. (2003). Design of SMB for a nonlinear amino acid system with mass-transfer effects. *AIChE Journal* **49** (11): 2850.

Xie, Y., Koo, Y.M., and Wang, N.H.L. (2001). Preparative chromatographic separation: simulated moving bed and modified chromatography methods. *Biotechnology and Bioprocess Engineering* **6** (6): 363.

Yamamoto, S., Nakanishi, K., and Matsuno, R. (1988). *Ion-Exchange Chromatography of Proteins*, Chromatographic Science. New York: Marcel Dekker.

Yu, W., Hidajat, K., and Ray, A.K. (2003). Optimal operation of reactive simulated moving bed and varicol systems. *Journal of Chemical Technology and Biotechnology* **78** (2–3): 287.

3

Shortcut Design of SMB Systems

In this chapter, we discuss the two major shortcut design methods for simulated moving bed (SMB) chromatography. In Section 3.1, we cover the general concepts and goals involved in designing any SMB system, as well as possible simplifications. Section 3.2 discusses the implications of some of the simplifications and how they lead to the so-called **triangle theory**. Before we cover the full theoretical derivation of triangle theory in Section 3.4, we apply the triangle theory to a simple problem in Section 3.3 so that we may get a feel of what the end result will look like. Section 3.5 covers the iterative **standing wave design**. Since we have already covered the triangle theory in great detail and many concepts of the standing wave design overlap, we jump right in and explain the full iterative process with reference to the concepts of triangle theory. After explaining both methods, we give the reader an opportunity to apply standing wave design to real problems and compare the results to triangle theory in Sections 3.6 and 3.7. Section 3.8 presents the conclusions, Section 3.9 gives the practice problems, and Section 3.10 presents the bibliography.

3.1 General Concepts

As discussed in Section 2.2 and illustrated in Figure 2.1, a perfect true moving bed (TMB) system would separate a feed stream containing two components dissolved in a solvent into two product streams, each containing one of the components in the same solvent. Essentially, we would want the situation depicted in Figure 3.1. The most adsorbed component travels in the direction of the solid stream, to the extract. The least adsorbed component travels in the direction of the liquid stream toward the raffinate.

In order to achieve this ideal scenario, each zone has a role to play. These roles are depicted in Figure 3.2.

Zone 1: Component 2 is desorbed from the solid adsorbent into the liquid phase. This ensures that all of component 2 is present in the extract by preventing the solid phase from carrying it through to zones 4 and 3.

Zone 2: Component 1 is desorbed from the solid adsorbent into the liquid phase. This ensures that there will not be any of this component in zone 1 and therefore the extract will be uncontaminated.

Design, Simulation, and Optimization of Adsorptive and Chromatographic Separations: A Hands-On Approach, First Edition. Kevin R. Wood, Y. A. Liu, and Yueying Yu.

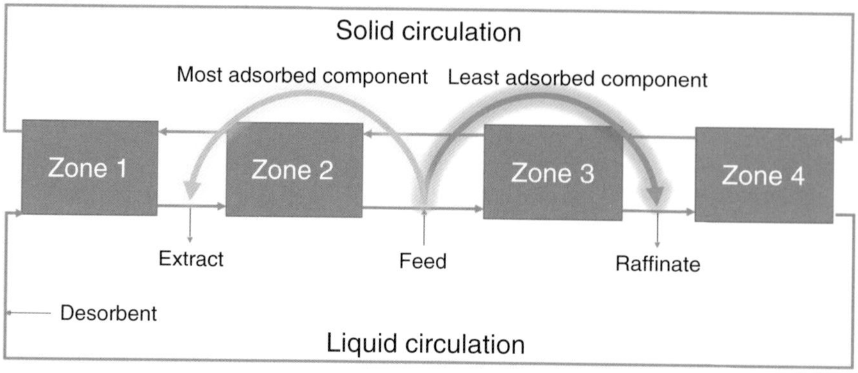

Figure 3.1 Ideal separation case.

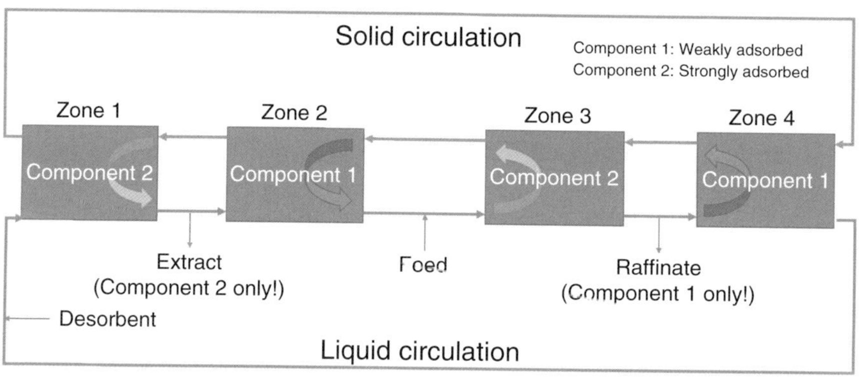

Figure 3.2 Zone roles in TMB separation.

Zone 3: Component 2 is adsorbed into the solid phase. This means the liquid will only contain component 1 so that the raffinate will remain uncontaminated.

Zone 4: The remainder of component 1 is adsorbed onto the solid. This ensures that component 1 will not be present in the liquid recycle, and will therefore not contaminate zone 1 and the extract.

Given those roles and our goals for the separation, we can assign variables as shown in Figure 3.3. Each stream is characterized by either a pair of concentrations (c_1, c_2) or solid loadings (w_1, w_2) and we know that some streams have zero concentrations.

In an ideal binary chromatographic system, the two components introduced in the feed stream would be perfectly separated into the extract and raffinate streams. Mass-transfer resistance would be absent, and the liquid phase would instantly reach equilibrium with the solid adsorbent. Under such conditions, we achieve the ideal situation illustrated in Figure 3.1. The net effect is that component 1 will travel through zone 3 in the liquid phase to be removed in the raffinate stream. Component 2, on the other hand, will move through zone 2 to be removed in the extract stream. In a real separation unit, the concentration profile that would achieve this outcome can look qualitatively as in Figure 3.4.

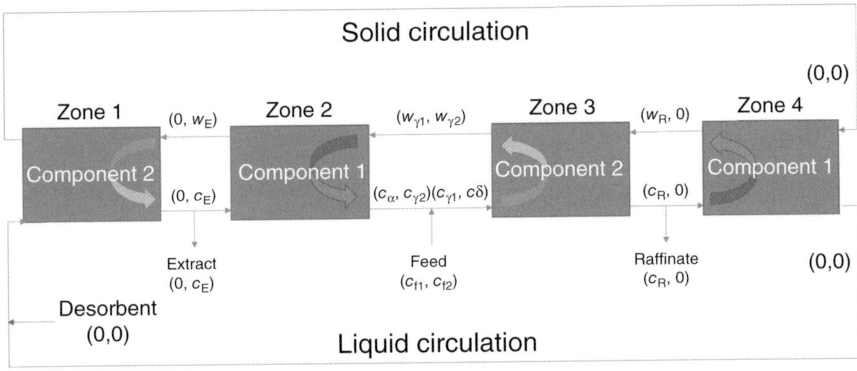

Figure 3.3 SMB unit under ideal separation conditions.

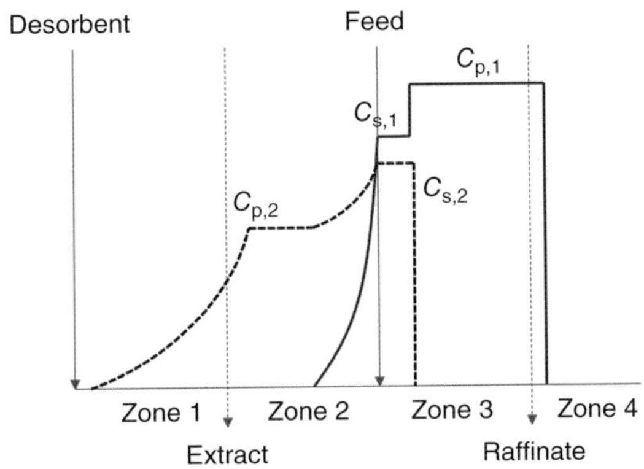

Figure 3.4 SMB concentration profile under ideal separation conditions.

Given the background we have discussed so far, the design problem now boils down to the question: "For a given system configuration (chemical and adsorbent choices, and column sizes), what set of flow rates allows us to achieve the ideal separation?" And, as discussed in Chapter 2, the flow rates are completely specified by the flow rate ratios or m-values, so that the question is equivalent to "Which m-values allow us to achieve the ideal separation?"

3.1.1 Mass Balances

Using the concepts and nomenclature developed in the previous sections, we can start discussing the mathematics of an adsorption bed. The obvious first step toward that end is to perform a mass balance on the system. Let us start by considering the case of a countercurrent bed, where the solid and liquid are both moving. A fixed bed will be a special case of this scenario, with the solid velocity equal to zero.

Figure 3.5 represents a column with solid particles moving downward and the liquid moving upward. In the figure, $\dot{M}_L$ is the mass flow rate of the liquid phase, while $\dot{M}_s$ is the mass flow rate of the solid phase. Keep in mind that "the solid phase" refers to both the solid particles and the liquid entrained in the particle

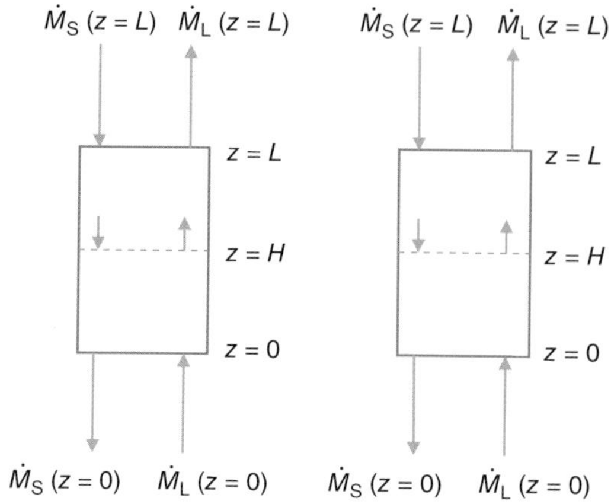

Figure 3.5 Diagram of a Countercurrent Adsorber.

pores. The overall steady-state mass balance is therefore

$$[\dot{M}_L(z = 0) + \dot{M}_s(z = L)] - [\dot{M}_L(z = L) + \dot{M}_s(z = 0)] = 0 \tag{3.1}$$

Unfortunately, without further assumptions, this mass balance is not very useful. We establish the following definitions:

1. Volumetric flow rates will be denoted $\dot{Q}_L$ for the liquid phase and $\dot{Q}_s$ for the solid phase.
2. Loading, which has the symbol q, is the mass of the liquid adsorbed per unit volume of the solid phase.
3. Concentration, which has the symbol C, is the mass of the liquid per unit volume in the liquid phase.
4. Linear velocities will be referred to as v for the liquid system and u_s for the solid system.

Assuming the following allows more useful statements about the system:

1. The volumetric flow rate of the solid phase is constant, and the actual solid part of the solid phase has a constant density, ρ_e.
2. The cross-sectional area of the column, A, is constant.
3. The fraction of the column's volume that is comprised of the upward-moving liquid is constant and equal to the external porosity, ε_i.
4. The superficial (or interstitial, or linear) velocity in the liquid phase is given by $v = \frac{\dot{Q}_L}{A\varepsilon_i}$; this can also be thought of as an assumption that the bulk porosity value holds true at differential length scales.

Thanks to definitions 1 and 3, we can immediately write

$$\dot{M}_L = C\dot{Q}_L \tag{3.2}$$

And definition 2 gives

$$\dot{M}_s = q\dot{Q}_s + \rho_s\dot{Q}_s = (q + \rho_e)\dot{Q}_s \tag{3.3}$$

Assumptions 2–4 allow us to substitute $vA\varepsilon_i$ for $\dot{Q}_L$ and $u_sA(1-\varepsilon_i)$ for $\dot{Q}_s$ to get

$$\dot{\boldsymbol{M}}_L = \boldsymbol{C}\dot{\boldsymbol{Q}}_L = \boldsymbol{C}\boldsymbol{v}\boldsymbol{A}\varepsilon_i \tag{3.4}$$

$$\dot{\boldsymbol{M}}_s = \boldsymbol{q}\dot{\boldsymbol{Q}}_s + \rho_s\dot{\boldsymbol{Q}}_s = (\boldsymbol{q} + \rho_e)\boldsymbol{u}_s\boldsymbol{A}(\mathbf{1} - \varepsilon_i) \tag{3.5}$$

Our overall mass balance is therefore

$$\begin{aligned}&[\boldsymbol{C}\boldsymbol{v}\boldsymbol{A}\varepsilon_i]_{z=0} + [(\boldsymbol{q} + \rho_e)\boldsymbol{u}_s\boldsymbol{A}(\mathbf{1} - \varepsilon_i)]_{z=L}\\ &\quad = [\boldsymbol{C}\boldsymbol{v}\boldsymbol{A}\varepsilon_i]_{z=L} + [(\boldsymbol{q} + \rho_e)\boldsymbol{u}_s\boldsymbol{A}(\mathbf{1} - \varepsilon_i)]_{z=0}\end{aligned} \tag{3.6}$$

We know that A, ε_i, and ρ_e are not functions of z, so their values are the same at $z=0$ and $z=L$ and we can treat them as constants. It is also generally assumed that the solid velocity u_s is constant.

$$\begin{aligned}&\boldsymbol{A}\varepsilon_i[\boldsymbol{C}\boldsymbol{v}]_{z=0} + \boldsymbol{u}_s\boldsymbol{A}(\mathbf{1} - \varepsilon_i)\rho_e + \boldsymbol{u}_s\boldsymbol{A}(\mathbf{1} - \varepsilon_i)[\boldsymbol{q}]_{z=L}\\ &\quad = \boldsymbol{A}\varepsilon_i[\boldsymbol{C}\boldsymbol{v}]_{z=L} + \boldsymbol{u}_s\boldsymbol{A}(\mathbf{1} - \varepsilon_i)\rho_e + \boldsymbol{u}_s\boldsymbol{A}(\mathbf{1} - \varepsilon_i)[\boldsymbol{q}]_{z=0}\end{aligned} \tag{3.7}$$

Simplifying by dividing throughout by $A\varepsilon_i$ and removing the $u_sA(1-\varepsilon_i)\rho_e$ term that appears on both sides gives

$$[\boldsymbol{C}\boldsymbol{v}]_{z=0} + \boldsymbol{u}_s\frac{(\mathbf{1} - \varepsilon_i)}{\varepsilon_i}[\boldsymbol{q}]_{z=L} = [\boldsymbol{C}\boldsymbol{v}]_{z=L} + \boldsymbol{u}_s\frac{(\mathbf{1} - \varepsilon_i)}{\varepsilon_i}[\boldsymbol{q}]_{z=0} \tag{3.8}$$

Before the final simplification, we can assume that the overall liquid concentration is constant. This is equivalent to assuming that the components being adsorbed are only present in small quantities. An immediate consequence is that the superficial velocity, v, is constant. Our final equation is therefore

$$\boldsymbol{v}(\boldsymbol{C}_L - \boldsymbol{C}_0) = \boldsymbol{u}_s\frac{(\mathbf{1} - \varepsilon_i)}{\varepsilon_i}(\boldsymbol{q}_L - \boldsymbol{q}_0) \tag{3.9}$$

3.1.2 Differential Equations

Now that we have established some background in equations used in adsorption, let us consider flow in packed beds, with adsorption. Starting with a differential volume element of the column depicted in Figure 3.5, we write the non-steady-state mass balance:

$$\text{Accumulation} = \text{In} - \text{Out}$$

Taking the net mass flow into the element as $[\dot{M}_L(z) + \dot{M}_s(z+\Delta z)]$, we find the total mass entering in a short period of time as $[\dot{M}_L(z) + \dot{M}_s(z+\Delta z)]\Delta t$. Likewise, the total mass leaving the differential element in that time period would be $[\dot{M}_L(z+\Delta z) + \dot{M}_s(z)]\Delta t$ so that we can write

$$M_{t+\Delta t} - M_t = [\dot{M}_L(z) + \dot{M}_s(z+\Delta z)]\Delta t - [\dot{M}_L(z+\Delta z) + \dot{M}_s(z)]\Delta t \tag{3.10}$$

We can now substitute the relationships we found in the mass balance section to find that

$$M_{t+\Delta t} - M_t = A\ u_s(1-\varepsilon_i)(q_{z+\Delta z} - q_z)\Delta t - A\varepsilon_i v(C_{z+\Delta z} - C_z)\Delta t \tag{3.11}$$

The mass of our differential volume section is $M = C\varepsilon_i A\Delta z + q(1-\varepsilon_i)A\Delta z$ so that the change over time is

$$M_{t+\Delta t} - M_t = (C_{t+\Delta t} - C_t)\varepsilon_i A\Delta z + (q_{t+\Delta t} - q_t)(1-\varepsilon_i)A\Delta z \tag{3.12}$$

Equations (3.10) and (3.11) have equal left-hand sides; we can therefore write

$$\begin{aligned}(C_{t+\Delta t} - C_t)\varepsilon_i A\Delta z &+ (q_{t+\Delta t} - q_t)(1-\varepsilon_i)A\Delta z \\ &= A\, u_s(1-\varepsilon_i)q_z\Delta t - A\, v\,(C_{z+\Delta z} - C_z)\Delta t\end{aligned} \tag{3.13}$$

Dividing both sides of Eq. (3.13) by $\varepsilon_i \Delta t \Delta z$, we end up with

$$\begin{aligned}\frac{(C_{t+\Delta t} - C_t)}{\Delta t} + \frac{(1-\varepsilon_i)}{\varepsilon_i}\frac{(q_{t+\Delta t} - q_t)}{\Delta t} &= u_s\frac{(1-\varepsilon_i)}{\varepsilon_i}\frac{(q_z - q_{z-\Delta z})}{\Delta z} \\ &\quad - \frac{v}{\varepsilon_i}\frac{(C_{z+\Delta z} - C_z)}{\Delta z}\end{aligned} \tag{3.14}$$

Taking the limit as Δz and Δt approach 0, we write Eq. (3.14) as a partial differential equation (PDE):

$$\frac{\partial C}{\partial t} + \frac{1-\varepsilon_i}{\varepsilon_i}\frac{\partial q}{\partial t} = u_s\frac{1-\varepsilon_i}{\varepsilon_i}\frac{\partial q}{\partial z} - v\frac{\partial C}{\partial z} \tag{3.15}$$

This PDE does not include an axial dispersion term. It represents the limit of TMB operation as the axial dispersion approaches 0. A derivation of the full differential equations is not provided here, but if we were to include mass transfer and dispersive effects, we would actually be left with a pair of differential-algebraic equations:

$$\frac{\partial C}{\partial t} = E_i\frac{\partial^2 C}{\partial z^2} - v\frac{\partial C}{\partial x} - k\frac{(1-\varepsilon_i)}{\varepsilon_i}(q - q^{\text{eq}}) \tag{3.16}$$

$$\epsilon_p\frac{\partial C^*}{\partial t} + (1-\epsilon_p)\frac{\partial q}{\partial t} = k\frac{(1-\varepsilon_i)}{\varepsilon_i}(C - C^*) + v\epsilon_p\frac{\partial C^*}{\partial x} + (1-\epsilon_p)v\frac{\partial q}{\partial x} \tag{3.17}$$

3.1.3 The Method of Characteristics

The method of characteristics is a well-known approach for solving first-order PDEs, that is, equations of the form (3.18).

$$A(x,t)\partial_t c + B(x,t)\partial_z c = C(x,t) \tag{3.18}$$

Specifically, the method requires us to find parametric curves, $(x(s), t(s))$, with a "characteristic" parameter s such that

$$\frac{\mathrm{d}t}{\mathrm{d}s} = A(x(s), t(s)) \tag{3.19}$$

$$\frac{\mathrm{d}x}{\mathrm{d}s} = B(x(s), t(s)) \tag{3.20}$$

$$\frac{\mathrm{d}c}{\mathrm{d}s} = C(x(s), t(s)) \tag{3.21}$$

Now we will notice that we can, in principle, integrate Eqs. (3.19)–(3.21) to find three distinct equations:

$$t(s) = \int A(x(s), t(s))\mathrm{d}s \tag{3.22}$$

$$x(s) = \int B(x(s), t(s))\mathrm{d}s \tag{3.23}$$

$$c(s) = \int C(x(s), t(s))\mathrm{d}s \tag{3.24}$$

We can subsequently combine Eqs. (3.22) through (3.24) with their initial conditions and completely eliminate s. In other words, the initial conditions let us uniquely determine a function such that $g(x, t) = s$. Consequently, because we know that $c(x, t) = c(s)$ and $s = g(x, t)$ we will obtain a unique solution of the form

$$c(g(x, t)) \tag{3.25}$$

Now, this is a simple and well-established process. For certain separation scenarios (i.e. noncompetitive adsorption isotherms), this analysis is sufficient to describe the separation behavior. However, in the case of competitive isotherms, this approach must be extended. This extension requires more advanced mathematical treatment, and we present a sketch of the more advanced process through application to the Langmuir isotherm in Section 3.4.

3.2 Triangle Theory

3.2.1 Notations

In this section, we use the following notations (Table 3.1).

3.2.2 Introduction

The triangle theory starts off by making the following simplifying assumptions:

- The system is a true moving bed.
- Mass transfer resistance and dispersion are negligible.
- Complete separation is achieved, as illustrated in Figure 3.3.
- The system is at steady state.

By carefully examining Figure 3.3, we can see that the "net effect" of the SMB column is to carry one component to the extract stream and the other component to the raffinate. In order to quantify this net effect, we can define a new term, $f_{i,j}^{\text{net}}$, which will be the net flow rate of component i in zone j. How can it be calculated? Let us again consider the mass balance discussed in Section 2.2.

Unfortunately, defining "the net flow through a zone" is easier to express in everyday language than in mathematics (Figure 3.6). How we quantify it crucially

Table 3.1 Notations used in the development of triangle theory.

Quantity	Symbol
Concentration of component i in state j	c_i^j
Bed length	L
Flow rate ratio	$m = \dfrac{1-\mu\epsilon_p}{\mu(1-\epsilon_p)}$
Solid phase loading of component i	n_i
Derivative of solid phase loading	$n_{i,j} = \dfrac{\partial n_i}{\partial c_j}$
Solid velocity	V_s
Liquid velocity	V
Phase ratio	$\nu^* = \dfrac{1-\epsilon_b-(1-\epsilon_b)\epsilon_p}{\epsilon_b+(1-\epsilon_b)\epsilon_p}$
Dimensionless position	$x = z/L$
States achieved during SMB operation between zones 2 and 3	α, β, γ
External or interparticle porosity	ε_b
Internal or intraparticle porosity	ε_p
Ratio of solid to fluid velocity	$\mu = \dfrac{(1-\epsilon_b)V_s}{\epsilon_b V}$
Bed void fraction	$\psi = \epsilon_b + (1-\epsilon_b)\epsilon_p$
Eigenvalues of the isotherm matrix	θ_1, θ_2
Eigenvalues of the F matrix	ϕ_1, ϕ_2
Eigenvalues of the G matrix	γ_1, γ_2
Eigenvalues of the H matrix	η_1, η_2

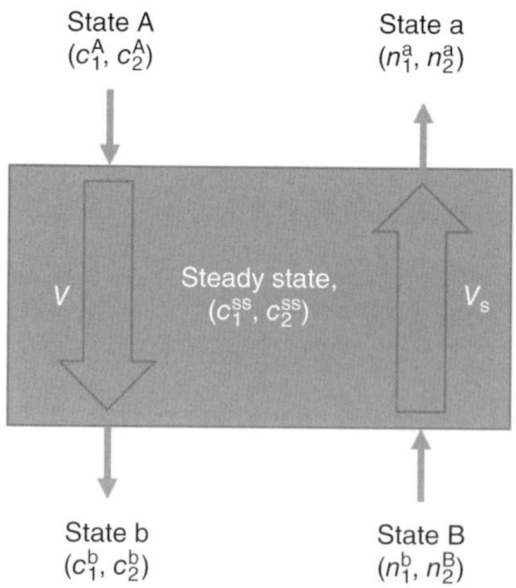

Figure 3.6 Net flow through a zone.

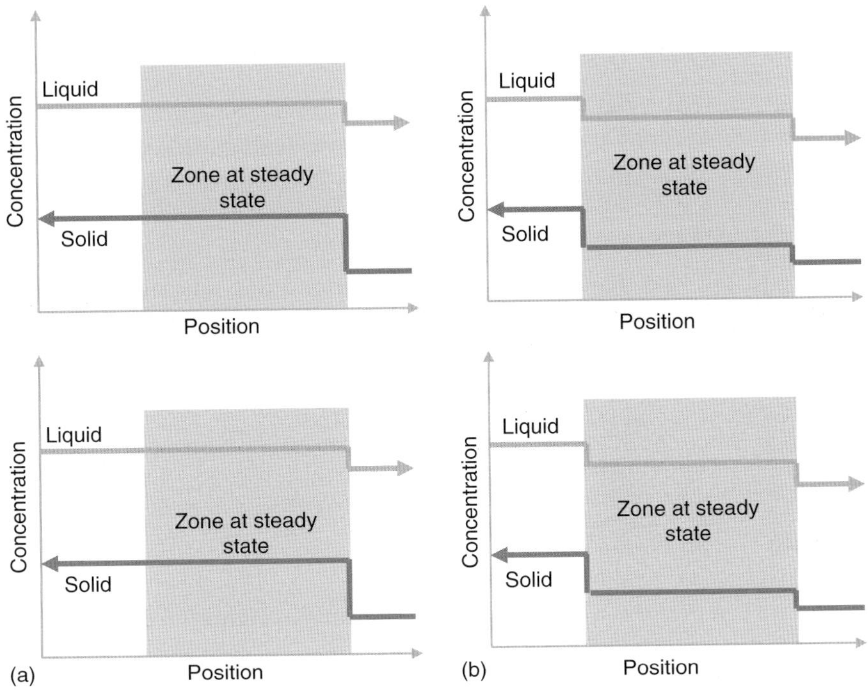

Figure 3.7 Examples of steady-state conditions.

depends on what happens at steady state inside the zone. For example, we could imagine that the flow rate of the components in the liquid phase is very large compared to the solid phase's capacity. In such a scenario, the solid phase would immediately become fully loaded, and the concentration inside the zone would be equal to the liquid inlet concentration. This scenario is sketched graphically in Figure 3.7a. On the other hand, if the liquid and solid capacities are balanced, then the zone may take on an intermediate state at steady state. This is drawn in Figure 3.7b.

Which state actually occurs depends on the various flow rates, concentrations, and isotherm. However, once we have that relationship, we can use the following definition for "net flow through the zone"

$$F_{i,j}^{\text{net}} = (Q_j - \varepsilon_p Q_s)c_i^{\text{SS}} - Q_s(1 - \varepsilon_p)n_i^{\text{SS}} \tag{3.26}$$

It is not too difficult to see how this equation expresses the concept of net flow. The first term is the net flow rate of each component in the liquid phase. Note that we subtract $\varepsilon_p Q_s$, which represents the flow rate of the *voids* in the solid stream. We can think of this as discounting the liquid that is entrained in the solid. The second term represents the movement of the components that have been adsorbed onto the solid. Note that the $(1 - \varepsilon_p)$ term converts the overall solid volume *including* voids to the solid volume *without* voids. It is also possible to write Eq. (3.26) in the following form:

$$F_{i,j}^{\text{net}} = \left(\frac{(Q_j - \varepsilon_p Q_s)}{Q_s(1-\varepsilon_p)} c_i^{\text{SS}} - n_i^{\text{SS}} \right) Q_s(1-\varepsilon_p) \tag{3.27}$$

$$F_{i,j}^{\text{net}} = (m_j c_i^{\text{SS}} - n_i^{\text{SS}}) Q_s(1-\varepsilon_p) \tag{3.28}$$

$$f_{i,j}^{\text{net}} = \frac{F_{i,j}^{\text{net}}}{Q_s(1-\varepsilon_p)} = m_j c_i^{\text{SS}} - n_i^{\text{SS}} \tag{3.29}$$

In Eq. (3.28), we have defined a very important quantity in SMB chromatography – specifically, **the flow rate ratio, m_j**, which represents the ratio of the net liquid flow rate to the net solid flow rate, as defined previously in Eq. (2.1):

$$\text{TMB} \qquad m_j = \frac{(Q_j - \varepsilon_p Q_s)}{Q_s(1-\varepsilon_p)}$$

$$\text{SMB} \qquad m_j = \frac{(Q_j t_{\text{sw}} - \varepsilon^* V_{\text{bed}})}{V_{\text{bed}}(1-\varepsilon^*)}$$

Finally, we can perform a mass balance on the zone in Figure 3.6 as follows:

$$\left(m + \frac{\varepsilon_p}{1-\varepsilon_p} \right) c_i^b = f_{i,j}^{\text{net}} + n_i^B + \frac{\varepsilon_p}{1-\varepsilon_p} c_i^B \tag{3.30}$$

$$\left(m + \frac{\varepsilon_p}{1-\varepsilon_p} \right) c_i^A = f_{i,j}^{\text{net}} + n_i^a + \frac{\varepsilon_p}{1-\varepsilon_p} c_i^a \tag{3.31}$$

Now that we have sketched out a few equations governing our system, we can begin the process of determining exactly which steady-state conditions prevail in each zone. This process is more easily said than done, because finding the answer requires delving into the PDEs. Before we do that, however, we will make some remarks.

It will turn out that the steady state that is achieved inside the bed is a function of the two inlet streams (liquid and solid) as well as the m value, that is, $f_{i,j}^{\text{net}} = f(c_1^A, c_2^A, n_1^B, n_2^B, m)$. Therefore, if we are able to calculate the net flow rate given values for each of those variables, we will also be able to calculate the values of the outlets at steady state.

3.2.3 Constraints on the System

In section 3.2.2, we discussed the concept of "net flow rate" through a bed. Going back to Figures 3.1 and 3.2, we can express the ideal separation in terms of net flow rate. Specifically, the only net flow rates that we want to be nonzero are the flow rates of component 2 through the second zone and the flow rate of component 1 in the third zone. Because we want 100% of component 1 to flow through zone 3 and into the raffinate, the net flow through zone 3 should be the same as the net flow of component 1 in the feed. Similarly, for component 2, we want the net flow through zone 2 to be equal to the feed of component 2, except that it is negative since the flow is moving in the direction of the solid phase.

$$F_1^{\mathrm{I}} = 0,\ F_2^{\mathrm{I}} = 0 \tag{3.32}$$

$$F_1^{\mathrm{II}} = 0,\ F_2^{\mathrm{II}} = -(Q_{\mathrm{feed}})c_2^f \tag{3.33}$$

$$F_1^{\mathrm{III}} = (Q_{\mathrm{feed}})c_1^f,\ F_2^{\mathrm{III}} = 0 \tag{3.34}$$

$$F_1^{\mathrm{IV}} = 0,\ F_2^{\mathrm{IV}} = 0 \tag{3.35}$$

If these are satisfied, we will have everything that we need for a perfect separation. We also notice that it is possible to express the feed flow rate as

$$\frac{Q_{\mathrm{feed}}}{Q_s(1-\varepsilon_p)} = m_3 - m_2 \tag{3.36}$$

which allows us to simply say

$$f_2^{\mathrm{II}} = -(m_3 - m_2)c_2^f, f_1^{\mathrm{III}} = (m_3 - m_2)c_1^f \tag{3.37}$$

If we look at the definition of net flow rate in Eq. (3.29), then we can immediately say something about the steady-state conditions in zones 2 and 3:

$$(m_3 - m_2)c_2^f = -(m_2\, c_2^{\mathrm{II}} - n_2^{\mathrm{II}}) \tag{3.38}$$

$$(m_3 - m_2)c_1^f = m_3\, c_1^{\mathrm{III}} - n_1^{\mathrm{III}} \tag{3.39}$$

where c^{II} and c^{III} are the steady-state conditions in zones 2 and 3, respectively.

Applying the mass balances described in Eqs. (3.30) and (3.31) to each zone gives

Zone 1:

$$\left(m_1 + \frac{\varepsilon_p}{1-\varepsilon_p}\right) c_2^{\mathrm{E}} = n_2^{\mathrm{ES}} + \frac{\varepsilon_p}{1-\varepsilon_p} c_2^{\mathrm{ES}} \tag{3.40}$$

Zone 2:

$$\left(m_2 + \frac{\varepsilon_p}{1-\varepsilon_p}\right) c_1^{\alpha} = n_1^{\gamma} + \frac{\varepsilon_p}{1-\varepsilon_p} c_1^{\gamma} \tag{3.41}$$

$$\left(m_2 + \frac{\varepsilon_p}{1-\varepsilon_p}\right) c_2^{\alpha} = -(m_2\, c_2^{\mathrm{II}} - n_2^{\mathrm{II}}) + n_2^{\gamma} + \frac{\varepsilon_p}{1-\varepsilon_p} c_2^{\gamma} \tag{3.42}$$

$$\left(m_2 + \frac{\varepsilon_p}{1-\varepsilon_p}\right) c_2^{E} = -(m_2\, c_2^{\mathrm{II}} - n_2^{\mathrm{II}}) + n_2^{\mathrm{ES}} + \frac{\varepsilon_p}{1-\varepsilon_p} c_2^{\mathrm{ES}} \tag{3.43}$$

Zone 3:

$$\left(m_3 + \frac{\varepsilon_p}{1-\varepsilon_p}\right) c_1^{R} = m_3\, c_1^{\mathrm{III}} - n_1^{\mathrm{III}} + n_1^{\mathrm{RS}} + \frac{\varepsilon_p}{1-\varepsilon_p} c_1^{\mathrm{RS}} \tag{3.44}$$

$$\left(m_3 + \frac{\varepsilon_p}{1-\varepsilon_p}\right) c_1^{\beta} = m_3\, c_1^{\mathrm{III}} - n_1^{\mathrm{III}} + n_1^{\gamma} + \frac{\varepsilon_p}{1-\varepsilon_p} c_1^{\gamma} \tag{3.45}$$

$$\left(m_3 + \frac{\varepsilon_p}{1-\varepsilon_p}\right) c_2^{\beta} = n_2^{\gamma} + \frac{\varepsilon_p}{1-\varepsilon_p} c_2^{\gamma} \tag{3.46}$$

Zone 4:

$$\left(m_4 + \frac{\varepsilon_p}{1-\varepsilon_p}\right) c_1^R = n_1^{RS} + \frac{\varepsilon_p}{1-\varepsilon_p} c_1^{RS} \tag{3.47}$$

We also know from the overall mass balances (or rearranging the previous equations) that

$$(m_3 - m_2)c_1^f = (m_3 - m_4)c_1^R \tag{3.48}$$

$$(m_3 - m_2)c_2^f = (m_3 - m_2)c_2^E \tag{3.49}$$

Unfortunately, this is about as far as we can go without knowing anything about the steady-state concentrations in zones 2 and 3. To calculate what those are, we must delve into the differential equations that govern those two zones.

3.3 Triangle Theory Workshop: Design of a System for the Separation of Amino Acids

Before we get any deeper into the detailed mathematical theory, let us jump ahead and assume we already know how to resolve all of these constraints, and are now interested in how to use triangle theory to design an SMB unit. Specifically, we will look at a binary separation of tryptophan and phenylalanine.

Step 1: Create a new Aspen Chromatography Flowsheet

Convert the component list to a set, and add phenylalanine and tryptophan as components (Figure 3.8). Save the file as **TriangleTheoryWorkshop.cra**.

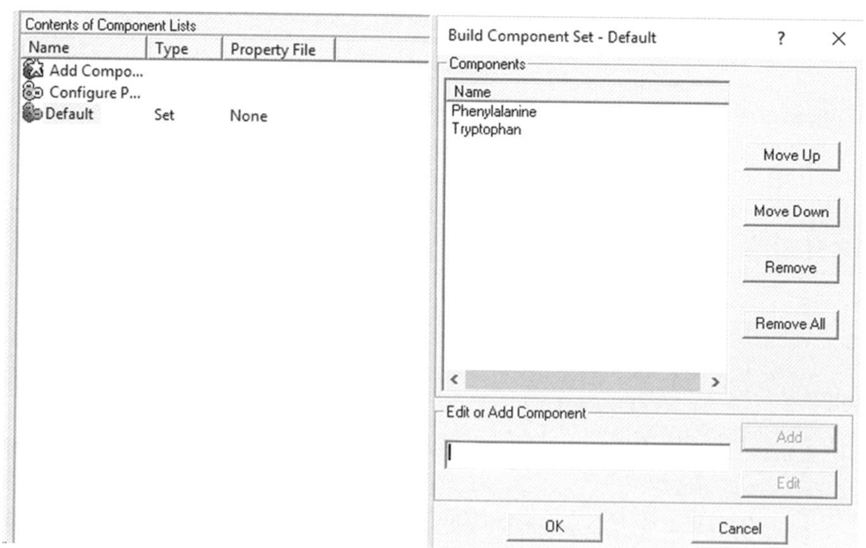

Figure 3.8 The workshop component set.

Step 2: Add a chrom_ccc_separator block, and connect the standard four inlets/outlets

Create the flowsheet as shown in Figure 3.9.

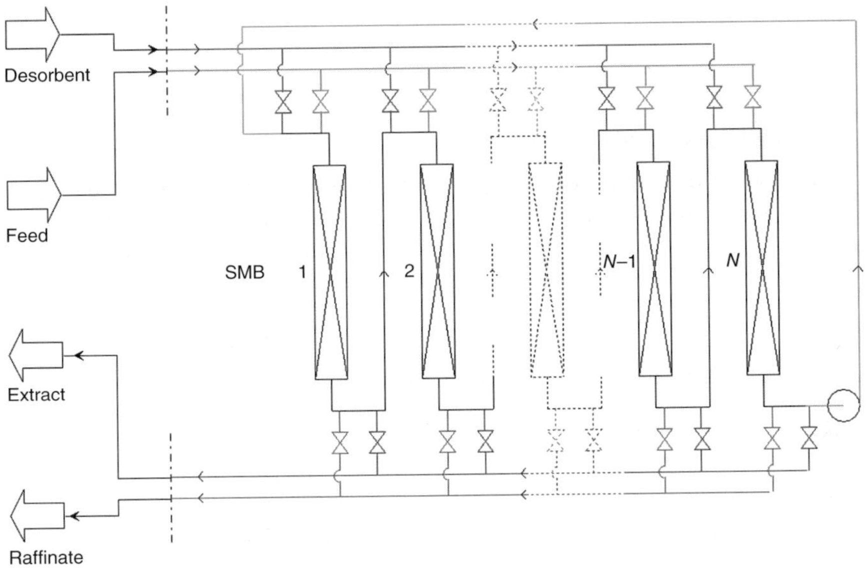

Figure 3.9 The workshop flowsheet connectivity.

Step 3: Set the column conditions

Configure the column to have 10 beds, as shown in Figure 3.10; then set the physical properties shown in Figure 3.11. As you can see from the available options, the dispersion and mass transfer coefficients should be constant, the isotherm should be "Extended Langmuir" based on bed volume, and the column isothermal.

Step 4: Open the Triangle Theory tool in Aspen Chromatography

To find Aspen's built-in triangle theory tool, look under "Forms" in the separator block (Figure 3.12). Once you have opened the form, fill in the information shown in Figure 3.13. Ensure that you actually set the feed concentration to match the numbers entered in the triangle theory form.

Step 5: Load the column configuration automatically

Fortunately, we do not need to manually fill in all the information as we did in the first tab. The "Column configuration" tab has a nice button that will automatically read the data we have already assigned to the column (Figure 3.14).

Also, take note of which isotherms are available in the triangle theory tool.

Only a small subset of the isotherms built into the separators can be used with Aspen's triangle theory form (Figure 3.15).

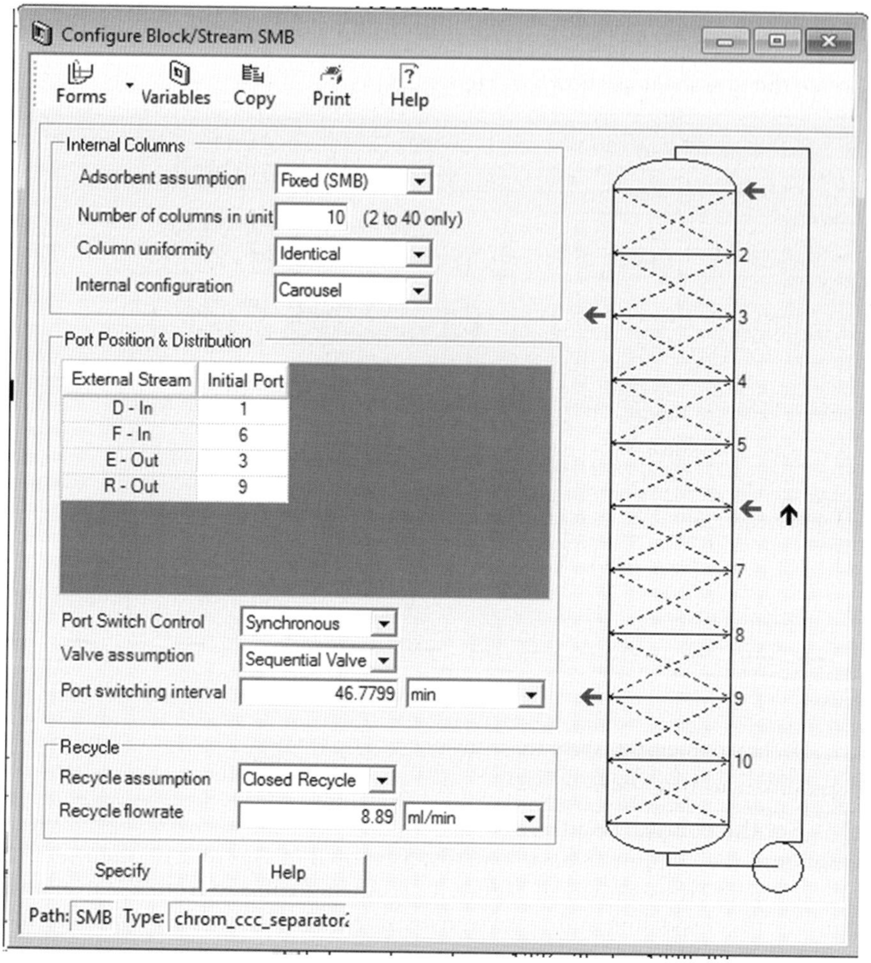

Figure 3.10 The column internal layout.

Step 6: View the separation plane

Open the separation tab to view the triangle that gives the triangle theory its name (Figure 3.16). You can move the red + around to select different m2,m3 coordinates. To reset the conditions to the optimum, click on the "update" button.

Step 7: Use the operating condition tab to choose the optimum operating conditions for a specific flow rate

The final tab allows you to specify a single operating condition (e.g. feed flow rate) and calculate the optimum operating conditions for the rest of the operating values. Set the feed flow rate to 20 ml min^{-1} and click "Calculate operating point." Once you are satisfied with the numbers, click on "Commit to flowsheet." See Figure 3.17.

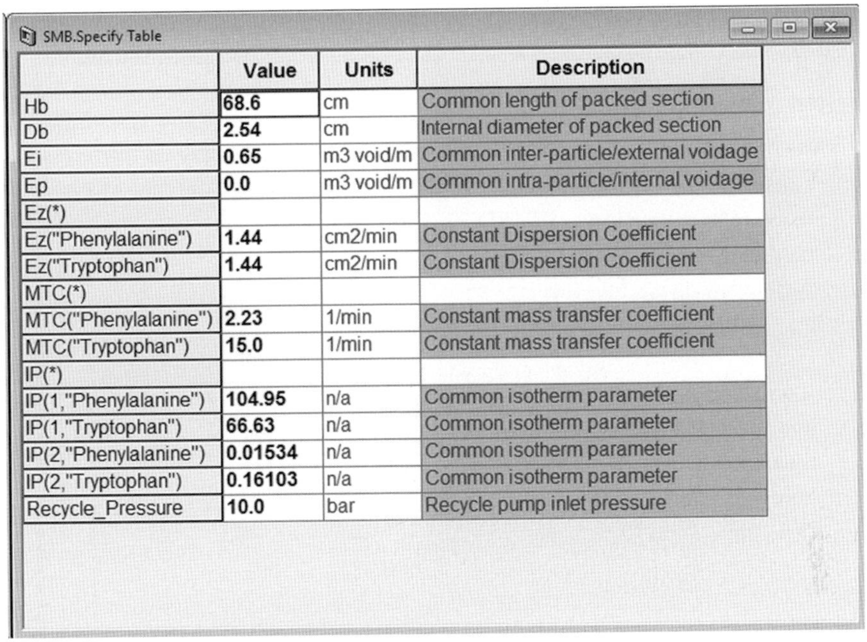

SMB.Specify Table

	Value	Units	Description
Hb	68.6	cm	Common length of packed section
Db	2.54	cm	Internal diameter of packed section
Ei	0.65	m3 void/m	Common inter-particle/external voidage
Ep	0.0	m3 void/m	Common intra-particle/internal voidage
Ez(*)			
Ez("Phenylalanine")	1.44	cm2/min	Constant Dispersion Coefficient
Ez("Tryptophan")	1.44	cm2/min	Constant Dispersion Coefficient
MTC(*)			
MTC("Phenylalanine")	2.23	1/min	Constant mass transfer coefficient
MTC("Tryptophan")	15.0	1/min	Constant mass transfer coefficient
IP(*)			
IP(1,"Phenylalanine")	104.95	n/a	Common isotherm parameter
IP(1,"Tryptophan")	66.63	n/a	Common isotherm parameter
IP(2,"Phenylalanine")	0.01534	n/a	Common isotherm parameter
IP(2,"Tryptophan")	0.16103	n/a	Common isotherm parameter
Recycle_Pressure	10.0	bar	Recycle pump inlet pressure

Figure 3.11 The column internal physical properties.

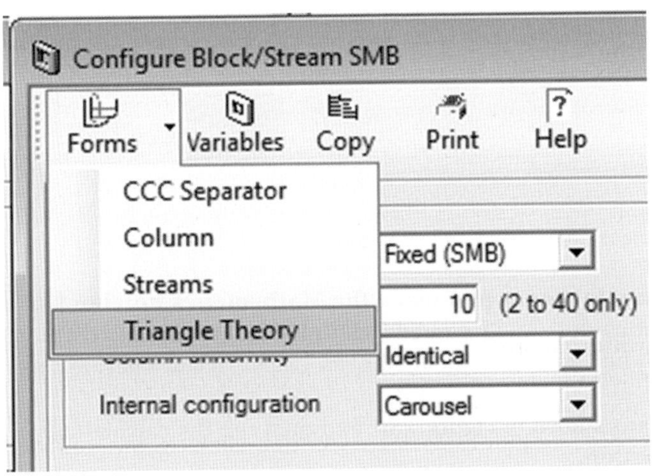

Figure 3.12 Opening the triangle theory form.

Step 8: Run the simulation

Set the run to pause at 1400 min, and run the simulation. Once it completes, create plots of the extract and raffinate concentrations. See Figures 3.18 and 3.19.

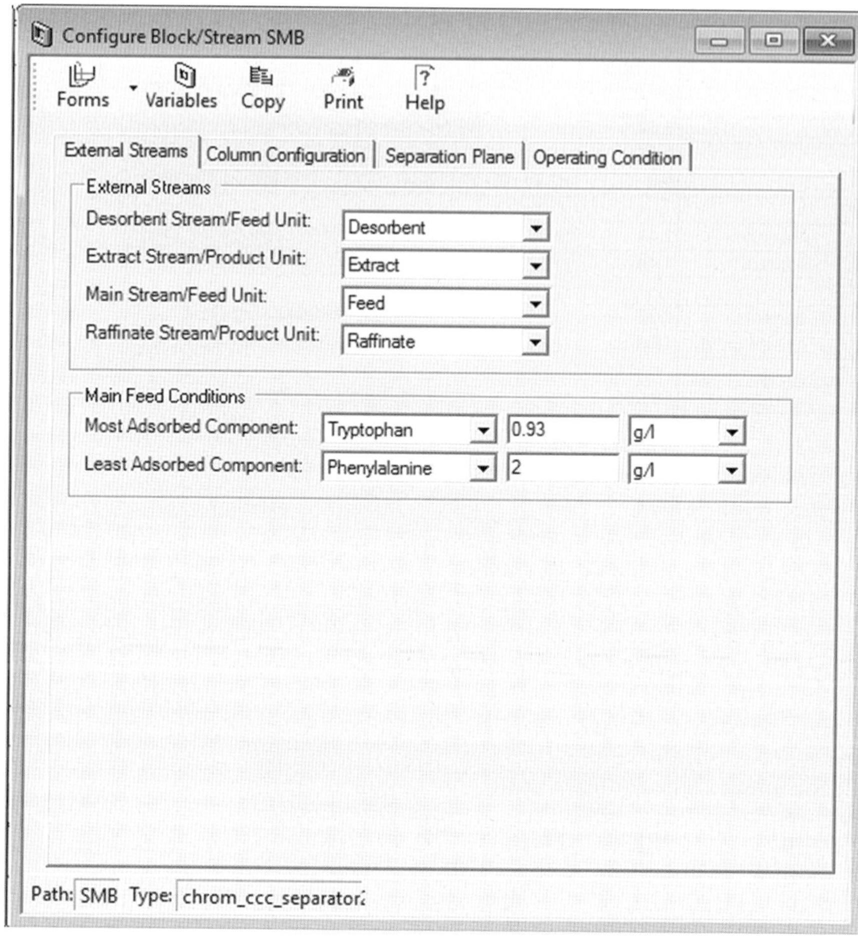

Figure 3.13 The triangle theory form.

Step 9: Create the triangle theory plot using the Mathematica notebooks.
Open the Mathematica document named **Triangle Theory Clean.cdf**. Note that CDF player can be legally downloaded from https://www.wolfram.com/cde-player/.

Apply the settings as shown in Figure 3.20, and then run the calculations by clicking the "Run" button.

Upon completing the calculations, you should see a triangular region very similar to the one from Aspen's tool, with the optimum condition clearly labeled. Indeed, we can check and see that the results from Mathematica match with Aspen's built-in tool (Figure 3.21).

If you review the Mathematica documents provided, you will find that they allow you to calculate the operating region and optimum operating condition for a much wider set of isotherms than the tool built into Aspen.

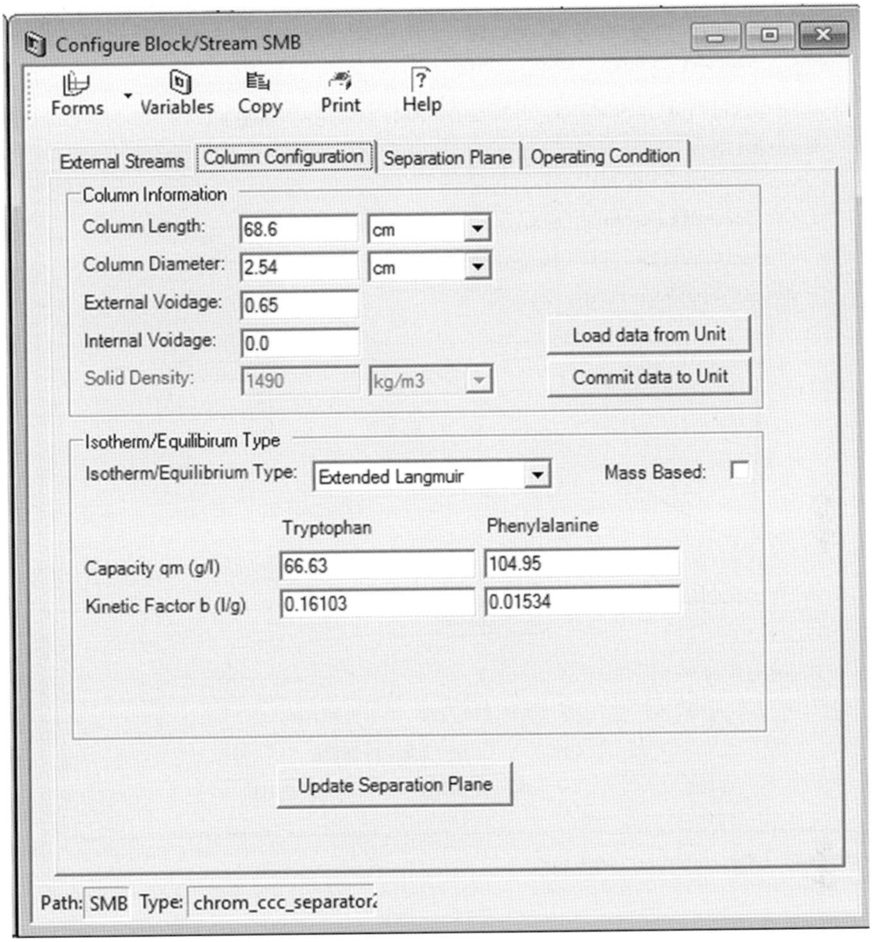

Figure 3.14 The imported data from clicking on "Load data from unit."

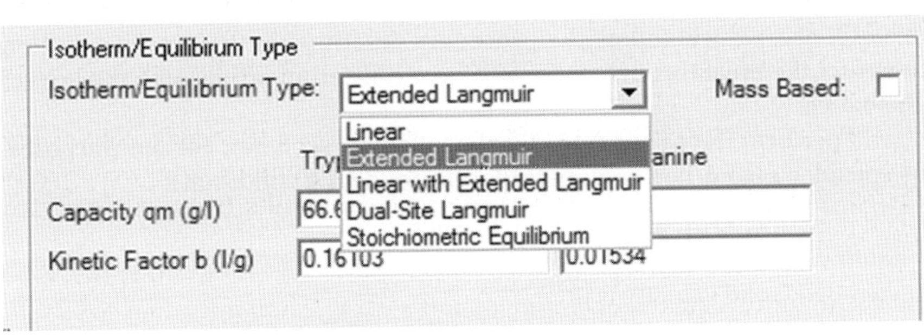

Figure 3.15 The isotherms available.

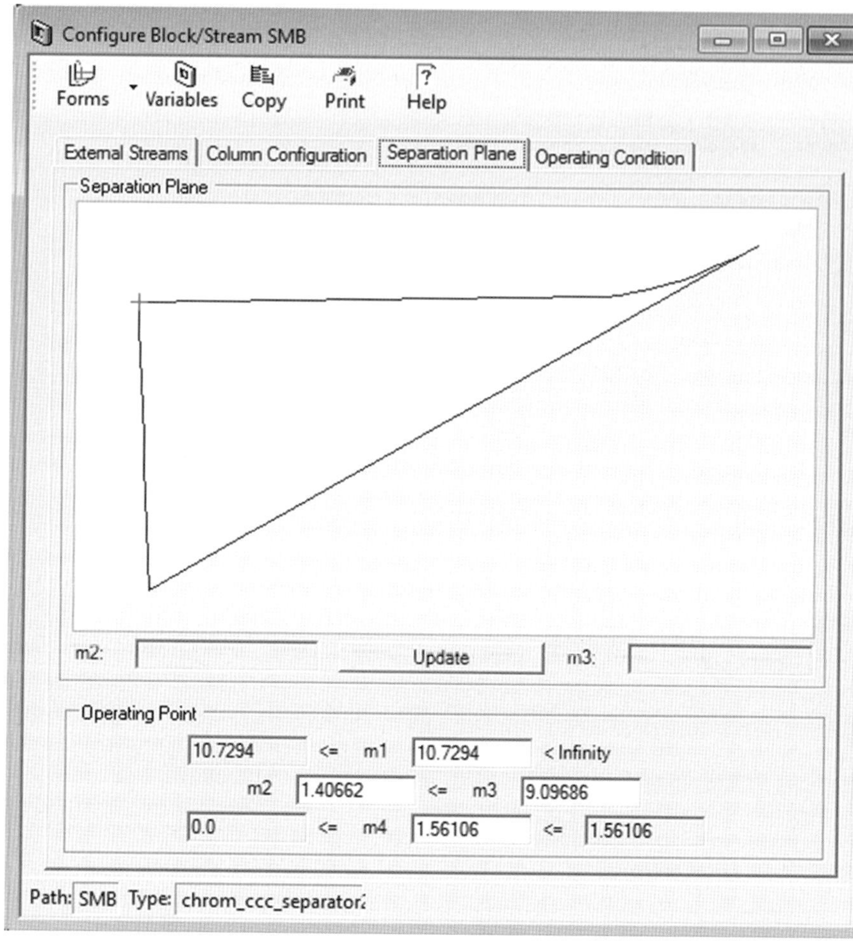

Figure 3.16 The separation plane.

3.4 Exercise 1: Calculating Transitions in a Fixed Bed Using Mathematica

In this exercise, we will follow the procedure described in the previous section to construct the breakthrough curve and a concentration profile with the help of Mathematica.

We will perform the calculations for the following scenario. For our purposes, let us consider a fixed-bed column under the following conditions:

- A bed porosity of 0.4, with no particle voids
- A liquid velocity of 0.5 units
- A fixed bed (solid velocity is 0)
- A Langmuir isotherm with $a_1 = 1$, $a_2 = 2$, $b_1 = 0.1$, $b_2 = 0.1$
- An initially empty column, $c_{i,\text{initial}} = 0$
- A feed with composition ($c_{1,\text{feed}} = 5$, $c_{2,\text{feed}} = 5$)

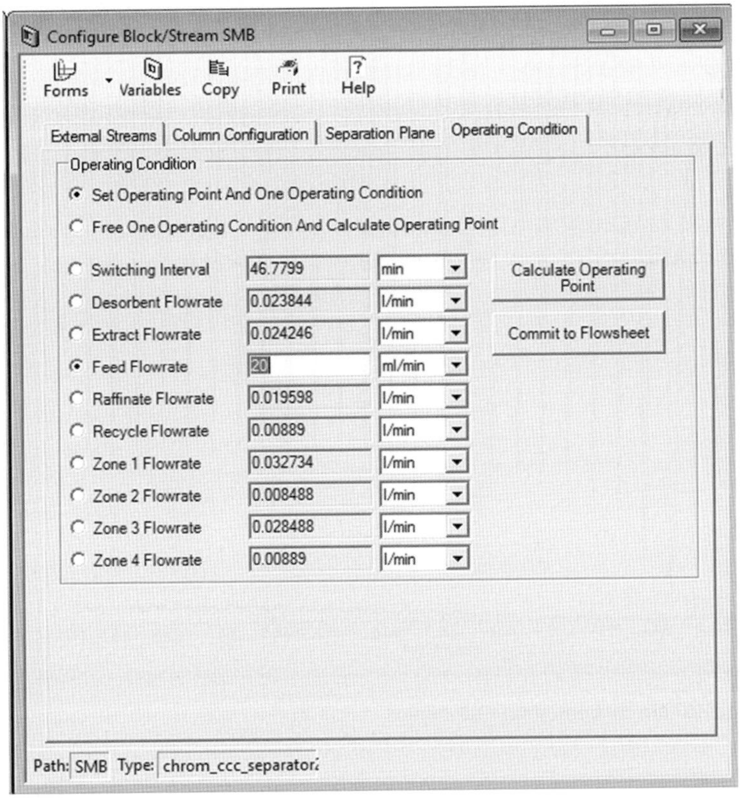

Figure 3.17 The optimum operating conditions for a feed flow rate of 20 ml min^{-1}

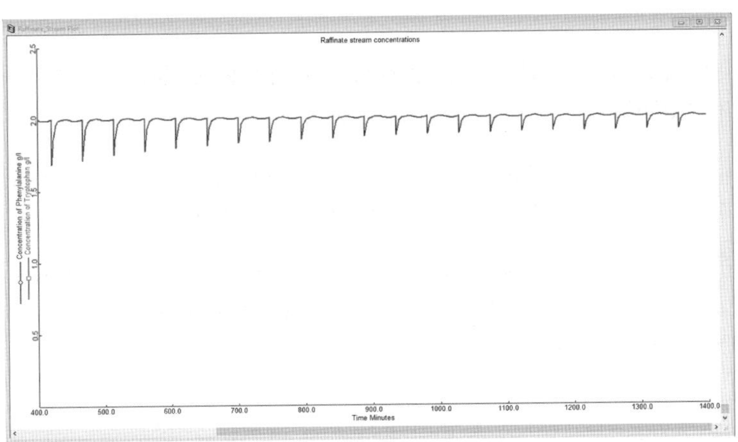

Figure 3.18 Raffinate history plot.

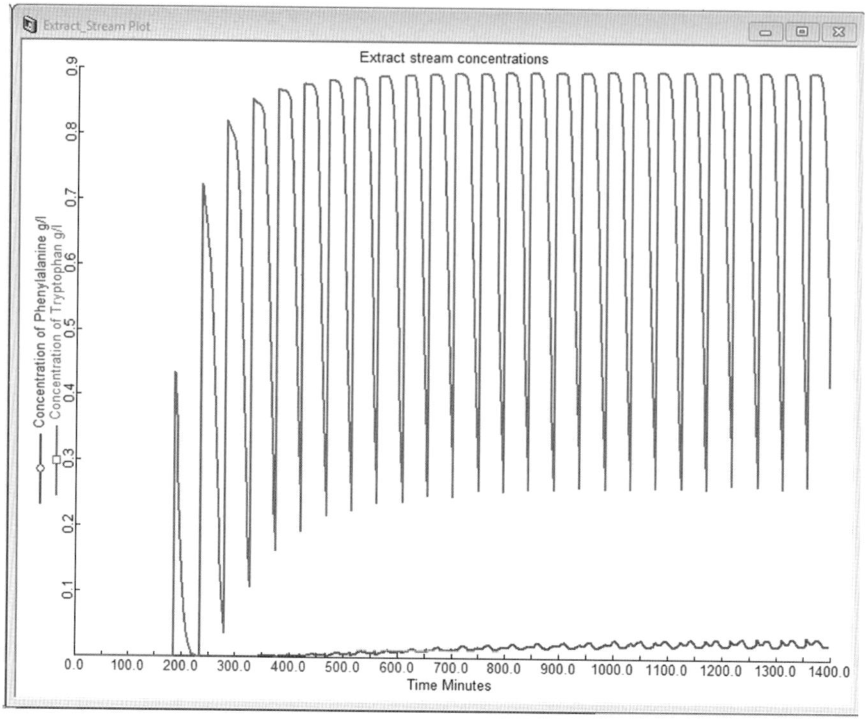

Figure 3.19 Extract stream concentration history plot.

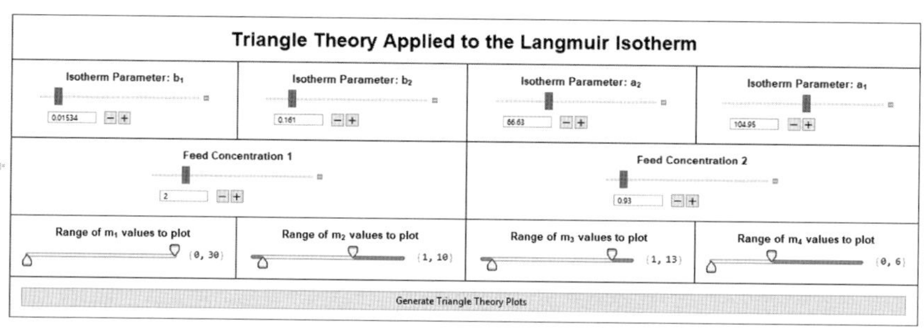

Figure 3.20 The interface for generating the Mathematica triangle theory plots.

3.4.1 Differential Equations – Analysis

To start off, let us define the simplified differential equation we use. Starting from the more general differential Eq. (3.16), we make the following assumptions:

- Negligible mass transfer resistance
  - The liquid and solid phases are in equilibrium at all times.
- No axial dispersion
- The density of the liquid does not depend on the component concentrations.

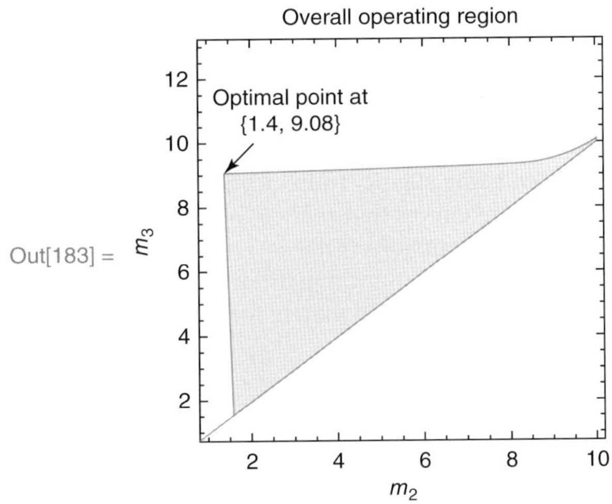

Figure 3.21 The output of the Mathematica triangle theory tool.

In this section, when the equation corresponds to an important calculation in the Triangle Theory Mathematica documents, **the Mathematica input required to generate the mathematical equation will be given alongside the equation, and designated (M#).**

With these assumptions, we can reduce the differential equation to Eq. (3.50) with $f_i = f_i(c_1 \ldots c_n)$ and $g_i = g_i(c_1 \ldots c_n)$

$$\frac{\partial g_i}{\partial x} + \frac{\partial f_i}{\partial \tau} = 0 \tag{3.50}$$

Specifically, f_i is given by

$$f_i = c_i + \varepsilon\, n_i \tag{3.51}$$

in the case of fixed beds. In moving beds,

$$f_i = \psi(c_i + v^*\, n_i) \tag{3.52}$$

```
f = Table[(ci + v* ni) ψ, {i, 1, 2}]
{ψ (c1 + v* n1), ψ (c2 + v* n2)}
```
(M1)

where n_i depends on the choice of isotherm.

We can also define g_i for moving beds as

$$g_i = ((1 - \mu \epsilon_p)c_i - \mu(1 - \epsilon_p)n_i) \tag{3.53}$$

```
g = Table[((1 - μ ϵp) ci - μ (1 - ϵp) ni), {i, 1, 2}]
{(1 - ϵp μ) c1 - (1 - ϵp) μ n1,
 (1 - ϵp μ) c2 - (1 - ϵp) μ n2}
```
(M2)

But in fixed beds,

$$g_i = c_i \tag{3.54}$$

To solve a PDE of this form, we can use the method of characteristics, as described in Section 3.1.3. Unfortunately, since we have multiple components, we need some slightly more advanced methods. The overall goal is the same: to find a set of curves that reduces our differential equations from PDEs to ordinary differential equations (ODEs).

The advanced solution method proceeds in the following way. To start with, we try to rewrite (3.50) in matrix form. We can use the chain rule to say

$$\frac{\partial f_i}{\partial \tau} = \frac{\partial f_i}{\partial c_1}\frac{\partial c_1}{\partial \tau} + \cdots + \frac{\partial f_i}{\partial c_n}\frac{\partial c_n}{\partial \tau} \tag{3.55}$$

$$\frac{\partial g_i}{\partial x} = \frac{\partial g_i}{\partial c_1}\frac{\partial c_1}{\partial x} + \cdots + \frac{\partial g_i}{\partial c_n}\frac{\partial c_n}{\partial x} \tag{3.56}$$

That means, we can write the system of equations defined in (3.50) as

$$\boldsymbol{g}\frac{\partial \vec{c}}{\partial x} + \boldsymbol{f}\frac{\partial \vec{c}}{\partial \tau} = 0 \tag{3.57}$$

where g and f are actually defined as the Jacobean matrix of the f and g functions defined in Eqs. (3.51) through (3.54):

$$\boldsymbol{f} = \begin{bmatrix} \frac{\partial f_1}{\partial c_1} & \cdots & \frac{\partial f_1}{\partial c_n} \\ \vdots & \ddots & \vdots \\ \frac{\partial f_n}{\partial c_1} & \cdots & \frac{\partial f_n}{\partial c_n} \end{bmatrix} \tag{3.58}$$

$$\boldsymbol{g} = \begin{bmatrix} \frac{\partial g_1}{\partial c_1} & \cdots & \frac{\partial g_1}{\partial c_n} \\ \vdots & \ddots & \vdots \\ \frac{\partial g_n}{\partial c_1} & \cdots & \frac{\partial g_n}{\partial c_n} \end{bmatrix} \tag{3.59}$$

```
In[41]:= jacobian[x_] := jacobian[x] = Simplify@D[x, {{c₁, c₂}}, NonConstants → {n₁, n₂}]
         jacobian[f]
Out[42]= {{ψ + v* ψ n1,1, v* ψ n1,2}, {v* ψ n2,1, ψ + v* ψ n2,2}}
```
(M3)

```
In[47]:= jacobian[g]
Out[47]= {{1 - ϵp μ + (-1 + ϵp) μ n1,1, (-1 + ϵp) μ n1,2},
          {(-1 + ϵp) μ n2,1, 1 - ϵp μ + (-1 + ϵp) μ n2,2}}
```
(M4)

Notice that Eq. (3.57) is a relatively simple matrix equation, and we can make it even simpler by taking an inverse:

$$\boldsymbol{g}^{-1}\boldsymbol{f} = \boldsymbol{h} \tag{3.60}$$

$$\frac{\partial \vec{c}}{\partial x} + \boldsymbol{g}^{-1}\boldsymbol{f}\frac{\partial \vec{c}}{\partial \tau} = \frac{\partial \vec{c}}{\partial x} + \boldsymbol{h}\frac{\partial \vec{c}}{\partial \tau} = 0 \tag{3.61}$$

In[49]:=

```
jacobian[h] = Simplify[Inverse[jacobian[g]].jacobian[f]];
Grid[%, Frame → All, Spacings → {3, 3}]
```

Out[50]=

$\left(-v^{*}\left(-1+\epsilon_p\right)\mu\psi n_{1,2}n_{2,1}+\left(\psi+v^{*}\psi n_{1,1}\right)\left(1-\epsilon_p\mu+\left(-1+\epsilon_p\right)\mu n_{2,2}\right)\right)/\left(-\left(-1+\epsilon_p\right)^2\mu^2 n_{1,2}n_{2,1}+\left(1-\epsilon_p\mu+\left(-1+\epsilon_p\right)\mu n_{1,1}\right)\left(1-\epsilon_p\mu+\left(-1+\epsilon_p\right)\mu n_{2,2}\right)\right)$	$\left(\left(\left(-1+\epsilon_p\right)\mu+v^{*}\left(-1+\epsilon_p\mu\right)\right)\psi n_{1,2}\right)/\left(\left(-1+\epsilon_p\right)^2\mu^2 n_{1,2}n_{2,1}+\left(-1+\epsilon_p\mu\right)\left(1-\epsilon_p\mu+\left(-1+\epsilon_p\right)\mu n_{2,2}\right)-\left(-1+\epsilon_p\right)\mu n_{1,1}\left(1-\epsilon_p\mu+\left(-1+\epsilon_p\right)\mu n_{2,2}\right)\right)$
$\left(\left(v^{*}+\mu-\epsilon_p\mu-v^{*}\epsilon_p\mu\right)\psi n_{2,1}\right)/\left(-\left(-1+\epsilon_p\right)^2\mu^2 n_{1,2}n_{2,1}-\left(-1+\epsilon_p\mu\right)\left(1-\epsilon_p\mu+\left(-1+\epsilon_p\right)\mu n_{2,2}\right)+\left(-1+\epsilon_p\right)\mu n_{1,1}\left(1-\epsilon_p\mu+\left(-1+\epsilon_p\right)\mu n_{2,2}\right)\right)$	$\left(-v^{*}\left(-1+\epsilon_p\right)\mu\psi n_{1,2}n_{2,1}+\left(1-\epsilon_p\mu+\left(-1+\epsilon_p\right)\mu n_{1,1}\right)\left(\psi+v^{*}\psi n_{2,2}\right)\right)/\left(-\left(-1+\epsilon_p\right)^2\mu^2 n_{1,2}n_{2,1}+\left(1-\epsilon_p\mu+\left(-1+\epsilon_p\right)\mu n_{1,1}\right)\left(1-\epsilon_p\mu+\left(-1+\epsilon_p\right)\mu n_{2,2}\right)\right)$

(M5)

Now this equation would be even more approachable if we were able to find some scalar multiple, σ, such that

$$\boldsymbol{h}\frac{\partial\vec{c}}{\partial\tau} = \sigma\frac{\partial\vec{c}}{\partial\tau} = \frac{\partial\vec{c}}{\partial x} \tag{3.62}$$

After all, we know that

$$\frac{\partial\vec{c}}{\partial\tau}\frac{\partial\tau}{\partial x} = \frac{\partial\vec{c}}{\partial x} \tag{3.63}$$

which would mean that

$$\sigma = \frac{\partial\tau}{\partial x} \tag{3.64}$$

But if we recall from linear algebra, the definition of eigenvalues is

$$\boldsymbol{h}\vec{v} = \sigma\vec{v} \tag{3.65}$$

which is exactly what we are trying to find! Moreover, finding eigenvalues is a fairly straightforward process; we can find the correct value of σ by calculating the eigenvalues of $\boldsymbol{h}$ with the equations defined in

$$|\boldsymbol{h} - \sigma\boldsymbol{I}| = 0 \tag{3.66}$$

The exact form of the eigenvalues will depend on the isotherm, and they may be functions of concentrations. But so far so good! We have identified how to calculate a curve in the x/τ plane. But what does that curve tell us? Not much, by itself, but we do know that each eigenvalue will have a corresponding eigenvector. For simplicity, let us consider the two-component case. When we identify the eigenvalues σ in Eq. (3.62), we also implicitly identify the corresponding eigenvector to be

$$\frac{\partial\vec{c}}{\partial\tau} = \left(\frac{\partial c_1}{\partial\tau}, \frac{\partial c_2}{\partial\tau}\right) = \text{Eigenvector}(\boldsymbol{h}) \tag{3.67}$$

In[110]:=
```
eigenSystem[x_] := eigenSystem[x] = Simplify@Eigensystem@jacobian[x]
eigenSystem[f]
```

Out[111]=

$$\left\{\left\{\frac{1}{2}\psi\left(2+v^{*}n_{1,1}+v^{*}n_{2,2}-v^{*}\sqrt{n_{1,1}^{2}+4n_{1,2}n_{2,1}-2n_{1,1}n_{2,2}+n_{2,2}^{2}}\right),\ \frac{1}{2}\psi\left(2+v^{*}n_{1,1}+v^{*}n_{2,2}+v^{*}\sqrt{n_{1,1}^{2}+4n_{1,2}n_{2,1}-2n_{1,1}n_{2,2}+n_{2,2}^{2}}\right)\right\},\ \left\{\left\{-\frac{-n_{1,1}+n_{2,2}+\sqrt{n_{1,1}^{2}+4n_{1,2}n_{2,1}-2n_{1,1}n_{2,2}+n_{2,2}^{2}}}{2n_{2,1}},1\right\},\ \left\{\frac{n_{1,1}-n_{2,2}+\sqrt{n_{1,1}^{2}+4n_{1,2}n_{2,1}-2n_{1,1}n_{2,2}+n_{2,2}^{2}}}{2n_{2,1}},1\right\}\right\}\right\} \qquad \text{(M6)}$$

This vector defines a second curve through its slope:

$$\frac{\partial c_2}{\partial \tau} / \frac{\partial c_1}{\partial \tau} = \frac{\partial c_2}{\partial c_1} \tag{3.68}$$

Now we have two curves:

- $\frac{\partial \tau}{\partial x}$ in the x, τ plane is defined by the eigenvalues of $\boldsymbol{h}$ and may be a function of concentration.
- $\frac{\partial c_2}{\partial c_1}$ in the c_1, c_2 plane is defined by the eigenvectors of $\boldsymbol{h}$ and is not a function of x or τ.

The eigenvalue, σ, will be constant along the x, τ curve, i.e. the x τ curve will be a straight line. However, the same is not true of the c_1, c_2 curve; it may not be a straight line. Regardless, the curves in the two planes allow us to find the complete solution to the differential equations. In Section 3.4.2, we show how to construct the solution from the equations we just derived.

Before we move on, however, it will turn out to be important to analyze how the eigenvalues of this $\boldsymbol{h}$ matrix are related to the eigenvalues of the Jacobean of the isotherm equation.

If we designate the eigenvalues according to Table 3.2, then we can derive Eqs. (3.69) through (3.71)

$$\phi_1 = \psi + v^*\psi\theta_1 \text{ and } \phi_2 = \psi + v^*\psi\theta_2 \tag{3.69}$$

$$\gamma_1 = 1 - \epsilon_p\mu + (-1+\epsilon_p)\mu\theta_1 \text{ and } \gamma_2 = 1 - \epsilon_p\mu + (-1+\epsilon_p)\mu\theta_2 \tag{3.70}$$

$$\eta_1 = \frac{\psi + v^*\psi\theta_1}{1 - \epsilon_p\mu + (-1+\epsilon_p)\mu\theta_1} \text{ and } \eta_2 = \frac{\psi + v^*\psi\theta_2}{1 - \epsilon_p\mu + (-1+\epsilon_p)\mu\theta_2} \tag{3.71}$$

Table 3.2 Eigenvalue nomenclature.

Matrix	Eigenvalue variable
Jacobean of the isotherm	θ
F	ϕ
G	γ
H	η

Now we should also note that, because the m-values can be written as

$$m = \frac{1 - \mu\varepsilon_p}{\mu(1 - \varepsilon_p)} \tag{3.72}$$

we can insert it into Eq. (3.71) to find

$$\eta_1 = \frac{\psi + v^*\psi\theta_1}{\mu(1 - \epsilon_p)(m - \theta_1)} \text{ and } \eta_2 = \frac{\psi + v^*\psi\theta_2}{\mu(1 - \epsilon_p)(m - \theta_2)} \tag{3.73}$$

which is fantastically useful. Note that the $(\psi + v^*\psi\theta)$ and $\mu(1 - \epsilon_p)$ terms in this equation are guaranteed to be positive. This means, the only way in which η can be less than zero is if $m - \theta$ is less than zero.

Thanks to Eqs. (3.67) and (3.68), we can write out explicit equations for $\frac{\partial c_2}{\partial c_1}$. If we integrate these equations with respect to c_1 (or integrate $\frac{\partial c_1}{\partial c_2}$ with respect to c_2), we can determine curves in the (c_1, c_2) plane that the solution will follow. In order to actually carry out this integration, we need initial conditions.

Because this exercise concerns a column-loading scenario, the column is initially loaded solely with solvent, i.e. $c_1^0 = c_2^0 = 0$. We also know that the column will eventually become fully loaded with components 1 and 2, so that it effectively stops separating the components, and the liquid inside the column will reach the feed conditions as a final state, i.e. ($c_1^{\text{final}} = c_1^{\text{feed}}$ and $c_2^0 = c_2^{\text{feed}}$).

Does this mean we can choose a single eigenvector, and then integrate from $c_{1,\text{initial}}$ to $c_{1,\text{final}}$ and have our answer? Unfortunately not – there is no guarantee that the $c_{2,\text{final}}$ value calculated this way will fall on the curve defined by the eigenvector. Instead, we integrate along the first eigenvector from the initial condition, then the second eigenvector from the final condition. These curves will intersect at an intermediate state.

One further note before we proceed is that we must keep careful track of which eigenvalues correspond to which eigenvectors. Along the (c_1, c_2) curve defined by the first eigenvector, the velocity in the (x, τ) plane will be defined by the first eigenvalue. Along the (c_1, c_2) curve defined by the second eigenvector, the velocity in the (x, τ) plane will be defined by the second eigenvalue.

This solution procedure is shown graphically in Figure 3.22. The five labeled points in (c_1, c_2) each have a corresponding velocity in the (x, τ) plane. Because all of the concentrations in the solution are fed in to the column at $x = 0$, $\tau = 0$, these velocities define the σ curves in the (x, τ) plane. Note that the intermediate state is special. It falls on both eigenvectors, and therefore has *two* velocity lines, one defined by σ_1 and one defined by σ_2.

While looking at this graph, let us consider for a second what it would mean for a (c_1, c_2) point to have a negative velocity in our bed. If the velocity is negative, then as we move forward in time (τ), that concentration point would try to move in the $-x$ direction. That is equivalent to saying that it would be "stuck" at the beginning of the column, and it would cause one of the "jumps" in concentration described in Figure 3.7.

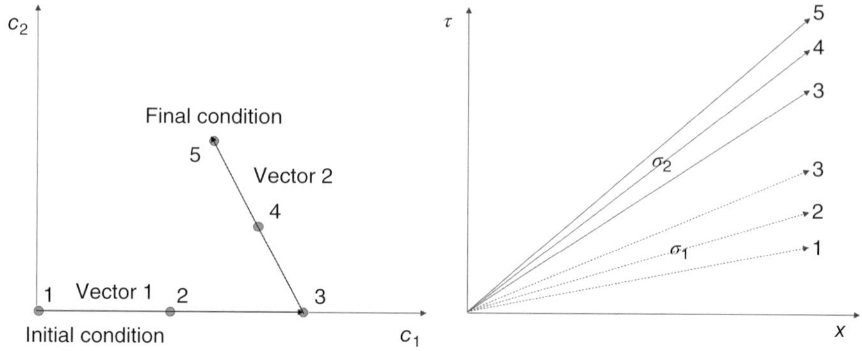

Figure 3.22 A sketch of the solution in the (c_1, c_2) and (x, τ) planes.

3.4.2 Constructing the Solution from Eigenvectors and Eigenvalues

To progress beyond the analysis in Section 3.4.1, we need to introduce boundary conditions. In a moving bed adsorber, these are most commonly feed concentrations and initial column loading. From those initial conditions, we can define the actual instances of the c_1, c_2 curves and construct the solution.

Figure 3.23 shows the three pairs of boundary conditions that are used in the solution. Specifically,

$$c_i = c_i^0 \text{ at } \tau = 0 \text{ and } 0 \leq x \leq 1 \tag{3.74}$$

$$c_i = c_{i,\text{Feed}} \text{ for all } \tau > 0 \text{ and } x = 0 \tag{3.75}$$

$$n_i = n_{i,\text{Feed}} \text{ for all } \tau > 0 \text{ and } x = 1 \tag{3.76}$$

Complete description of this situation requires that we find a solution for the top of the column and the bottom of the column separately, at least until they begin interacting. Once that happens, we need to describe the evolution of the interacting systems until they reach steady state. All of that is possible, but it is not necessary to describe that process here. It turns out that "all roads lead to Rome," because the eventual steady state is independent of the initial column state; it is characterized by Eqs. (3.74) and (3.75) alone. Therefore, in the following analysis, we assume that the initial column state is the same as (i.e. is in equilibrium with) the solid stream's state. This simplifies the process, since we only need to consider the transition taking place at the top of the column, and we will get the correct steady state.

To construct the solution, we follow this procedure:

1. Calculate the ***g***, ***f***, and ***h*** matrices defined in Eqs. (3.58)–(3.60).
2. Calculate the eigenvalues and eigenvectors of these matrices.
3. Find the path that the transition will take in the (c_1, c_2) plane. This is done by
   a. calculating the two curves defined by the eigenvectors at the initial condition;
   b. calculating the two curves defined by the eigenvectors at the final condition;

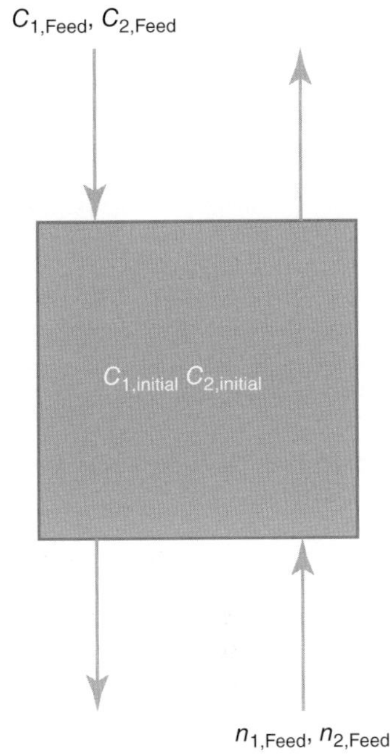

Figure 3.23 Boundary conditions in a chromatographic bed.

   c. the transition path follows the second eigenvector curve from the initial condition until it intersects the first eigenvector curve from the final condition. See Figure 3.8.
4. At each point along the transition, calculate the value of the corresponding eigenvalue. Thanks to Eq. (3.64), we know $\tau = \sigma\, x$, so for each concentration pair in a given transition curve, we have an eigenvalue and therefore a curve in the x, τ plane.
   a. Along the curve defined by the second eigenvector, use the second eigenvalue.
   b. Along the curve defined by the first eigenvector, use the first eigenvalue.
5. Take a slice of the x, τ plane curves at x = 1 to generate breakthrough curves for our bed.
6. Take a slice of the x, τ plane curves at $\tau = \tau_i$ to generate the concentration profile in the bed at time τ_i.

3.4.3 Use the Steady-State Information to Constrain Operating Conditions

At last, we are now able to return to the constraints we were discussing in Section 3.2.3. Armed with the fresh knowledge about transitions in the c_1, c_2 plane, we are now able to make some statements about the operating conditions in the SMB unit. As it turns out, we can place the entire SMB unit onto the same type of (c_1, c_2) plot that we used in the transition exercise. Specifically, the separation can be laid out in this way (Figure 3.24).

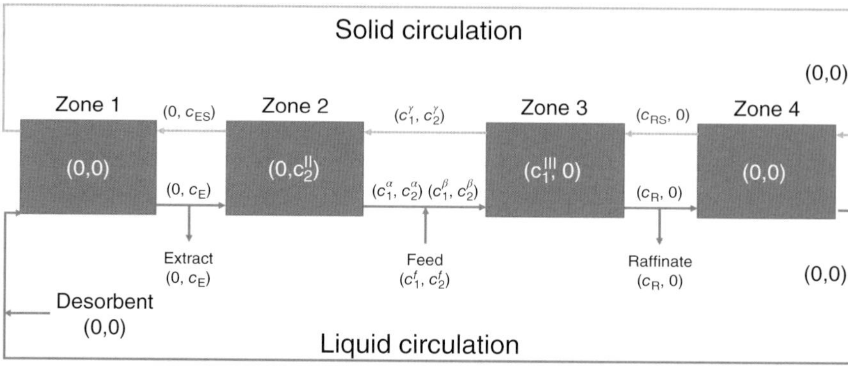

Figure 3.24 TMB operation with zone steady states labeled.

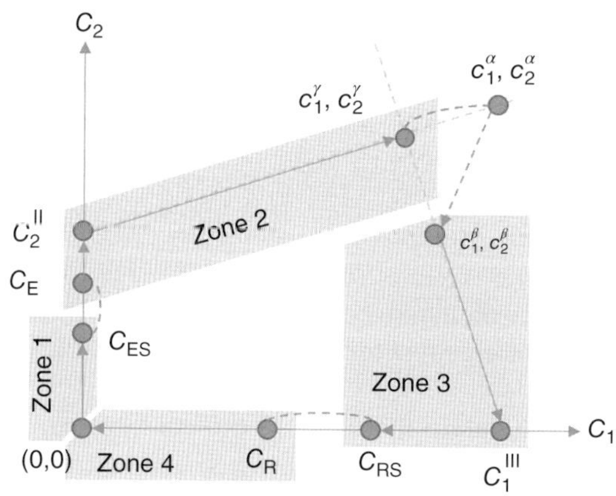

Figure 3.25 Transitions in a TMB column.

Figure 3.24 shows what states will be achieved in a TMB column. Figure 3.25 shows how those states are laid out in the c_1, c_2 plane. The *most important* feature of these diagrams is to notice that all the states labeled in Figure 3.24 fall on the nice curves defined by the eigenvectors (i.e. the dashed lines in Figure 3.25). It is also important to note that the states achieved in zones 2 and 3 are intermediate states, and the concentration of one component is zero. Taken together, those features suggest a line of attack on this problem. Thanks to the exercise in section 3.3, we already have a general method for calculating c_2^{II} and c_1^{III} given (c_1^γ, c_2^γ).

Once we have calculated c_2^{II} and c_1^{III} from (c_1^γ, c_2^γ), we can use Eqs. (3.38) and (3.39) to solve for (c_1^γ, c_2^γ) as a function of the feed concentrations, m_2, m_3, and the isotherm parameters. Ensuring that these concentrations are physically meaningful (i.e. real and positive) constitutes one set of constraints on this system, and it is important to note that these constraints are **not** functions of the raffinate or extract flow rates, and also do not depend on m_1 or m_4.

The second set of constraints comes from Eq. (3.73). Notice that the velocity of the concentrations in each zone is proportional to $(\theta - m)$, where θ is an eigenvector of the isotherm Jacobean. In order to actually achieve the intermediate state, we need to ensure that all states past the intermediate state have positive velocities, so $(\theta_{\text{state}} - m) \geq 0$, and that all states before the intermediate state have negative velocities, so $(\theta_{\text{state}} - m) \leq 0$. In practice, what this means is that in order to achieve the intermediate state, we need to choose an m-value in such a way that

$$\sigma_2^{\text{Int}} \geq 0 \text{ and } \sigma_1^{\text{Int}} \leq 0 \tag{3.77}$$

$$\theta_2^{\text{Int}} \geq m \geq \theta_1^{\text{Int}} \tag{3.78}$$

We can see how this works in Figure 3.26. As in Exercise 1, the intermediate state is flanked by the two transitions, with boundaries defined by the two eigenvalues. If we decreased the m value, eventually σ_1^{Int} will become 0; when that happens, the column will achieve the intermediate state at steady state.

3.4.4 Calculate the Curves Defined by the Eigenvectors

The equations for the eigenvectors are somewhat complex, and they contain a problematic feature: there is a ratio whose denominator is proportional to c_2. This means that in all cases where $c_2 = 0$, we are going to have problems with division by zero. To understand what is going on, let us consider the eigenvectors at the origin, $c_1 = 0$, $c_2 = 0$. We will ask Mathematica to calculate the direction of the eigenvectors as c_2 approaches zero. To do this, we will first normalize the eigenvectors:

$$\text{vectorDirections} = |\text{isotherm eigenvector}| \tag{3.79}$$

In[389]:=
```
generalVectorDirections = Simplify[Normalize /@ eigenvectors[isotherm],
  Assumptions → {a₁ < a₂, a₁ > 0, a₂ > 0, c₁ ≥ 0, c₂ ≥ 0}]
```

Out[389]=

$$\left\{\left\{\frac{-a_1 + a_2 + a_2 b_1 c_1 - a_1 b_2 c_2 + \sqrt{-4 a_1 a_2 (1 + b_1 c_1 + b_2 c_2) + (a_2 (1 + b_1 c_1) + a_1 (1 + b_2 c_2))^2}}{\sqrt{4 + \text{Abs}\left[\frac{a_2 (1+b_1 c_1) - a_1 (1+b_2 c_2) + \sqrt{-4 a_1 a_2 (1+b_1 c_1+b_2 c_2) + (a_2 (1+b_1 c_1) + a_1 (1+b_2 c_2))^2}}{a_2 b_1 c_2}\right]^2}\, a_2 b_1 c_2},\right.\right.$$
$$\left.\frac{2}{\sqrt{4 + \text{Abs}\left[\frac{a_2 (1+b_1 c_1) - a_1 (1+b_2 c_2) + \sqrt{-4 a_1 a_2 (1+b_1 c_1+b_2 c_2) + (a_2 (1+b_1 c_1) + a_1 (1+b_2 c_2))^2}}{a_2 b_1 c_2}\right]^2}}\right\},$$
$$\left\{\frac{-a_1 + a_2 + a_2 b_1 c_1 - a_1 b_2 c_2 - \sqrt{-4 a_1 a_2 (1 + b_1 c_1 + b_2 c_2) + (a_2 (1 + b_1 c_1) + a_1 (1 + b_2 c_2))^2}}{\sqrt{4 + \text{Abs}\left[\frac{-a_2 (1+b_1 c_1) + a_1 (1+b_2 c_2) + \sqrt{-4 a_1 a_2 (1+b_1 c_1+b_2 c_2) + (a_2 (1+b_1 c_1) + a_1 (1+b_2 c_2))^2}}{a_2 b_1 c_2}\right]^2}\, a_2 b_1 c_2},\right.$$
$$\left.\left.\frac{2}{\sqrt{4 + \text{Abs}\left[\frac{-a_2 (1+b_1 c_1) + a_1 (1+b_2 c_2) + \sqrt{-4 a_1 a_2 (1+b_1 c_1+b_2 c_2) + (a_2 (1+b_1 c_1) + a_1 (1+b_2 c_2))^2}}{a_2 b_1 c_2}\right]^2}}\right\}\right\} \tag{M7}$$

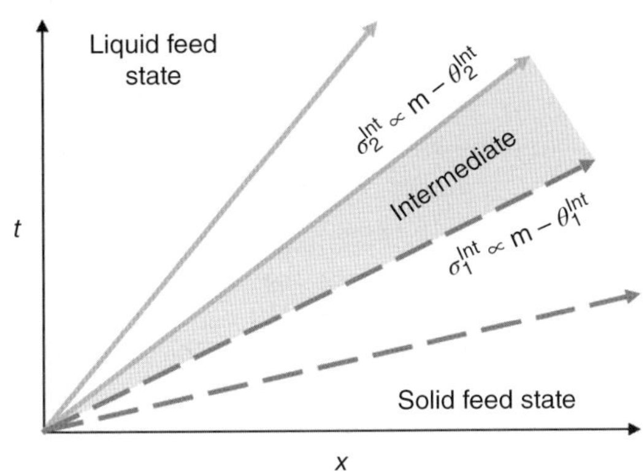

Figure 3.26 Velocities are proportional to $\boldsymbol{m - \theta}$.

The next step is to take the limit as c_2 goes to 0. It turns out that the result of this limit depends on which component is more strongly adsorbed. Therefore, we restrict our consideration to the scenario where component 2 is more strongly adsorbed.

$$\text{originVectors} = \text{vectorDirections}|_{c_1=0,c_2=0} \tag{3.80}$$

```
In[390]:= originVectorDirections = Limit[generalVectorDirections, c₂ → 0,
             Assumptions → {a₁ < a₂, a₁ > 0, a₂ > 0, c₁ == 0, b₁ ∈ Reals}]

Out[390]= {{Abs[b₁]/b₁, 0}, {0, 1}}
```
(M8)

Given these results, we have the situation depicted in Figure 3.27.

With this in mind, we will assume that $a_1 < a_2$ for the rest of this exercise.

Now, we can return to the main task: we need to calculate the curves in the eigenvector field from the initial and final conditions. To start with, we remember that we have already calculated the vector directions discussed in Eq. (3.68) when we calculated the eigenvectors.

```
slopeEq1=vectors[[1,2]]/vectors[[1,1]]//Simplify
slopeEq2=vectors[[2,1]]/vectors[[2,2]]//Simplify
```

Note that we have treated the first and the second vector differently. This is because of what we know from Figure 3.27. Specifically, we have said that

$$\text{slopeEq1} = \frac{\partial c_2}{\partial \tau} / \frac{\partial c_1}{\partial \tau} = \frac{\partial c_2}{\partial c_1} \tag{3.81}$$

$$\text{slopeEq2} = \frac{\partial c_1}{\partial \tau} / \frac{\partial c_2}{\partial \tau} = \frac{\partial c_1}{\partial c_2} \tag{3.82}$$

This is necessary because $\frac{\partial c_1}{\partial c_2}$ is infinite along the first vector, while $\frac{\partial c_2}{\partial c_1}$ is infinite along the second vector.

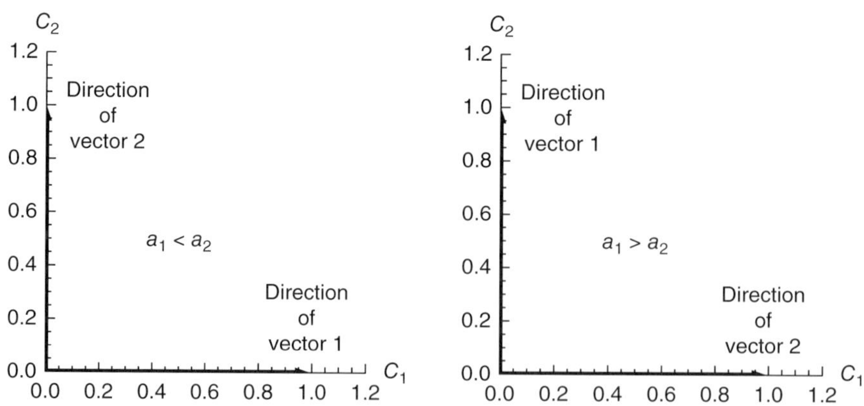

Figure 3.27 The eigenvector directions depend on which component is more strongly adsorbed.

Now that we have these two slopes in the c_1, c_2 plane, we can ask Mathematica to do the following:

$$\text{DSolve}[c_2'[c_1] == (\text{slopeEq1}/c_2 \rightarrow c_2[c_1]), c_2[c_1], c_1]$$

DSolve is Mathematica's differential equation solver. The above command asks Mathematica to find an analytic solution to the differential equation we have specified so far. In other words, if Mathematica can, it will output the function, $c_2[c_1]$, in terms of c_1. Since we did not specify an initial condition, the resulting equation would contain an unknown integration constant, but we could use our initial conditions to find the value. Unfortunately, Mathematica's DSolve function is currently unable to solve this differential equation in a reasonable amount of time, so we must resort to more manual methods.

For our manual method, it turns out that the curves we are looking for are linear. We can prove this to ourselves in the following way. First, we tell Mathematica that along the first curve, c_2 is a function of c_1; along the second curve, c_1 is a function of c_2:

```
funcSlopes1=slopeEq1/.c2->c2[c1]
funcSlopes2=slopeEq2/.c1->c1[c2]
```

Then, we take the derivative of those slopes:

$$\text{secondDerivative1} = \text{Simplify}[\partial_{c_1}\text{funcSlopes1}]$$
$$\text{secondDerivative2} = \text{Simplify}[\partial_{c_2}\text{funcSlopes2}]$$

And finally, we tell Mathematica to replace the first derivatives with the first derivatives we already knew.

```
Simplify[secondDerivative1/.c2'[c1]-> funcSlopes1]
Simplify[secondDerivative2/.c1'[c2]-> funcSlopes2]
                                       0
                                       0
```

Since the output is zero for both cases, we have effectively demonstrated that the second derivatives are zero, and therefore the curves are linear. This is explained in Eqs. (3.83) and (3.84):

$$\frac{\partial \text{slopeEq1}}{\partial c_1} = \frac{\partial^2 c_2}{\partial {c_1}^2} = 0 \tag{3.83}$$

$$\frac{\partial \text{slopeEq2}}{\partial c_2} = \frac{\partial^2 c_1}{\partial {c_2}^2} = 0 \tag{3.84}$$

which implies that the curves we are looking for are linear. This means that the slope of the curves is constant at all points including the initial conditions. Therefore, the slope can be calculated once and for all at the initial and final states of the column. We can easily calculate these slopes, and subsequently the equations of the lines in the c_1, c_2 plane.

```
initialSlope1 = slopeEq1/.{c1->c01,c2->c02}
initialSlope2 = slopeEq2/.{c1->c01,c2->c02}
line1 = Simplify[c2==c1*initialSlope1+(c02-c01*initialSlope1)]
line2 = Simplify[c1==c2*initialSlope2+(c01-c02*initialSlope2)]
```

Finally, we are ready to insert the values for this problem and plot the result. Figure 3.28 shows a plot of the lines that we have calculated, once we substitute in the isotherm parameters and operating conditions. The lines represent the analytic equations, while the x marks denote the results of a detailed numeric simulation. Clearly, we are not just wasting our time, and the eigenvector transitions are able to correctly describe the loading of a packed bed.

3.4.5 Calculate the Eigenvalues along the Transition

As shown in Figure 3.28, we now have information about how the column gets loaded in the concentration plane. However, we have no information about how that plot corresponds to position and time. This information comes from the eigenvalues, as we saw in Eq. (3.64). For reasons that will be discussed later, we will calculate the eigenvalues of the isotherm matrix first (θ_i), which will then be used to calculate the eigenvalues of the other matrices using Eqs. (3.69)–(3.71). So to get started, we will set up a temporary variable containing the eigenvalues we calculated in step 2. We will also create replacement rules by solving the line equations we produced in step 3.

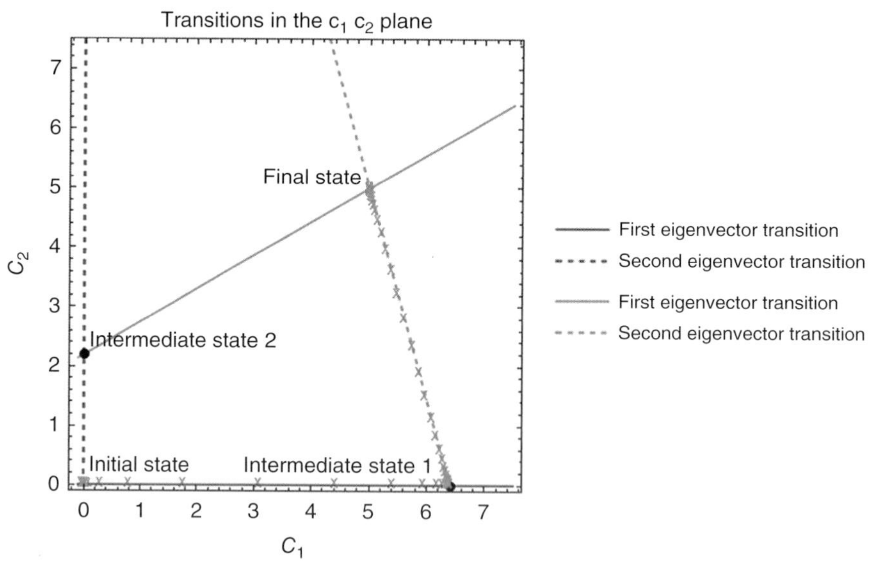

Figure 3.28 Transitions in the c_1,c_2 plane; the eigenvector prediction versus numerical simulation in Aspen Chromatography. The dark gray lines are the eigenvectors through the initial state, light gray lines are the eigenvectors through the final state. The forward transition simulated in Aspen proceeds through intermediate state 1 (X marks), while the reverse transition would proceed through intermediate state 2.

```
eigenvalues=eigensIso[[1]]
line1Replacement=Solve[line1,c2]//First
line2Replacement=Solve[line2,c1]//First
```

Now we must apply the replacement rules to the corresponding eigenvalue. Along line1, the curve defined by the first eigenvector, we must use the first eigenvalue. The second eigenvalue is used along the second curve.

```
eigenvalueEquation1=Simplify[eigenvalues[[1]]/.
   line1Replacement]
eigenvalueEquation2=Simplify[eigenvalues[[2]]/.
   line2Replacement]
```

This results in two equations, the first one in terms of c_1 and the second in terms of c_2. These equations allow us to calculate the eigenvalues at a particular point along the transition curves. For example, if we wanted to calculate the θ eigenvalue at the point (4.5, 0) in Figure 3.28, we would do the following:

```
eigenvalueEquation1/.{c01->0, c02->0, c1->4.5, a1->1,
a2->2, b1->0.1, b2->0.1}
0.475624
```

As we see in Figure 3.28, the first transition proceeds from (0,0) to (6.5,0). We can therefore create a nice plot of these slopes using our eigenvalue equations, as seen in Figure 3.29.

If we pay close attention, we will notice that these transitions do not quite make sense. For example, why would the σ_1 corresponding to the intermediate condition 1 be adjacent to the initial state instead of the intermediate state? The answer is that these transitions actually form "**shock waves**," which we will discuss after we try to calculate the breakthrough curves in the next step.

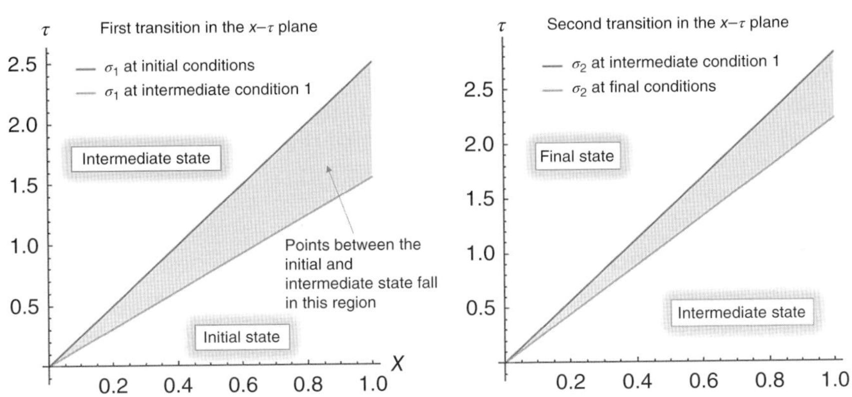

Figure 3.29 Transitions in the $x - \tau$ plane.

3.4.6 Calculate the Concentrations in Time and Space

Now that each concentration pair in the transition has a curve in the $\tau - x$ plane, we can choose a constant position ($x = 1$) and calculate the time corresponding to each concentration pair. To do this, we start by calculating the τ equations at $x = 1$:

```
τprofile1=1*ϕ1/.ϕSolutions/.θ1->eigenvalueEquation1
τprofile2=1*ϕ2/.ϕSolutions/.θ2->eigenvalueEquation2
```

These two equations tell us τ as a function of c_1 or c_2. We could immediately make a plot of this, but having τ on the vertical axis is somewhat awkward. We also need a way to plot both concentrations at all times, but these equations only involve one of the components each.

To resolve this, we will set ourselves up to create a parametric plot.

```
parametricCurves1={{τprofile1,c1},{τprofile1,(c2/.
   line1Replacement)}}
parametricCurves2={{τprofile2,c1/.line2Replacement},
   {τprofile2,c2}}
```

What we have just done is to define two pairs of parametric curves. Specifically, parametricCurve1 is now defined as

$$\text{parametricCurve1} = \begin{pmatrix} \tau[c_1] & c_1 \\ \tau[c_1] & c_2[c_1] \end{pmatrix} \tag{3.85}$$

The first row of that matrix is a (τ, c_1) pair, where τ is a function of c_1. The second row is a (τ, c_2) pair, where both τ and c_2 are functions of c_1. This means that we can make a parametric plot where we vary c_1 from $c_{1,\text{ initial}}$ to $c_{1,\text{ intermediate state 1}}$. The exact same procedure applies to the second parametricCurve2, except that we vary c_2 from $c_{2,\text{ intermediate state 1}}$ to $c_{2,\text{ final}}$. When we do this, we find the curves shown in Figure 3.30.

Clearly, something has gone wrong – the curves look to be reversed in time. The first transition looks like it starts from the intermediate condition, (6.4,0), and ends at the initial condition (0,0). The second curve transitions from the final state (5,5) to the intermediate state. Why are the transitions backward? The answer, which we might have noticed earlier, is that these transitions form shock waves.

3.4.7 Account for Shock Waves

Clearly, our solution in step 5 is not physically meaningful because we have not accounted for shock wave formation. What are shock waves and how can we calculate a solution that takes them into account?

Shock waves occur when the state at the end of a transition moves through the column faster than the state at the start. This process is shown in Figure 3.31,

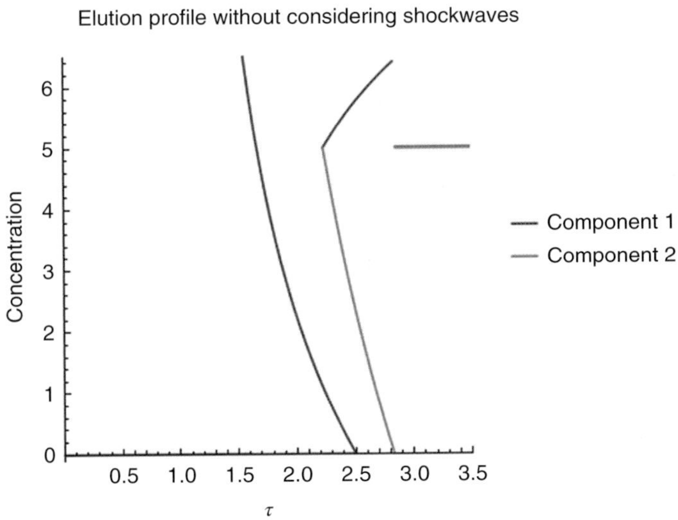

Figure 3.30 The breakthrough curve defined by the curves in Figure 3.22, without accounting for shocks.

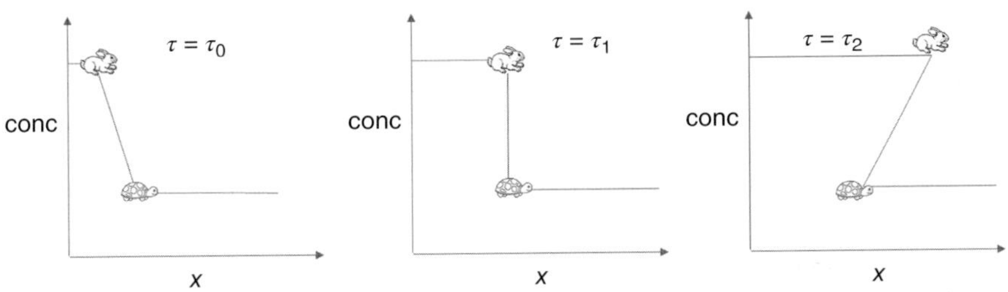

Figure 3.31 Formation of shock waves.

where the fast final state is labeled with a rabbit, while the slow initial state is labeled with a turtle.

As we can see, the situation in Figure 3.31 causes the same kind of backward transitions that we calculated in Figure 3.30. The backward state shown in the $\tau = \tau_2$ plot is not physically meaningful or achievable. Instead, the $\tau = \tau_1$ state puts an end to the speed difference between the initial and the final states. The instantaneous transition in that state is called *a shock wave*, and it has its own velocity that is calculated differently. To calculate the velocity of a transition that has formed a shock wave, we replace the θ values in Eqs. (3.69)–(3.71) with a value based on the magnitude of the jump in the shock wave, here denoted by the brackets.

$$\theta_i|_{\text{shock}} = \left.\frac{[n_i]}{[c_i]}\right|_{\text{shock}} \tag{3.86}$$

For example, $[c_1] = |c_{1,\text{final}} - c_{1,\text{intermediate}}|$. This is our reason for using θ_i, the eigenvalue of the isotherm matrix defined in Table 3.2, instead of just working with the explicit eigenvalues originally calculated by Mathematica. After calculating the θ value using Eq. (3.86), we apply Eq. (3.71) to find the velocity of the shock wave.

In Mathematica, we can calculate this easily:

$$\theta_{\text{shocks}} = \frac{(\text{isotherm}/.\{c_1 \rightarrow \text{cF}_1, c_2 \rightarrow \text{cF}_2\}) - (\text{isotherm}/.\{c_1 \rightarrow \text{c0}_1, c_2 \rightarrow \text{c0}_2\})}{\{\text{cF}_1 - \text{c0}_1, \text{cF}_2 - \text{c0}_2\}}$$

With these θ values, we can generate a new version of the plots in Figure 3.29 that takes into account the shock wave formation. To do this, we calculate the τ profiles just like we did in step 5.

```
τprofile1shock=ϕ1/.ϕSolutions/.θ1->θshocks[[1]]
τprofile2shock=ϕ2/.ϕSolutions/.θ2->θshocks[[2]]
```

Then we plot these two shock transitions:

Pay attention to the differences between Figures 3.29 and 3.32. In Figure 3.29, we plot the two transitions separately, and the transitions contain a range of slopes in the $x - \tau$ plane. Figure 3.29 fails to account for shocks, and so the plot put the slopes corresponding to initial conditions next to the final conditions; the results of this show up in Figure 3.30. In Figure 3.32, the transitions are represented by a single slope, because the entire transition occurs instantly. Figure 3.33 shows what these shockwaves look like as a breakthrough curve.

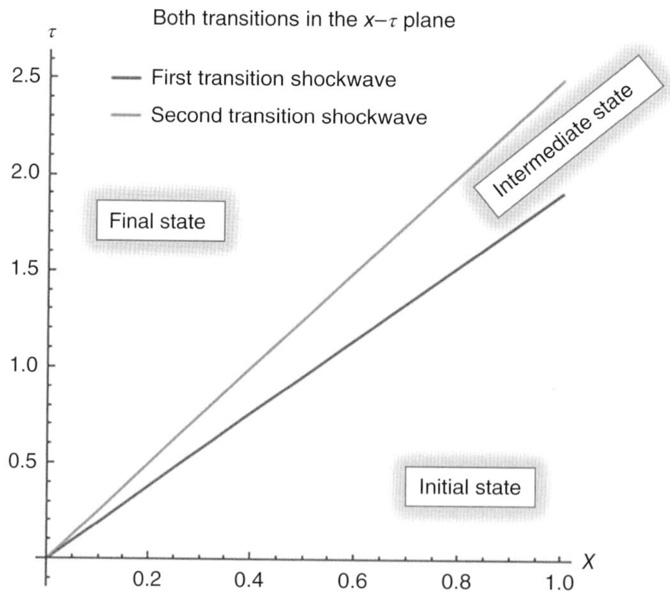

Figure 3.32 Transitionstransitions in the $x - \tau$ plane with shock waves accounted for.

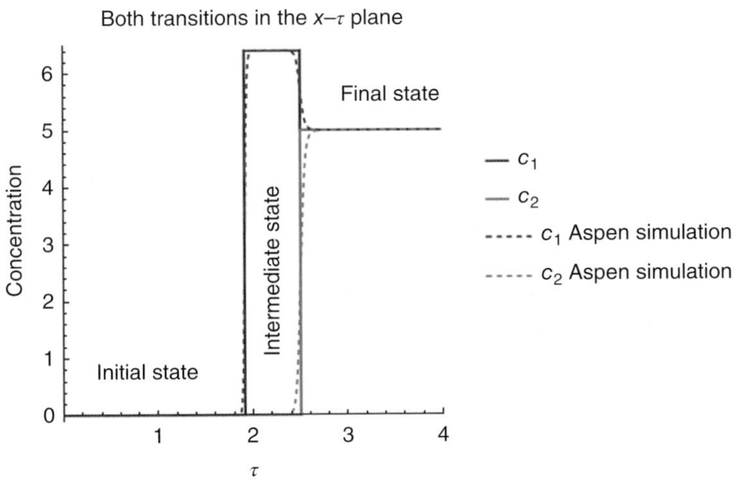

Figure 3.33 Breakthrough curves with shock waves accounted for.

3.5 Exercise 2: Constructing the Constraints on the TMB System in Mathematica

Step 1: Define the zone 2 and 3 constraints

To do this, we simply need to tell Mathematica to use Eq. (3.78). That can be easily done with the following code:

```
thetas=eigensIso[[1]]
zone2state={c_1->0,c_2->cII_2}
zone2thetas=Simplify[thetas/.zone2state]
zone2constraint={ zone2thetas[[1]]∈Reals &&
   zone2thetas[[2]]∈Reals && zone2thetas[[1]]
   <=m_2<=zone2thetas[[2]]}
```

Here we set up the zone2constraint, which consists of the inequality in (3.78) as well as a test to ensure that the θ values are real. We can do the same thing for zone 3, although we need to remember that as we saw in Section 3.4.7, the transitions in this zone will be shock waves. Therefore, we need to use the θ_{shocks} instead of the eigenvalues.

```
zone3states={{cF_1->cIII_1,cF_2->0},{c0_1->cβ_1,c0_2->cβ_2}}
zone3theta=θ_shocks[[1]]/.Flatten[zone3states]
zone3constraint={zone3theta∈Reals&&m_3<=zone3theta}
```

Fortunately, Mazzotti [44] has proved that the zone 3 constraints are always satisfied when the various other constraints are satisfied, and so there will be no need to use these relationships any further.

Step 2: Define the feed and c^γ constraints

The feed concentrations provide some constraints on the c^γ values indirectly. Specifically, the intermediate concentrations inside zones 2 and 3 cannot exceed the values predicted from the feed concentration. Therefore, c^γ values leading to steady-state concentrations that are too large are forbidden, and we have an upper bound on the c^γ values. We can tell Mathematica this by calculating c_2^{II} and c_1^{III} from the feed concentrations and setting the limit as follows:

```
relevantIntersections2[{0,0},{cF1,cF2}]
feedConstraint=Simplify[{cIII1<=c1,cII2<=c2}]/.Diagonal[%]
```

Naturally, the c^γ values must also be real and positive:

```
cγConstraints={cγ1∈Reals && cγ1>0, cγ2∈Reals && cγ2>0}
```

Step 3: Calculate the intersection concentrations as functions of c^γ

This step is fairly self-explanatory. We use the relevant intersection function to calculate the steady-state concentrations for zones 2 and 3. The diagonal function simply selects the c_1 and c_2 functions we care about.

```
intersectionSolutions={cIII1->c1,cII2->c2}/.
Diagonal@relevantIntersections2[{0,0},{cγ1,cγ2}]
```

Step 4: Calculate the intersections as functions of the flow rate ratios, using mass balances

Now we can calculate the values a second way, through the mass balances on zones 2 and 3. This is very simply done with

$$\text{balanceEq1} = -(m_3 - m_2)\text{cF}_2 == m_2\text{cII}_2 - \frac{a_2\text{cII}_2}{1 + b_2\text{cII}_2};$$

$$\text{balanceEq2} = (m_3 - m_2)\text{cF}_1 == m_2\text{cIII}_1 - \frac{a_1\text{cIII}_1}{1 + b_1\text{cIII}_1};$$

$$\text{balSolutions} = \text{Simplify@Solve}[\{\text{balanceEq1}, \text{balanceEq2}\}, \{\text{cII}_2, \text{cIII}_1\}]$$

Here, we simply solve the quadratic equations for $\mathrm{c}_2^{\mathrm{II}}$ and $\mathrm{c}_1^{\mathrm{III}}$. Unfortunately, we get four possible solutions to these equations, which we will have to keep track of.

Step 5: Combine the results of steps 3 and 4 to solve for c^γ

Now, we have two separate equations for $\mathrm{c}_2^{\mathrm{II}}$ and $\mathrm{c}_1^{\mathrm{III}}$ that we can combine to find an equation for c^γ in terms of the flow rate ratios.

```
combinedEquations={
(cIII1/.intersectionSolutions)==cIII1,
(cII2/.intersectionSolutions)==cII2
}/.balSolutions;
```

In this equation, we are starting from the equation $\text{cIII}_1 == \text{cIII}_1$, but replacing the left-hand side with the solutions we got from the eigenvector curves, and the right-hand side with the solutions from the mass balances. From this point, we need to simplify the resulting equations, and then solve them for c^{γ}.

```
simplifiedEquations=FullSimplify[combinedEquations,
Assumptions->{a1<a2,a2>0}];
cγSolutions=First@Simplify@Solve[#,{cγ1,cγ2}]&/
    @simplifiedEquations
```

This input uses the advanced Mathematica technique called *pure functions.* Since we have four possible solutions from our mass balances, our SimplifiedEquations variable contains four different pairs of equations. We want to solve each pair, so we define a function using # and the &:

First@Simplify@Solve[#,{cγ_1,cγ_2}]&

This function is "pure" because we have not assigned it a name. Whatever we pass into this function will be placed in the position of the hashtag # symbol, and then the function will solve (and simplify, and take the first answer) that input. Since we want Mathematica to solve each pair of equations, we can do so by telling Mathematica to pass the pairs into the function one at a time with the code:

/@simplifiedEquations

The result will be four separate solutions for $\{c_1^{\gamma}, c_2^{\gamma}\}$.

Step 6: View the regions specified by each constraint

First, we have to enumerate all the constrains on the system, and apply the relationships we derived in steps 3 and 4.

```
allConstraints={Simplify[zone2constraint/.balSolutions],
feedConstraint/.balSolutions,
cγConstraints/.cγSolutions};
```

Once we have these, we can plot them using Mathematica's RegionPlot function. First, we can define our own custom plot function with

```
plotRegion[r_,label_]:=RegionPlot[r,{m2,0.5,2.5},{m3,0.5,2.5},
PlotPoints->100,
PlotLabel->label,
FrameLabel->{m2,m3}
]
```

We can then assume some set of system parameters and plot the various regions. For convenience, we will use the same parameters as were chosen in Mazzotti [44], so that we have a straightforward basis for comparison.

The results of our coding are shown in Figure 3.34. Each row in the table corresponds to a solution for c_2^{II} and c_1^{III} while each column represents one of

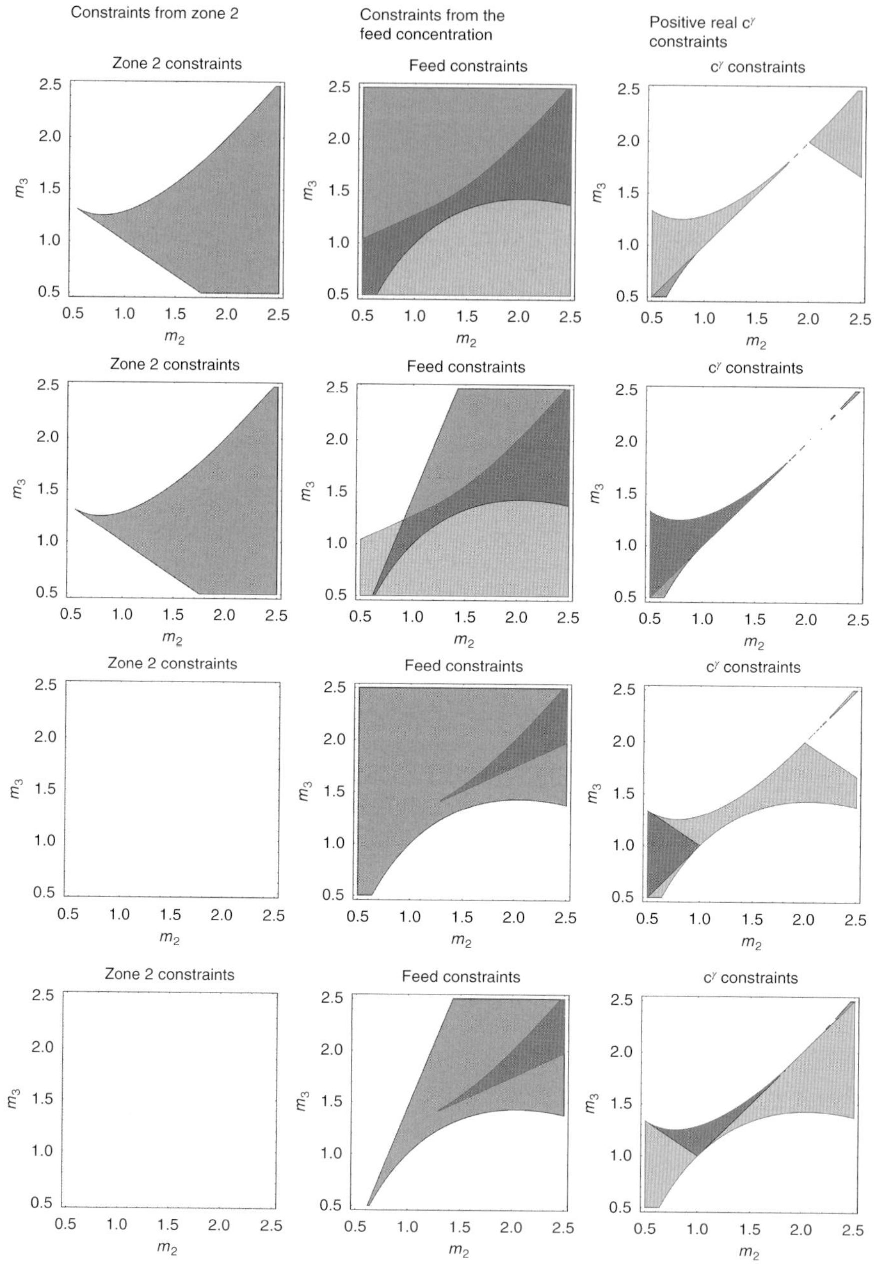

Figure 3.34 Constraints on column operation by the four solutions for $\mathbf{c}_2^{II}$ and $\mathbf{c}_1^{III}$. Heavy and light black denote regions where a constraint is satisfied. When both colors are present, operation is only possible where the regions intersect.

the three constraints on the flow rate ratios. As we can see, the second row is the only solution where all three constraints overlap, and therefore represents our true solution.

Therefore, we can apply our plotting function to all three constraints at once, and get the final operating region.

With this result, we have completed the derivation of the triangle theory for the Langmuir isotherm. Given this diagram, it is easy to select the m_2, m_3 point, which gives both the maximum feed flow rate (since the feed flow rate is proportional to $m_3 - m_2$) and complete separation (lies inside the region in Figure 3.35).

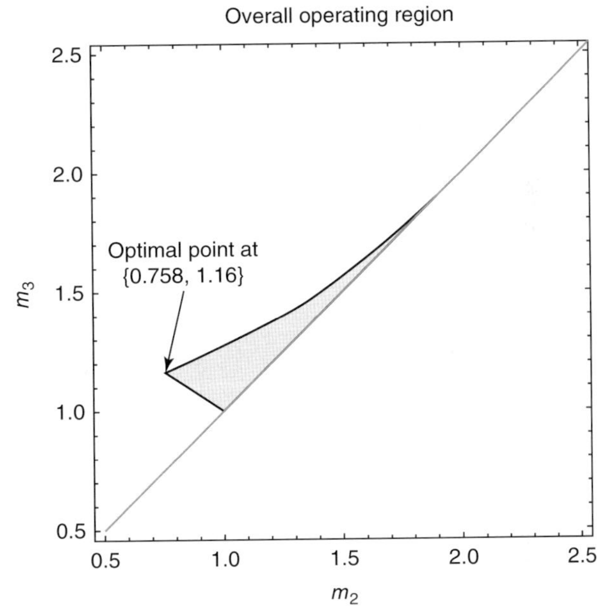

Figure 3.35 Region where all constraints are satisfied.

3.6 Standing Wave Design

Now that we have covered in detail the mathematics behind triangle theory, we can turn our attention to the alternative method known as "standing wave design." Fortunately, standing wave design (SWD) leans heavily on the same mathematical principles as the triangle theory, so there is no need to reinvent the wheel. At the fundamental level, SWD attempts to solve the same problem as triangle theory, and is constrained by the same mathematics. Given some SMB system, how can we predict the optimal flow rates, or the flow rates that give complete separation? SWD only really diverges from the triangle theory at the very end of the analysis. Instead of using the relationships we described in Section 3.2 to put analytical bounds on the m values, the authors of the SWD approach opted to employ an iterative solution approach.

3.6.1 Standing Wave Design in a Nonlinear Ideal System

We start the explanation of SWD in the case of a Langmuir isotherm without significant mass transfer effects. This will allow us to easily see the parallels between the SWD approach and the ideas we developed in Sections 3.2–3.5.

Figure 3.36 shows the standing wave design procedure for an ideal (no mass-transfer resistance), nonlinear isotherm system.

Before proceeding, it is important to inventory the differences in notations and in algorithm between this approach and the triangle theory approach discussed in Section 3.2. When the SWD procedure discusses "plateau concentrations," they are talking about the steady-state concentrations reached in zones 2 and 3 that we discussed in the triangle theory derivation. Moreover, they refer to the slopes of the characteristic curves (eigenvectors) as γ values. They have also calculated the eigenvalues at the "corners" of the separation region in the c_1, c_2 plane, and called these values "retention factors." The notation used for these concepts is explained in Table 3.3.

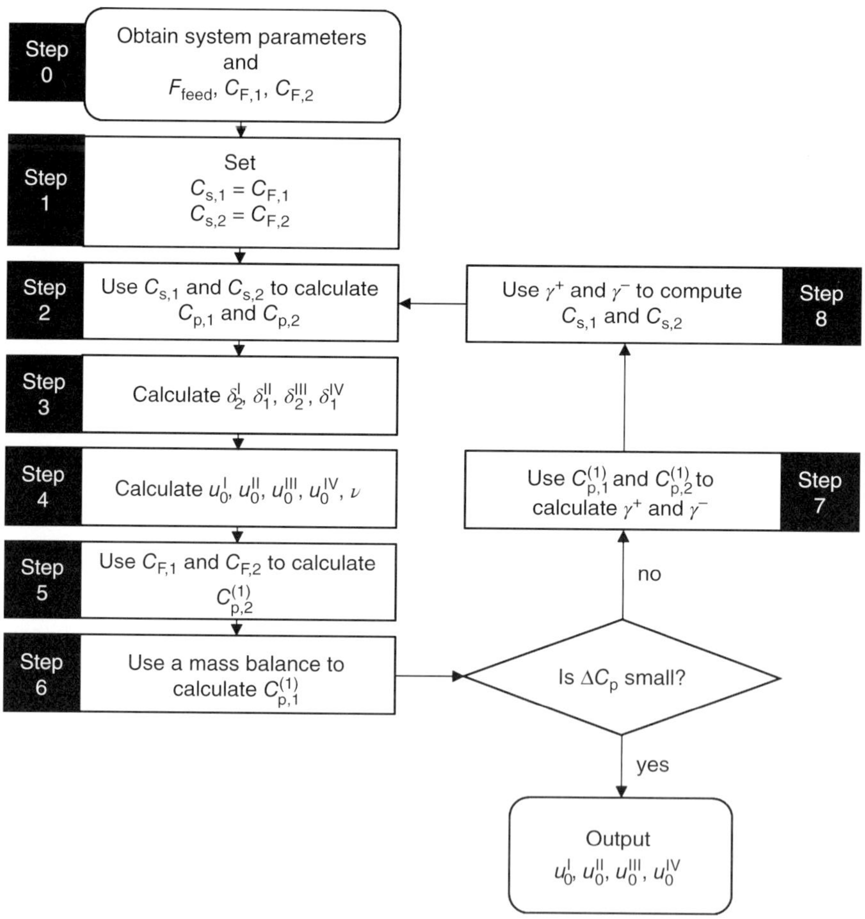

Figure 3.36 Simple Standing Wave Design Algorithm.

Table 3.3 Notations used in standing wave design.

Quantity	Symbol
Column cross section	A
Intercept of γ_+/γ_- characteristic	A_+/A_-
First Langmuir isotherm parameter, maximum loading	a_i
Second Langmuir isotherm parameter	b_i
Concentration	C
Plateau concentrations in constant regions	$C_{\text{p},i}$
Dilution concentration	$C_{\text{s},i}$
Concentration of component i	c_i
Equilibrium liquid phase concentration	c_i^{eq}
Axial dispersion	E_i
Volumetric flow rate of stream i	F_i
Henry's constant for component i	H_i
Mass transfer parameter	k_i
Column length	L
Flow rate ratio in zone k	m_k
Mass flow rate of the liquid	$\dot{M}_\text{L}$
Mass flow rate of the solid	$\dot{M}_\text{s}$
Solid phase loading	q
Volumetric flow rate of the liquid	$\dot{Q}_\text{L}$
Volumetric flow rate of the solid	$\dot{Q}_\text{s}$
Time	t
SMB switching time	t^*
Solid linear velocity	u_s
Liquid interstitial velocity in zone k	u_0^k
Target yield of component i	Y_i
Liquid linear velocity	v
Dead volume	V_D
Volume of bed k	V_k
Column axial position	z
Retention factor of component i in zone k	δ_i^k
Bulk density	ρ_b
Skeletal or solid density	ρ_s
Envelope density	ρ_e
External or inter-particle porosity	ε_i
Internal or intra-particle porosity	ε_p
Total porosity	ε_t
Slope of the characteristic curves in the hodograph plane	$\gamma_+\ \gamma_-$

Step 1: Calculate an initial guess of the "dilution concentrations"

To begin the procedure, an initial guess of C_s is required, and different methods for making this guess have been proposed [45–47]. In [48], the authors propose that the feed concentration be used. Another proposed method [46] is to iteratively increase the initial guesses, starting from values that were much smaller than the feed concentrations. The reason these approaches work is because of the feed concentration limit mentioned in step 2. The feed state represents a boundary on the zone 2 intermediate state, and this is shown visually in Figure 3.37.

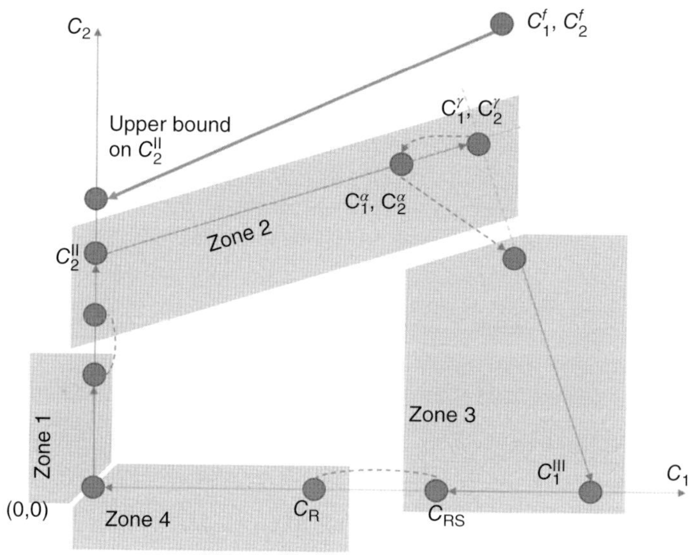

Figure 3.37 Feed state imposes upper bound on c_2^{II}.

Therefore, if we were trying to find some concentration state that satisfied our conditions, we could start from (0,0) and work upward, or we could start from c^f and work inward. It turns out that the feed concentration is always a better idea.

Step 2: Calculate the slopes of the characteristics in the (c_1, c_2) plane, $\gamma_{+,-}$

This step introduces a procedure that is used here and again in step 5. In this step, the roots of the equation

$$\frac{\partial q_2}{\partial C_1}\gamma^2 + \left(\frac{\partial q_2}{\partial C_2} - \frac{\partial q_1}{\partial C_1}\right)\gamma - \frac{\partial q_1}{\partial C_2} = 0 \tag{3.87}$$

```
characteristicPolynomial = Inactivate[ γ^2 ∂_{cA} qB + (∂_{cB} qB - ∂_{cA} qA) γ - ∂_{cB} qA == 0] /. isotherm
γ^2*∂_{cA} (aB cB)/(1 + bA cA + bB cB) + (∂_{cB} (aB cB)/(1 + bA cA + bB cB) + -1*∂_{cA} (aA cA)/(1 + bA cA + bB cB))*γ + -1*∂_{cB} (aA cA)/(1 + bA cA + bB cB) == 0
```
(M9)

are found and designated γ_+ and γ_-. This is the SWD nomenclature for the slopes of the characteristic curves in the (c_1, c_2) plane. We have already been able to calculate these in the triangle theory section, and there is nothing new in this formulation. The only difference is the nomenclature, with $q_1$and q_2 being the isotherm equations.

In this step, $C_{s,1}$ and $C_{s,2}$ are used as the basis point for calculating the characteristic curves. As a reminder, these curves represent the transitions from the (c_{s1}, c_{s2}) state to the $(c_{p,1}, 0)$ and $(0, c_{p,2})$ states [49, 50]. Therefore, finding the x and y intercepts of those lines allows us to come up with an estimate for $c_{p,1}$ and $c_{p,2}$. This procedure is conceptually identical to the first step of Exercise 2 in the triangle theory section, although care should be taken not to conflate the c^{γ} state from triangle theory with the C_s state from SWD; the C_s state is actually the SWD name for the c^{β} state. This is why the $c_{p,2}$ value is merely an estimate.

In the SWD literature, the intersections of the characteristic curves are found with the following procedure. First, two parameters, A_+ and A_- are found with the equations

$$A_+ = \frac{(a_1 - a_2)}{a_2 b_1 + \dfrac{a_1 b_2}{\gamma_+}} \tag{3.88}$$

$$A_- = \frac{(a_1 - a_2)}{a_2 b_1 + \dfrac{a_1 b_2}{\gamma_-}} \tag{3.89}$$

In[37]:=
```
{A₊, A₋} = FullSimplify[(a_A - a_B)/(a_B b_A + a_A b_B / γ) /. γPolynomialSolutions]
```

Out[37]=

$$\left\{\frac{a_A - a_B + a_B b_A c_A + a_A b_B c_B + \sqrt{4 a_A a_B b_A b_B c_A c_B + (a_A - a_B (1 + b_A c_A) + a_A b_B c_B)^2}}{2 a_B b_A}, \frac{a_A - a_B + a_B b_A c_A + a_A b_B c_B - \sqrt{4 a_A a_B b_A b_B c_A c_B + (a_A - a_B (1 + b_A c_A) + a_A b_B c_B)^2}}{2 a_B b_A}\right\} \tag{M10}$$

Finally these equations update the C_p values:

$$C_{p,2} = -\frac{A_+}{\gamma_+} \tag{3.90}$$

$$C_{p,1} = A_- \tag{3.91}$$

In[43]:=
```
peakEstimate = FullSimplify[{c_A^P → A₋, c_B^P → -A₊ / γ₊} /. γValues]
```

Out[43]=

$$\left\{c_A^P \to \frac{a_A - a_B + a_B b_A c_A + a_A b_B c_B - \sqrt{4 a_A a_B b_A b_B c_A c_B + (a_A - a_B (1 + b_A c_A) + a_A b_B c_B)^2}}{2 a_B b_A}, c_B^P \to \frac{-a_A + a_B + a_B b_A c_A + a_A b_B c_B + \sqrt{4 a_A a_B b_A b_B c_A c_B + (a_A - a_B (1 + b_A c_A) + a_A b_B c_B)^2}}{2 a_A b_B}\right\} \tag{M11}$$

As we have mentioned previously, these explicit equations are only possible because of the linearity of the Langmuir isotherm's characteristic curves. Regardless of the shape of the curves, however, the principle of finding the x and y intercepts of the two characteristic curves coming from the (c_{s1}, c_{s2}) point is consistent across isotherms, and *the Mathematica algorithms can find these intercepts even for nonlinear characteristics.*

Step 3: Calculate the "retention factors" in each zone

Now we begin to estimate wave velocities. In SWD, this is done by calculating δ values, which are referred to as retention factors for each zone. For a Langmuir isotherm, they are calculated with the following equations [47]:

$$\delta_2^{\mathrm{I}} = \varepsilon_\mathrm{p} + (1 - \varepsilon_\mathrm{p})a_2 \tag{3.92}$$

$$\delta_1^{\mathrm{II}} = \varepsilon_\mathrm{p} + (1 - \varepsilon_\mathrm{p})\frac{a_1}{1 + b_2 C_{\mathrm{p},2}} \tag{3.93}$$

$$\delta_2^{\mathrm{III}} = \varepsilon_\mathrm{p} + (1 - \varepsilon_\mathrm{p})\frac{a_2}{1 + b_1 C_{\mathrm{s},1} + b_2 C_{\mathrm{s},2}} \tag{3.94}$$

$$\delta_1^{\mathrm{IV}} = \varepsilon_\mathrm{p} + (1 - \varepsilon_\mathrm{p})\frac{a_1}{1 + b_1 C_{\mathrm{p},1}} \tag{3.95}$$

In[30]:=

```
idealRetention = {
  δB^"I" → ϵp + (1 - ϵp) aB,
  δA^"II" → ϵp + (1 - ϵp) aA / (1 + bB cB^p),
  δB^"III" → ϵp + (1 - ϵp) aB / (1 + bA cA^s + bB cB^s),
  δA^"IV" → ϵp + (1 - ϵp) aA / (1 + bA cA^p)}
```

(M12)

Out[30]=

$$\left\{\delta_B^{I} \to a_B (1 - \epsilon_p) + \epsilon_p,\ \delta_A^{II} \to \frac{a_A (1 - \epsilon_p)}{1 + b_B c_B^p} + \epsilon_p,\ \delta_B^{III} \to \frac{a_B (1 - \epsilon_p)}{1 + b_A c_A^s + b_B c_B^s} + \epsilon_p,\ \delta_A^{IV} \to \frac{a_A (1 - \epsilon_p)}{1 + b_A c_A^p} + \epsilon_p\right\}$$

Retention factors can be derived in a very straightforward way for any isotherm [49]. Specifically, as we will see in the next step, they are a simple function of the eigenvalues of the isotherm matrix. In other words, we have already calculated these values by calculating θ_1 and θ_2 during our triangle theory derivation.

Step 4: Calculate the interstitial velocities in each zone

To find the zone flow rates and interstitial velocity, we have to solve a system of five equations with five unknowns. Specifically, we have

$$u_0^{\mathrm{I}} = v\left(1 + \frac{1 - \varepsilon_\mathrm{b}}{\varepsilon_\mathrm{b}}\delta_2^{\mathrm{I}}\right) \tag{3.96}$$

$$u_0^{\mathrm{II}} = v\left(1 + \frac{1 - \varepsilon_\mathrm{b}}{\varepsilon_\mathrm{b}}\delta_1^{\mathrm{II}}\right) \tag{3.97}$$

$$u_0^{\mathrm{III}} = v\left(1 + \frac{1-\varepsilon_\mathrm{b}}{\varepsilon_\mathrm{b}}\delta_2^{\mathrm{III}}\right) \tag{3.98}$$

$$u_0^{\mathrm{IV}} = v\left(1 + \frac{1-\varepsilon_\mathrm{b}}{\varepsilon_\mathrm{b}}\delta_1^{\mathrm{IV}}\right) \tag{3.99}$$

$$\frac{F_\mathrm{feed}}{\varepsilon_\mathrm{b}\pi r^2} = u_0^{\mathrm{III}} - u_0^{\mathrm{II}} \tag{3.100}$$

(M13)

```
In[39]:= idealZoneVelocityEstimates =
  Thread[{u0"I", u0"II", u0"III", u0"IV"} → (v (1 + phaseRatio * #) & /@ {δB"I", δA"II", δB"III", δA"IV"})]
```

$$\text{Out[39]=}\ \left\{u_0^{I} \to v\left(1+\frac{\delta_B^{I}(1-\epsilon_b)}{\epsilon_b}\right),\ u_0^{II} \to v\left(1+\frac{\delta_A^{II}(1-\epsilon_b)}{\epsilon_b}\right),\ u_0^{III} \to v\left(1+\frac{\delta_B^{III}(1-\epsilon_b)}{\epsilon_b}\right),\ u_0^{IV} \to v\left(1+\frac{\delta_A^{IV}(1-\epsilon_b)}{\epsilon_b}\right)\right\}$$

It is no accident that these equations resemble Eq. (3.69). Indeed, these equations expose the retention factors as simple functions of the eigenvalues. However, care should be taken, since in zones where shock waves occur (e.g. zone 3 for the Langmuir isotherm), we still need to replace the proper eigenvalue equation with the difference equation from Eq. (3.86).

Step 5: Calculate an estimate for the peak concentration based on feed concentration

In this step, we compute a new value for $C_{\mathrm{p},2}$ by estimating it from the feed concentration, using the same procedure as in step 2. As we mentioned earlier, the $C_{\mathrm{p},2}$ value calculated from the feed concentration represents an upper bound, and so we take that upper bound as our estimate here.

Step 6: Calculate a mass balance based estimate for the peak concentration

In this step, we compute a new value for $C_{\mathrm{p},1}$ based on the mass balance according to Eq. (3.81). There is nothing too remarkable here, except that the consequences of our decisions in the previous step have been made clear. By allowing the contamination of the extract and raffinate streams, we can no longer ignore the concentrations of the extract and raffinate streams, as we did in the triangle theory derivation. Impurities mean that all the zones are coupled, and we cannot get away with mathematically isolating zones 2 and 3 from zones 1 and 4.

$$C_{\mathrm{p},1} = \frac{C_{\mathrm{F},1}F_\mathrm{feed}Y_1}{F_\mathrm{raf}} \tag{3.101}$$

(M14)

$$\text{In[47]:=}\ \texttt{massBalanceEstimate} = \left\{c_A^P \to \frac{c_A^{feed}\ f_{feed}\ y_A}{\epsilon_b\ \texttt{crossSection}\ (u_0^{III} - u_0^{IV})}\right\}$$

$$\text{Out[47]=}\ \left\{c_A^P \to \frac{c_A^{feed}\ f_{feed}\ y_A}{\texttt{crossSection}\ \epsilon_b\ (u_0^{III} - u_0^{IV})}\right\}$$

Step 7: Perform a Tolerance Check

At this point, we can compare how much the $C_{\mathrm{p},1}$ and $C_{\mathrm{p},2}$ values have changed by taking the difference between the values found in step 2, and the values found in steps 5 and 6. If the change is small enough (i.e. our estimates were close enough), we can stop iterating; otherwise we continue to step 8.

Step 8: Calculate the slopes of the characteristics at our new peak concentration values

Using our new estimates for the peak concentration, we calculate a new pair of γ values with a different method from step 2. These are once again the slopes of the transitions, except this time we are interested in the transition from $(C_{\mathrm{p},1}, C_{\mathrm{p},2})$ to $(C_{\mathrm{s},1}, C_{\mathrm{s},2})$. For a Langmuir system, the slopes are

$$\gamma_+ = \frac{a_2 - a_1 - a_1 b_2 C_{\mathrm{p},2}}{a_2 b_1 C_{\mathrm{p},2}} \tag{3.102}$$

$$\gamma_- = \frac{a_1 b_2 C_{\mathrm{p},1}}{a_1 - a_2 - a_2 b_1 C_{\mathrm{p},1}} \tag{3.103}$$

In[38]:= $\texttt{secondary}\gamma\texttt{Values} = \left\{\gamma_+ \to \dfrac{a_B - a_A - a_A\, b_B\, c_B^p}{a_B\, b_A\, c_B^p},\ \gamma_- \to \dfrac{a_A\, b_B\, c_A^p}{a_A - a_B - a_B\, b_A\, c_A^p}\right\}$

Out[38]= $\left\{\gamma_+ \to \dfrac{-a_A + a_B - a_A\, b_B\, c_B^p}{a_B\, b_A\, c_B^p},\ \gamma_- \to \dfrac{a_A\, b_B\, c_A^p}{a_A - a_B - a_B\, b_A\, c_A^p}\right\}$ (M15)

These can easily be derived by simplifying the eigenvectors when one component goes to zero.

Step 9: Calculate the intersections of the two characteristics

With our new γ values from step 7, we can now estimate a new pair of C_s values using the following equations, derived for the Langmuir isotherm. *This is the most difficult step for isotherms with nonlinear characteristics,* because it involves finding the intersection of two curves, the γ_+ characteristic coming from $(0, C_{\mathrm{p},2})$ and the γ_- characteristic coming from $(C_{\mathrm{p},1}, 0)$. Fortunately, it is almost always possible to solve this numerically if analytic solutions are not forthcoming. Note, however, that the intersection is only part of this equation. If we did simply calculate the intersection, then when we returned to step 2, the γ values calculated there would be identical, and lead us back to the exact same C_p state.

$$C_{\mathrm{s},1} = \frac{a_2 b_1 \gamma_+ \gamma_- (a_1 - a_2)}{a_1 a_2 b_1 b_2 (\gamma_+ + \gamma_-) + {a_1}^2 {b_2}^2 + {a_2}^2 {b_1}^2 \gamma_+ \gamma_-} \tag{3.104}$$

$$C_{\mathrm{s},2} = \frac{a_1 b_2 (a_2 - a_1)}{a_1 a_2 b_1 b_2 (\gamma_+ + \gamma_-) + {a_1}^2 {b_2}^2 + {a_2}^2 {b_1}^2 \gamma_+ \gamma_-} \tag{3.105}$$

In[45]:= $\texttt{denominator} = a_A\, a_B\, b_A\, b_B\, (\gamma_+ + \gamma_-) + {a_A}^2\, {b_B}^2 + {a_B}^2\, {b_A}^2\, \gamma_+\, \gamma_-;$

$\texttt{dilutionEstimate} = \left\{c_A^s \to \dfrac{a_B\, b_A\, \gamma_+\, \gamma_-\, (a_A - a_B)}{\texttt{denominator}},\ c_B^s \to \dfrac{a_A\, b_B\, (a_B - a_A)}{\texttt{denominator}}\right\}$

Out[46]= $\left\{c_A^s \to \dfrac{(a_A - a_B)\, a_B\, b_A\, \gamma_-\, \gamma_+}{a_A^2\, b_B^2 + a_B^2\, b_A^2\, \gamma_-\, \gamma_+ + a_A\, a_B\, b_A\, b_B\, (\gamma_- + \gamma_+)},\ c_B^s \to \dfrac{a_A\, (-a_A + a_B)\, b_B}{a_A^2\, b_B^2 + a_B^2\, b_A^2\, \gamma_-\, \gamma_+ + a_A\, a_B\, b_A\, b_B\, (\gamma_- + \gamma_+)}\right\}$ (M16)

From this point, we return to step 2.

3.6.2 Standing Wave Design in a System with Nonlinear Isotherm and Significant Mass Transfer Effects

So far, we have not really seen the power of the standing wave design system. SWD really shines in systems with significant mass transfer and dispersive effects. When those are present, the algorithm expands to include iterative estimates of the velocities, as shown in Figure 3.38.

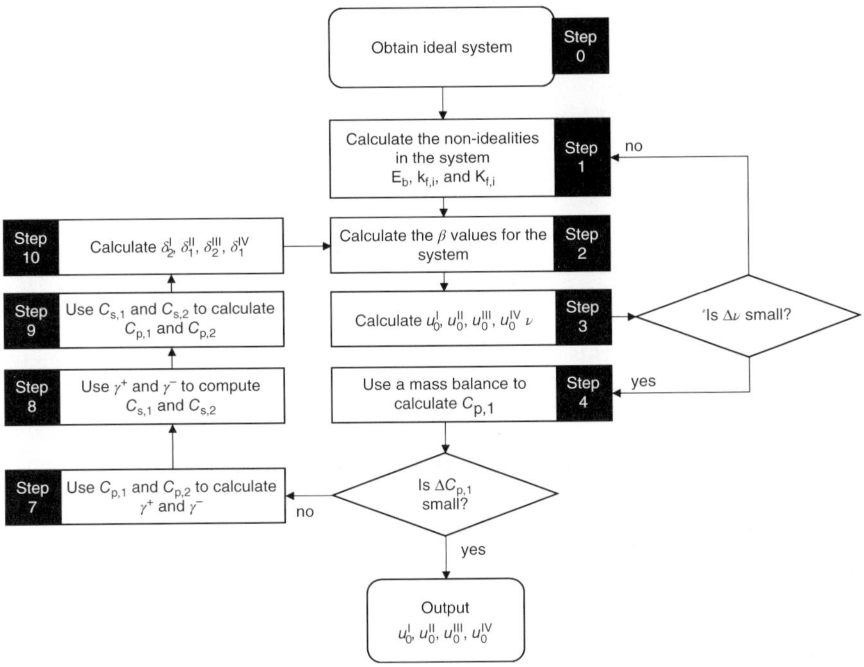

Figure 3.38 Standing wave design in a nonlinear and nonideal system.

In this expanded version of the algorithm, steps 4 through 10 are exactly the same as their ideal counterparts, i.e. the system without mass transfer resistance. We will therefore only cover steps 0–3 here.

Step 1: Perform an "ideal" SWD
In this step, we carry out the SWD as though the system were ideal, to give us a starting estimate of the mass transfer resistance.

Step 2: Estimate the value of the parameters related to mass transfer resistance
It is assumed that the mass transfer resistances and dispersion coefficients may be functions of the fluid velocities. If that is the case, then in this step, we use the estimates of the fluid velocities to calculate them.

Step 3: Calculate the "decay factors" for each zone
This is the beginning of the key contribution of the standing wave design. What happens here is that instead of sticking with the negligible dispersion assumption that we used very early on, we now turn our attention to a

steady-state analysis of the differential equations. Specifically, we attempt to solve the equation

$$(u_0 - (1 + P\delta)v)\frac{dc_i}{dx} = \left(E_b + \frac{P\, v^2\, \delta^2}{K_f}\right)\frac{d^2c_i}{dx^2} \tag{3.106}$$

```
In[229]:= zoneVelocityEstimates = {
   u0"I" → (1 + ϕ * δB"I") v + βB"I"/Lz (EB"I" + ϕ (v δB"I")^2 / KfB"I"),
   u0"II" → (1 + ϕ * δA"II") v + βA"II"/Lz (EA"II" + ϕ (v δA"II")^2 / KfA"II"),
   u0"III" → (1 + ϕ * δB"III") v - βB"III"/Lz (EB"III" + ϕ (v δB"III")^2 / KfB"III"),
   u0"IV" → (1 + ϕ * δA"IV") v - βA"IV"/Lz (EA"IV" + ϕ (v δA"IV")^2 / KfA"IV")}
```
(M17)

As we might expect, this equation has a solution that looks like an exponential decay in the x direction. Because of this, it is possible to write the solution in the form of

$$\frac{c_i|_{x=0}}{c_i|_{x=L}} = \exp\left[\frac{(1 + P\,\delta)v - u_0}{E_b + \frac{P\, v^2\, \delta^2}{K_f}}L\right] \tag{3.107}$$

Each zone has its own special exponential decay, but they can all be written in this form. Therefore, the velocity difference that we care about (i.e. $(1 + P\,\delta)v - u_0$) has a relationship with $\ln\left[\frac{c_i|_{x=0}}{c_i|_{x=L}}\right]$. This is where the β values come into play. Because the exponential decay functions only decrease along a column, the concentration ratio inside the natural log will end up being the ratio of the largest and smallest concentrations. This is where the β values come into play. By using the information we get from the characteristic transitions, we already have estimates of the smallest and largest concentrations in each column. The β values in each zone are therefore chosen to correspond to the component whose wave we want to have stand, and are defined as follows:

$$\beta_2^{\mathrm{I}} = \ln\left(\frac{C_{E,2}}{C_2^{\mathrm{I}}}\right) \tag{3.108}$$

$$\beta_1^{\mathrm{II}} = \ln\left(\frac{C_{s,1}^{*}}{C_{E,1}}\right) \tag{3.109}$$

$$\beta_2^{\rm III} = \ln\left(\frac{C_{\rm s,2}}{C_{\rm R,2}}\right) \tag{3.110}$$

$$\beta_1^{\rm IV} = \ln\left(\frac{C_{\rm R,1}}{C_2^{\rm I}}\right) \tag{3.111}$$

In[32]:=

```
decayFactors = {
  β_B^"I" → Log[ce2/c12],
  β_A^"II" → Log[css1/ce1],
  β_B^"III" → Log[cs2/cr2],
  β_A^"IV" → Log[cr1/c41]}
```

(M18)

Out[32]=

```
{β_B^I → Log[ce2/c12], β_A^II → Log[css1/ce1],
 β_B^III → Log[cs2/cr2],
 β_A^IV → Log[cr1/c41]}
```

Step 4: Estimate the interstitial velocity of the fluid in each zone

In this step, we leverage the β values together with (3.107) to generate the equations that allow us to calculate the interstitial liquid velocities that make this new concentration wave – which takes into account nonidealities – stand still where we want it to. Those equations are as follows:

$$u_0^{\rm I} = (1 + P\,\delta_2^{\rm I})v + \frac{\beta_2^{\rm I}}{L^{\rm I}}\left(E_b^I + \frac{P\,v^2\,(\delta_2^{\rm I})^2}{K_{\rm f}^{\rm I}}\right) \tag{3.112}$$

$$u_0^{\rm II} = (1 + P\,\delta_1^{\rm II})v + \frac{\beta_1^{\rm II}}{L^{\rm II}}\left(E_b^{\rm II} + \frac{P\,v^2\,(\delta_1^{\rm II})^2}{K_{\rm f}^{\rm II}}\right) \tag{3.113}$$

$$u_0^{\rm I} = (1 + P\,\delta_2^{\rm III})v + \frac{\beta_2^{\rm I}}{L^{\rm III}}\left(E_{\rm b}^{\rm III} + \frac{P\,v^2\,(\delta_2^{\rm III})^2}{K_{\rm f}^{\rm III}}\right) \tag{3.114}$$

$$u_0^{\rm I} = (1 + P\,\delta_1^{\rm IV})v + \frac{\beta_1^{\rm IV}}{L^{\rm IV}}\left(E_{\rm b}^{\rm IV} + \frac{P\,v^2\,(\delta_1^{\rm IV})^2}{K_{\rm f}^{\rm IV}}\right) \tag{3.115}$$

In[40]:=

$$\text{zoneVelocityEstimates} = \Bigg\{$$

$$u_0^{\text{"I"}} \rightarrow (1 + \text{phaseRatio} * \delta_B^{\text{"I"}})\, v + \frac{\beta_B^{\text{"I"}}}{L_z}\left(E_B^{\text{"I"}} + \frac{\text{phaseRatio}\,(v\,\delta_B^{\text{"I"}})^2}{Kf_B^{\text{"I"}}}\right),$$

$$u_0^{\text{"II"}} \rightarrow (1 + \text{phaseRatio} * \delta_A^{\text{"II"}})\, v + \frac{\beta_A^{\text{"II"}}}{L_z}\left(E_A^{\text{"II"}} + \frac{\text{phaseRatio}\,(v\,\delta_A^{\text{"II"}})^2}{Kf_A^{\text{"II"}}}\right),$$

$$u_0^{\text{"III"}} \rightarrow (1 + \text{phaseRatio} * \delta_B^{\text{"III"}})\, v - \frac{\beta_B^{\text{"III"}}}{L_z}\left(E_B^{\text{"III"}} + \frac{\text{phaseRatio}\,(v\,\delta_B^{\text{"III"}})^2}{Kf_B^{\text{"III"}}}\right),$$

$$u_0^{\text{"IV"}} \rightarrow (1 + \text{phaseRatio} * \delta_A^{\text{"IV"}})\, v - \frac{\beta_A^{\text{"IV"}}}{L_z}\left(E_A^{\text{"IV"}} + \frac{\text{phaseRatio}\,(v\,\delta_A^{\text{"IV"}})^2}{Kf_A^{\text{"IV"}}}\right)\Bigg\}$$

Out[40]=

$$\Bigg\{u_0^{I} \rightarrow v\left(1 + \frac{\delta_B^{I}(1-\epsilon_b)}{\epsilon_b}\right) + \frac{\beta_B^{I}\left(E_B^{I} + \frac{v^2(\delta_B^{I})^2(1-\epsilon_b)}{Kf_B^{I}\,\epsilon_b}\right)}{L_z},$$

$$u_0^{II} \rightarrow v\left(1 + \frac{\delta_A^{II}(1-\epsilon_b)}{\epsilon_b}\right) + \frac{\beta_A^{II}\left(E_A^{II} + \frac{v^2(\delta_A^{II})^2(1-\epsilon_b)}{Kf_A^{II}\,\epsilon_b}\right)}{L_z},$$

$$u_0^{III} \rightarrow v\left(1 + \frac{\delta_B^{III}(1-\epsilon_b)}{\epsilon_b}\right) - \frac{\beta_B^{III}\left(E_B^{III} + \frac{v^2(\delta_B^{III})^2(1-\epsilon_b)}{Kf_B^{III}\,\epsilon_b}\right)}{L_z},$$

$$u_0^{IV} \rightarrow v\left(1 + \frac{\delta_A^{IV}(1-\epsilon_b)}{\epsilon_b}\right) - \frac{\beta_A^{IV}\left(E_A^{IV} + \frac{v^2(\delta_A^{IV})^2(1-\epsilon_b)}{Kf_A^{IV}\,\epsilon_b}\right)}{L_z}\Bigg\}$$

(M19)

Step 5: Calculate the peak concentrations from a mass balance

In this step, we once again use a mass balance to calculate new values for the peak concentration. This step is virtually identical to step 6 in the ideal case; mass transfer resistances do not change how you do a mass balance.

Step 6: Tolerance Check

This is a simple tolerance check, and works the same way as the ideal case.

Step 7: Output Results

If we succeed in our tolerance check, then we simply output the zone velocities we have calculated thus far.

Step 8: Calculate the slope of the characteristics

This step is identical to step 7 from the ideal case. We once again calculate the slope of the characteristics by using the modified slope formula.

Step 9: Find the dilution concentrations

The slopes of the characteristics are used to calculate the dilution concentrations by finding their intercepts, just like in step 8 of the ideal case.

Step 10: Use the dilution concentrations to calculate the peak concentrations

The dilution concentrations are used to create a new estimate of the peak concentrations, using the characteristic polynomial method as in step 2 of the ideal case.

Step 11: Calculate the retention factors, with additional nonideal contributions

We calculate the retention factors; however they are modified to account for dead volume:

$$\delta_2^{I} = \varepsilon_p + (1-\varepsilon_p)a_2 + \frac{V_{dead}}{pV_{bed}\varepsilon_b} \tag{3.116}$$

$$\delta_1^{II} = \varepsilon_p + (1-\varepsilon_p)\frac{a_1}{1 + b_2 C_{p,2}} + \frac{V_{dead}}{pV_{bed}\varepsilon_b} \tag{3.117}$$

$$\delta_2^{\mathrm{III}} = \varepsilon_{\mathrm{p}} + (1-\varepsilon_{\mathrm{p}})\frac{a_2}{1+b_1C_{\mathrm{s},1}+b_2C_{\mathrm{s},2}} + \frac{V_{\mathrm{dead}}}{pV_{\mathrm{bed}}\varepsilon_{\mathrm{b}}} \tag{3.118}$$

$$\delta_1^{\mathrm{IV}} = \varepsilon_{\mathrm{p}} + (1-\varepsilon_{\mathrm{p}})\frac{a_1}{1+b_1C_{\mathrm{p},1}} + \frac{V_{\mathrm{dead}}}{pV_{\mathrm{bed}}\varepsilon_{\mathrm{b}}} \tag{3.119}$$

In[67]:=

$$\texttt{nonidealRetention} = \left\{ \delta_B^{\text{"I"}} \to \epsilon_p + (1-\epsilon_p)\, a_B + \frac{\mathrm{DV}}{\texttt{phaseRatio}\, L\, A_c\, \epsilon_b},\ \delta_A^{\text{"II"}} \to \epsilon_p + (1-\epsilon_p)\frac{a_A}{1+b_B c_B^p} + \frac{\mathrm{DV}}{\texttt{phaseRatio}\, L\, A_c\, \epsilon_b},\ \delta_B^{\text{"III"}} \to \epsilon_p + (1-\epsilon_p)\left(a_B / \left(1+b_A c_A^s + b_B c_B^s\right)\right) + \frac{\mathrm{DV}}{\texttt{phaseRatio}\, L\, A_c\, \epsilon_b},\ \delta_A^{\text{"IV"}} \to \epsilon_p + (1-\epsilon_p)\frac{a_A}{1+b_A c_A^p} + \frac{\mathrm{DV}}{\texttt{phaseRatio}\, L\, A_c\, \epsilon_b} \right\} \tag{M20}$$

Out[67]=

$$\left\{ \delta_B^{I} \to \frac{\mathrm{DV}}{A_c L (1-\epsilon_b)} + a_B(1-\epsilon_p) + \epsilon_p,\ \delta_A^{II} \to \frac{\mathrm{DV}}{A_c L (1-\epsilon_b)} + \frac{a_A(1-\epsilon_p)}{1+b_B c_B^p} + \epsilon_p,\ \delta_B^{III} \to \frac{\mathrm{DV}}{A_c L (1-\epsilon_b)} + \frac{a_B(1-\epsilon_p)}{1+b_A c_A^s + b_B c_B^s} + \epsilon_p,\ \delta_A^{IV} \to \frac{\mathrm{DV}}{A_c L (1-\epsilon_b)} + \frac{a_A(1-\epsilon_p)}{1+b_A c_A^p} + \epsilon_p \right\}$$

Once these retention factors are found, the procedure returns to step 2 and they are used to calculate new β values.

3.7 Standing Wave Design Workshop: Calculating the Operating Conditions for an Ideal and a Nonideal System

Design an SMB system for separating Trogers base enantiomers with the following properties:

- A racemic feed, with an overall concentration of $1\ \mathrm{g\ l^{-1}}$.
- Able to handle a feed flow rate of $5.7\ \mathrm{ml\ min^{-1}}$ while achieving complete separation of the enantiomers.
- Constructed from stock adsorbent beds that have a diameter of 2.56 cm and a height of 96 cm.
- The beds have an interparticle void fraction of 0.45, and no intraparticle voids.
- Has no dead volume, and negligible mass transfer effects.

Step 1: Open the Mathematica CDF File: "SWD Ideal Langmuir.cdf"
Step 2: Execute the entire notebook

When you open the file, Mathematica does not load the equations and variables in the notebook into its "kernel." Using the evaluation menu as shown in Figure 3.39, we can perform all initialization calculations, as well as load any variable values into the kernel (Figure 3.40).

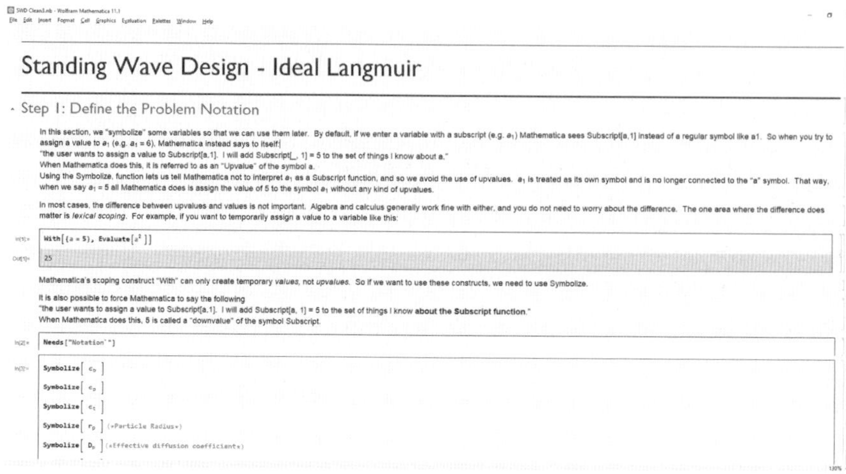

Figure 3.39 Ideal Langmuir CDF file.

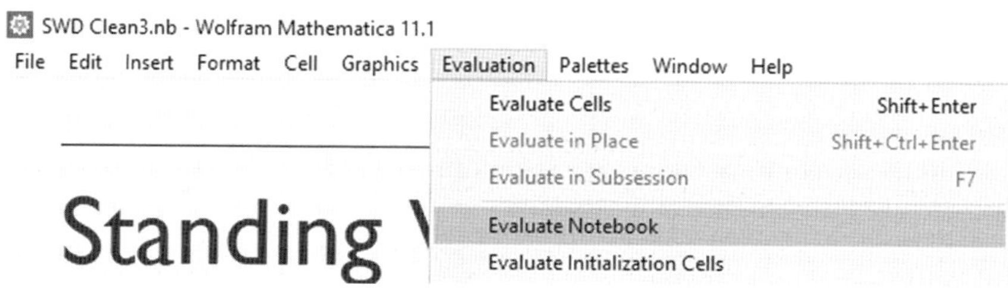

Figure 3.40 Evaluating all the equations defined in the notebook.

Step 3: Scroll to the end of the document

Collapse the sections by clicking on either the small arrow to the left of the section heading (#1 in Figure 3.41) or the square bracket to the right of the section heading (#2 in Figure 3.41). This way, you can see only the "final result" that we are interested in.

Figure 3.41 Collapsing the section headings.

Once the headings are collapsed, it is much easier to find the section containing the user interface, as shown in Figure 3.42.

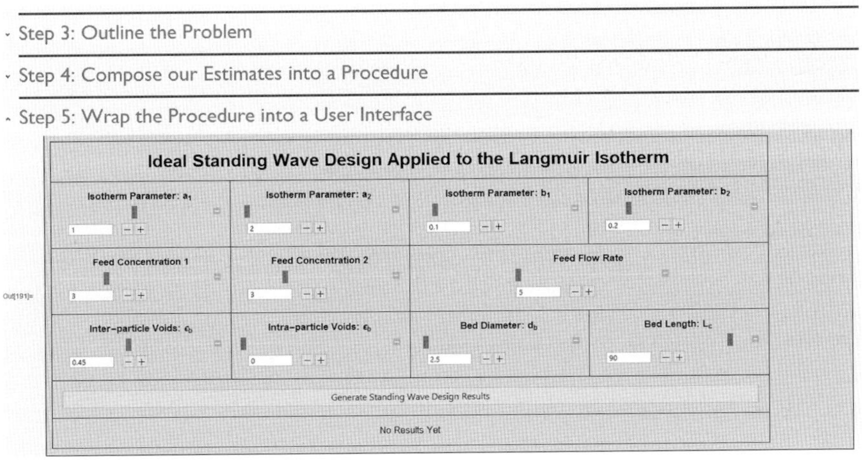

Figure 3.42 The SWD user interface.

Step 4: Enter the problem's values into the boxes provided, then click "Generate" (Figure 3.43)

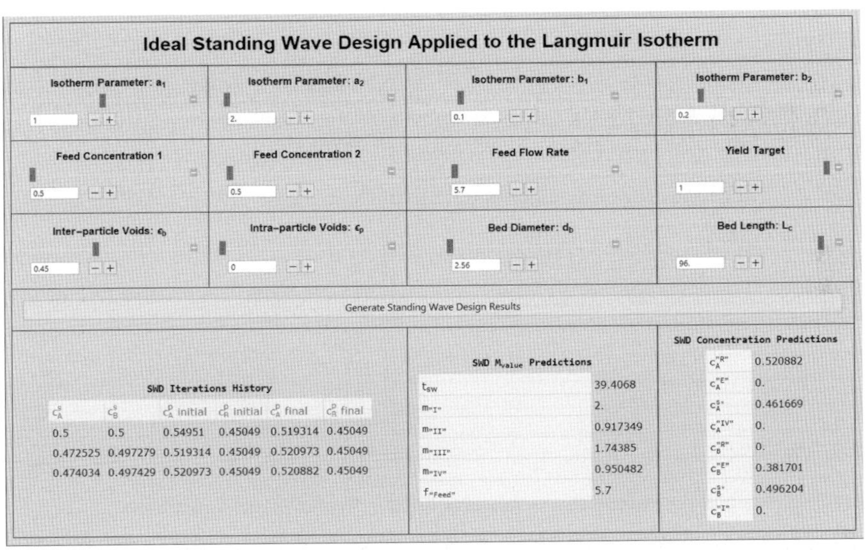

Figure 3.43 The SWD user interface.

Step 5: Create a copy of the Aspen file from Workshop 1.7 and modify it to match this problem statement.

Adjust the bed size and conditions, increase the mass transfer coefficients to 500, and then open the Flowsheet Constraints. Because the standing wave design predicts a switching time for us, we no longer need the pressure drop constraint to fully specify the flow rates. Delete (or comment out) these lines from the flowsheet constraints (Figure 3.44):

```
//M VALUE SPECIFICATIONS

M1 as RealVariable(fixed);
M2 as RealVariable(fixed);
M3 as RealVariable(fixed);
M4 as RealVariable(fixed);

//MaxPdrop as Pressure_Change(fixed);

ColVol as c_Volume(free);
EffectVol as c_Volume(free);

ColVol = SMB.Hb * 3.1415926 * (SMB.Db / 2) ^ 2;

//The 1000 terms are unit conversions, we want to be working in cm^3, but the volumes flow rates are in L
EffectVol * ColVol = SMB.Hb ^ 3 * (1 * DESORBENT.Flowrate - 2 * FEED.Flowrate + 4 * RAFFINATE.Flowrate + 6 * SMB.Recycle_Flowrate) * 1000;

//MaxPdrop * SMB.Hb = 0.1 * EffectVol;

M1 * ColVol * (SMB.Ei - 1) = ColVol * SMB.Ei - SMB.Switch_Time * 1000 * (DESORBENT.Flowrate                  + SMB.Recycle_Flowrate) + 0.23;
M2 * ColVol * (SMB.Ei - 1) = ColVol * SMB.Ei - SMB.Switch_Time * 1000 * (RAFFINATE.Flowrate - FEED.Flowrate + SMB.Recycle_Flowrate) + 0.23;
M3 * ColVol * (SMB.Ei - 1) = ColVol * SMB.Ei - SMB.Switch_Time * 1000 * (RAFFINATE.Flowrate                  + SMB.Recycle_Flowrate) + 0.23;
M4 * ColVol * (SMB.Ei - 1) = ColVol * SMB.Ei - SMB.Switch_Time * 1000 * (                                      SMB.Recycle_Flowrate) + 0.23;
```

Figure 3.44 The Updated Flowsheet Constraints.

```
MaxPdrop as Pressure_Change(fixed);

MaxPdrop * SMB.Hb = 0.1 * EffectVol;
```

Once the new constraints are compiled, we also need to modify the Switch_Time variable to be fixed. Open the M_Specced form and change the variable specification (Figure 3.45).

M_Specced Table

	Value	Units	Spec
SMB.Recycle_Flowrate	12.1977	ml/min	Free
SMB.Switch_Time	**39.4068**	min	Fixed
DESORBENT.Flowrate	7.231	ml/min	Free
EXTRACT.Flowrate	7.46	ml/min	Free
FEED.Flowrate	5.7	ml/min	Free
RAFFINATE.Flowrate	5.471	ml/min	Free
SMB.Hb	**96.0**	cm	Fixed
SMB.Db	**2.56**	cm	Fixed
SMB.Ei	**0.45**	m3 void/m3 b	Fixed
M1	**1.99898**		Fixed
M2	**0.917277**		Fixed
M3	**1.74378**		Fixed
M4	**0.950482**		Fixed
ColVol	494.13	cm3	Free

Figure 3.45 Assigning values to the variables defined in the flowsheet constraints.

Step 6: Run the simulation

Once you have updated the simulation, open the axial concentration plot and run the simulation in **steady state mode**. You will see the result shown in Figure 3.46.

Compare the concentrations on this plot to the predictions made by the SWD calculations (Figure 3.43). Also compare the extract and raffinate predictions.

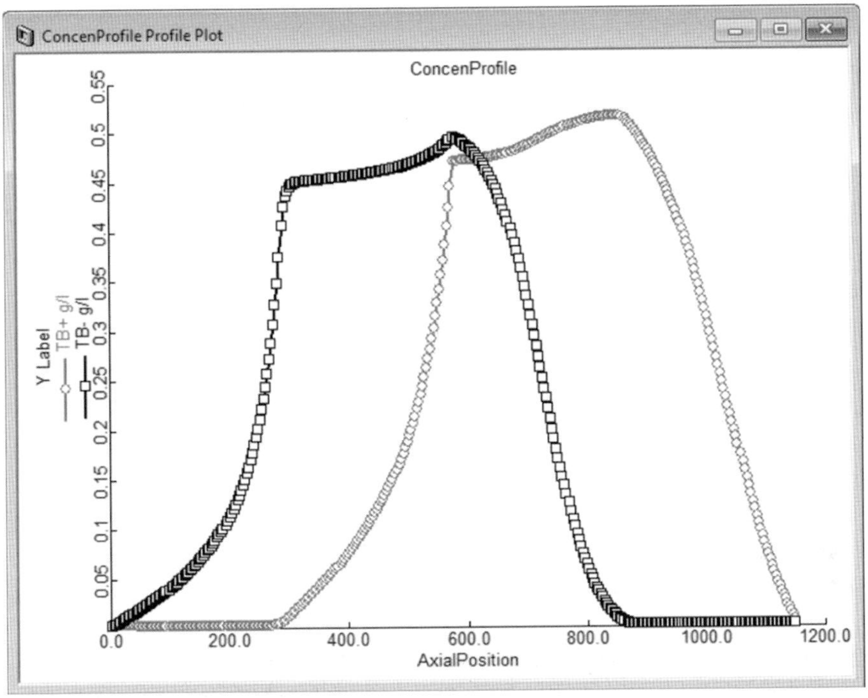

Figure 3.46 The Separation Performance.

Step 7: Return the mass transfer effects to normal

Earlier, we increased the mass transfer coefficient to simulate a system with no mass transfer resistance. Now, let us observe the effects of large mass transfer resistance. Change the mass transfer coefficients of the system to the values shown in Figure 3.47.

MTC(*)			
MTC("TB+")	1.e-003	1/s	Constant mass transfer coefficient
MTC("TB-")	0.003	1/s	Constant mass transfer coefficient
IP(*)			

Figure 3.47 A system with significant mass transfer resistance.

With this large mass transfer resistance, the system will no longer behave as predicted by the ideal model equations. Instead, the concentration peaks will be spread out and overlap. This means the purity and recovery of both components will be compromised (Figure 3.48).

To design a system to work with this high mass transfer scenario, we will upgrade to the more detailed SWD calculations described in step 8.

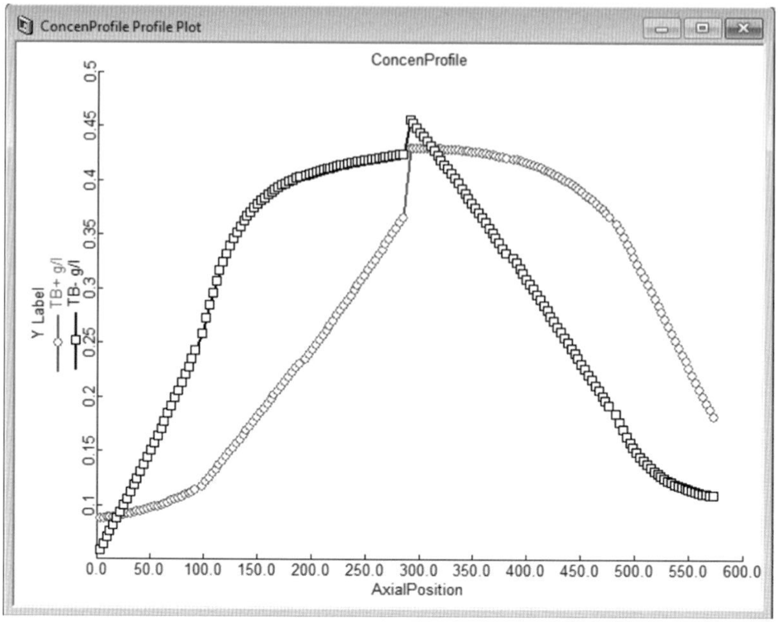

Figure 3.48 Separation performance with high mass transfer resistance.

Step 8: Calculating the operating conditions for a system with significant mass transfer resistance

Open the Mathematica CDF File: "SWD NonIdeal Langmuir.cdf" and scroll to the bottom to find the GUI. Enter the updated mass transfer values as shown in Figure 3.49.

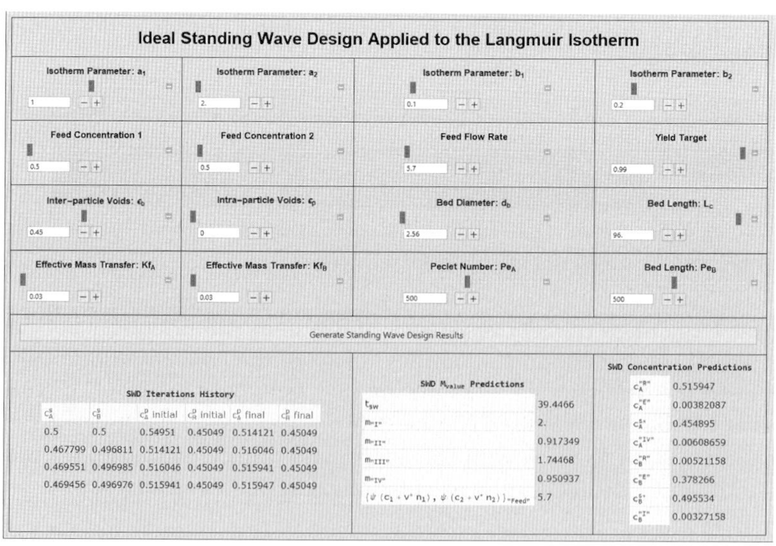

Figure 3.49 Entering the Mass Transfer Coefficients.

Press the "Generate" button to view the results.

Step 9: Run the Aspen simulation using the values from SWD

The updated results are shown in Figure 3.50.

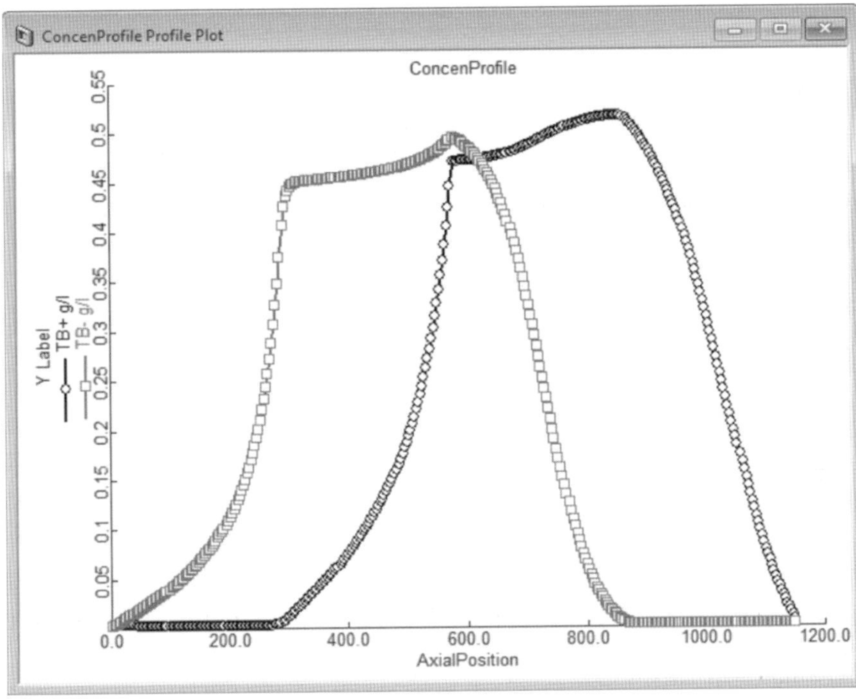

Figure 3.50 The Separation Performance of the High Mass Transfer Resistance Case.

3.8 Conclusions

In this chapter, we have covered the theory behind the two principal design methods for SMB systems: triangle theory and standing wave design. Specifically, we have learned about

1. the simplifications applied to the differential equations that allow us to put the problem into a matrix form;
2. the significance of the eigenvectors and eigenvalues of the matrices;
3. how these eigenvalues and eigenvectors are used differently between standing wave design and triangle theory.

In addition to the theory, we have also learned how to apply them to real problems using tools developed in Mathematica. Specifically we learned the following:

1. How to use the triangle theory tool built into Aspen chromatography
2. How to use Mathematica tools to implement triangle theory for isotherms not available in Aspen
3. How to use Mathematica tools to implement standing wave design in situations where we expect mass transfer resistance to be significant
4. How to convert the results of standing wave design into Aspen input

3.9 Practice Problems

3.9.1 Use the Triangle Theory Tool and the Standing Wave Design Tool to Create an Aspen Simulation of the Separation of 1-phenol-1-propanol on Tribenzoate

This enantioseparation was studied by Kaczmarski et al. [47] and the adsorption behavior was found to follow the Toth isotherm. Using the user interface generated in Mathematica, we can easily generate the triangle theory plots for this Toth isotherm. The isotherm parameters for 1-phenol-1-propanol on chircacel OB are given in Table 3.4.

You may notice that Aspen does not have a Toth isotherm model. In order to specify our own custom isotherm, choose "User Submodel" when selecting the isotherm assumption. Afterwards, enter the following code into the flowsheet constraints:

```
CONSTRAINTS
//Flowsheet variables and equations...

Within SMB
 For cc In PortSet Do//For each bed number
  Within Column_(cc)//Use the variables inside the bed
   For ii In z.InteriorSet Do//For each node in the
   column
    For jj In AdsorbedComps Do//For each chemical
    component
     Ws(ii,jj) * (1 + (IP(2,"R-PP") * C(ii,"R-PP")
      + IP(2,"S-PP") * C(ii,"S-PP"))^IP(3,jj))^(1/IP(3,jj))
      = C(ii,jj) * IP(1,jj) * IP(2,jj);
    EndFor
   EndFor
  EndWithin
 EndFor
EndWithin
END
```

Table 3.4 Toth isotherm parameters.

Parameter	Value
q_s	71
b_1	0.06
b_2	0.09
v	0.77

Source: Kaczmarski et al. 2006 [47]. Reproduced with permission of Elsevier.

Table 3.5 Adsorbent characteristics.

Adsorbent characteristics	Value
Bed void fraction: ε_b	0.35
Particle void fraction: ε_p	0.561

Table 3.6 Mass transfer effects in the 1-phenol-1-propanol separation.

Parameter	Value
Molecular diffusion coefficient ($cm^2\ min^{-1}$)	1.7e-3
Effective diffusion coefficient ($cm^2\ min^{-1}$)	1.5e-5
Dispersion coefficient ($cm^2\ min^{-1}$)	3.2e-3

This assumes that your components are named "R-PP" and "S-PP" and that your SMB block is named "SMB." Compile the flowsheet to complete the Toth isotherm specification.

In addition to these parameters, we need to specify a feed concentration. Since the Kaczmarski paper derived the isotherm parameters from racemic mixtures with concentrations of less than 5 $g\ l^{-1}$ we will select a feed concentration of $c_1 = c_2 = 5\ g\ l^{-1}$ for our hypothetical SMB column. The adsorbent characteristics are given in Table 3.5.

Using the triangle theory tool, generate an initial simulation without any mass transfer resistance. Next, use the standing wave design tool to incorporate mass transfer resistance into the column design. The mass transfer resistance parameters are shown in Table 3.6.

The triangle theory region for this separation is shown in Figure 3.51.

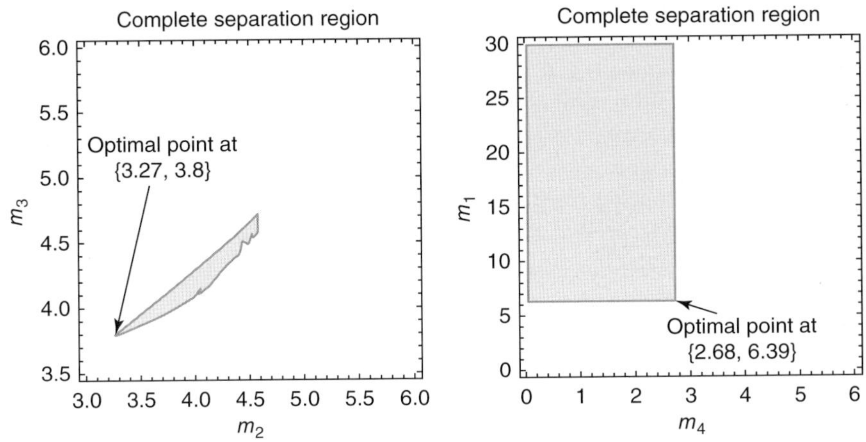

Figure 3.51 Optimum Operating Regions for the Toth isotherm.

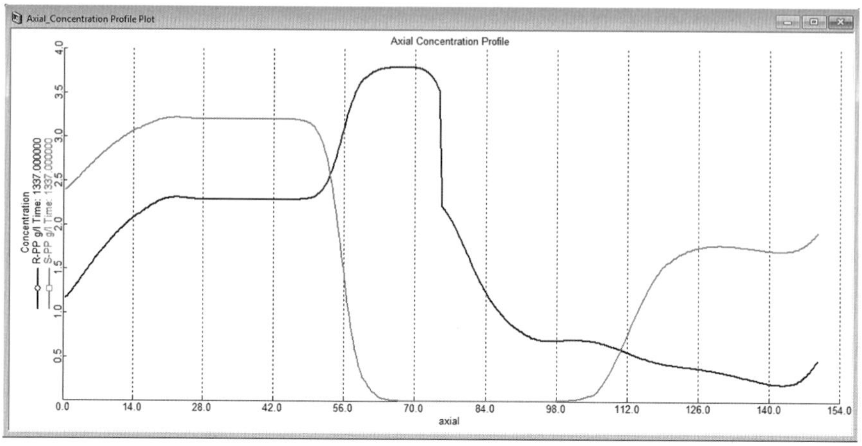

Figure 3.52 The separation performance of our 1-phenol-1-propanol separator.

The Aspen simulation of this separation will produce the interesting axial concentration profile shown in Figure 3.52.

Bibliography

Triangle Theory

Abel, S., Erdem, G., Amanullah, M. et al. (2005). Optimizing control of simulated moving beds—experimental implementation. *Journal of Chromatography A* **1092** (1): 2–16.

Abel, S., Erdem, G., Mazzotti, M. et al. (2004). Optimizing control of simulated moving beds—linear isotherm. *Journal of Chromatography A* **1033** (2): 229–239.

Abel, S., Mazzotti, M., and Morbidelli, M. (2002). Solvent gradient operation of simulated moving beds: I. Linear isotherms. *Journal of chromatography A* **944** (1): 23–39.

Araújo, J.M.M., Rodrigues, R.C.R., and Mota, J.P.B. (2006). Optimal design and operation of a certain class of asynchronous simulated moving bed processes. *Journal of Chromatography A* **1132** (1–2): 76–89.

Azevedo, D. and Rodrigues, A.E. (1999). Design of a simulated moving bed in the presence of mass-transfer resistances. *AIChE journal* **45** (5): 956–966.

Azevedo, D. and Rodrigues, A.E. (2001). Fructose–glucose separation in a SMB pilot unit: modeling, simulation, design, and operation. *AIChE journal* **47** (9): 2042–2051.

Azevedo, D. and Rodrigues, A.E. (2001). Design methodology and operation of a simulated moving bed reactor for the inversion of sucrose and glucose–fructose separation. *Chemical Engineering Journal* **82** (1): 95–107.

Bae, Y.-S., Kim, K.-M., Moon, J.-H. et al. (2008). Effects of flow-rate ratio on startup and cyclic steady-state behaviors of simulated moving bed under linear conditions. *Separation and Purification Technology* **62** (1): 148–159.

Biressi, G., Mazzotti, M., and Morbidelli, M. (2002). Experimental investigation of the behavior of gas phase simulated moving beds. *Journal of Chromatography A* **957** (2): 211–225.

Buhlert, K., Lehr, M., and Jungbauer, A. (2009). Construction and development of a new single-column simulated moving bed system on the laboratory scale. *Journal of Chromatography A* **1216** (50): 8778–8786.

Chiang, A.S. (1998). Equilibrium theory for simulated moving bed adsorption processes. *AIChE journal* **44** (11): 2431–2441.

Ching, C.B., Arlt, W., Lisso, M., and Wozny, G. (2000). Kinetic and equilibrium study of the enantioseparation of fenoprofen in a batch setup. *Industrial & Engineering Chemistry Research* **39** (11): 4365–4369.

Ching, C.B., Hidajat, K., Ho, C., and Ruthven, D.M. (1987). Modelling of a simulated counter-current adsorption process. *Reactive Polymers, Ion Exchangers, Sorbents* **6** (1): 15–20.

Denet, F., Hauck, W., Nicoud, R.M. et al. (2001). Enantioseparation through Supercritical Fluid Simulated Moving Bed (SF-SMB) Chromatography. *Industrial & Engineering Chemistry Research* **40** (21): 4603–4609.

Dunnebier, G., Weirich, I., and Klatt, K.U. (1998). Computationally efficient dynamic modelling and simulation of simulated moving bed chromatographic processes with linear isotherms. *Chemical Engineering Science* **53** (14): 2537–2546.

Erdem, G., Abel, S., Morari, M. et al. (2004). Automatic control of simulated moving beds. *Industrial & Engineering Chemistry Research* **43** (2): 405–421.

Erdem, G., Abel, S., Morari, M. et al. (2004). Automatic control of simulated moving beds II: nonlinear isotherm. *Industrial & Engineering Chemistry Research* **43** (14): 3895–3907.

Erdem, G., Morari, M., Amanullah, M. et al. (2006). Optimizing control of an experimental simulated moving bed unit. *AIChE Journal* **52** (4): 1481–1494.

Fricke, J. and Schmidt-Traub, H. (2003). A new method supporting the design of simulated moving bed chromatographic reactors. *Chemical Engineering and Processing: Process Intensification* **42** (3): 237–248.

García, M.-S.G., Balsa-Canto, E., Banga, J.R., and Vande Wouwer, A. (2006). Dynamic optimization of a simulated moving bed (SMB) chromatographic separation process. *Industrial & Engineering Chemistry Research* **45** (26): 9033–9041.

Gentilini, A., Migliorini, C., Mazzotti, M., and Morbidelli, M. (1998). Optimal operation of simulated moving-bed units for non-linear chromatographic separations: II. Bi-Langmuir isotherm1Part I of this series is Mazzotti et al., 1997 [1]. 1. *Journal of Chromatography A* **805** (1): 37–44.

Giacobello, S., Storti, G., and Tola, G. (2000). Design of a simulated moving bed unit for sucrose–betaine separations. *Journal of Chromatography A* **872** (1): 23–35.

Gomes, P.S., Zabkova, M., Zabka, M. et al. (2009). Separation of chiral mixtures in real SMB units: the FlexSMB-LSRE®. *AIChE Journal*, NA-NA.

Gonçalves, C.V., Carpes, M.J.S., Correia, C.R.D., and Santana, C.C. (2008). Purification of *n*-boc-Rolipram racemate on chiral stationary phase using simulated moving bed chromatography under linear conditions. *Biochemical Engineering Journal* **40** (3): 526–536.

Grossmann, C., Amanullah, M., Morari, M. et al. (2007). Optimizing control of simulated moving bed separations of mixtures subject to the generalized Langmuir isotherm. *Adsorption* **14** (2–3): 423–432.

Grossmann, C., Langel, C., Mazzotti, M. et al. (2010). Experimental implementation of automatic “cycle to cycle” control to a nonlinear chiral simulated moving bed separation. *Journal of Chromatography A* **1217** (13): 2013–2021.

Heuer, C., Küsters, E., Plattner, T., and Seidel-Morgenstern, A. (1998). Design of the simulated moving bed process based on adsorption isotherm measurements using a perturbation method. *Journal of Chromatography A* **827** (2): 175–191.

Houwing, J., van Hateren, S.H., Billiet, H.A., and van der Wielen, L.A. (2002). Effect of salt gradients on the separation of dilute mixtures of proteins by ion-exchange in simulated moving beds. *Journal of Chromatography A* **952** (1): 85–98.

Hur, J.S. and Wankat, P.C. (2005). New design of simulated moving bed (SMB) for ternary separations. *Industrial & Engineering Chemistry Research* **44** (6): 1906–1913.

Huthmann, E. and Juza, M. (2005). Less common applications of simulated moving bed chromatography in the pharmaceutical industry. *Journal of Chromatography A* **1092** (1): 24–35.

Jermann, S., Katsuo, S., and Mazzotti, M. (2012). Intermittent simulated moving bed processes for chromatographic three-fraction separation. *Organic Process Research & Development* **16** (2): 311–322.

Juza, M. (1999). Development of an high-performance liquid chromatographic simulated moving bed separation from an industrial perspective. *Journal of Chromatography A* **865** (1): 35–49.

Katsuo, S., Langel, C., Schanen, P., and Mazzotti, M. (2009). Extra-column dead volume in simulated moving bed separations: theory and experiments. *Journal of Chromatography A* **1216** (7): 1084–1093.

Katsuo, S. and Mazzotti, M. (2010). Intermittent simulated moving bed chromatography: 1. Design criteria and cyclic steady-state. *Journal of Chromatography A* **1217** (8): 1354–1361.

Katsuo, S. and Mazzotti, M. (2010). Intermittent simulated moving bed chromatography: 2. Separation of Tröger’s base enantiomers. *Journal of Chromatography A* **1217** (18): 3067–3075.

Keßler, L.C. and Seidel-Morgenstern, A. (2008). Improving performance of simulated moving bed chromatography by fractionation and feed-back of outlet streams. *Journal of Chromatography A* **1207** (1–2): 55–71.

Khattabi, S., Cherrak, D.E., Mihlbachler, K., and Guiochon, G. (2000). Enantioseparation of 1-phenyl-1-propanol by simulated moving bed under linear and nonlinear conditions. *Journal of Chromatography A* **893** (2): 307–319.

Kurup, A.S., Subramani, H.J., Hidajat, K., and Ray, A.K. (2005). Optimal design and operation of SMB bioreactor for sucrose inversion. *Chemical Engineering Journal* **108** (1–2): 19–33.

Langel, C., Grossmann, C., Jermann, S. et al. (2010). Experimental optimizing control of the simulated moving bed separation of Tröger’s base enantiomers. *Industrial & Engineering Chemistry Research* **49** (23): 11996–12003.

Lee, A.-L. and Lim, Y.-I. (2009). Performance evaluation of the conservation element and solution element method in SMB process simulation. *Chemical Engineering and Processing: Process Intensification* **48** (4): 878–884.

Lee, H.-J., Xie, Y., Koo, Y.-M., and Wang, N.-H.L. (2008). Separation of lactic acid from acetic acid using a four-zone SMB. *Biotechnology Progress* **20** (1): 179–192.

Lorenz, H., Sheehan, P., and Seidel-Morgenstern, A. (2001). Coupling of simulated moving bed chromatography and fractional crystallisation for efficient enantioseparation. *Journal of Chromatography A* **908** (1): 201–214.

Mazzotti, M. and Rajendran, A. (2013). Equilibrium theory–based analysis of nonlinear waves in separation processes. *Annual Review of Chemical and Biomolecular Engineering* **4** (1): 119–141.

Mazzotti, M., Storti, G., and Morbidelli, M. (1997). Optimal operation of simulated moving bed units for nonlinear chromatographic separations. *Journal of Chromatography A* **769** (1): 3–24.

Mazzotti, M. (2006). Local equilibrium theory for the binary chromatography of species subject to a generalized Langmuir isotherm. *Industrial & Engineering Chemistry Research* **45** (15): 5332–5350.

Mazzotti, M. (2006). Design of simulated moving bed separations: generalized Langmuir isotherm. *Industrial & Engineering Chemistry Research* **45** (18): 6311–6324.

Mazzotti, M. (2006). Equilibrium theory based design of simulated moving bed processes for a generalized Langmuir isotherm. *Journal of Chromatography A* **1126** (1–2): 311–322.

Migliorini, C., Fillinger, M., Mazzotti, M., and Morbidelli, M. (1999). Analysis of simulated moving-bed reactors. *Chemical engineering science* **54** (13): 2475–2480.

Migliorini, C., Gentilini, A., Mazzotti, M., and Morbidelli, M. (1999). Design of simulated moving bed units under nonideal conditions. *Industrial & engineering chemistry research* **38** (6): 2400–2410.

Migliorini, C., Mazzotti, M., and Morbidelli, M. (1998). Continuous chromatographic separation through simulated moving beds under linear and nonlinear conditions. *Journal of Chromatography A* **827** (2): 161–173.

Migliorini, C., Mazzotti, M., and Morbidelli, M. (2000). Design of simulated moving bed multicomponent separations: Langmuir systems. *Separation and purification technology* **20** (1): 79–96.

Mihlbachler, K., Fricke, J., Yun, T. et al. (2001). Effect of the homogeneity of the column set on the performance of a simulated moving bed unit: I. Theory. *Journal of Chromatography A* **908** (1): 49–70.

Mihlbachler, K., Jupke, A., Seidel-Morgenstern, A. et al. (2002). Effect of the homogeneity of the column set on the performance of a simulated moving bed unit: II. Experimental study. *Journal of Chromatography A* **944** (1): 3–22.

Mihlbachler, K., Seidel-Morgenstern, A., and Guiochon, G. (2004). Detailed study of Tröger's base separation by SMB process. *AIChE journal* **50** (3): 611–624.

Miller, L., Grill, C., Yan, T. et al. (2003). Batch and simulated moving bed chromatographic resolution of a pharmaceutical racemate. *Journal of Chromatography A* **1006** (1): 267–280.

Minceva, M., Silva, V.M.T., and Rodrigues, A.E. (2005). Analytical solution for reactive simulated moving bed in the presence of mass transfer resistance. *Industrial & Engineering Chemistry Research* **44** (14): 5246–5255.

Molnár, Z., Nagy, M., Aranyi, A. et al. (2005). Separation of amino acids with simulated moving bed chromatography. *Journal of Chromatography A* **1075** (1–2): 77–86.

Mun, S. (2010). Enhanced separation performance of a five-zone simulated moving bed process by using partial collection strategy based on alternate opening and closing of a product port. *Industrial & Engineering Chemistry Research* **49** (19): 9258–9270.

Mun, S. (2011). Effect of a partial-feeding application on product purities and throughput of a five-zone simulated moving bed process for the separation of a ternary nucleoside mixture. *Process Biochemistry* **46** (4): 977–986.

Nicolaos, A., Muhr, L., Gotteland, P. et al. (2001). Application of equilibrium theory to ternary moving bed configurations (four+ four, five+ four, eight and nine zones): I. Linear case. *Journal of Chromatography A* **908** (1): 71–86.

Palacios, J.G., Kramer, B., Kienle, A., and Kaspereit, M. (2011). Experimental validation of a new integrated simulated moving bed process for the production of single enantiomers. *Journal of Chromatography A* **1218** (16): 2232–2239.

Palacios, J.G., Kaspereit, M., and Kienle, A. (2011). Integrated simulated moving bed processes for production of single enantiomers. *Chemical Engineering & Technology* **34** (5): 688–698.

Pedeferri, M., Zenoni, G., Mazzotti, M., and Morbidelli, M. (1999). Experimental analysis of a chiral separation through simulated moving bed chromatography. *Chemical Engineering Science* **54** (17): 3735–3748.

Peper, S., Johannsen, M., and Brunner, G. (2007). Preparative chromatography with supercritical fluids. *Journal of Chromatography A* **1176** (1–2): 246–253.

Polzer, W.L., Rao, M.G., Fuentes, H.R., and Beckman, R.J. (1992). Thermodynamically derived relationships between the modified Langmuir isotherm and experimental parameters. *Environmental Science & Technology* **26** (9): 1780–1786.

Rajendran, A. (2008). Equilibrium theory-based design of simulated moving bed processes under reduced purity requirements. *Journal of Chromatography A* **1185** (2): 216–222.

Ruthven, D.M. and Ching, C.B. (1989). Counter-current and simulated counter-current adsorption separation processes. *Chemical Engineering Science* **44** (5): 1011–1038.

Sá Gomes, P., Lamia, N., and Rodrigues, A.E. (2009). Design of a gas phase simulated moving bed for propane/propylene separation. *Chemical Engineering Science* **64** (6): 1336–1357.

Sá Gomes, P. and Rodrigues, A.E. (2007). Outlet streams swing (OSS) and MultiFeed operation of simulated moving beds. *Separation Science and Technology* **42** (2): 223–252.

Saska, M., Wu, M.D., Clarke, S.J., and Iqbal, K. (1992). Continuous separation of sugarcane molasses with a simulated moving-bed adsorber. Adsorption equilibria, kinetics, and application. *Separation Science and Technology* **27** (13): 1711–1732.

Seidel-Morgenstern, A., Keßler, L.C., and Kaspereit, M. (2008). New developments in simulated moving bed chromatography. *Chemical Engineering & Technology* **31** (6): 826–837.

Silva, V.M.T., Minceva, M., and Rodrigues, A.E. (2004). Novel analytical solution for a simulated moving bed in the presence of mass-transfer resistance. *Industrial & Engineering Chemistry Research* **43** (16): 4494–4502.

Song, I.-H., Amanullah, M., Erdem, G. et al. (2006). Experimental implementation of identification-based optimizing control of a simulated moving bed process. *Journal of Chromatography A* **1113** (1–2): 60–73.

Song, I.-H., Lee, S.-B., Rhee, H.-K., and Mazzotti, M. (2006). Identification and predictive control of a simulated moving bed process: purity control. *Chemical Engineering Science* **61** (6): 1973–1986.

Song, I.-H., Lee, S.-B., Rhee, H.-K., and Mazzotti, M. (2006). Optimization-based predictive control of a simulated moving bed process using an identified model. *Chemical Engineering Science* **61** (18): 6165–6179.

Ströhlein, G., Mazzotti, M., and Morbidelli, M. (2005). Optimal operation of simulated-moving-bed reactors for nonlinear adsorption isotherms and equilibrium reactions. *Chemical Engineering Science* **60** (6): 1525–1533.

Thome, B.M. and Ivory, C.F. (2007). Increasing the scale of true moving bed electrophoretic separations using filtration to reduce solvent volumetric flows between sections II and III. *Journal of Chromatography A* **1138** (1–2): 291–300.

Toumi, A., Engell, S., Diehl, M. et al. (2007). Efficient optimization of simulated moving bed processes. *Chemical Engineering and Processing: Process Intensification* **46** (11): 1067–1084.

Van Duc Long, N., Le, T.-H., Kim, J.-I. et al. (2009). Separation of D-psicose and D-fructose using simulated moving bed chromatography. *Journal of Separation Science* **32** (11): 1987–1995.

Vu, T.D., Seidel-Morgenstern, A., Grüner, S., and Kienle, A. (2005). Analysis of ester hydrolysis reactions in a chromatographic reactor using equilibrium theory and a rate model. *Industrial & Engineering Chemistry Research* **44** (25): 9565–9574.

Wei, F., Shen, B., and Chen, M. (2006). From analytical chromatography to simulated moving bed chromatography: resolution of omeprazole enantiomers. *Industrial & Engineering Chemistry Research* **45** (4): 1420–1425.

Wei, F. and Zhao, Y. (2008). Separation of capsaicin from capsaicinoids by simulated moving bed chromatography. *Journal of Chromatography A* **1187** (1–2): 281–284.

Weon Lee, J., Kim, J., Ho Lee, C. et al. (2007). Design and sensitivity analysis of simulated moving bed chromatography to separate sugar alcohols. *Journal of Liquid Chromatography & Related Technologies* **30** (13): 1859–1877.

Wu, D.J., Xie, Y., Ma, Z., and Wang, N.H. (1998). Design of simulated moving bed chromatography for amino acid separations. *Industrial & engineering chemistry research* **37** (10): 4023–4035.

Wu, J., Peng, Q., Arlt, W., and Minceva, M. (2009). Model-based design of a pilot-scale simulated moving bed for purification of citric acid from fermentation broth. *Journal of Chromatography A* **1216** (50): 8793–8805.

Xie, Y., Wu, D., Ma, Z., and Wang, N.-H.L. (2000). Extended standing wave design method for simulated moving bed chromatography: linear systems. *Industrial & Engineering Chemistry Research* **39** (6): 1993–2005.

Zhang, L., Gedicke, K., Kuznetsov, M.A. et al. (2007). Application of an eremomycin-chiral stationary phase for the separation of dl-methionine using simulated moving bed technology. *Journal of Chromatography A* **1162** (1): 90–96.

Zhang, Z., Mazzotti, M., and Morbidelli, M. (2004). Continuous chromatographic processes with a small number of columns: comparison of simulated moving bed with Varicol, PowerFeed, and ModiCon. *Korean Journal of Chemical Engineering* **21** (2): 454–464.

Zhong, G. and Guiochon, G. (1996). Analytical solution for the linear ideal model of simulated moving bed chromatography. *Chemical Engineering Science* **51** (18): 4307–4319.

Zhong, G. and Guiochon, G. (1998). Steady-state analysis of simulated moving-bed chromatography using the linear, ideal model. *Chemical Engineering Science* **53** (6): 1121–1130.

Zhong, G., Smith, M.S., and Guiochon, G. (1997). Effect of the flow rates in linear, ideal, simulated moving-bed chromatography. *AIChE Journal* **43** (11): 2960–2969.

Ziomek, G., Kaspereit, M., Jeżowski, J. et al. (2005). Effect of mobile phase composition on the SMB processes efficiency. *Journal of Chromatography A* **1070** (1–2): 111–124.

Standing Wave Design

Berninger, J.A., Whitley, R.D., Zhang, X., and Wang, N.-H.L. (1991). A versatile model for simulation of reaction and nonequilibrium dynamics in multicomponent fixed-bed adsorption processes. *Computers & Chemical Engineering* **15** (11): 749–768.

Cauley, F.G., Xie, Y., and Wang, N.-H.L. (2004). Optimization of SMB systems with linear adsorption isotherms by the standing wave annealing technique. *Industrial & Engineering Chemistry Research* **43** (23): 7588–7599.

Helfferich, F.G. and Carr, P.W. (1993). Non-linear waves in chromatography. *Journal of Chromatography A* **629** (2): 97–122.

Hritzko, B.J., Xie, Y., Wooley, R.J., and Wang, N.-H.L. (2002). Standing-wave design of tandem SMB for linear multicomponent systems. *AIChE Journal* **48** (12): 2769–2787.

Lee, H.-J., Xie, Y., Koo, Y.-M., and Wang, N.-H.L. (2008). Separation of lactic acid from acetic acid using a four-zone SMB. *Biotechnology Progress* **20** (1): 179–192.

Lee, K.B., Chin, C.Y., Xie, Y. et al. (2005). Standing-wave design of a simulated moving bed under a pressure limit for enantioseparation of phenylpropanolamine. *Industrial & Engineering Chemistry Research* **44** (9): 3249–3267.

Lee, K.B., Kasat, R.B., Cox, G.B., and Wang, N.-H.L. (2008). Simulated moving bed multiobjective optimization using standing wave design and genetic algorithm. *AIChE Journal* **54** (11): 2852–2871.

Lee, K.B., Mun, S., Cauley, F. et al. (2006). Optimal standing-wave design of nonlinear simulated moving bed systems for enantioseparation. *Industrial & Engineering Chemistry Research* **45** (2): 739–752.

Lohrmann, R. (1975). Formation of nucleoside 5′-polyphosphates from nucleotides and trimetaphosphate. *Journal of Molecular Evolution* **6** (4): 237–252.
Ma, Z. and Wang, N.-H.L. (1997). Standing wave analysis of SMB chromatography: linear systems. *AIChE Journal* **43** (10): 2488–2508.
Ma, Z., Katti, A., Lin, B., and Guiochon, G. (1990). Simple wave effects in two-component nonlinear liquid chromatography: application to the measurement of competitive adsorption isotherms. *The Journal of Physical Chemistry* **94** (17): 6911–6922.
Mallmann, T., Burris, B.D., Ma, Z., and Wang, N.H.L. (1998). Standing wave design of nonlinear SMB systems for fructose purification. *AIChE journal* **44** (12): 2628–2646.
Mun, S., Chin, C., Xie, Y., and Wang, N.-H.L. (2006). Standing wave design of carousel ion-exchange processes for the removal of zinc ions from a protein mixture. *Industrial & Engineering Chemistry Research* **45** (1): 316–329.
Rodrigues, A.E. and Pais, L.S. (2005). Design of SMB chiral separations using the concept of separation volume. *Separation Science and Technology* **39** (2): 245–270.
Xie, Y., Mun, S., Kim, J., and Wang, N.-H.L. (2002). Standing wave design and experimental validation of a tandem simulated moving bed process for insulin purification. *Biotechnology Progress* **18** (6): 1332–1344.
Xie, Y., Farrenburg, C.A., Chin, C.Y. et al. (2003). Design of SMB for a nonlinear amino acid system with mass-transfer effects. *AIChE Journal* **49** (11): 2850–2863.
Xie, Y., Wu, D., Ma, Z., and Wang, N.-H.L. (2000). Extended standing wave design method for simulated moving bed chromatography: linear systems. *Industrial & Engineering Chemistry Research* **39** (6): 1993–2005.

4

Operational Modes of SMB Processes

This chapter contains a detailed discussion of five "operational modes" in SMB chromatography. Sections 4.1 and 4.2 provide a high-level overview of operational modes, covering the motivations and considerations involved. Sections 4.3–4.6 cover Varicol, PowerFeed, ModiCon, and their combinations. We provide heuristics for each of those operational modes and discuss how to implement them in an Aspen chromatography simulation. Section 4.7 describes the "parallel two-zone" configuration and its application, as well as its implementation in Aspen. Section 4.8 summarizes the conclusions. Problems that allow readers to apply the practices described in this chapter are offered in Section 4.9, and section 4.10 presents the bibliography.

4.1 Overview

As we discussed in Chapter 3, engineers typically design simulated moving bed (SMB) systems using some idealized assumptions. Specifically, the design equations rely on the fact that the SMB system will approximate a true moving bed (TMB) system once the SMB unit reaches the cyclic steady state. The more beds an SMB unit has, the better this assumption is. Large-scale industrial separations can have over 20 beds to ensure that they perform as close as possible to TMB units and the corresponding design equations. Naturally, as the number of beds increases, so does the up-front cost of the unit. Smaller scale units commonly have eight or fewer beds. In these kinds of units, there are frequent deviations from the TMB approximation, and this led researchers to seek ways to achieve better performance from units that do not approximate a TMB separator. These methods are collectively called **operational modes**.

Operational modes involve changing various operating conditions during the SMB switching time. For example, in the PowerFeed operational mode, the feed flow rate may change several times between each port switching event. Doing this breaks the ties between SMB and TMB in a fundamental way. In a TMB unit, there is no such thing as "between" the switching events, and so no design method based on TMB behavior can capture the effect these changes have on SMB behavior.

Design, Simulation, and Optimization of Adsorptive and Chromatographic Separations: A Hands-On Approach, First Edition. Kevin R. Wood, Y. A. Liu, and Yueying Yu.

In this chapter, we will provide a discussion of the various operational modes, and provide guidance for selecting operational modes, estimating their benefits, and implementing them in Aspen chromatography.

- Guidance on the selection of operational modes (Section 4.2)
- Details and unique characteristics of Varicol processes (Section 4.3)
- Introduction to, and heuristic methods for, the PowerFeed operational mode (Section 4.4)
- Details and heuristic methods for the ModiCon Operational mode (Section 4.5)
- Summary of methods for combining operational modes (Section 4.6)
- Discussion of the use of parallel two-zone columns for de-bottlenecking (Section 4.7)

4.2 Selection of Operational Modes

Past research has shown that there are definite winners and losers among the various operational modes presented in this chapter. As we cover the various operational modes, we will discuss their potential benefits and applications. Before we dive into the detailed explanations, Table 4.1 summarizes the operational modes that this chapter covers.

4.3 Varicol

The key difference between the Varicol operation and the basic SMB is an asynchronous shift of the inlet/outlet ports in a multicolumn system. The zone lengths

Table 4.1 Summary of operational modes for binary separation.

Operational mode	Operating condition changed	Ease of implementation	Potential benefits	Best case scenario
Varicol	Port switching time	Moderate	Low	Column with 4–5 beds
ModiCon	Feed concentration	Difficult	High	Column with 4–8 beds
PowerFeed	Feed flow rate	Easy	Low	—
I-SMB	Zone connectivity	Easy	Moderate	Low selectivity separations
Solvent gradient	Solvent composition	Difficult	High	Separations near solubility limits
7+ zone	Dead volume composition	Difficult	Moderate	Systems with large dead volumes

are varied in time, but the variation is periodic so that the column recovers its initial position after a whole switching cycle. Thus, the operation is cyclic. The fact that the zone lengths vary over time means that effective solid flow rate is not a meaningful concept. The change in the relative position of the ports suggests that we cannot say that each zone moves past one bed per switching time as in the case of the basic SMB.

Figure 4.1 illustrates the asynchronous shifting of inlet/outlet ports in a Varicol process within a 4-zone and 6-bed SMB column. We denote the number of beds from zone 1 to zone 4, namely, (1, 2, 2, 1) by the initial column configuration denoted by $\chi = (1\text{-}2\text{-}2\text{-}1)$. We define the asynchronous shifting ratio (AS) as a fraction of the switching time t_{sw} at which the inlet or outlet port moves to its next port position. Specifically, AS_D, AS_E, AS_F, and AS_R with values between 0 and 1 represent, respectively, the asynchronous shifting ratios for the desorbent-in port, extract-out port, feed-in port, and raffinate-out port. For example, $AS_F = AS_E = 0.5$, together with $AS_D = AS_R = 1$ means that at $t = 0.5^* t_{sw}$, both feed-in port and extract-out port switch to the next bed, and the column configuration becomes $\chi = (2\text{-}2\text{-}1\text{-}1)$. Then, at $t = t_{sw}$, desorbent-in port and raffinate-out port move to their next beds. At $t = t_{sw}$, all four ports have switched to their next beds, and the column configuration returns to $\chi = (1\text{-}2\text{-}2\text{-}1)$.

The effective number of beds is a useful parameter for describing a specific Varicol configuration. This number describes the effect of the asynchronous switching ratios on the length of a given zone. Given in equation form, we calculate the effective numbers of beds, Z_1 to Z_4, for a 4-zone Varicol system by Eqs. (4.1) through (4.4):

$$Z_1 = N_1 + AS_D - AS_E \tag{4.1}$$

$$Z_2 = N_2 + AS_E - AS_F \tag{4.2}$$

$$Z_3 = N_3 + AS_F - AS_R \tag{4.3}$$

$$Z_4 = N_4 + AS_R - AS_D \tag{4.4}$$

where N_1, N_2, N_3, and N_4 are the bed numbers in the four zones, and Z_1, Z_2, Z_3, and Z_4 are the effective zone lengths.

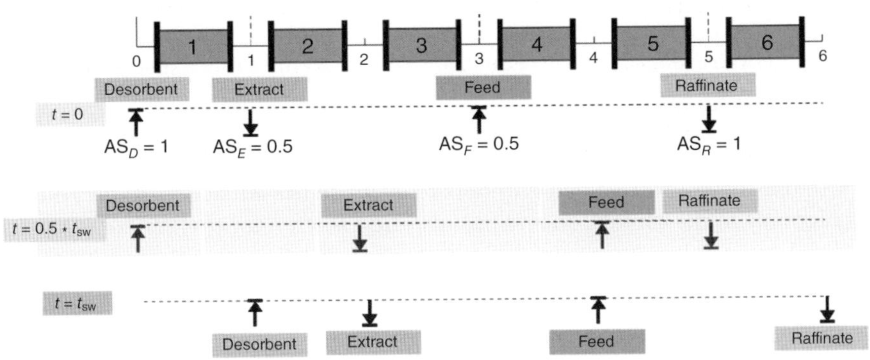

Figure 4.1 Diagram of the asynchronous shift of inlet/outlet ports in a Varicol process.

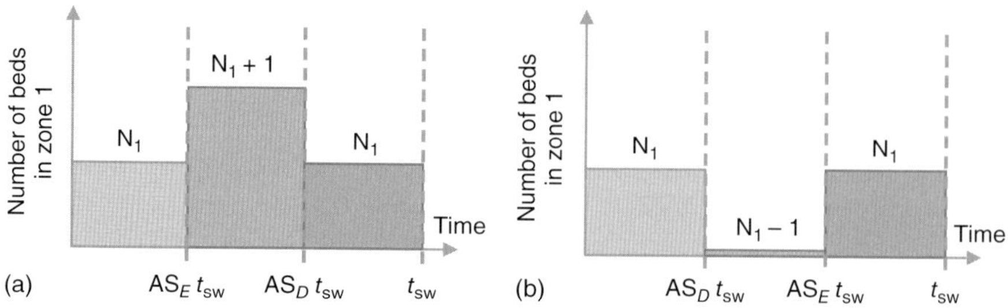

Figure 4.2 Illustration of the effective number of beds in the first zone of a Varicol system. (a) Scenario #1 and (b) Scenario #2.

The derivation of the equation for zone 1 is provided in Figure 4.2 as an example.

The starting point is the base configuration N_1, i.e. the initial number of beds in the zone. When the desorbent port is switched at time AS_D, the length of the zone decreases by the number of beds that the desorbent port moves past. When the extract port is switched at time AS_E, the length of the zone increases by the number of beds that the extract port moves past. Our ports only move by one bed at a time, so the change in the number of beds is always exactly equal to 1. We therefore have two possible scenarios, either the extract port switches before the desorbent port (Figure 4.2a) or the desorbent port switches before the extract port (Figure 4.2b). When they switch at exactly the same time, the two scenarios are equivalent.

We find the average number of beds in zone 1 by averaging the number of beds over a switching interval. Since we have three rectangular regions, we can immediately write down the equation for the weighted average:

Scenario #1

$$Z_1 = \frac{N_1(AS_E)(t_{sw}) + (N_1 + 1)(AS_D - AS_E)(t_{sw}) + N_1(1 - AS_D)(t_{sw})}{t_{sw}} \tag{4.5}$$

Scenario #2

$$Z_1 = \frac{N_1(AS_D)(t_{sw}) + (N_1 - 1)(AS_E - AS_D)(t_{sw}) + N_1(1 - AS_E)(t_{sw})}{t_{sw}} \tag{4.6}$$

Simple algebra reveals that both equations reduce to the same expression:

$$Z_1 = N_1 + 1(AS_D) - 1(AS_E) \tag{4.7}$$

We identify individual Varicol configurations as (Z_1-Z_2-Z_3-Z_4) for the study. We can use these effective bed numbers to evaluate the effect on product purity and recovery in the evaluation and optimization section for the Varicol.

4.3.1 Design Heuristics and Examples

To start modeling and optimizing a Varicol process, all the process and operating conditions are the same as those for the basic SMB, except for the new asynchronous port switching. Yu et al. [38] present a simple and effective evaluation

Figure 4.3 The Varicol evaluation and improvement steps.

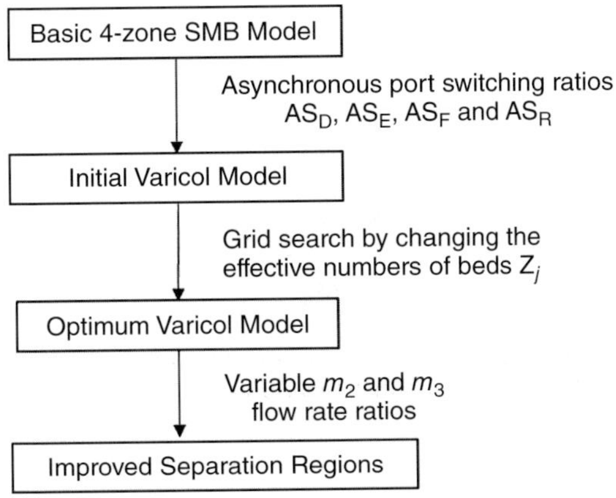

procedure using the rigorous models developed in Aspen chromatography to demonstrate how the Varicol can achieve superior separation performance over the basic SMB. We want to maximize the purity and recovery of the process by changing the effective bed numbers Z_j for each zone. This means that we need to maximize the lengths of the separation regions (zones 2 and 3), while making the non-separating regions as small as possible without causing impurity contamination. Figure 4.3 shows the strategy for the Varicol evaluation.

Figure 4.4 illustrates the general trends by looking at a contour plot of A and B purities in the (Z_1, Z_3) plane at $Z_2 = 2.4$ for the separation of Trögers' base enantiomers presented in Section 2.4. Here, the system contains two components, TB+ (the most adsorbed component) and TB− (the least adsorbed component). In the following, we call TB+ as component A and TB− as component B. The figure shows that there is no overlap between the regions where both A and B purities exceed 99% or 99.9%. We find that if we improve A purity to 99.9%, then B purity will drop, and this is observed at all Z_2 values. Therefore, we can conclude that the Varicol operational mode is not able to achieve greater than 99% or 99.9% purity in both components at the optimization point. However, it is possible to improve on the base case in some of performance indices significantly, while only slightly sacrificing other separation performance indices.

We see a nice change in the concentration profiles between the Varicol and the basic SMB column in Figure 4.5. The concentration of B (curve 2 for SMB and curve 4 for Varicol) moves to the right a little, so the regions containing components A and B overlap less; this means we are getting a better separation, although the improvement is not that much.

4.3.2 Workshop 1: Apply Varicol to the 4-Zone SMB Model

As mentioned in the introduction, the Varicol operation applies an asynchronous shift of the inlet/outlet ports upon the SMB column, where the zone lengths are varied in time. In the Varicol model, the asynchronous shifting ratio (AS) for each inlet or outlet port is a fraction of the switching time t_{sw} at which it moves to its next port position. Specifically, AS_D, AS_E, AS_F, and AS_R represent, respectively, the asynchronous shifting ratios for the desorbent-in port, extract-out port,

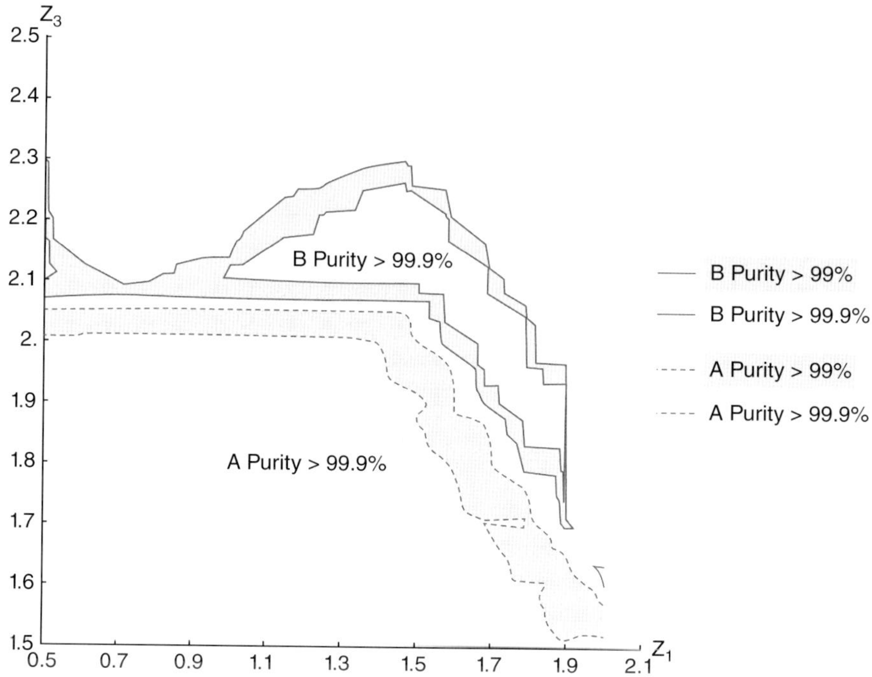

Figure 4.4 Example contour plot of A and B purities in the (Z_1, Z_3) plane with $Z_2 = 2.4$.

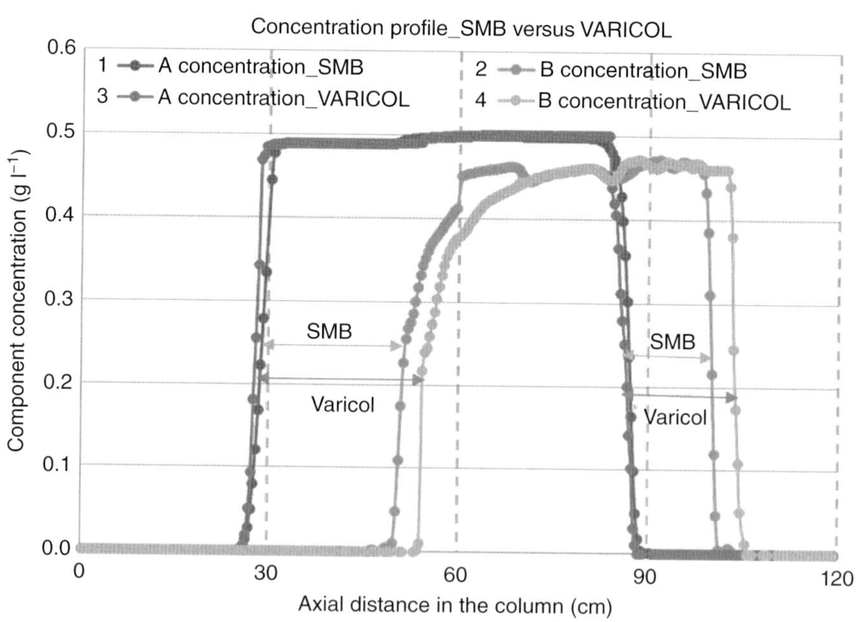

Figure 4.5 Concentration profiles between SMB and Varicol.

feed-in port, and raffinate-out port. Their values are all between 0 and 1. For the Varicol modeling, all the process and operating conditions are the same as the basic SMB, except for the new asynchronous port switching. In this workshop, we demonstrate how to apply the Varicol operation to the current 4-zone SMB model developed in Section 2.4. Open the "**4-zone Basic SMB model.cra**" and save it as "**4-zone Varicol model.cra.**"

Open the column configuration form and review the settings (Figure 4.6).

Step 1: Port Switch Control → Asynchronous.

We use the Port Switch Control box to specify whether the ports are shifted synchronously or asynchronously. In the SMB model, it is synchronous port

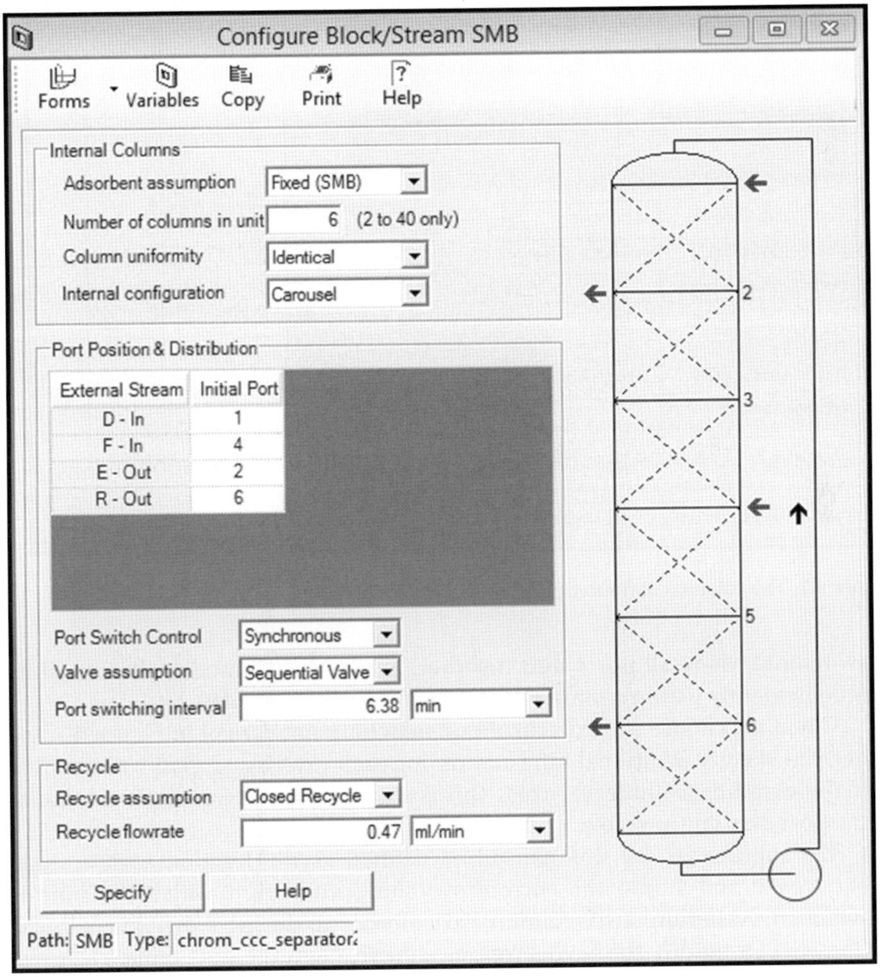

Figure 4.6 The column configuration block in the SMB model.

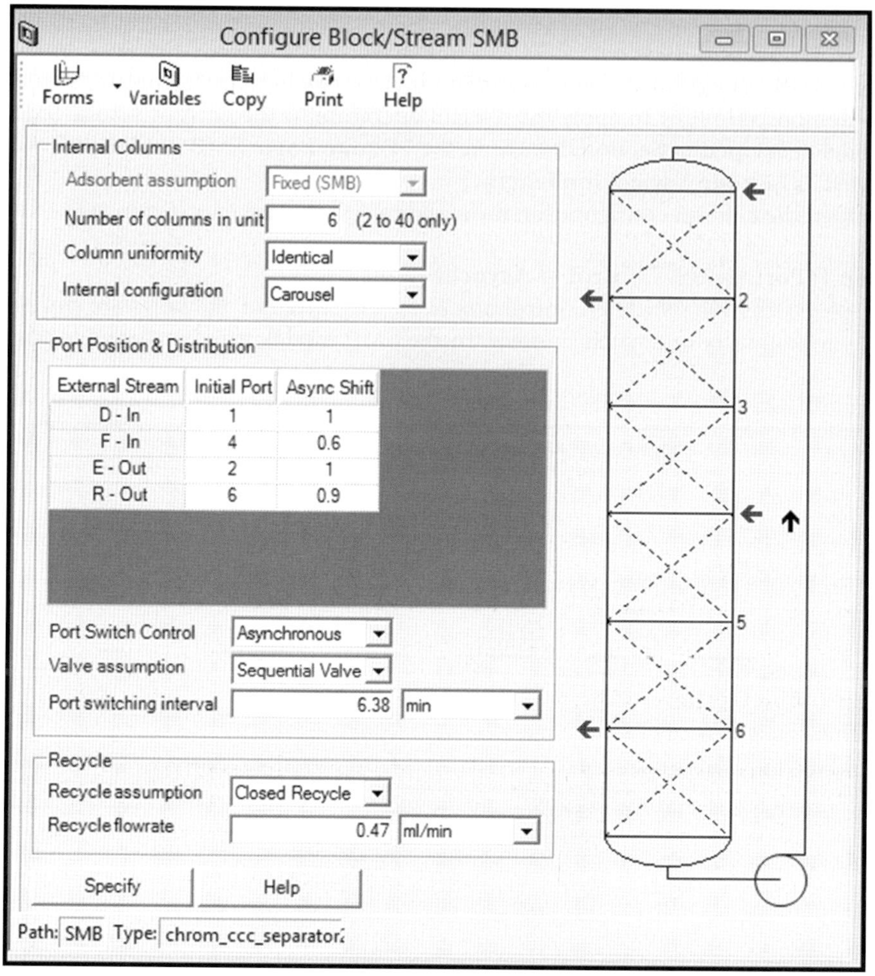

Figure 4.7 The column configuration block in the Varicol model.

switching, where all ports shift together; in the Varicol model, all ports shift independently within a master switching interval.

When we choose the asynchronous switching for Varicol in Figure 4.7, we need to specify additional information for each external stream with respect to the switching subinterval used. This is entered in the Async Shift column of the port distribution table.

The subinterval, or shift period, is defined as the fraction with a value between 0 and 1 into the current switching interval at which the port is switched. Assuming a six-column Varicol system, during a dynamic simulation we can modify the asynchronous shift through the port table. Here, we give an example of $AS_D = 1$, $AS_E = 1$, $AS_F = 0.6$, and $AS_R = 0.9$ (Figure 4.7).

Step 2: Configure Block → Initialize (Figure 4.8).

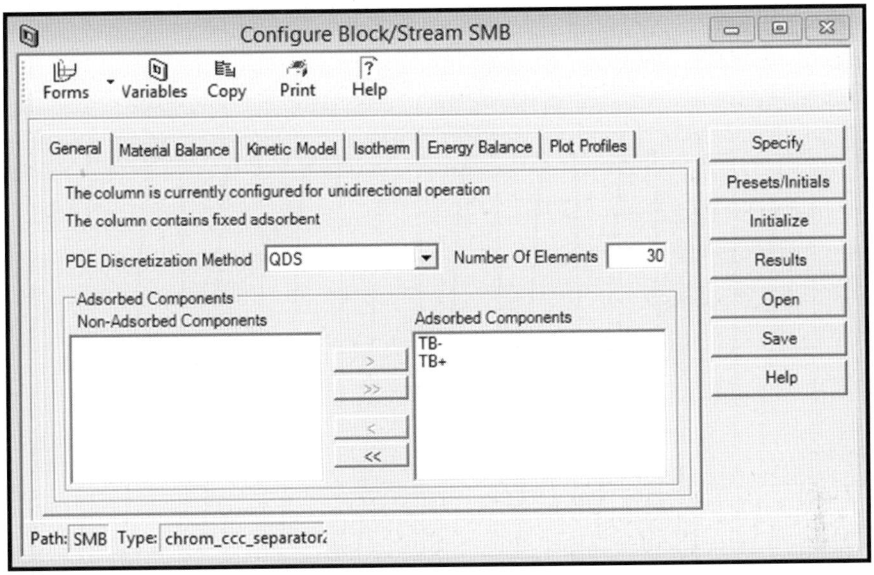

Figure 4.8 The initialization of the Varicol column.

Step 3: Run Options → Initialization (Figure 4.9).
The model is ready to initialize.

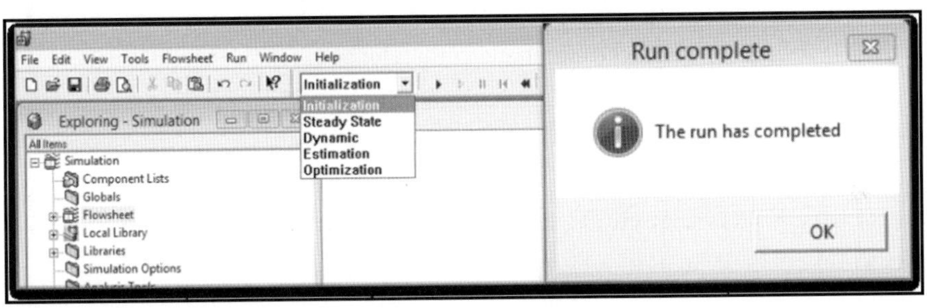

Figure 4.9 The initialization of the Varicol model.

Step 4: Run Options → Dynamic.
The model is ready for the dynamic run after reaching the pause time (382.8 min) we set (Figure 4.10).

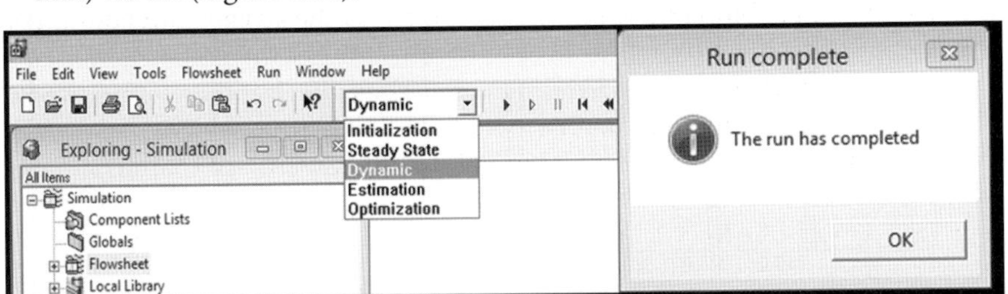

Figure 4.10 The dynamic run of the Varicol model.

Step 5: Results → Plots.

This step presents the extract concentration plot, raffinate concentration plot, and the column concentration profile (Figures 4.11–4.13).

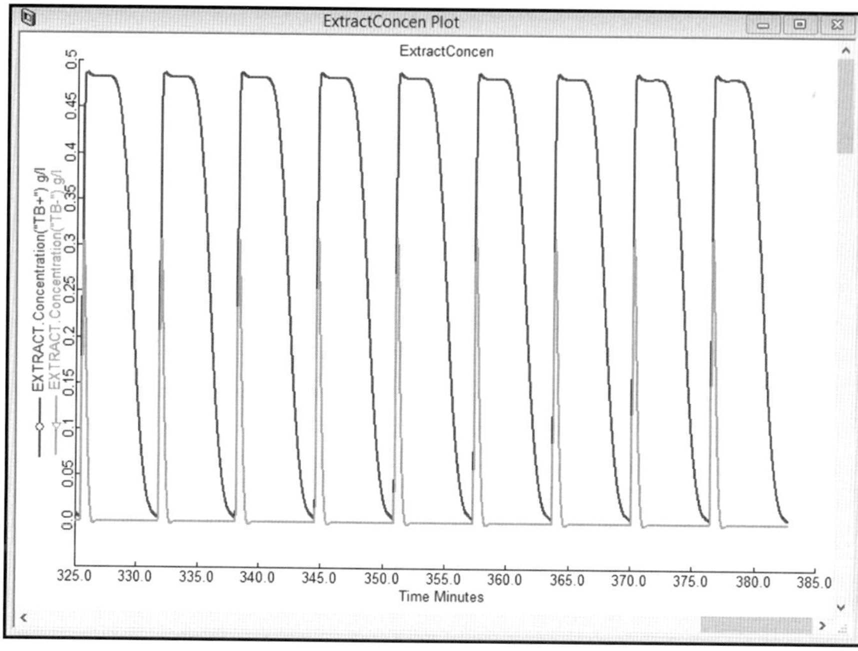

Figure 4.11 The resulting extract concentration results for both components in Varicol.

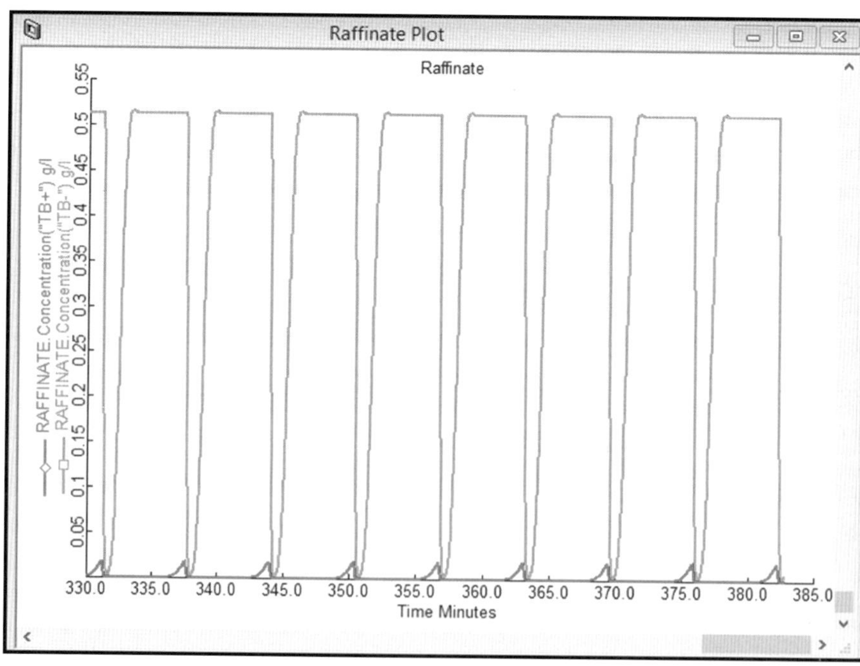

Figure 4.12 The resulting raffinate concentration results for both components in Varicol.

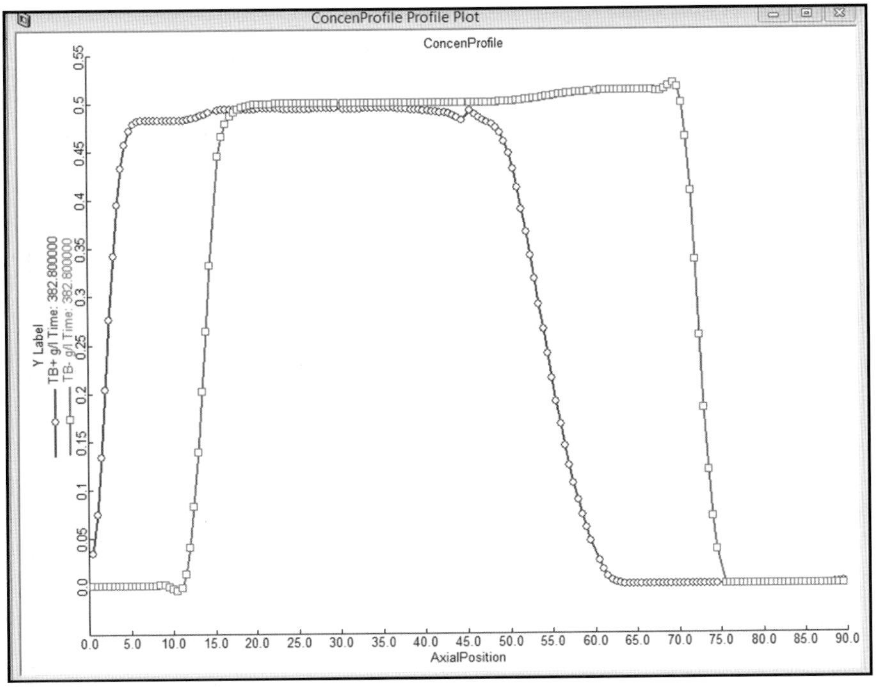

Figure 4.13 The resulting concentration profile within the Varicol column.

4.4 PowerFeed

The PowerFeed operational mode involves changing the feed flow rate between SMB switching times. In order to stay within the pressure drop constraint, we also adjust the recycle flow rates within the switching subintervals. Once again, we will consider the separation of Trögers' base enantiomers presented in Section 2.4 as a model system to explore the PowerFeed mode. Following the procedure of Yu et al. [38], we divide the switching time t_{sw} into three subintervals, and vary the feed flow rates (F_1, F_2, and F_3) from one subinterval to the next. In order to satisfy the maximum pressure drop criterion, we vary the recycled flow rates (Recy_1, Recy_2, and Recy_3) during each operation cycle, as illustrated in Figure 4.14. Figure 4.15 compares the feed flow rate policies of a basic SMB and the PowerFeed operation in three switching subintervals. We keep the average feed flow rate the same as the feed flow rate in our basic SMB model. In this modified PowerFeed with three switching subintervals, $t = 0{\sim}t_1$, $F = F_1$, $\text{Recy} = \text{Recy}_1$; $t = t_1{\sim}t_2$, $F = F_2$, $\text{Recy} = \text{Recy}_2$; $t = t_2{\sim}t_3$, $F = F_3$, $\text{Recy} = \text{Recy}_3$; $t = t_{sw,}$ $F = F_1$, $\text{Recy} = \text{Recy}_1$.

We note that partial feeding [51] is a special case of the PowerFeed operation when the feed stream and the raffinate stream are active only during certain subintervals within a switching time.

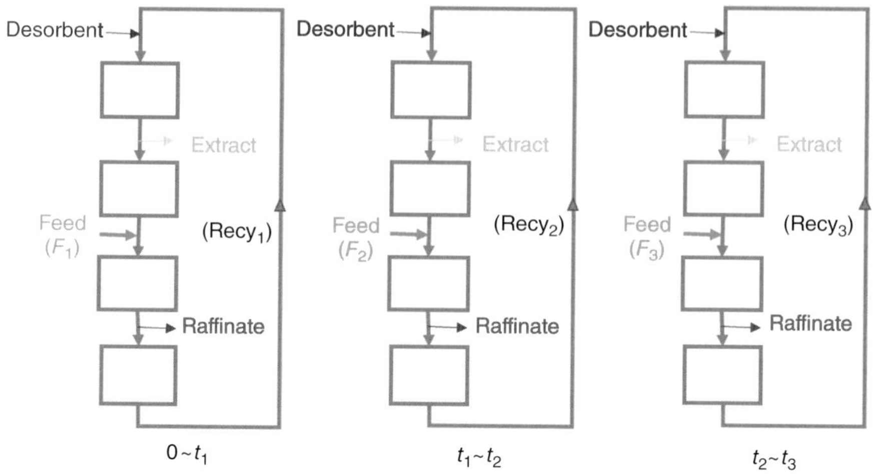

Figure 4.14 An illustration of variable feed and recycle flow rates within subintervals in the modified PowerFeed process.

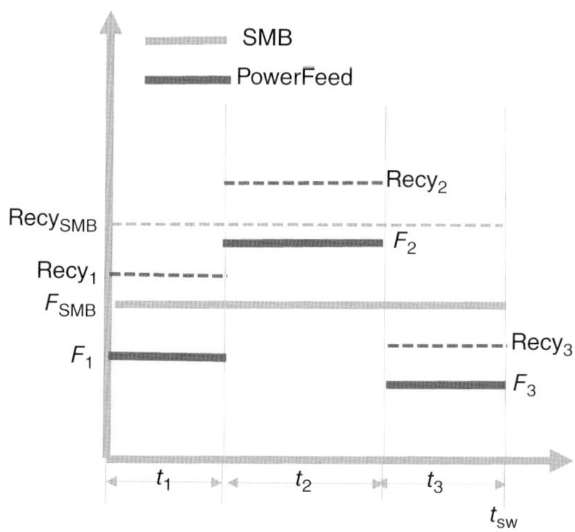

Figure 4.15 A comparison of feed and recycle flow rates policies of the basic SMB and PowerFeed in three switching subintervals.

4.4.1 Design Heuristics and Examples

For the binary separation, we start with the same basic 4-zone SMB. We choose the average m_i values under PowerFeed operation to be equal to those of the original SMB, and apply the analysis scheme shown in Figure 4.16 and the feed and recycle flow rate policies of Figure 4.16. We analyze the effects of this PowerFeed system on the product purity and recovery, and seek an improved PowerFeed model with two or three subintervals. We then adjust feasible m_2 and m_3 flow rate ratios to make the (m_2−m_3) separation plane for providing some useful guidelines of the PowerFeed.

In step 1, we evaluate the PowerFeed with two subintervals, as shown in Figure 4.17. In order to see if a front feed loading (larger F_1) or a rear feed

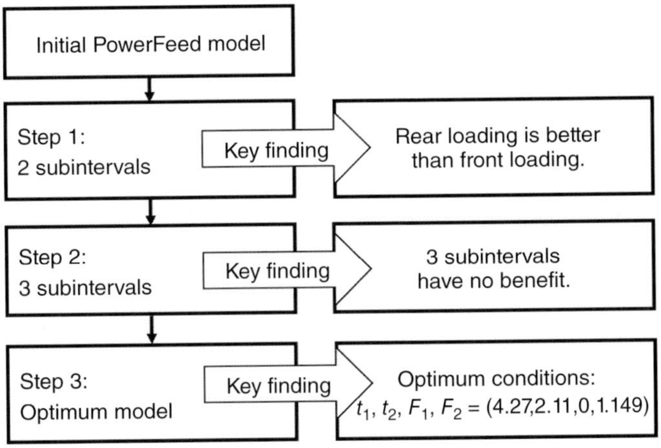

Figure 4.16 The PowerFeed evaluation and improvement steps.

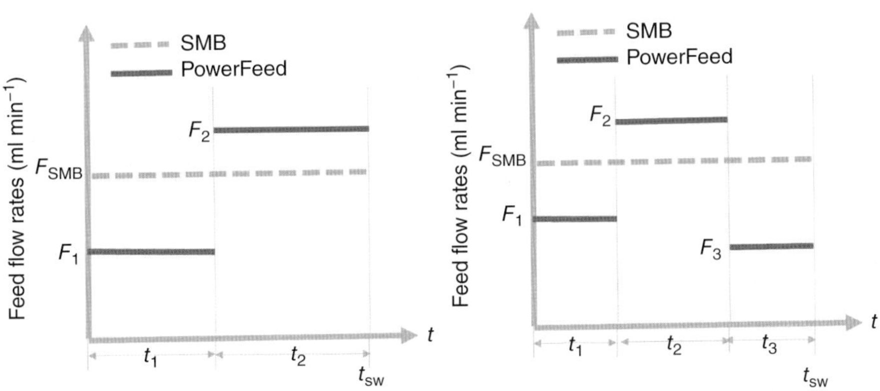

Figure 4.17 PowerFeed with two and three subintervals.

loading (Larger F_2) gives a better separation performance, we evaluate the performance by changing the subinterval feed flow rates and find that rear loading is better than front loading.

From step 2 of evaluating the PowerFeed with three subintervals in Figure 4.17, we find that there is no benefit to the PowerFeed system with three subintervals.

In step 3, we evaluate the partial feeding [51] with two switching subintervals to find the best manipulated variables for PowerFeed. We identify the second subinterval feed flow rate F_2 as the most important manipulated variable. We consider the B purity as the most important performance index, because the A purity, and the recovery of the products A and B essentially do not change much when we increase the F_2 value.

Figure 4.18 shows the effect of increasing the second subinterval feed flow rate F_2 on B purity for three different switching times in the PowerFeed with two subintervals. Compared to the SMB results, A purity and both A and B recoveries do not change much throughout the simulations.

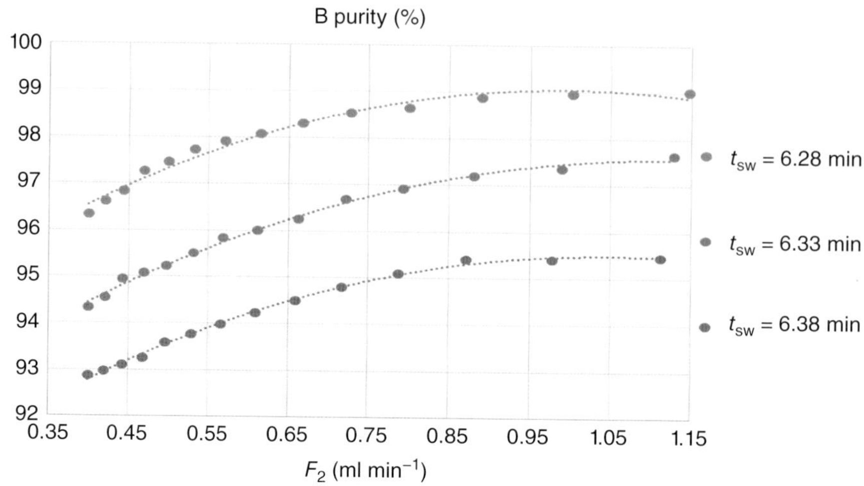

Figure 4.18 B purity simulation results in the PowerFeed model with two subintervals.

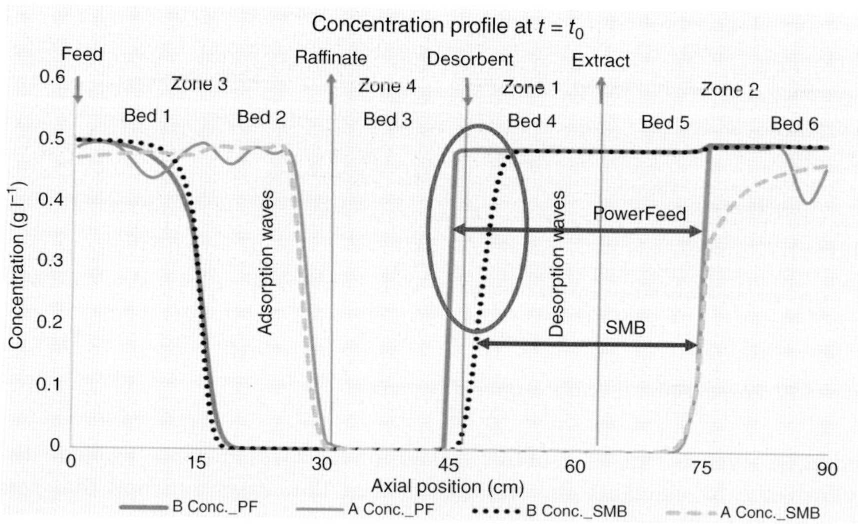

Figure 4.19 Concentration profiles for PowerFeed (PF) and SMB.

A comparison of the column concentration profiles of the PowerFeed model and the basic SMB model reveals the reasons for this trend. Figure 4.19 shows the steady-state concentration profiles for both PowerFeed and SMB. Solid curves represent concentration values from the PowerFeed model, while dashed curves refer to the SMB model. We find that the adsorption wave in zone 3 for component A in PowerFeed drops more quickly and earlier than the one in SMB, which results in more of component B being left in the flow coming into the raffinate stream. This explains the improved raffinate purity; the raffinate contains a high concentration of B for a longer period of time. The desorption waves in zones 1 and 2 are sharper in PowerFeed than those in the SMB model. In other words, the components are desorbed over a shorter distance in PowerFeed, which may improve the A recovery a little. The arrowed gap between two components in PowerFeed is wider than the one in SMB, suggesting that components separate better under PowerFeed operation.

4.4.2 Workshop 2: Apply PowerFeed to the Four-Zone SMB Model

In this workshop, we demonstrate how to create a PowerFeed model based on the basic SMB model. Since PowerFeed adjusts both the feed and recycle flow rates within the switching subintervals, we need to introduce the concept of the cycle organizer in Aspen chromatography to define the cyclic operations. We have already covered the cycle organizer in Aspen adsorption, and the cycle organizer in Aspen chromatography is nearly identical.

Many adsorption processes operate in a cyclic manner. Each cycle is described by a series of one or more sequential steps or discrete events. When simulating a cyclic process, you must be able to specify when certain events are going to occur.

The cycle organizer lets us rapidly create the steps that define a cyclic process. Use it to

- create any number of steps
- define the step termination event
- manipulate flowsheet variables for a given step
- generate a cyclic task based on the task language
- distribute cycle information to other flowsheet blocks through global variables
- store multiple cycle definitions
- control variable recording and automated snapshots
- execute VB (Visual Basic) scripts for additional calculations and control.

By way of review, the major features of the cycle organizer are as follows:

- The main Configure form gives the status of the system, as well as the active state of the cyclic task.
- All entered data are stored in the block on the flowsheet. This allows the data to be saved with the flowsheet input file.
- Only one instance of the Cycle Organizer is allowed on the flowsheet.
- When we configure the flowsheet for cyclic operation, it is advisable to configure it as if it is about to execute the first step of the cycle.
- On addition of a new step, we are asked two questions:
  - Is the new step to be placed before or after the currently selected step?
  - Is the information to be copied from the currently selected step into the newly created one (to act as a template)?

We give more details on how to create a cycle organizer in the PowerFeed model in the following steps.

Step 1: Open the "4-zone Basic SMB model.cra", and save it as "4-zone PowerFeed model.cra"

In this workshop, we apply the PowerFeed with a partial feeding. In Figure 4.20, we display the feed flow rate policies of the PowerFeed with two switching subintervals, at $t = 0 \sim t_1$, $F = F_1 = 0$; at $t = t_1 \sim t_2$, $F = F_2$.

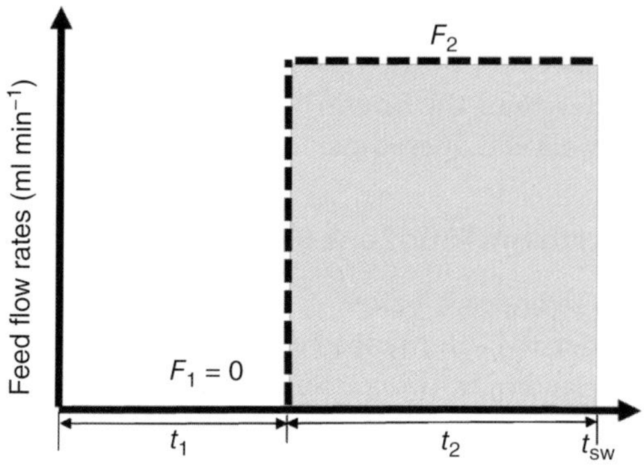

Figure 4.20 The feed flow rate policies in the PowerFeed with two subintervals.

Step 2: Add the pressure drop constraints.

In PowerFeed, if we only change the feed flow rates, it may cause the system to exceed the maximum pressure drop. For example, we might choose the Carman–Kozeney pressure drop correlation, Eq. (4.8):

$$\Delta P_j = \frac{-150\mu(1-\varepsilon)^2}{(2r_p\psi)^2\varepsilon^3}\frac{L^* Q_j}{S} \tag{4.8}$$

In the equation, ΔP_j = pressure drop in each zone; L = length of each zone; S = cross-sectional area of the bed; ε = overall bed voidage; μ = fluid viscosity; r_p = particle radius; and ψ = particle shape factor.

If we only increase the subinterval feed flow rate Q_F, then Q_3 ($Q_3 = Q_2 + Q_F$) could be very high, and ΔP_3 would exceed the maximum pressure drop of the system in that subinterval. In this workshop, we adjust the recycle flow rate to satisfy the pressure drop constraint.

In this step, we write a program to define and record the pressure drop through the whole column, and make sure that the total pressure drop is below the maximum pressure drop (40 bar). We define PreDrop as the total pressure drop variable for all the six beds in the column. Figure 4.21 shows the resulting value for the total pressure drop after each simulation run.

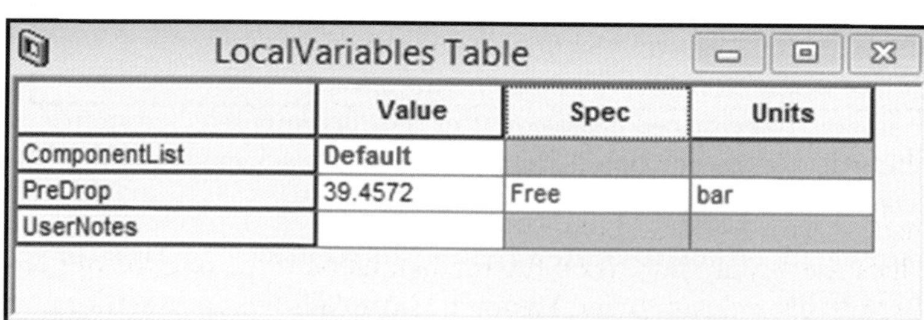

	Value	Spec	Units
ComponentList	Default		
PreDrop	39.4572	Free	bar
UserNotes			

Figure 4.21 The resulting value for the total pressure drop after each simulation run.

PreDrop as Pressure_Change;
PreDrop=SMB.Column_(1).Pdrop+SMB.Column_(2).Pdrop+SMB.Column_(3).Pdrop+SMB.Column_(4).Pdrop+SMB.Column_(5).Pdrop+SMB.Column_(6).Pdrop;

Step 3: Tool → Cycle Organizer.

To access the Cycle Organizer, we click the Cycle Organizer from the Tools menu, as shown in Figure 4.22. If a Cycle Organizer block does not exist on the flowsheet, one is automatically placed on the flowsheet and the Cycle Organizer window appears. Once a Cycle Organizer block is present on the flowsheet, you can open it either using the Tools menu, or by double-clicking on the flowsheet block. The Cycle Organizer toolbar gives access to the various fields and controls our specifications of a cyclic task.

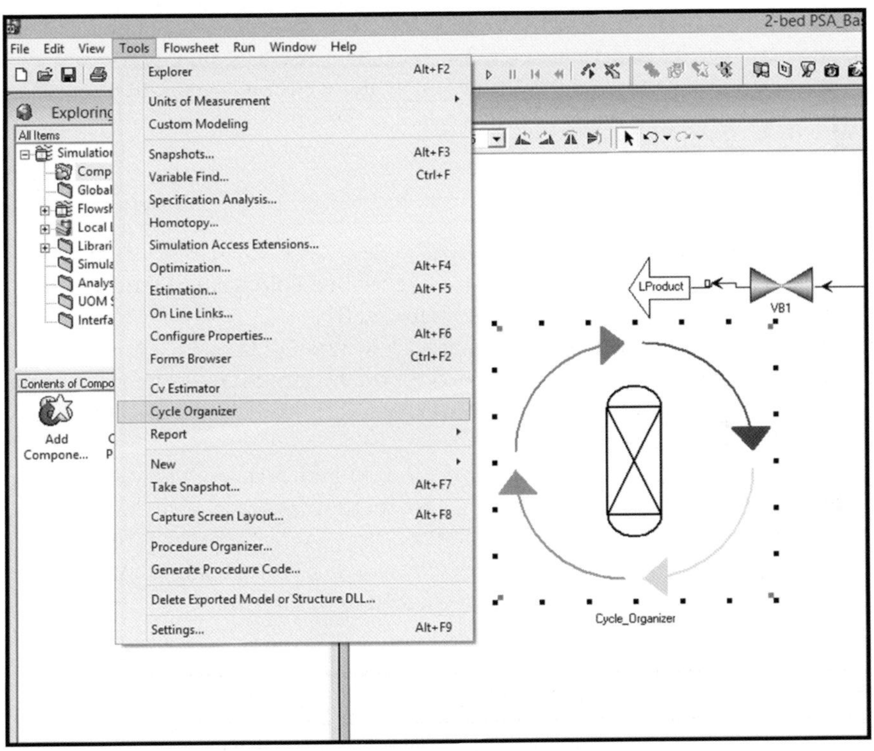

Figure 4.22 The Cycle Organizer block in the PowerFeed model.

Step 4: Cycle Organizer → Cycle Options.

Table 4.2 lists the "Cycle" button on the toolbar, its purpose, and the options available on the drop-down menu. Figure 4.23 displays the cycle settings for the process in PowerFeed operation. We set the maximum cycle number to be 100.

Table 4.2 The "Cycle" button on the toolbar in the Cycle Organizer.

Toolbar button	Purpose	Options
Cycle	Cycle controls, such as creating and activating cycles	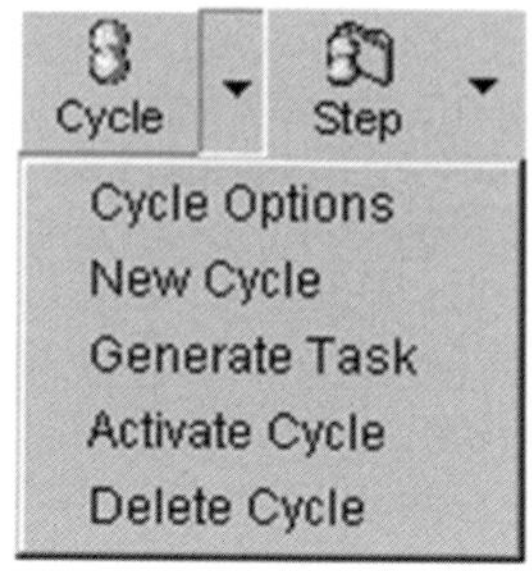

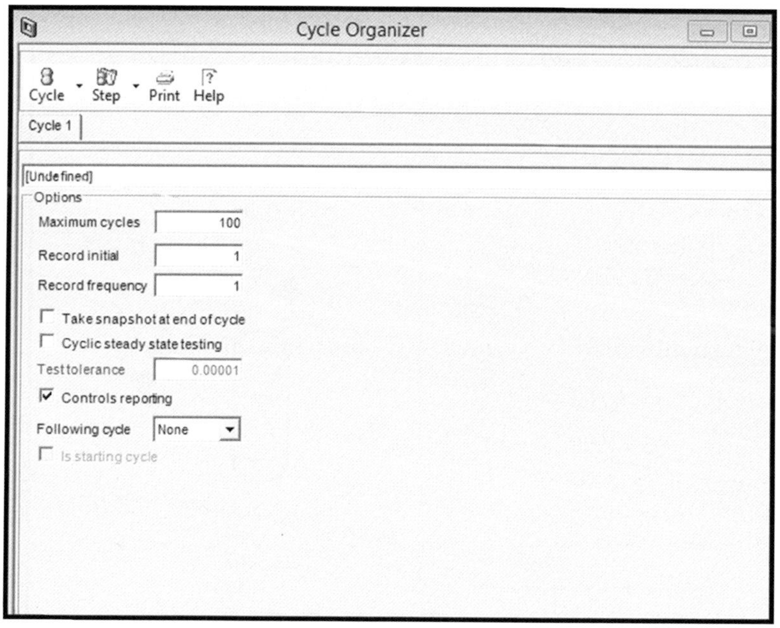

Figure 4.23 The cycle options in the Cycle Organizer of the PowerFeed model.

Step 5: Cycle Organizer → Step Control → Step 1: the first subswitching interval.

Table 4.3 shows the "Step" button on the toolbar, its purpose, and the options available on the drop-down menu.

Table 4.4 lists the cycle organizer settings for the subinterval switching times, and feed and recycle flow rates in the PowerFeed model. We get all these optimized subinterval conditions from our simulation and evaluation strategy. The detailed calculations for the values of Table 4.4 are as follows:

From looking at the effect of the second subinterval feed flow rate on the B purity in Figure 5.6, we achieve the best performance when $F_1 = 0$ and

Table 4.3 The "Step" button on the toolbar in the Cycle Organizer.

Toolbar button	Purpose	Options
Step	Step controls, such as modifying and inserting steps	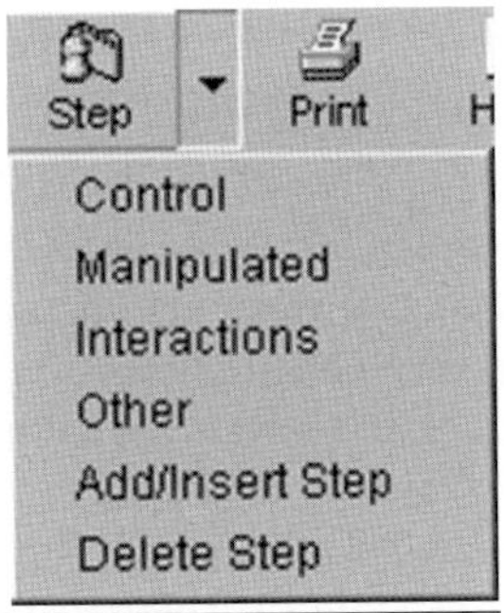

Table 4.4 The cycle organizer conditions for the PowerFeed model.

	Step 1	Step 2
Subinterval switching time	4.27 min	2.11 min
Subinterval feed flow rates	0	1.149 ml min^{-1}
Subinterval recycle flow rates	0.344 ml min^{-1}	0.727 ml min^{-1}

$F_2 = 0.149$ ml min^{-1} in the PowerFeed operation with two subintervals.

$$F_{\text{SMB}}{}^{*}t_{\text{sw}} = F_1{}^{*}t_1 + F_2{}^{*}t_2 \tag{4.9}$$

In this case, $F_{\text{SMB}} = 0.38$ ml min^{-1}; $t_1 + t_2 = t_{\text{sw}} = 6.38$ min

Then, $t_1 = 4.27$ min and $t_2 = 2.11$ min

For the calculation of the recycle flow rates, we combine the following equations:

$$m_j = \frac{Q_j t_{\text{sw}} - V_{\text{bed}}\varepsilon - V_j^D}{V_{\text{bed}}(1-\varepsilon)} \tag{4.10}$$

$$t_{\text{sw}} = \frac{\phi L^2}{\Delta P_{\max}} \sum_{j=1}^{j=4} n_j(m_j(1-\varepsilon)+\varepsilon) \tag{4.11}$$

$$Q_4 = Q_{\text{recycle}} \tag{4.12}$$

$$Q_3 = Q_2 + Q_{\text{F}} \tag{4.13}$$

$$Q_2 = Q_1 - (Q_{\text{F}} + Q_{\text{D}} - Q_{\text{R}}) \tag{4.14}$$

$$Q_1 = Q_4 + Q_D \tag{4.15}$$

We can therefore derive Eq. (4.16):

$$Q_{\text{recycle}} = \frac{V_{\text{bed}}\Delta\text{Pmax} - L^2\phi(n_1 Q_{\text{D}} - n_2 Q_{\text{F}} + (n_2+n_3)Q_{\text{R}})}{L^2\phi(n_1+n_2+n_3+n_4)} \tag{4.16}$$

Since $V_{bed} = 2.493\ cm^3$; $\Delta Pmax = 40$ bar; $L = 15$ cm; $\phi = 0.1$, $n_1 = 1$; $n_2 = 2$; $n_3 = 2$; $n_4 = 1$; $Q_D = 0.65\ ml\ min^{-1}$; $Q_R = 0.43\ ml\ min^{-1}$

we get the final recycle flow rate equation based on the feed flow rate:

$$Q_{recycle} = 0.344 + 0.333Q_F \tag{4.17}$$

Since $Q_{F_1} = 0\ ml\ min^{-1}$; $Q_{F_2} = 0.149\ ml\ min^{-1}$

we get $Q_{recy_1} = 0.344\ ml\ min^{-1}$; $Q_{recy_2} = 0.727\ ml\ min^{-1}$

Then, press the step control and follow Figures 4.24 and 4.25 to edit steps 1 and 2. In this case, we choose the time-driven method, which is the most common step control method. Here, the Step Control is a fixed elapsed time; for example, the step is set to terminate after 4.27 min. The step time remains constant from cycle to cycle.

Table 4.5 lists the main buttons on the toolbar, their purposes, and the options available on their drop-down menus of the Variables. Within each step of a cycle, we may modify different variables. These variables may control, for example,

- feed condition
- valve opening
- heater duty.

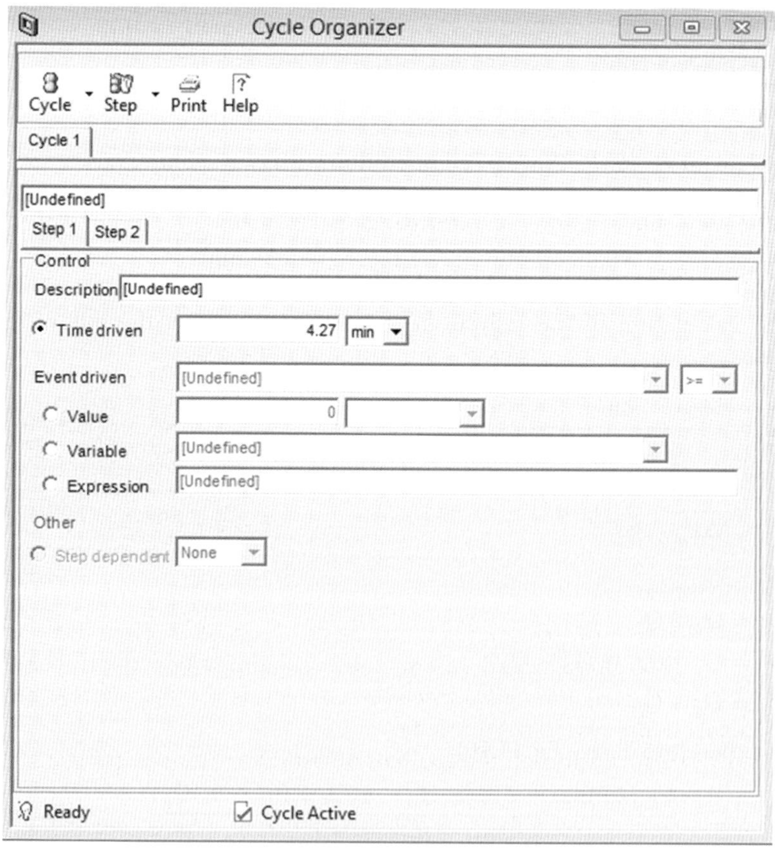

Figure 4.24 The step control setting for step 1 in the PowerFeed model.

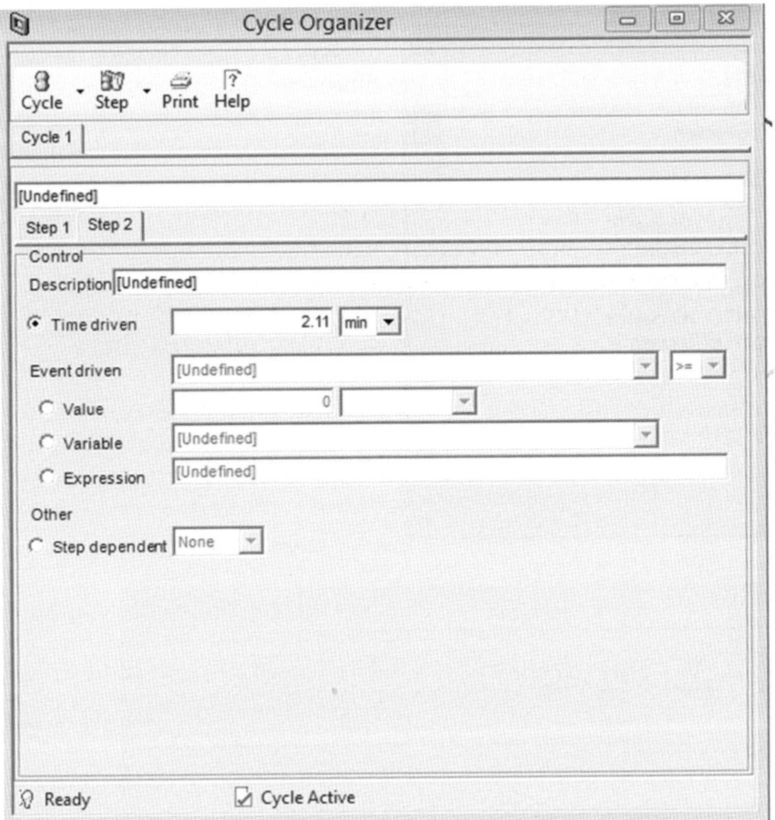

Figure 4.25 The step control setting for step 2 in the PowerFeed model.

Table 4.5 The "Variable" button on the toolbar in the Cycle Organizer.

Toolbar button	Purpose	Options
Variable (available only if you selected Manipulated from the Step menu)	Adding or deleting variables	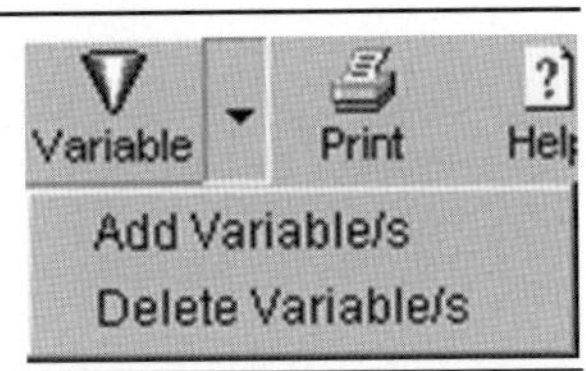

To access the list of manipulated variables: From the Step button's drop-down list, click Manipulated. In our case, we mainly adjust the feed and recycle flow rates during different subswitching intervals.

To add a new manipulated step variable:

1) Click the Variable button on the toolbar; or from its drop-down list, click Add Variable/s.

   The Variable Selector dialog box appears, which lists the available fixed and initial variables that have not already been selected in the current step (Figure 4.26).

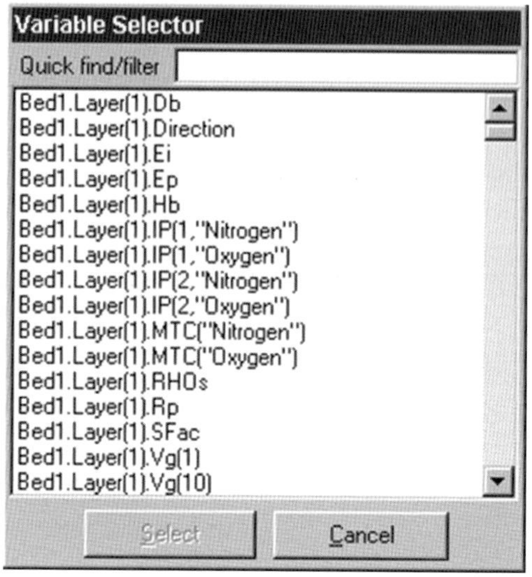

Figure 4.26 The variable selector window for variable finding in the Cycle Organizer.

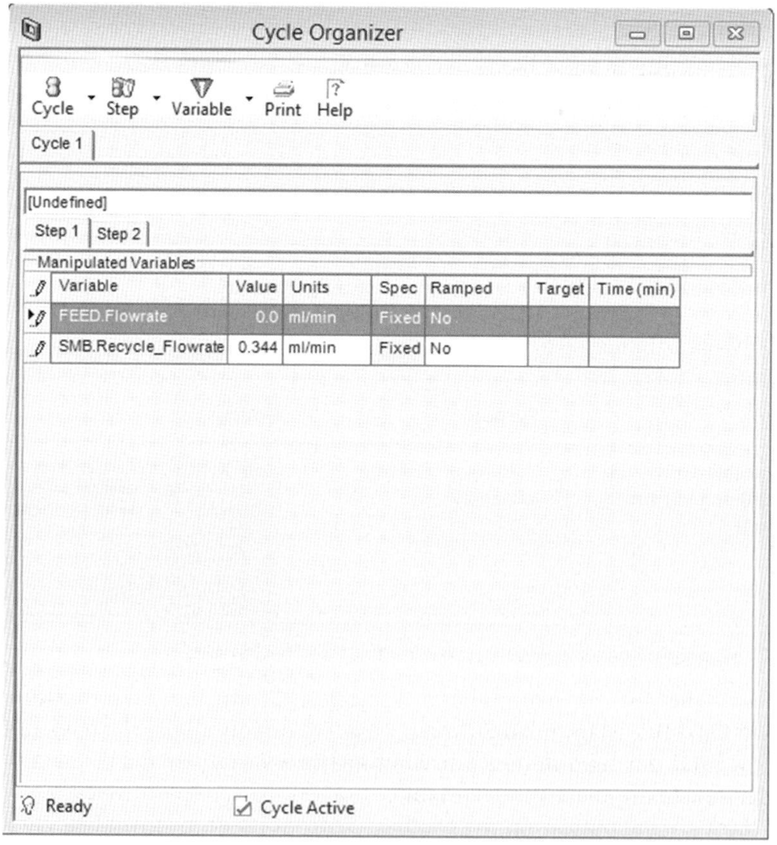

Figure 4.27 The manipulated variables for step 1 in the PowerFeed model.

2) Select a variable using one of these actions:
   a. Double-click on the variable in the list.
   b. Type the name of the variable in the text box at the top of the dialog box (a dynamic search will take place during typing).
   c. Select multiple variables, using either the SHIFT or CTRL key.
   d. Use wildcards in the text box to reduce the list size and then select.

Figure 4.27 is the final look of step 1 and shows the subinterval conditions in step 1:

Step 6: Cycle Organizer → Step Manipulated → Step 2: second subswitching interval (Figure 4.28).

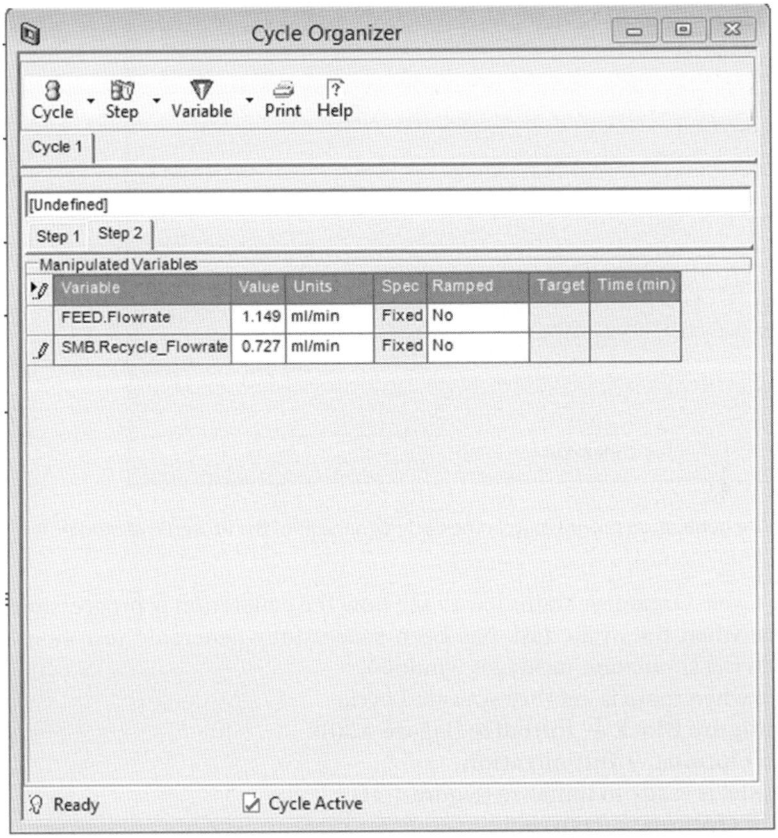

Figure 4.28 The manipulated variables for step 2 in the PowerFeed model.

Step 7: Cycle Options → Generate Task.

Once a new cycle has been defined, or changes made to an existing definition, you must generate the cyclic task before the simulation can be run. To generate the cyclic task

- on the Cycle button's drop-down menu, click Generate Task, as shown in Figure 4.29.

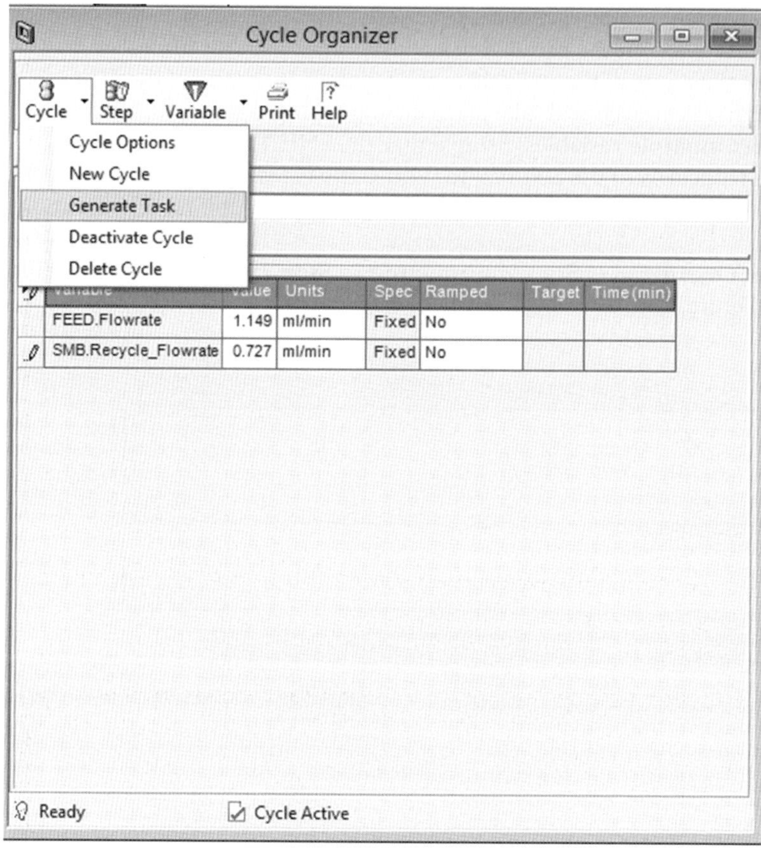

Figure 4.29 The generation progressing in the Cycle Organizer of the PowerFeed model.

View the Cycle Organizer status bar to see how the generation is progressing:

- You see when the cyclic task has been successfully generated (errors are noted in the simulation messages window).
- You see when there is another activated cycle.

Step 8: Configure Block → Initialize (Figure 4.30).

Step 9: Run Options → Initialization.

The model is ready to initialize (Figure 4.31).

Step 10: Run Options → Dynamic.

The model is ready for the dynamic run after reaching the pause time (382.8 min) we set (Figure 4.32).

Step 11: Results → Plots.

This step presents the extract concentration plot, raffinate concentration plot, and the column concentration profile (Figures 4.33–4.35).

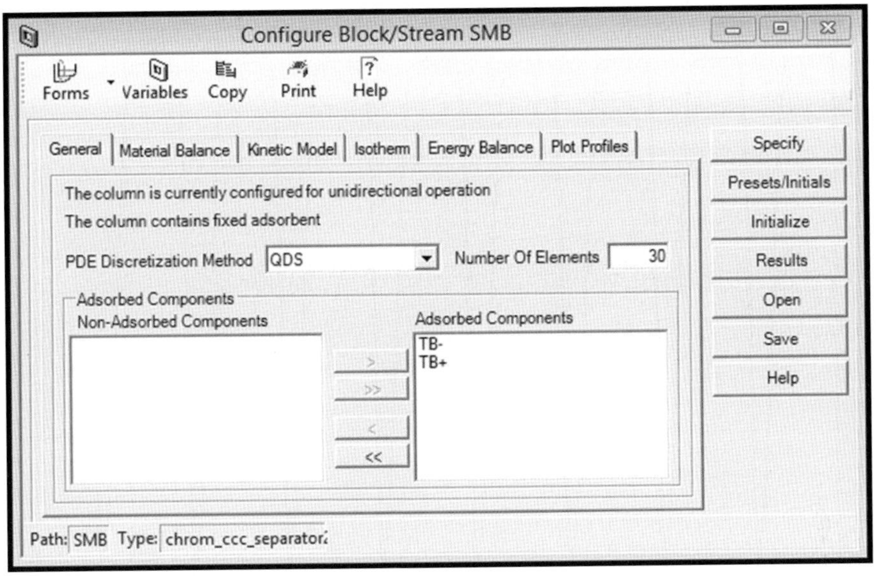

Figure 4.30 The initialization of the PowerFeed column.

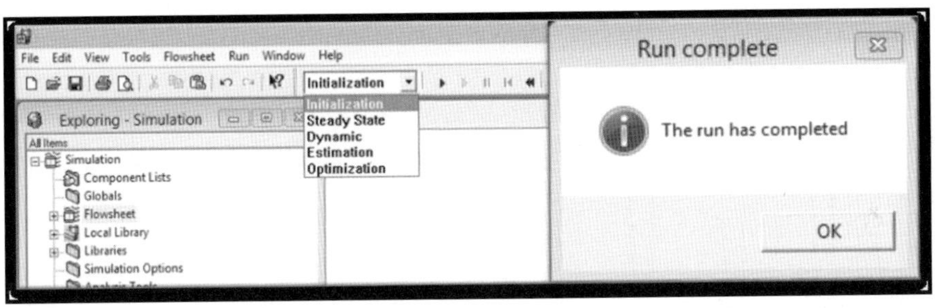

Figure 4.31 The initialization of the PowerFeed model.

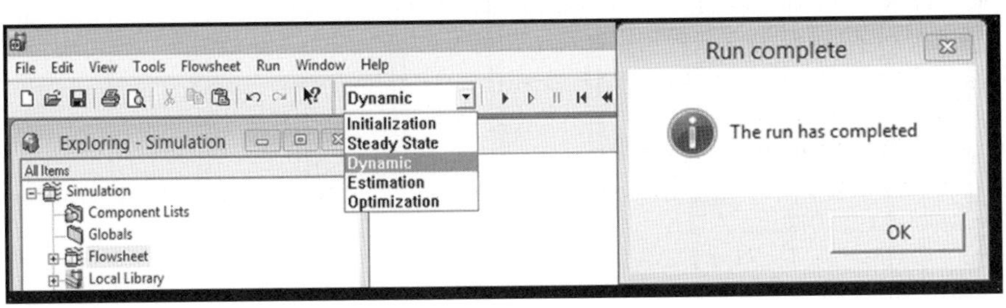

Figure 4.32 The dynamic run of the PowerFeed model.

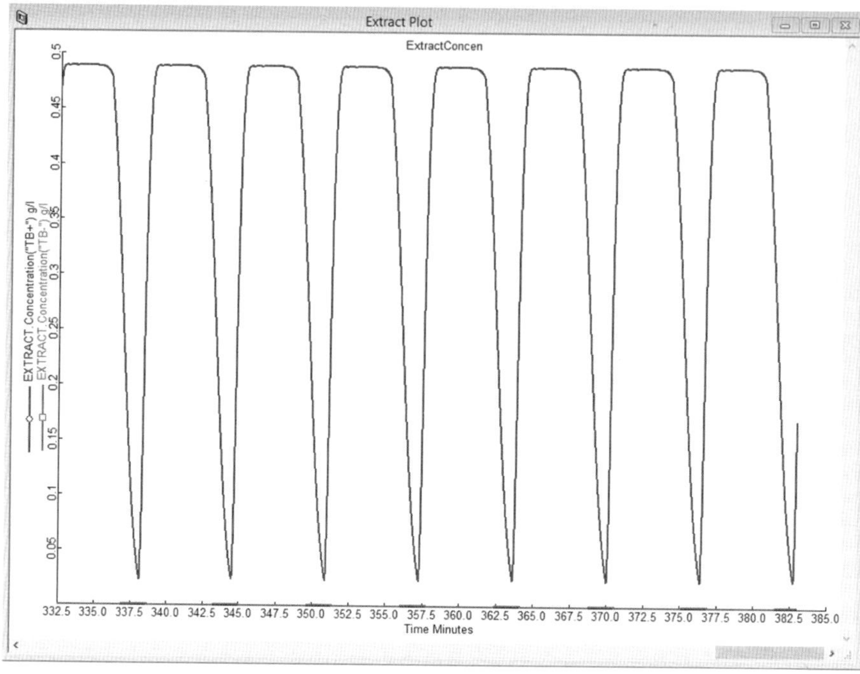

Figure 4.33 The resulting extract concentration results for both components in PowerFeed.

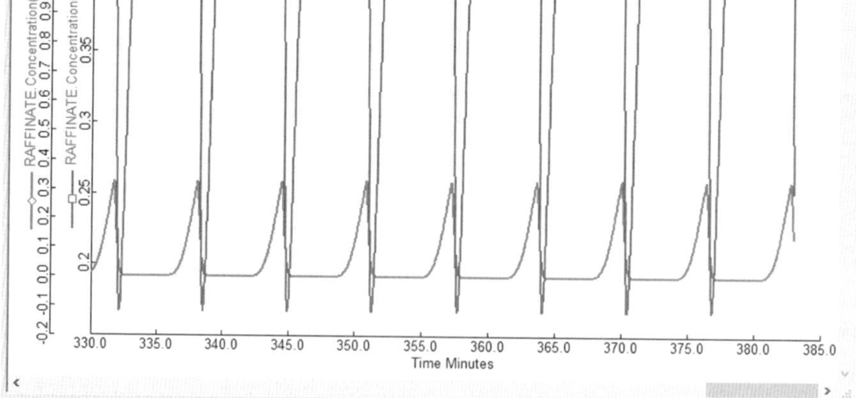

Figure 4.34 The resulting raffinate concentration results for both components in PowerFeed.

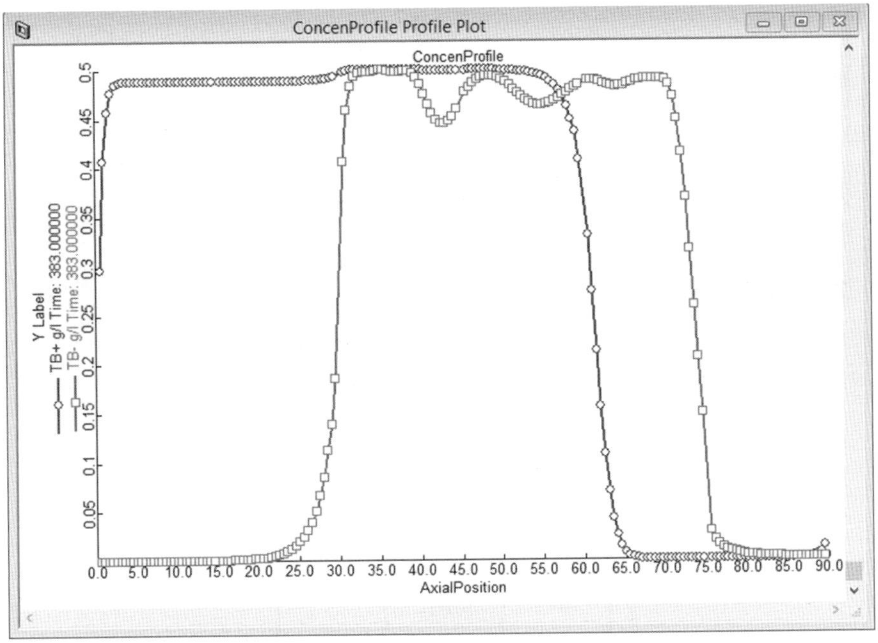

Figure 4.35 The resulting concentration profile within the PowerFeed column.

4.5 ModiCon

The ModiCon operational mode is similar to PowerFeed in concept: instead of varying the feed flow rate, we vary the feed concentration. This can cause significant improvement in separation performance, but it is more difficult to implement. Not every process has the luxury of choosing the feed concentration, and the ones that do may require additional piping to accomplish the switching.

4.5.1 Design Heuristics and Examples

ModiCon varies the feed concentration (C_1, C_2, and C_3) of the SMB process. It keeps the feed flow rate and the column configuration the same as the basic SMB model. Unlike PowerFeed, adding a third subinterval to ModiCon operation generally results in an improved performance. Figure 4.36 defines the three different ModiCon patterns within three switching subintervals: higher front, higher middle, and higher later feed concentration.

If we once again choose amino acid separation as a model, we find that the configuration with greater middle area is the best, and the second subinterval feed concentration C_2 is the most effective manipulated variable in improving separation performance.

Figure 4.37 shows that the larger C_2 values give higher B purity and A recovery; both B purity and A recovery remain constant after 1.6 g l^{-1} total feed concentration. We observe diminishing returns as C_2 increases beyond 1.6 times the original concentration. Therefore, we select a C_2 value of 1.6 g l^{-1} as our

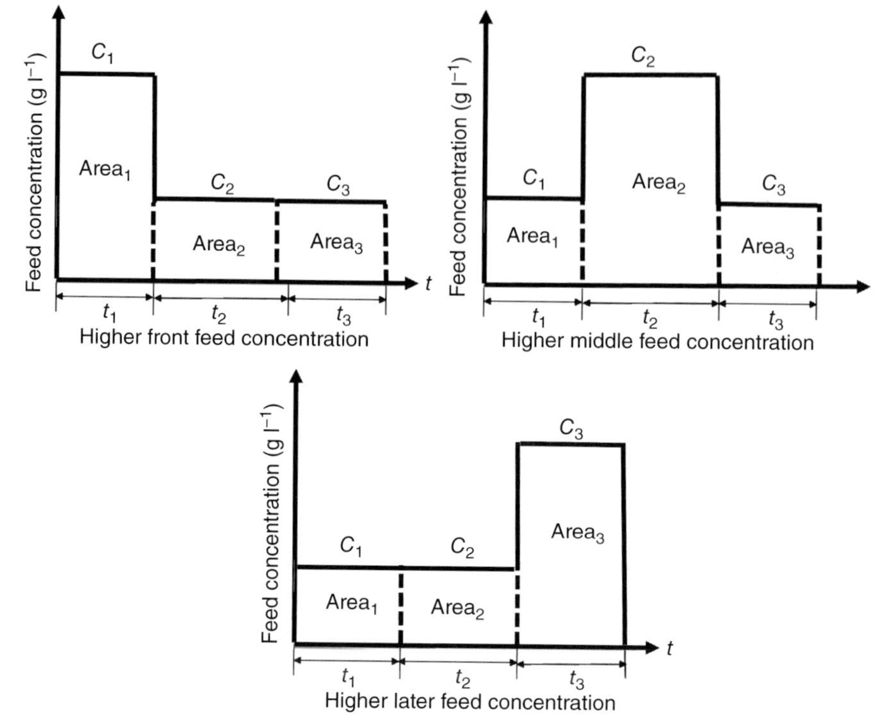

Figure 4.36 The definition of three ModiCon patterns.

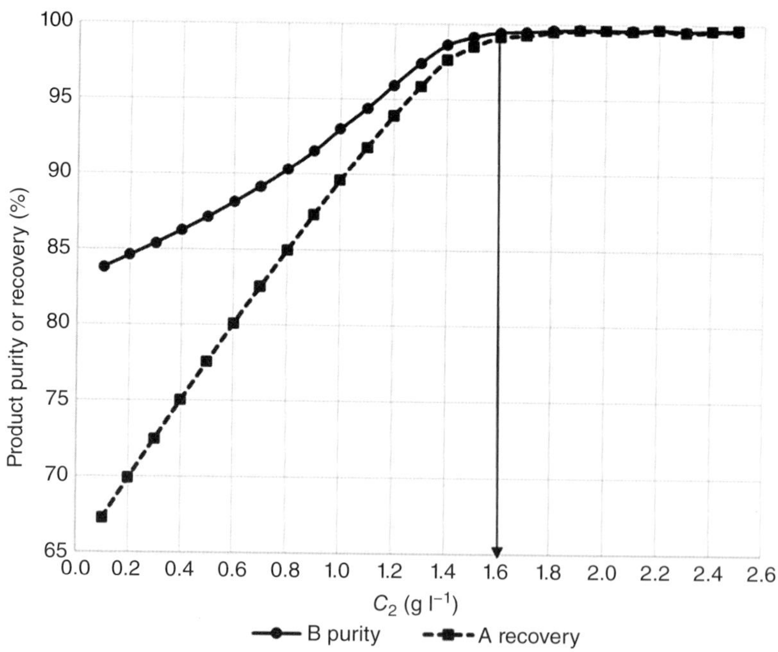

Figure 4.37 The ModiCon simulation results for improved purity and recovery.

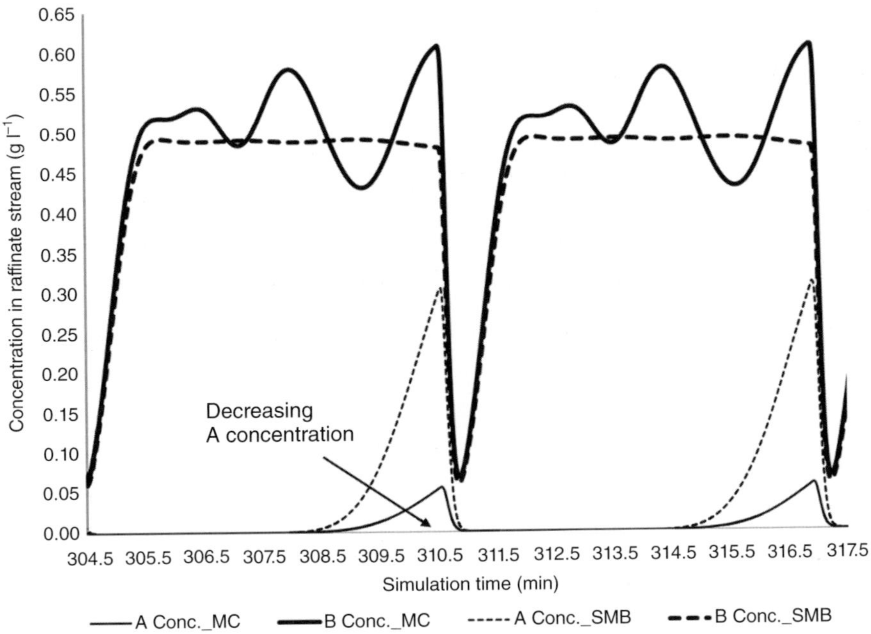

Figure 4.38 Raffinate concentrations over time in SMB and ModiCon (MC).

improved model. ModiCon achieves an improved separation performance, an 11.3% increase from that in the basic SMB model.

Figures 4.38 and 4.39 explain why ModiCon performs better than the basic SMB for both B purity and A recovery. Figure 4.39 displays that the concentration of A in the raffinate for the ModiCon is lower and the average concentration of B is higher than those in the basic SMB, which increases B purity in the raffinate. Figure 4.39 shows the improved A concentration in the extract stream when component B concentrations are effectively zero in both SMB and ModiCon models, which increases A recovery in the extract.

The column concentration profiles in Figure 4.39 show how modifying the subinterval feed concentrations in ModiCon achieves a better separation performance. The figure shows the steady-state concentration profiles at $t = t_1 + t_2$ within one switching time for both ModiCon and SMB. When the subinterval feed concentration in ModiCon increases, the adsorption wave of component B in zones 3 and 4 moves slowly through the column. As a result, more component B reaches the raffinate stream over the course of the switching time, which results in a higher concentration of B in the raffinate. We find that in Figure 4.41, the A concentration in zone 3 is much lower in the ModiCon model than in the SMB model, because the components are accumulating in zones 1 and 2, which is good for collecting component A in the extract stream. This explains the improved A recovery in the extract.

In order to study the system's oscillation tendency shown in Figure 4.40 further, we have validated our ModiCon simulations, showing that 40 and 80 grid points all give similar simulation results. We show the concentration profiles with 80 grid points after steady state in Figure 4.42, which displays a very similar

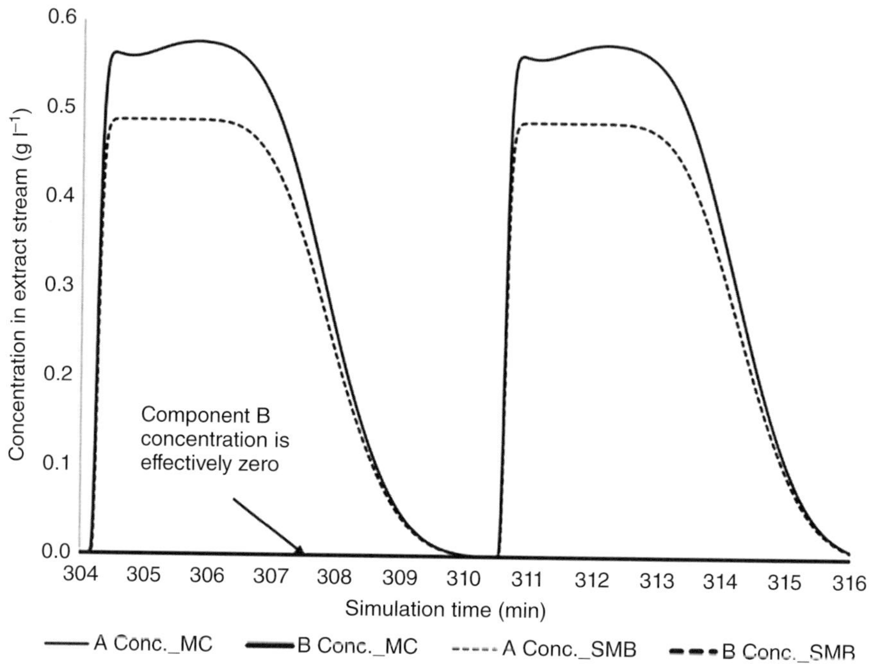

Figure 4.39 Extract concentrations over time in SMB and ModiCon (MC).

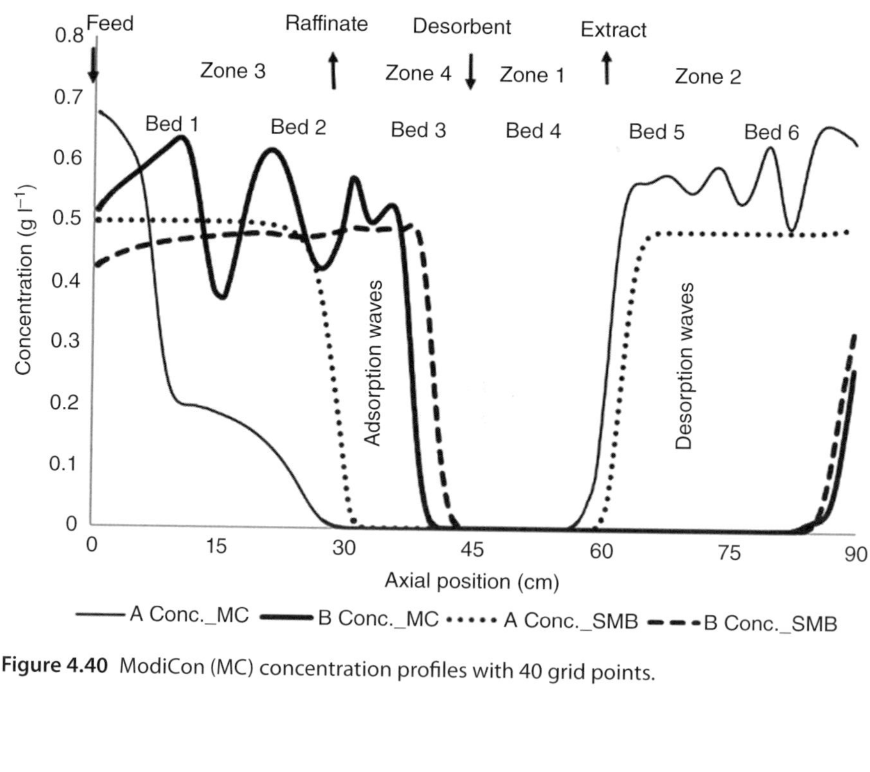

Figure 4.40 ModiCon (MC) concentration profiles with 40 grid points.

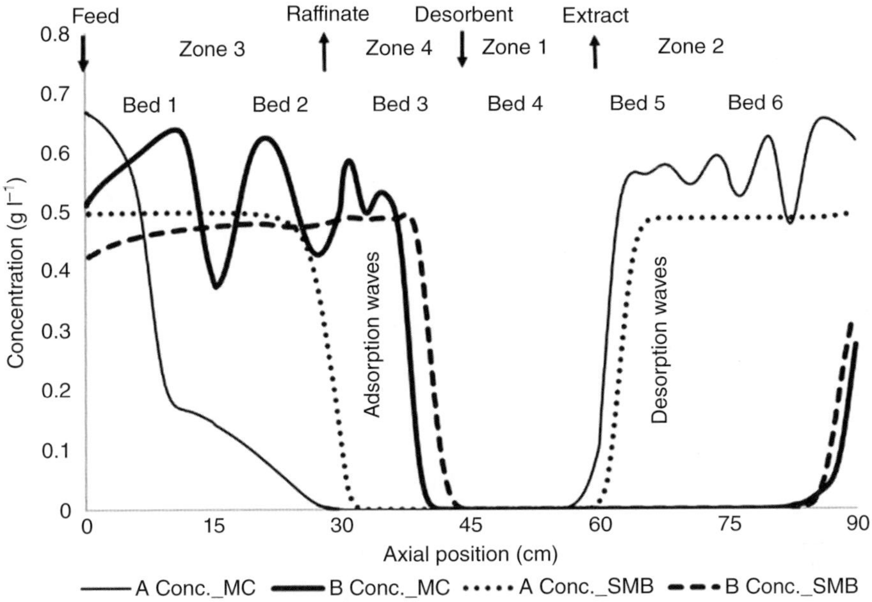

Figure 4.41 ModiCon (MC) concentration profiles with 80 grid points.

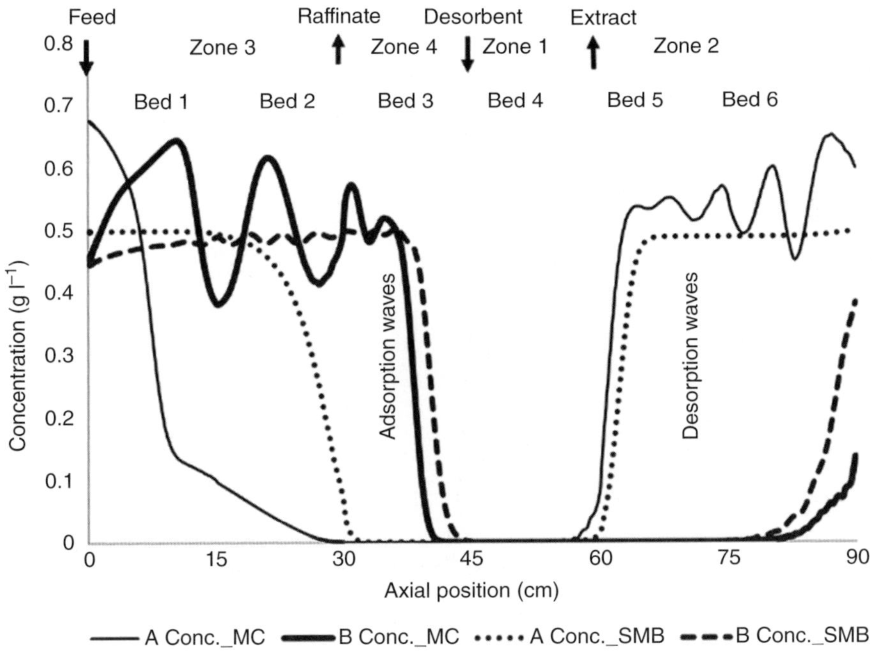

Figure 4.42 ModiCon (MC) concentration profiles with OCFE.

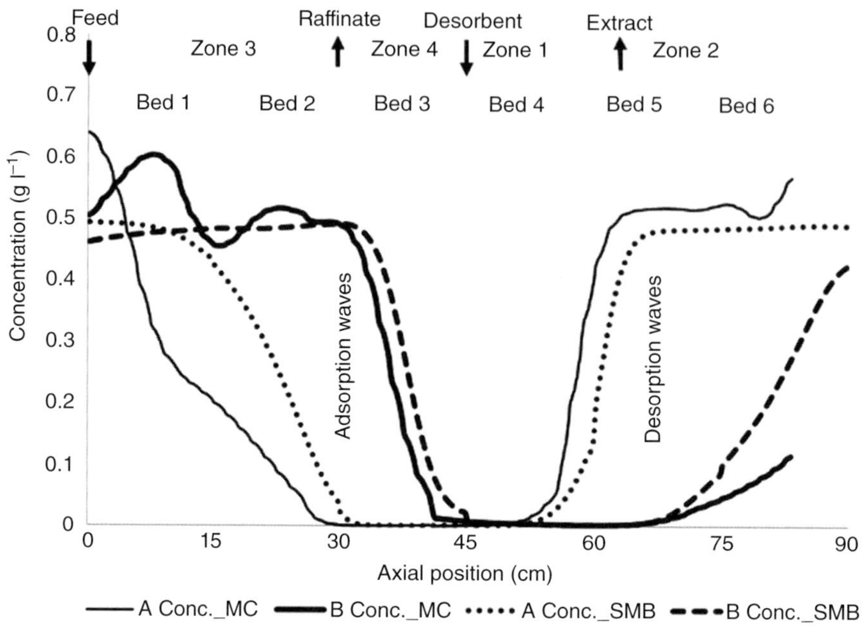

Figure 4.43 ModiCon (MC) concentration profiles with Peclet numbers ($Pe = 50$).

oscillatory tendency. The oscillations are also observed in simulations with small Peclet numbers ($Pe = 50$) and in simulations that employ an orthogonal collocation on finite elements (OCFE) in Figures 4.42 and 4.43. Therefore, we believe that the oscillations are actually a feature of the ModiCon operational mode; moreover, the ModiCon study of Schramm et al. [52] also shows similar oscillations in

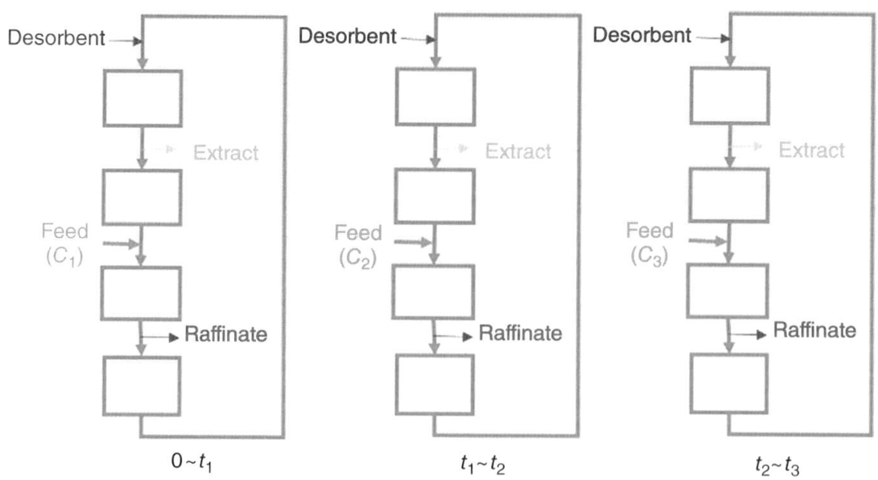

Figure 4.44 An illustration of variable feed concentrations within three subintervals in the ModiCon process.

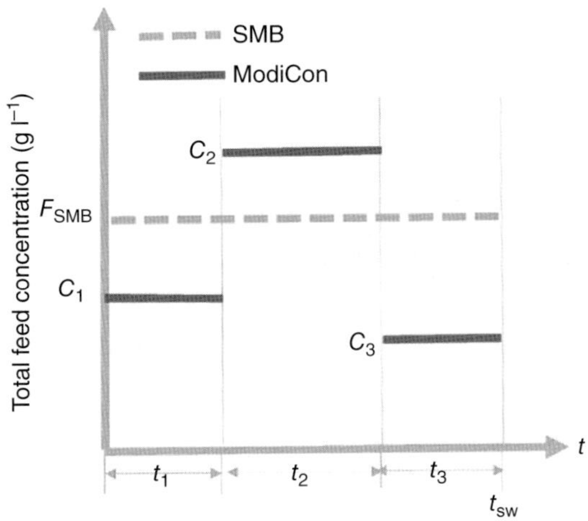

Figure 4.45 A comparison of feed concentration policies of the basic SMB and ModiCon within three switching subintervals.

their concentration profiles. The main reason for the concentration oscillation is the feed concentration switching in combination with the dead volumes.

ModiCon varies the feed concentration (C_1, C_2, and C_3) of the SMB process within three switching subintervals, as illustrated in Figures 4.44 and 4.45. It keeps the feed flow rate and the column configuration unchanged as the basic SMB model. In this ModiCon with three switching subintervals, $t = 0{\sim}t_1$, $C = C_1$; $t = t_1{\sim}t_2$, $C = C_2$; $t = t_2{\sim}t_3$, $C = C_3$; $t = t_{sw,}$ $C = C_1$.

4.5.2 Workshop 3: Apply ModiCon to the 4-Zone SMB Model

In this workshop, we demonstrate how to create a ModiCon model from the basic 4-zone SMB model. ModiCon adjusts the feed concentration, instead of feed flow rates, within the switching subintervals; we still need to add a Cycle Organizer in the ModiCon model to control the change of the cyclic feed concentrations. We keep the feed flow rate and the column configuration the same as in the basic SMB model. Basically, we follow pretty similar steps as we did for the PowerFeed to build a ModiCon model.

Step 1: Open the "4-zone Basic SMB model" and save it as "4-zone ModiCon model" (Figure 4.46).

Step 2: Tool → Cycle Organizer. We add a new Cycle Organizer block into the ModiCon model (Figure 4.47).

Step3: Cycle Organizer → Step Control → Step 1: first subswitching interval. Table 4.6 lists the cycle organizer settings for the subinterval switching times, and feed concentrations in the ModiCon model with three subintervals. We get all these optimized subinterval conditions from our simulation and evaluation strategy in Section 4.5.1. We show the detailed calculations for the values of Table 4.6 here:

From looking at the effect of the second subinterval feed concentration on the B purity and A recovery in Figure 4.39, we achieve the best performance when $C_1 = C_3 = 0.166\,\text{g l}^{-1}$ and $C_2 = 1.6\,\text{g l}^{-1}$ in the ModiCon operation with

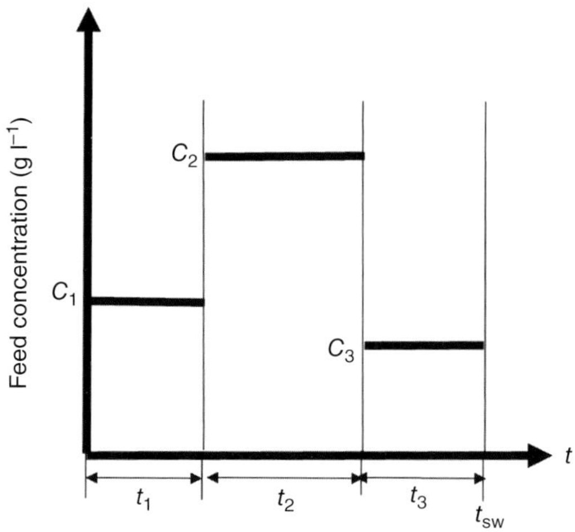

Figure 4.46 The feed flow rate policies in the ModiCon with three subintervals.

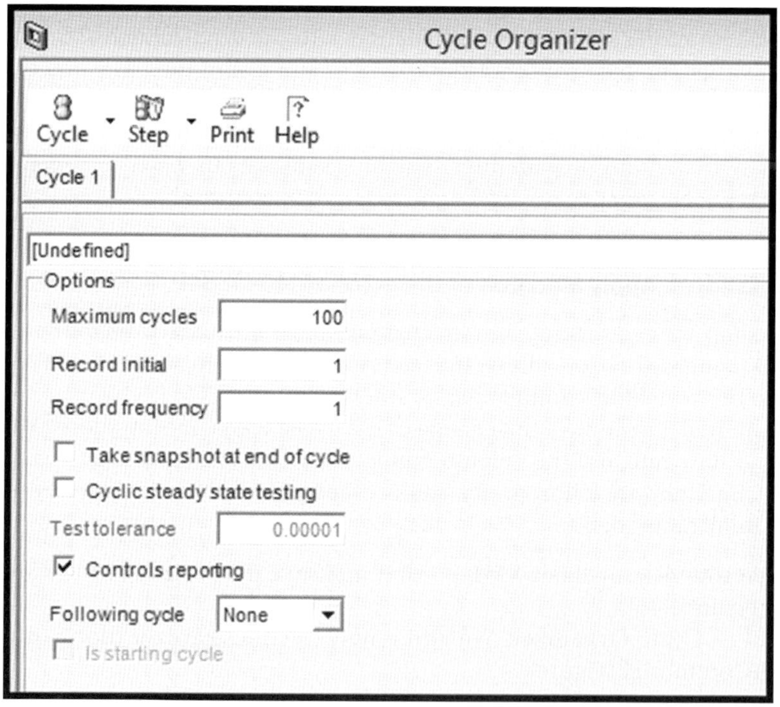

Figure 4.47 The cycle options in the Cycle Organizer of the ModiCon model.

Table 4.6 The cycle organizer conditions for the ModiCon model.

	Step 1	Step 2	Step 3
Subinterval switching time	1.3 min	3.78 min	1.3 min
Subinterval feed concentrations	0.166 g l^{-1}	1.6 g l^{-1}	0.166 g l^{-1}

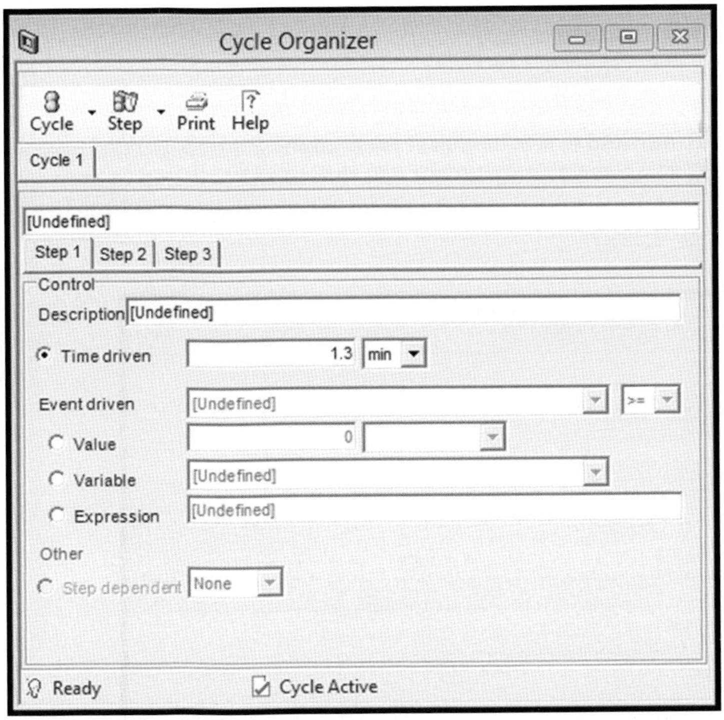

Figure 4.48 The step control setting for step 1 in the ModiCon model.

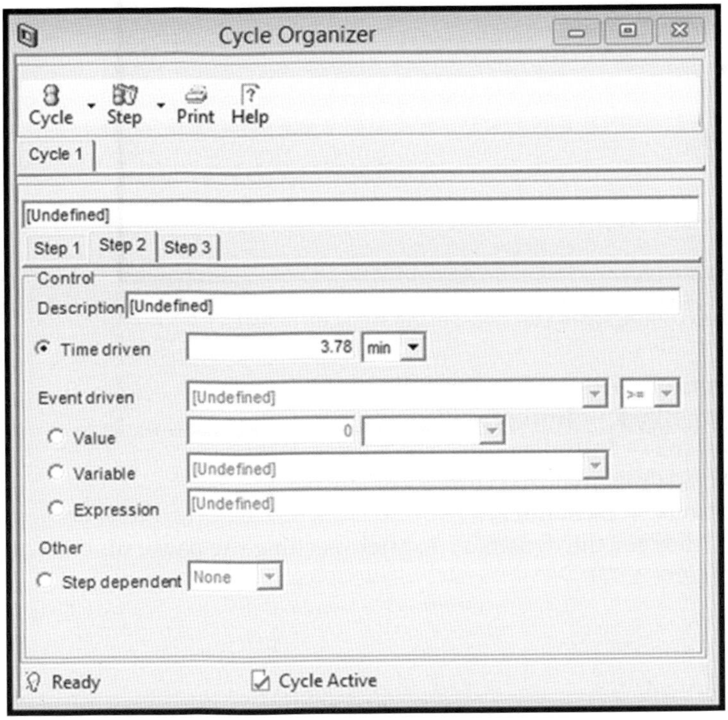

Figure 4.49 The step control setting for step 2 in the ModiCon model.

three subintervals.

$$C_{SMB}{}^{*}t_{sw} = C_1{}^{*}t_1 + C_2{}^{*}t_2 + C_3{}^{*}t_3 \tag{4.18}$$

In this case, $C_{SMB} = 1\ g/L$; $t_1 + t_2 + t_3 = t_{sw} = 6.38$ min

Then, $t_1 = 1.3$ min and $t_2 = 3.78$ min; $t_3 = 1.3$ min

We follow the Figures 4.48 and 4.49 to edit the steps 1 to 3 step by step.

Step 4: Cycle Organizer → Step Control → Step 2: second subswitching interval (Figures 4.50–4.53).

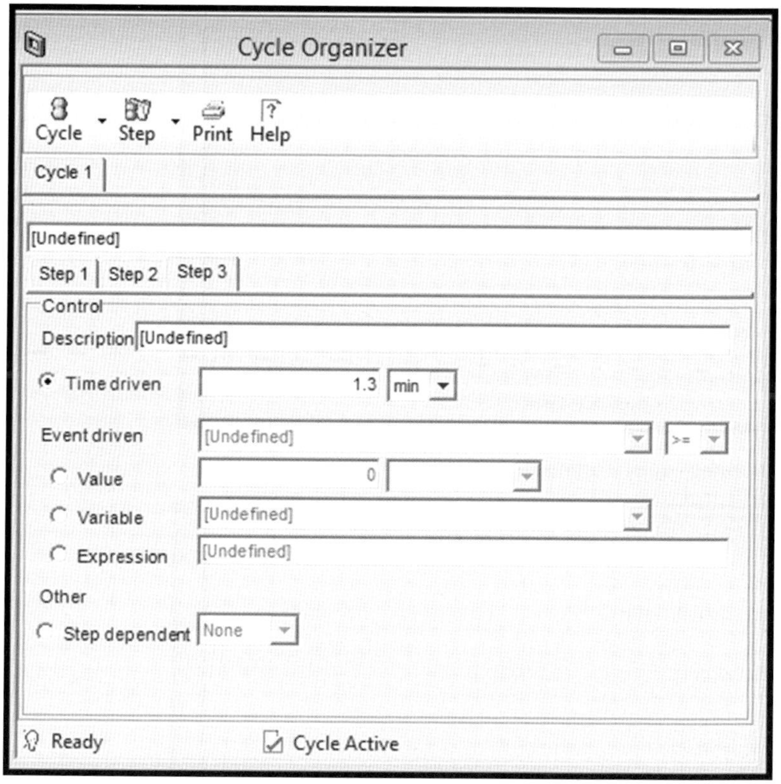

Figure 4.50 The step control setting for step 3 in the ModiCon model.

Step 5: Cycle Options → Generate Task (Figure 4.54).

Step 6: Configure Block → Initialize (Figure 4.55).

Step 7: Run Options → Initialization.

The model is ready to initialize (Figure 4.56).

Step 8: Run Options → Dynamic.

The model is ready for the dynamic run after reaching the pause time (382.8 min) we set (Figure 4.57).

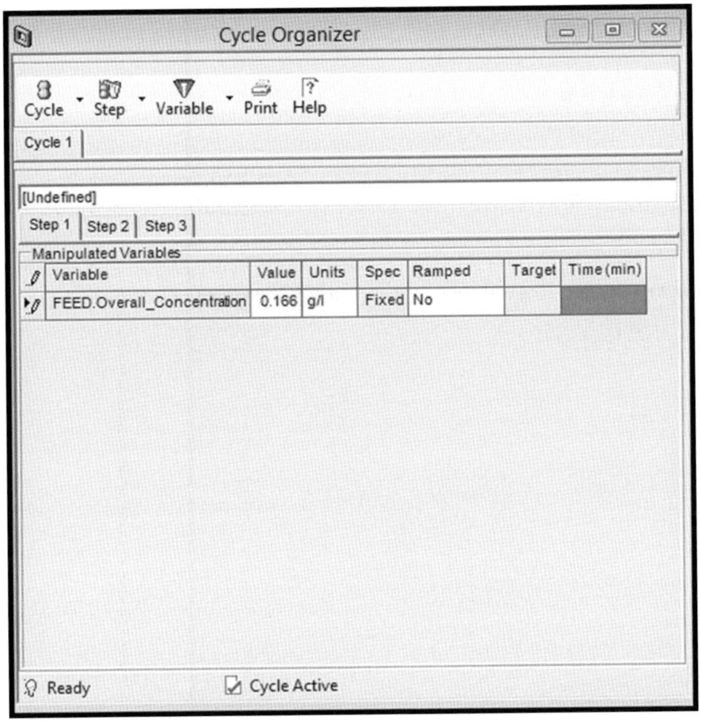

Figure 4.51 The manipulated variables for step 1 in the ModiCon model.

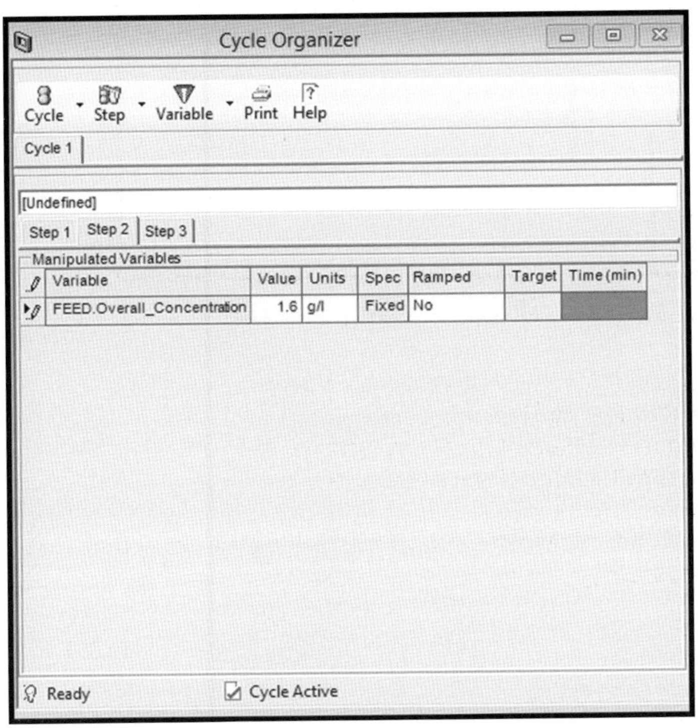

Figure 4.52 The manipulated variables for step 2 in the ModiCon model.

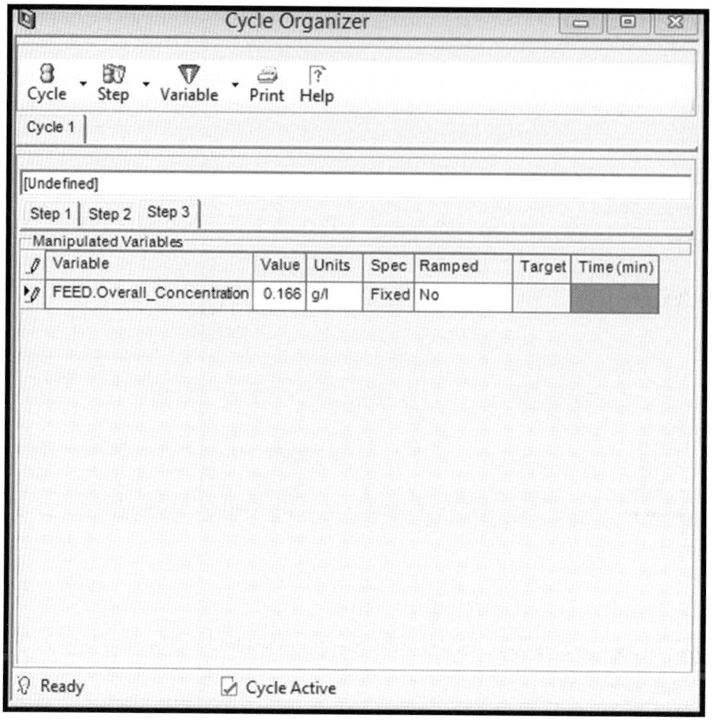

Figure 4.53 The manipulated variables for step 3 in the ModiCon model.

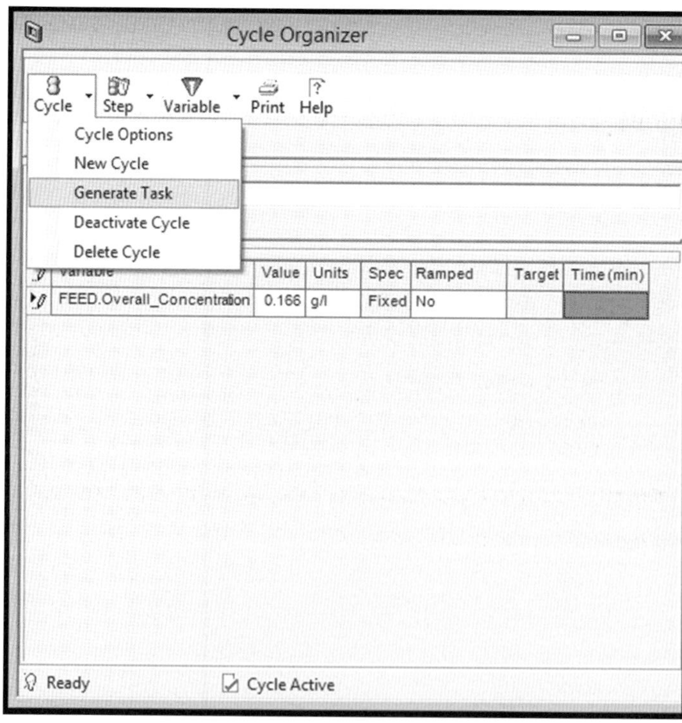

Figure 4.54 The generation progressing in the Cycle Organizer of the ModiCon model.

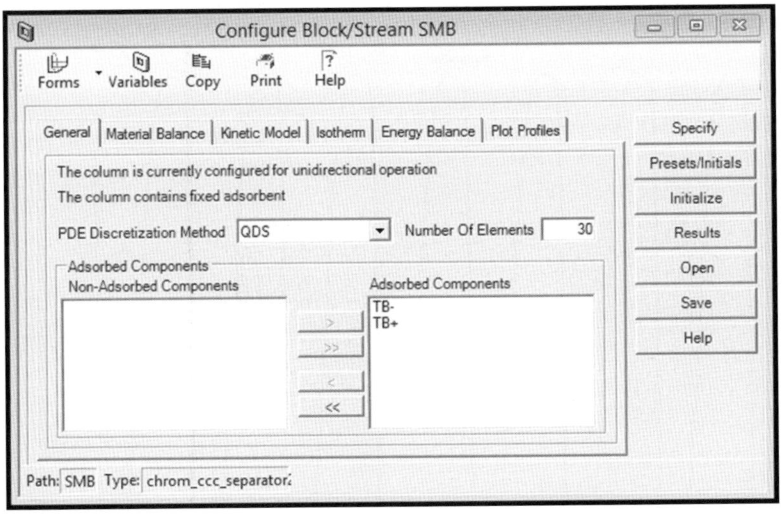

Figure 4.55 Initialize the configure block.

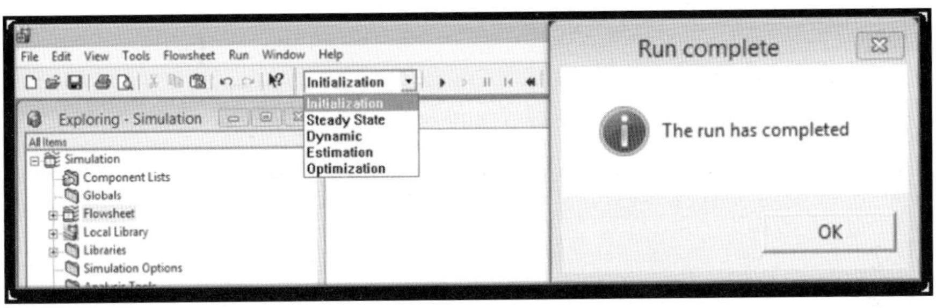

Figure 4.56 The initialization of the ModiCon model.

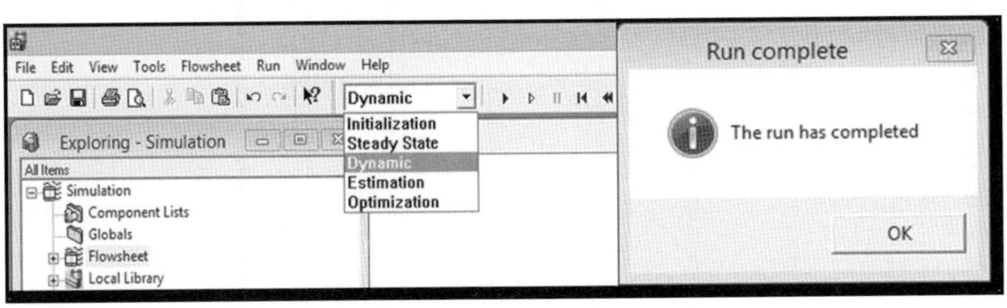

Figure 4.57 The *dynamic run* of the ModiCon model.

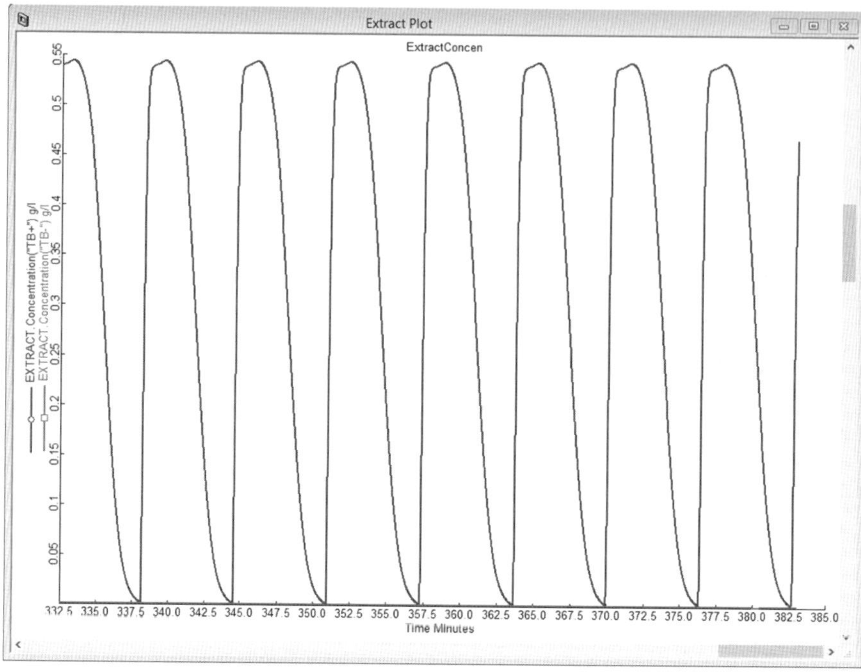

Figure 4.58 The resulting extract concentration results for both components in ModiCon.

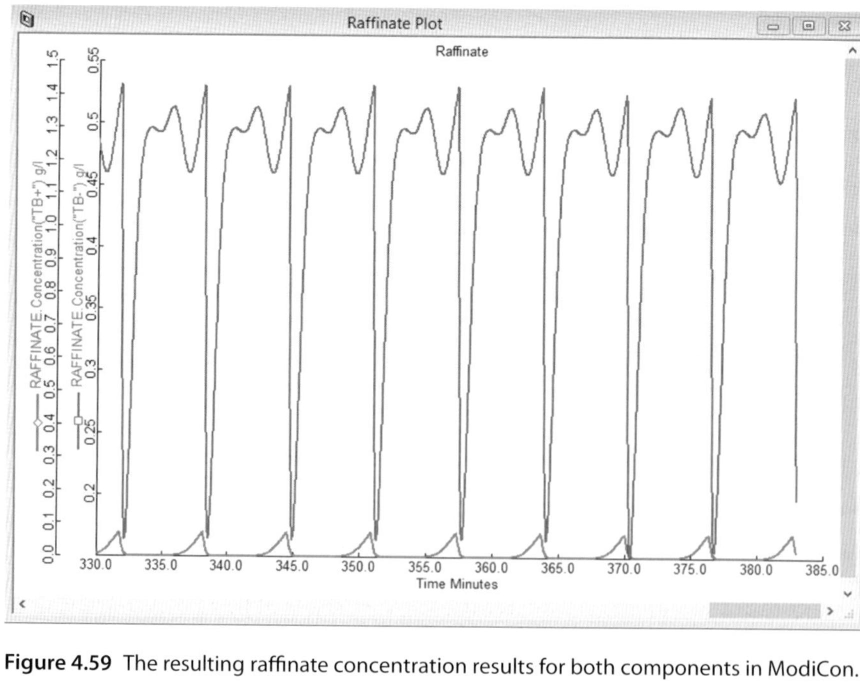

Figure 4.59 The resulting raffinate concentration results for both components in ModiCon.

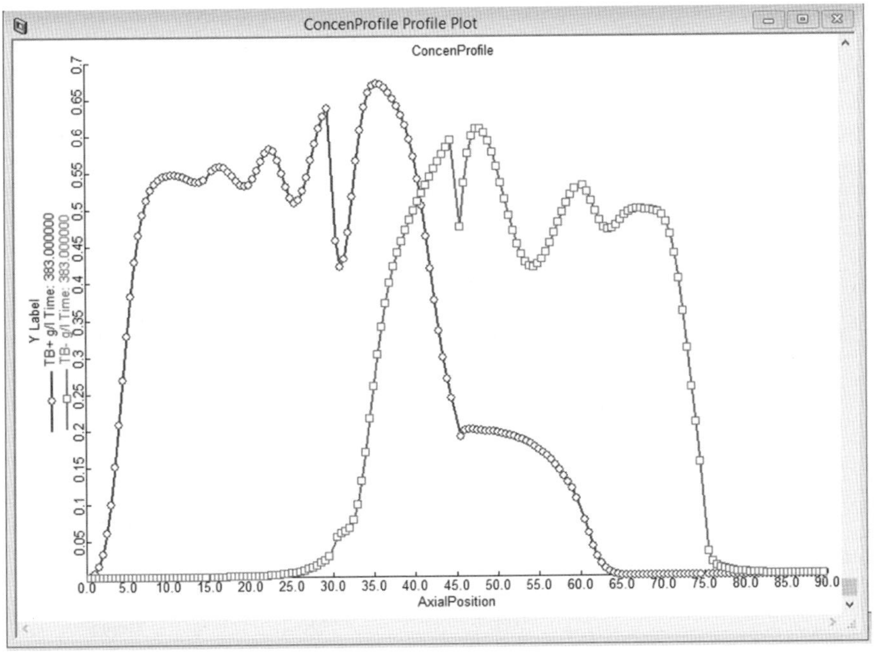

Figure 4.60 The resulting concentration profile within the ModiCon column.

Step 9: Results → Plots (Figures 4.58–4.60).
This step presents the extract concentration plot, raffinate concentration plot, and the column concentration profile.

4.6 Combined Modes

The previous operational modes have each changed different operating conditions, and so can freely be combined. This combination has the potential to greatly increase not only separation performance but also the complexity. Indeed, combining the modes also increases the number of optimization variables, making optimization much more time consuming. Yu et al. [38] found that for our model separation, PowerFeed and ModiCon worked well together, while combinations involving Varicol saw little improvement.

4.6.1 Workshop 4: Extend Previously Created Flowsheets

In this workshop, we combine two operational modes as a new combined operational mode to see if the integrated system will perform better than either one individually. We consider the combination of two operational modes, PowerFeed and ModiCon, and refer to it as PF&MC. We vary the feed flow rates, recycle flow

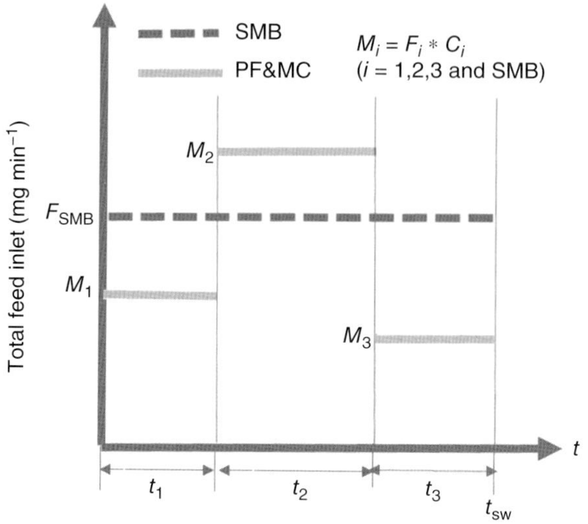

Figure 4.61 The feed inlet policies in the PF&MC model with three subintervals.

Table 4.7 The cycle organizer conditions for the combined PF&MC model.

	Step 1	Step 2	Step 3
Subinterval switching time	1.117 min	4.147 min	1.117 min
Subinterval feed concentrations	0.1 g l^{-1}	1.751 g l^{-1}	0.1 g l^{-1}
Subinterval feed flow rates	0.494 ml min^{-1}	0.319 ml min^{-1}	0.494 ml min^{-1}
Subinterval recycle flow rates	0.508 ml min^{-1}	0.45 ml min^{-1}	0.508 ml min^{-1}

rates, and feed concentrations within three switching subintervals, as shown in Figure 4.61 and Table 4.7.

Step 1: Open the "4-zone Basic SMB model", and save it as "4-zone PF&MC model".

Figure 4.61 shows the feed inlet policies in the PF&MC model with three subintervals.

Step 2: Tool → Cycle Organizer.

Once again, we add a cycle organizer into the basic SMB model and edit the cycle based on the conditions in Table 4.7.

Table 4.7 lists the cycle organizer settings for the subinterval switching times, feed concentrations, and feed and recycle flow rates in the PF&MC model. We get all these optimized subinterval conditions from our simulation and evaluation strategy in Sections 4.4.1 and 4.5.1. We show the detailed calculations for the values of Table 4.7 here.

From evaluation of the PF&MC operation on the product purity and recovery, we achieve the best performance when $C_1 = C_3 = 0.1$ g l^{-1} and

$C_2 = 1.751\ \text{g l}^{-1}$, and $F_1 = F_3 = 0.494\ \text{ml min}^{-1}$ and $F_2 = 0.319\ \text{ml min}^{-1}$ in the PowerFeed operation with two subintervals.

$$C_{\text{SMB}}{}^* F_{\text{SMB}}{}^* t_{\text{sw}} = C_1{}^* F_1{}^* t_1 + C_2{}^* F_2{}^* t_2 + C_3{}^* F_3{}^* t_3 \tag{4.19}$$

In this case, $F_{\text{SMB}} = 0.38\ \text{ml min}^{-1}$; $C_{\text{SMB}} = 1\ \text{g l}^{-1}$; $t_1 + t_2 + t_3 = t_{\text{sw}} = 6.38\ \text{min}$
Then, $t_1 = 1.117\ \text{min}$ and $t_2 = 4.147\ \text{min}$; $t_3 = 1.117\ \text{min}$

For the calculation of the recycle flow rates, we use the same equation:

$$Q_{\text{recycle}} = 0.344 + 0.333Q_{\text{F}} \tag{4.20}$$

Since $Q_{F_1} = 0.494\ \text{ml min}^{-1}$; $Q_{F_2} = 0.319\ \text{ml min}^{-1}$; $Q_{F_3} = 0.494\ \text{ml min}^{-1}$ we get $Q_{\text{recy}_1} = 0.508\ \text{ml min}^{-1}$; $Q_{\text{recy}_2} = 0.45\ \text{ml min}^{-1}$; $Q_{\text{recy}_3} = 0.508\ \text{ml min}^{-1}$

Step 3: Cycle Organizer → Cycle Options. We still choose 100 as the maximum cycle number.

Step 4: Cycle Organizer → Step Control → Step 1: first subswitching interval.

Figures 4.62–4.67 demonstrate how to edit the steps for the PF&MC mode.

Step 5: Cycle Options → Generate Task (Figure 4.68).

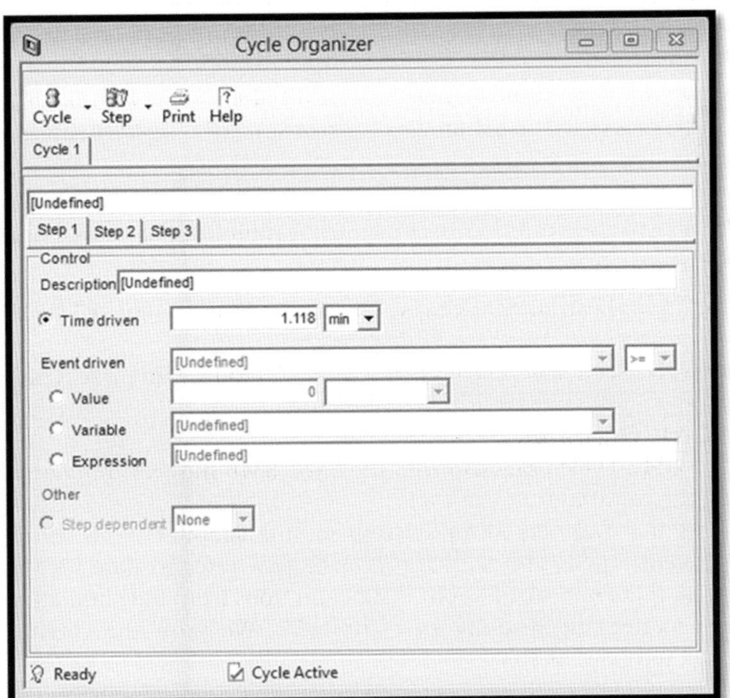

Figure 4.62 The step control setting for step 1 in the PF&MC model.

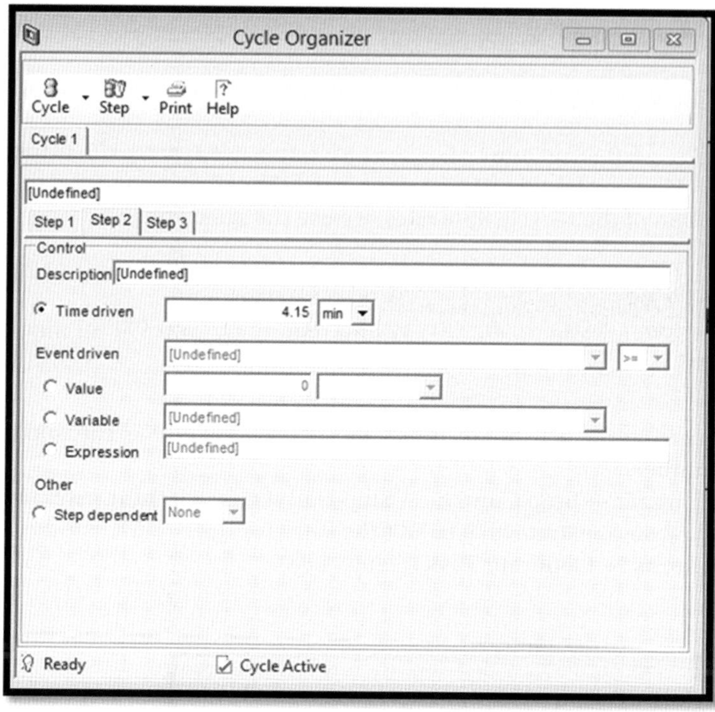

Figure 4.63 The step control setting for step 2 in the PF&MC model.

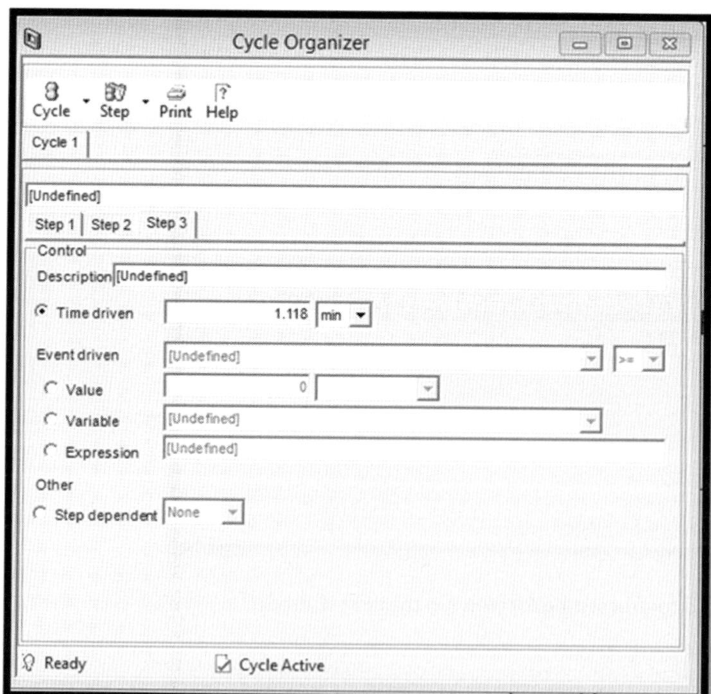

Figure 4.64 The step control setting for step 3 in the PF&MC model.

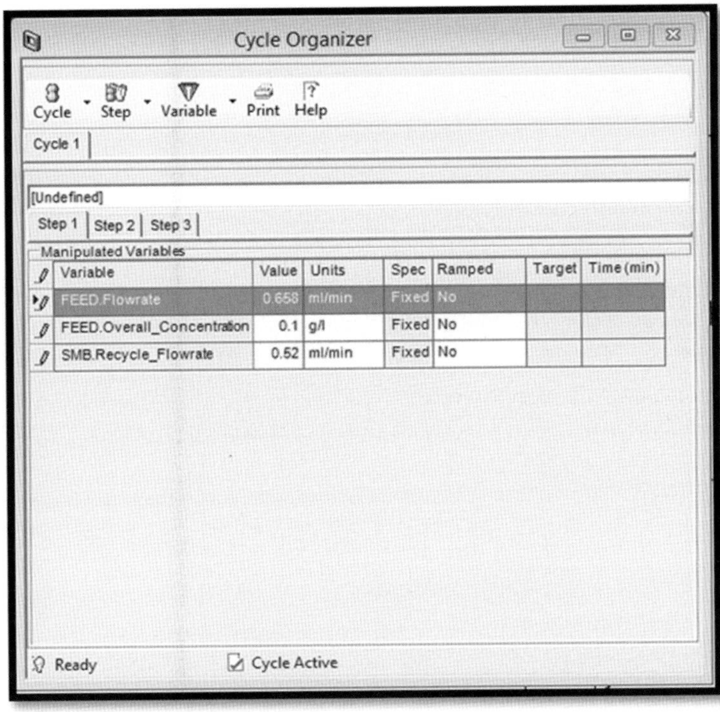

Figure 4.65 The manipulated variables for step 1 in the PF&MC model.

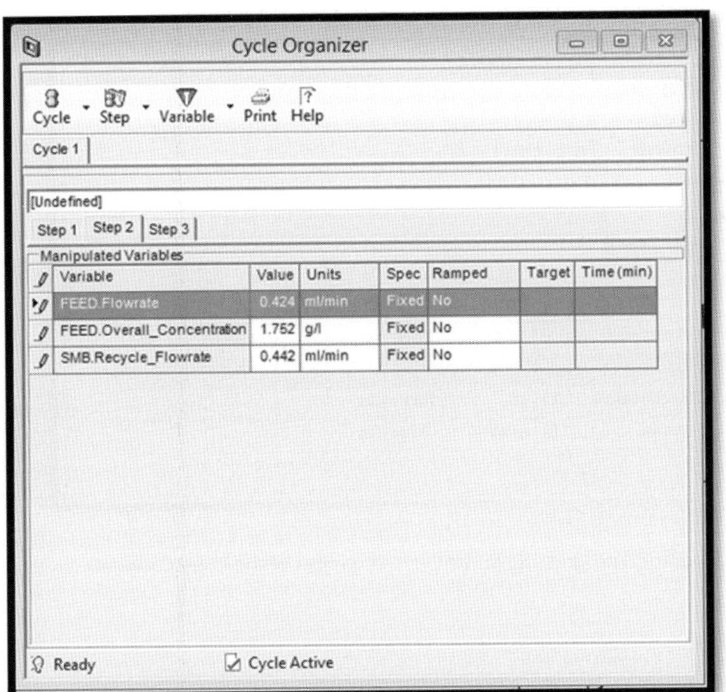

Figure 4.66 The manipulated variables for step 2 in the PF&MC model.

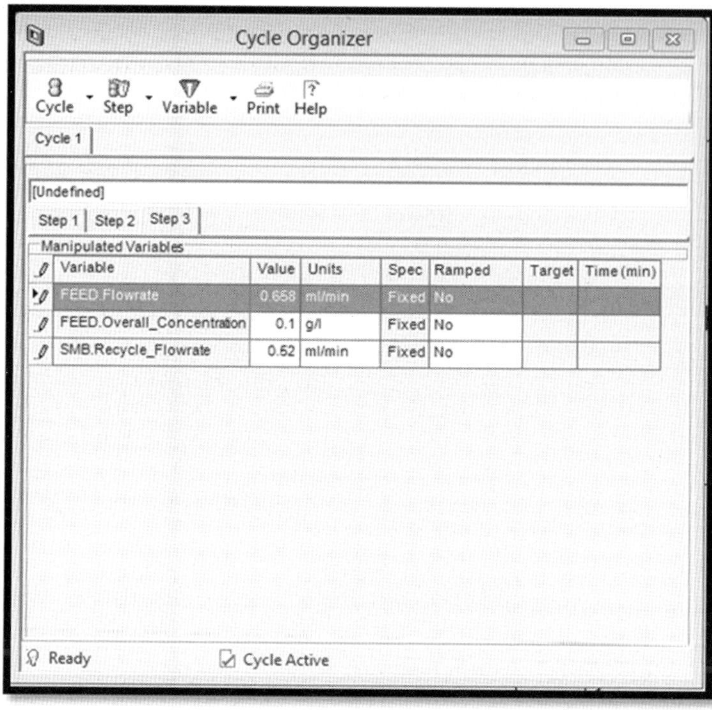

Figure 4.67 The manipulated variables for step 3 in the PF&MC model.

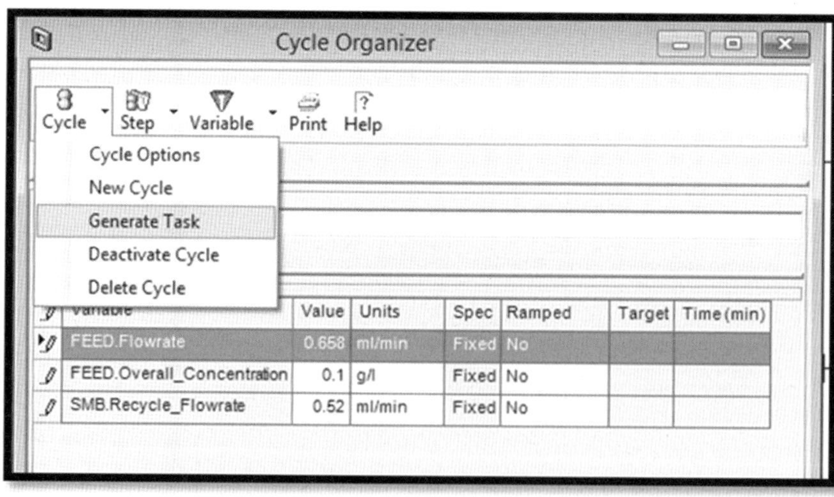

Figure 4.68 The generation progressing in the Cycle Organizer of the PF&MC model.

Step 6: Configure Block → Initialize (Figure 4.69).

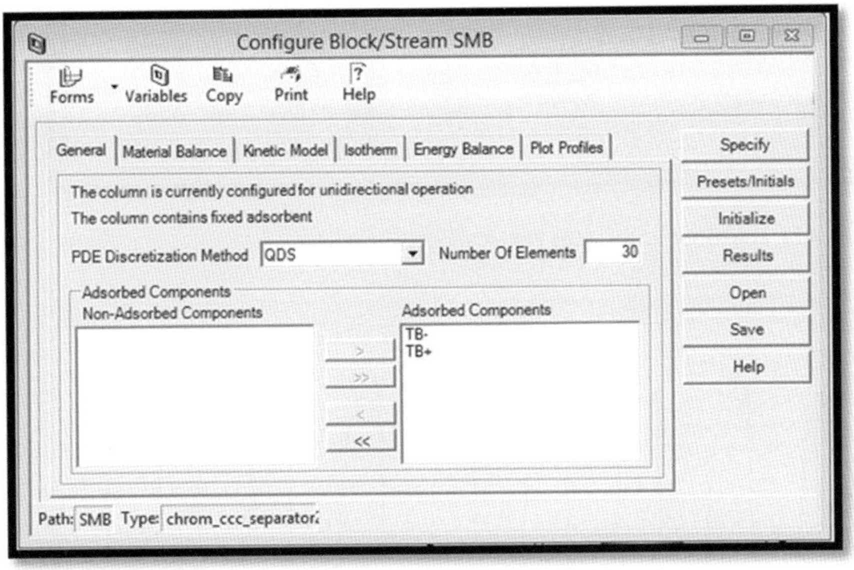

Figure 4.69 The initialization of the PF&MC column.

Step 7: Run Options → Initialization.

The model is ready to initialize (Figure 4.70).

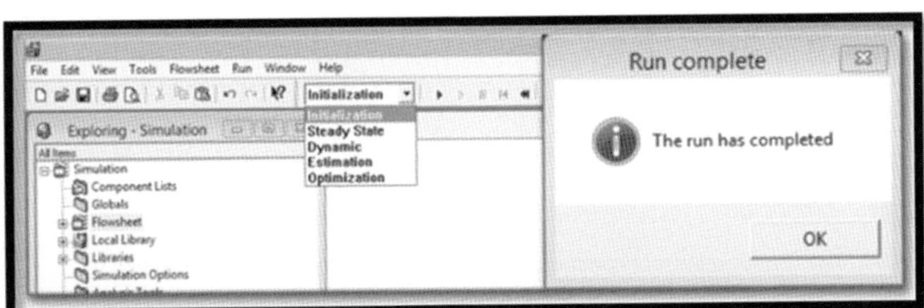

Figure 4.70 The initialization of the PF&MC model.

Step 8: Run Options → Dynamic.

The model is ready for the dynamic run until reaching the pause time (382.8 min) we set (Figure 4.71).

Step 9: Results → Plots.

This step presents the extract concentration plot, raffinate concentration plot, and column concentration profile (Figures 4.72–4.74).

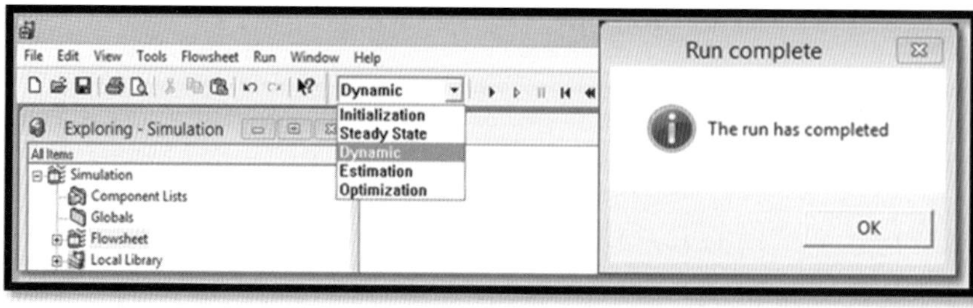

Figure 4.71 The dynamic run of the PF&MC model.

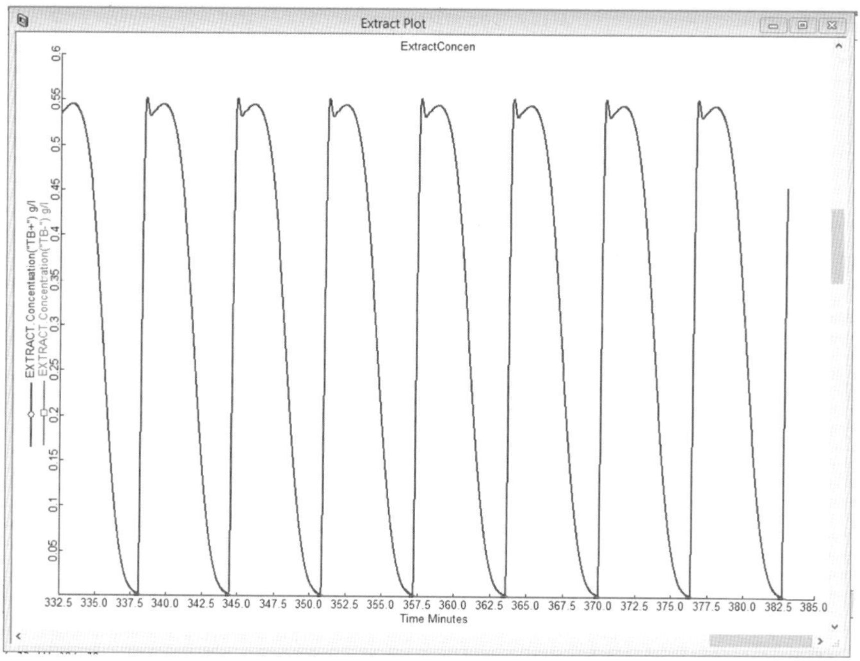

Figure 4.72 The resulting extract concentration results for both components in PF&MC.

4.7 Parallel Two Zones

4.7.1 Introduction to Parallel Two Zones

In 2007, Jin and Wankat developed the concept of a parallel two-zone separator [39]. Their motivation for doing so was the bottleneck that exists in typical xylene SMB units. Specifically, the low concentration of *p*-xylene in the feed to conventional SMB units limits their maximum productivity. By finding an efficient way of increasing the feed concentration to the SMB unit, the productivity of the unit could be dramatically increased. By using the parallel two-zone process, they were able to raise the feed concentration and achieve a 100% increase in productivity for a conventional SMB-based xylene separator.

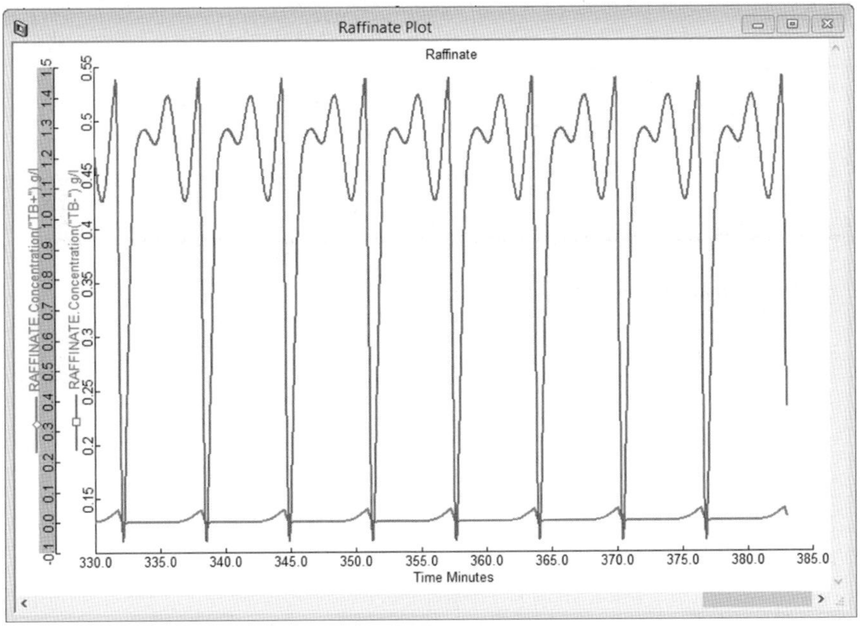

Figure 4.73 The resulting raffinate concentration results for both components in PF&MC.

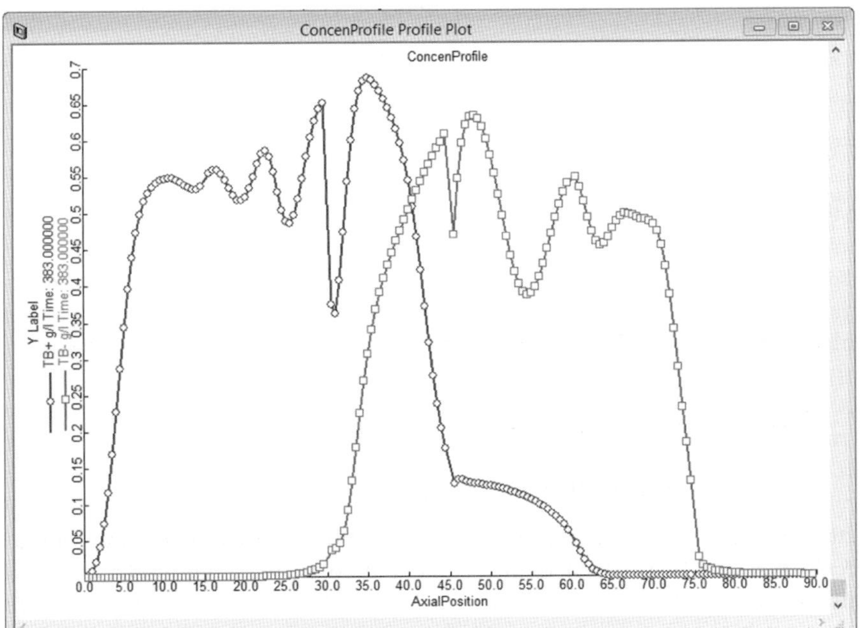

Figure 4.74 The resulting concentration profile within the PF&MC column.

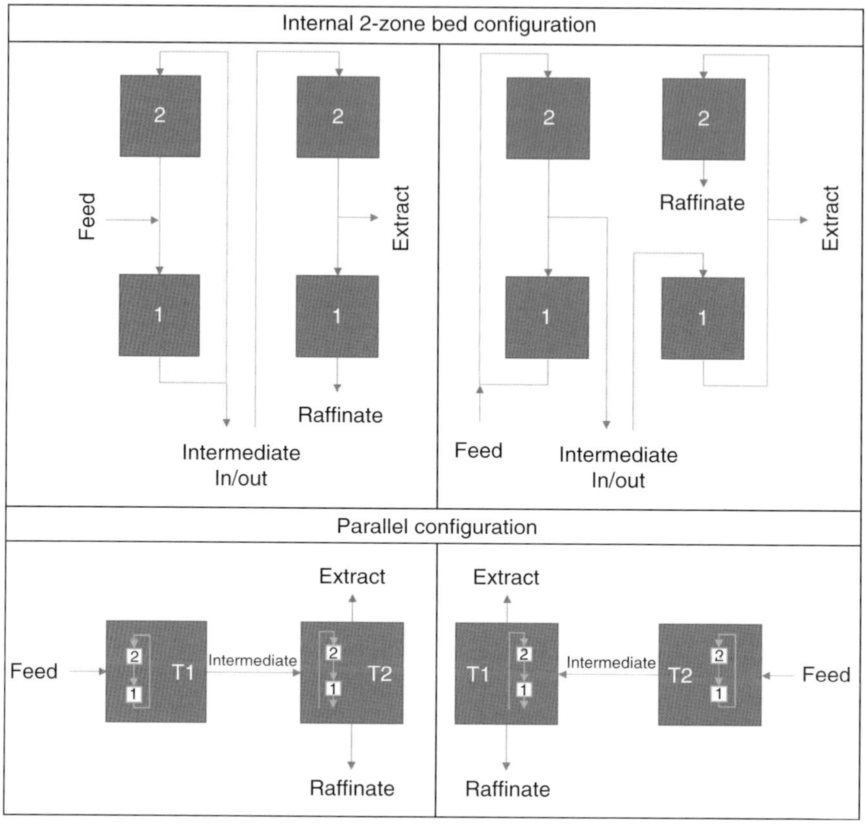

Figure 4.75 The parallel two-zone bed configuration.

In this workshop, we will learn how a parallel two-zone separator works, as well as how to build one in Aspen chromatography.

As the name implies, the parallel two-zone (P2Z) separator consists of columns operated in parallel, and each column has two zones at any given time. In Jin and Wankat's paper, they used columns with two beds each, configured as shown in Figure 4.75.

As we can see, this is a rather complex configuration. If we look specifically at Train 1, we see that in the first switching interval (steps 1 and 2,) the output from bed 1 may either be recycled or sent to the raffinate product. Likewise, the output of the second bed may be mixed with the feed stream, or a fraction may be taken as the extract. In the second switching interval (steps 3 and 4), these roles reverse; and bed 2 is either recycled or raffinate, while the outlet of bed 1 encounters either the feed or the extract. If we look at the second train, we notice that it is operated "out of phase" with Train 1, in such a way that it always lags one step behind Train 1.

How can we model in Aspen? The Chrom_CCC separator we have used up until now does not seem like a good fit for simulating the P2Z separator for the following reasons:

- If we try to model both trains in one CCC separator, the ports do not switch in order. The CCC separator will not be able to move the feed stream from Train 1 all the way to Train 2 and back again without an extensive set of valves.
- If we try to give the trains one Chrom_CCC each, the ports switch more regularly, but there are other problems:
  - The Chrom_CCC separator requires more than three beds.
  - You cannot connect multiple inlets or outlets to the same point in the column.
  - You would need to explicitly provide the recycle, or toggle the recycle on and off manually.
- If we try to model the problem by using a Chrom_r_column for each bed, the flowsheet will become very complex and time consuming to create.
  - Each of the four beds needs to be connected to four different inlets and four different outlets. They would also each require a number of pumps.

The simplest way to perform this modeling is by using a Chrom_r_carousel. The carousel model is very similar to the CCC separator, but slightly more flexible in that it allows for trains of only two beds. We will use one carousel per train and construct the trains separately.

Step 1: Create a new Aspen document to contain Train 1.

Add the xylene components shown in Figure 4.76 and use the Peng-Robinson property method to model them. Ensure that they are all added to the flowsheet.

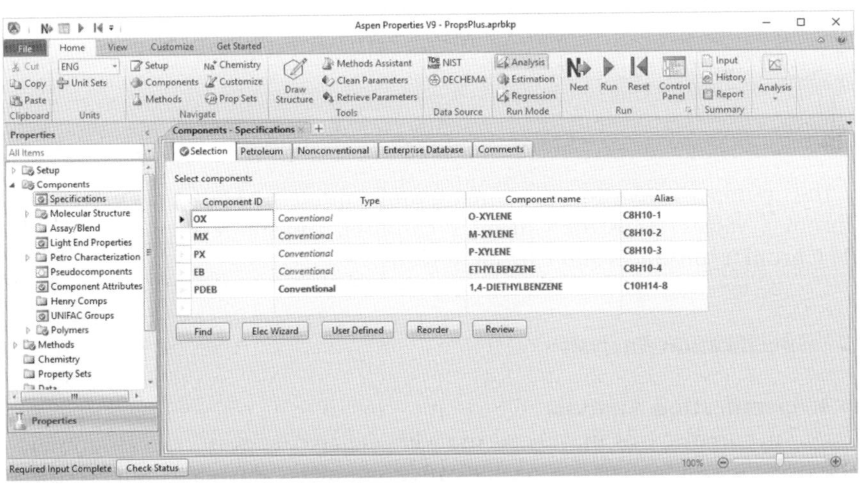

Figure 4.76 The components of a xylene separation.

Step 2: Specify the layout of Train 1.

Using ALL reversible blocks, create the flowsheet shown in Figure 4.77. **This includes the streams.** Make sure that you select Chrom_r_Material connectors while connecting the blocks.

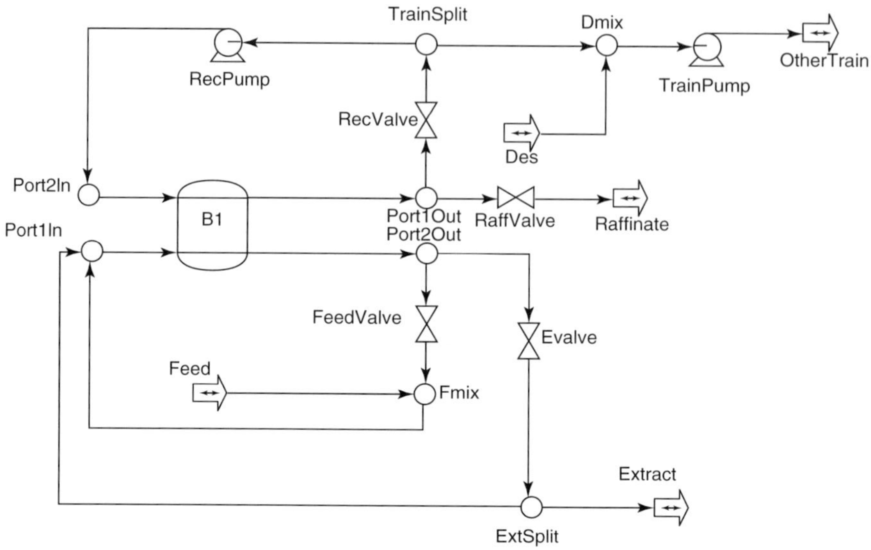

Figure 4.77 The flowsheet for Train 1.

The circles are Chrom_r_nodes. You should pay close attention while connecting these blocks to ensure that you always choose the Process_r_inlet or Process_r_Outlet arrows, as shown in Figure 4.78.

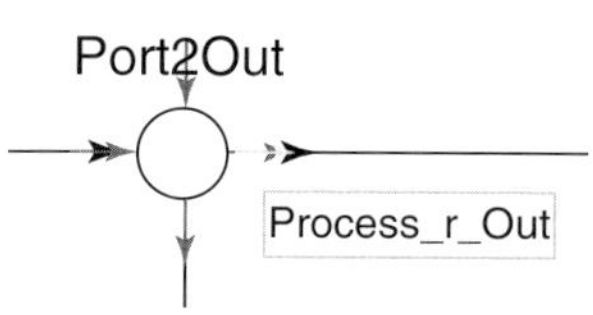

Figure 4.78 Selecting the reversible ports on a Chrom_r_node.

The pump icons are Chrom_r_pump models, and the valves are Chrom_r_valves. B1 is the Chrom_r_carousel, and the feed/product blocks are Chrom_r_products and Chrom_r_feeds.

4.7.2 Specification Analysis

Step 3: Specification Analysis

Once you complete the flowsheet, you will need to change the specifications of a number of variables. Create a table containing all the flow rates and pressures from the feed and product blocks, as shown in Figure 4.79. Set the values shown in that figure, and change the specification if needed.

Now we will also set the valves as though Train 1 is in step 1. To do this, create a table containing all the valve "Active_specification" variables and set them as shown in Figure 4.80.

At this point, the flowsheet will still be underspecified. We do not have any other flow or pressure settings to change, so what might be the problem?

OperatingCond Table

	Value	Units	Spec
Des.Flowrate	**4.62**	l/min	Fixed
Feed.Flowrate	**4.62**	l/min	Fixed
OtherTrain.Flowrate	9.24	l/min	Free
Raffinate.Flowrate	**0.0**	l/min	Fixed
RecPump.Flowrate	**1.59**	l/min	Fixed
TrainPump.Flowrate	9.24	l/min	Free
Des.Pressure	19.9379	bar	Free
Feed.Pressure	**20.0**	bar	Fixed
OtherTrain.Pressure	20.0	bar	Free
Raffinate.Pressure	20.0	bar	Free
RecPump.Pressure	20.008	bar	Free
TrainPump.Pressure	**20.0**	bar	Fixed

Figure 4.79 Setting the flowrate specifications.

valves Table

	Value	Units	Spec
Evalve.Active_Specification	**0.0**	n/a	Fixed
FeedValve.Active_Specification	**1.0**	n/a	Fixed
RaffValve.Active_Specification	**0.0**	n/a	Fixed
RecValve.Active_Specification	**1.0**	n/a	Fixed

Figure 4.80 Setting the valve specifications.

Step 4: Run a specification analysis

Double click on the small "under-specified" triangle at the bottom of the Aspen window. This will open the specification status window that helpfully informs us that we need to fix five additional variables to be able to run the simulation. Click on "analyze" to see which variables Aspen thinks we should change from free to fixed. At first, Aspen will recommend variables such as concentrations, as shown in Figure 4.81. However, we clearly want Aspen to calculate those for us.

Select all the concentration variables and press "reject" and then reanalyze the problem. Repeat this until Aspen suggests that you fix some "reverse concentration variables" (Figure 4.82).

Reverse concentration variables will not matter in our simulation, because the flow never actually does reverse. Therefore, we can safely "accept" these five variables to complete our specification.

Step 5: Run an initialization simulation.

Save your file as **Train1.ada**, and then ensure that your flowsheet is correctly connected and configured by running an initialization.

Step 6: Create the second train flowsheet.

Using exactly the same procedure as steps 1–5, create a second flowsheet containing the second train. Notice that it is quite similar to the first, although the outlets of the bed connect to different sinks (Figure 4.83).

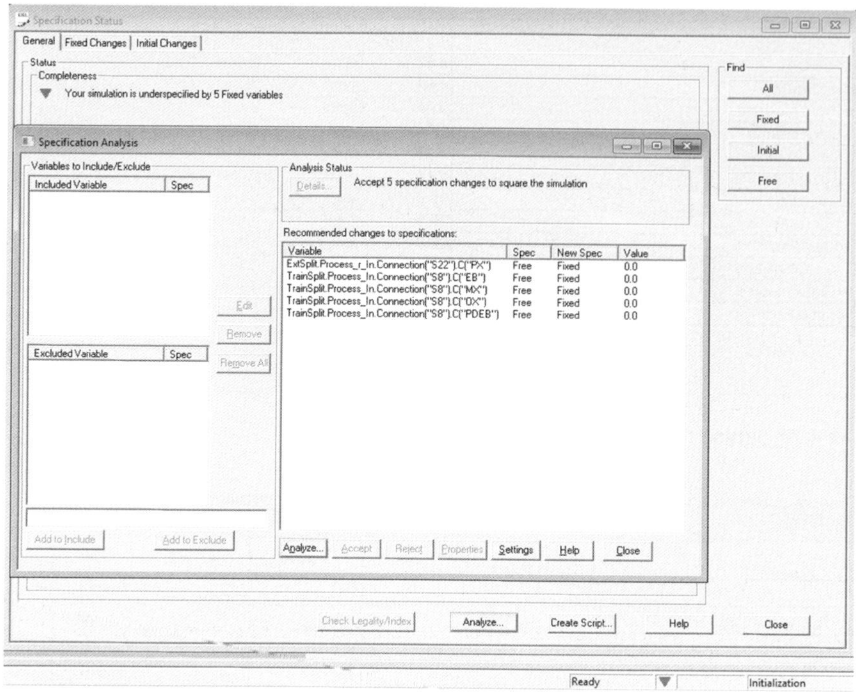

Figure 4.81 The specification analysis.

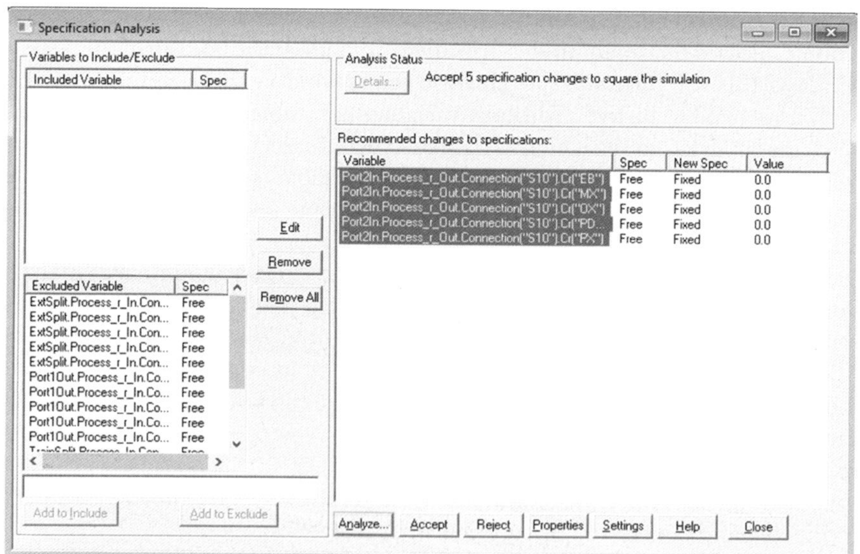

Figure 4.82 Aspen suggesting "reverse concentrations."

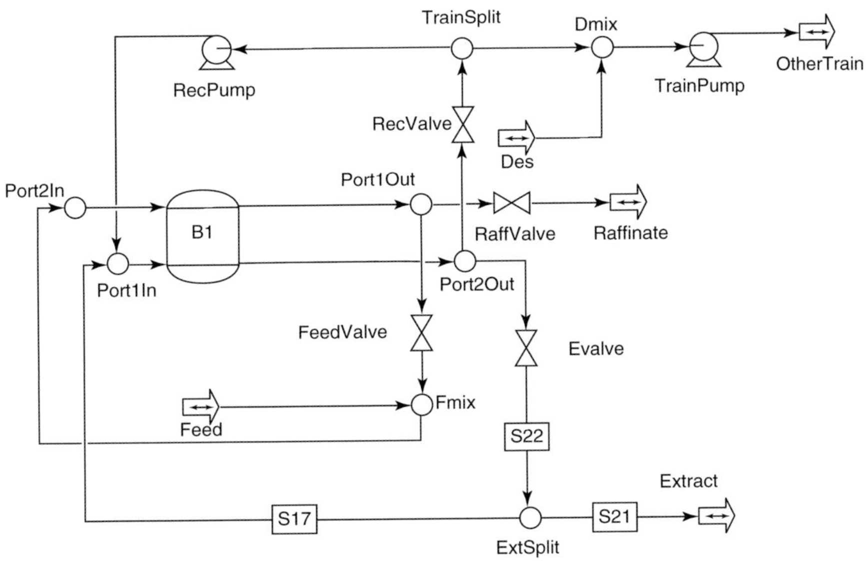

Figure 4.83 The flowsheet for Train 2.

Step 7: Run the initialization on Train 2, and save the document as Train 2.ada.

4.7.3 Importing Flowsheets

Step 8: With Train 2 open, import Train 1 using the file menu (Figure 4.84). While importing the flowsheet, add the prefix T1 to all the blocks that come from Train 1.

Step 9: Rearrange the mess.

All the blocks are imported, but the organization is a mess. Rearrange them so that they look like Figure 4.85.

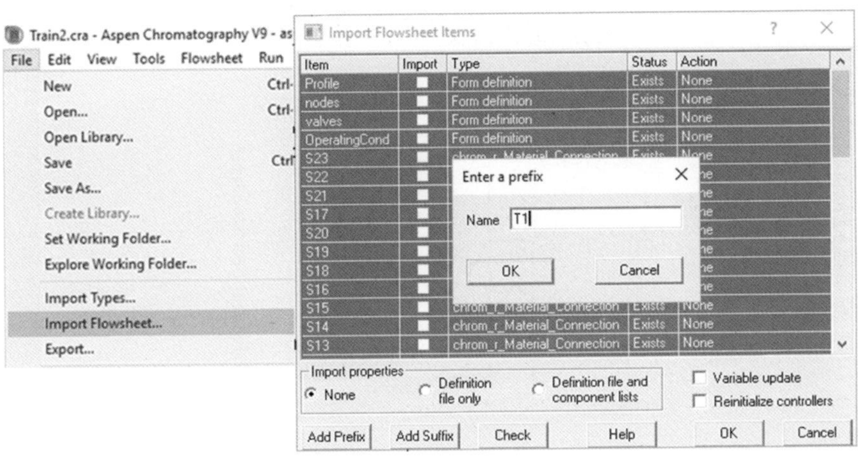

Figure 4.84 Importing a flow sheet.

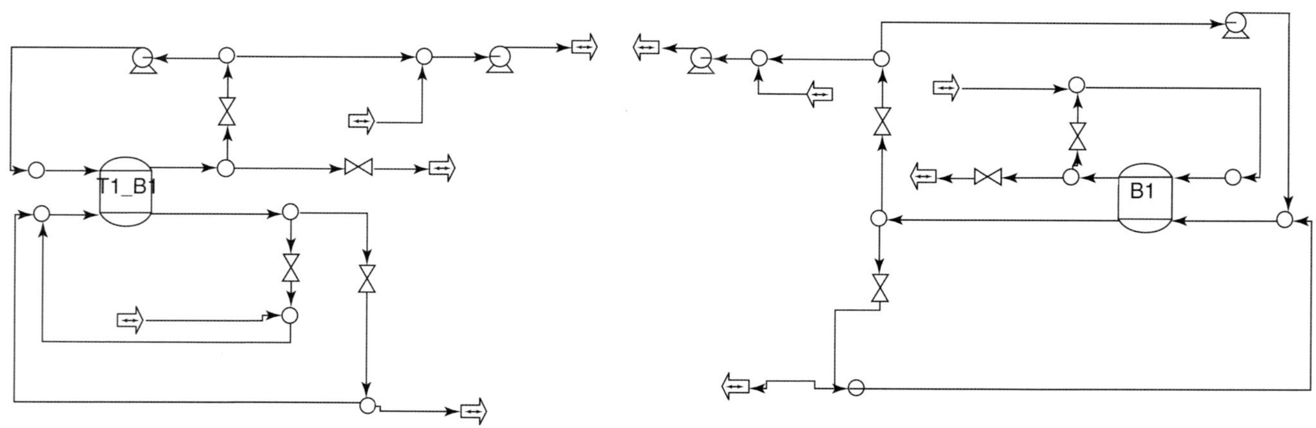

Figure 4.85 Rearranged flowsheet.

Step 10: Specify the beds.
Apply the settings shown in Figures 4.86 and 4.87 to both columns.

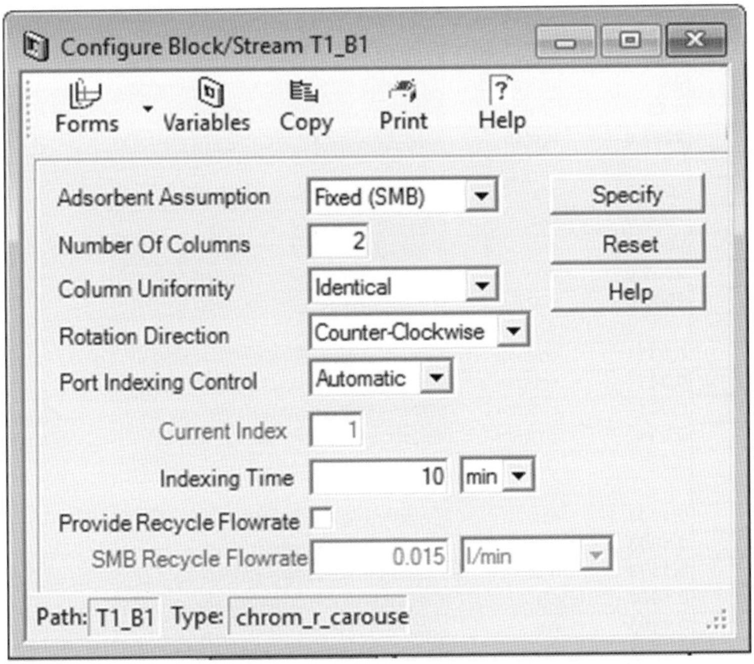

Figure 4.86 T1_B1 settings.

Step 11: Run an initialization.
With both flowsheets still separate, ensure that your simulation does initialize. Notice that some specification changes are lost on import. Reapply the specification changes and save the result as "Combined."
Step 12: Join the Trains together.
Using Chrom_r_nodes, Chrom_r_pumps, and reversible connectors, join the trains together as shown in Figure 4.88.
Step 13: Specify the operating conditions (Figure 4.89).
And the valves (Figure 4.90).
And run an initialization. Save the result as **P2Zworkshop.ada.**
Step 14: Perform a dynamic run.
Change the run mode to dynamic and run the simulation for 20 min.
Create a profile plot of the *p*-Xylene concentration in both trains and observe the behavior (Figure 4.91).
Step 15: Define the valve relationships.
With the flowsheet specified thus far, the carousel blocks already handle the internal port switching for us. However, we need to specify the valve switching

T1_B1.Specify Table

	Value	Units	Description
Hb	113.5	cm	Common length of packed section
Db	411.7	cm	Internal diameter of packed section
Ei	0.39	m3 void/m3	Common inter-particle/external voidag
Ep	0.0	m3 void/m3	Common intra-particle/internal voidag
Rp	0.46	cm	Particle radius of adsorbent
RHOs	2253.0	kg/m3	Common apparent density of adsorbe
Pe(*)			
Pe("EB")	2000.0	n/a	Peclet number for dispersion
Pe("MX")	2000.0	n/a	Peclet number for dispersion
Pe("OX")	2000.0	n/a	Peclet number for dispersion
Pe("PDEB")	2000.0	n/a	Peclet number for dispersion
Pe("PX")	2000.0	n/a	Peclet number for dispersion
Kf(*)			
kf("EB")	0.031	cm/min	Effective mass transfer coefficient
kf("MX")	0.031	cm/min	Effective mass transfer coefficient
kf("OX")	0.031	cm/min	Effective mass transfer coefficient
kf("PDEB")	0.031	cm/min	Effective mass transfer coefficient
kf("PX")	0.031	cm/min	Effective mass transfer coefficient
IP(*)			
IP(1,"EB")	0.1303	n/a	Common isotherm parameter
IP(1,"MX")	0.1303	n/a	Common isotherm parameter
IP(1,"OX")	0.1303	n/a	Common isotherm parameter
IP(1,"PDEB")	0.1077	n/a	Common isotherm parameter
IP(1,"PX")	0.1303	n/a	Common isotherm parameter
IP(2,"EB")	0.3067	n/a	Common isotherm parameter
IP(2,"MX")	0.2299	n/a	Common isotherm parameter
IP(2,"OX")	0.1884	n/a	Common isotherm parameter
IP(2,"PDEB")	1.2935	n/a	Common isotherm parameter
IP(2,"PX")	1.0658	n/a	Common isotherm parameter

Figure 4.87 T1_B1 specifications.

behavior. To do this, we will use a combination of scripts and flowsheet constraints. First, enter the following flowsheet constraints:

```
CONSTRAINTS
  //Flowsheet variables and equations...
  if E1_out.Active_Specification == 1 then
    E2_out.Active_Specification = 0;
  else
    E2_out.Active_Specification = 1;
  endif

  if E2_out.Active_Specification == 1 then
    Evalve.Active_Specification = 1;
  else
    Evalve.Active_Specification = 0;
  endif
```

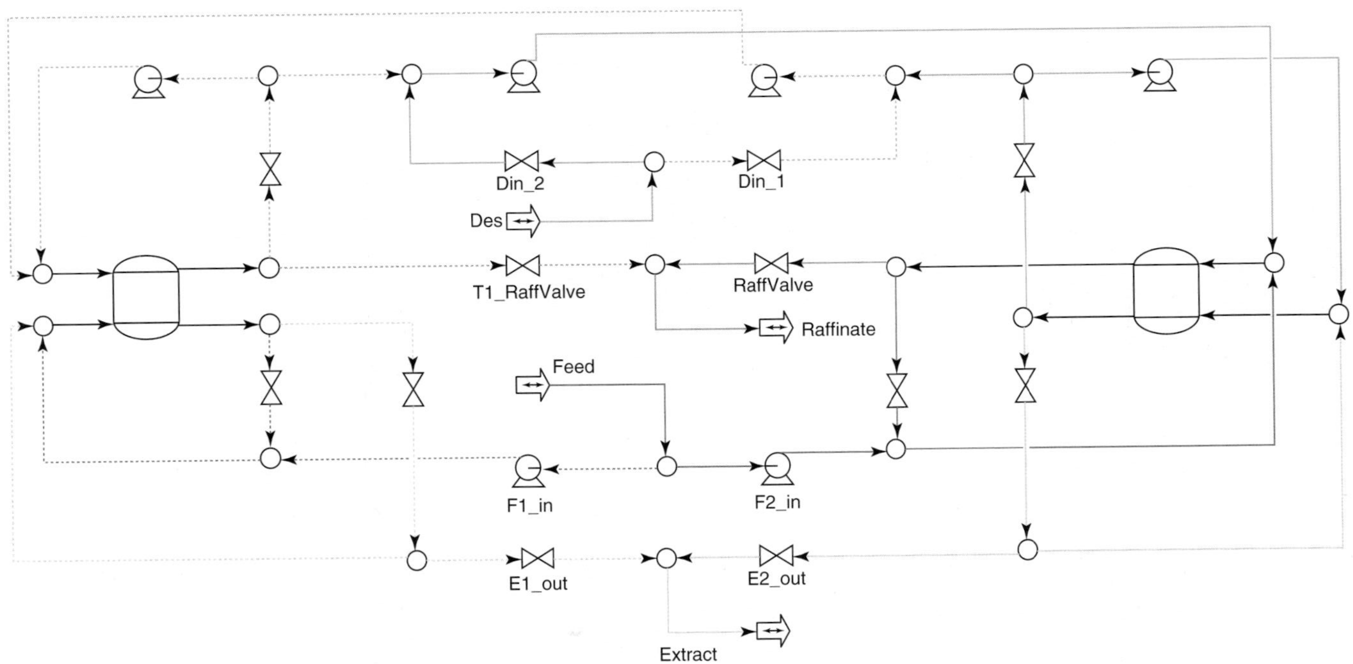

Figure 4.88 The combined trains.

OperatingConditions Table

	Value	Units	Spec
Raffinate.Flowrate	3.82	m3/hr	Free
Feed.Flowrate	**4.62**	m3/hr	Fixed
Extract.Flowrate	**5.42**	m3/hr	Fixed
Des.Flowrate	**4.62**	m3/hr	Fixed
T1_RecPump.Flowrate	**1.59**	m3/hr	Fixed
F1_in.Flowrate	4.62	m3/hr	Free
F2_in.Flowrate	**0.0**	l/min	Fixed
RecPump.Flowrate	**0.0**	l/min	Fixed
T1_TrainPump.Flowrate	9.24	m3/hr	Free
TrainPump.Flowrate	0.0	m3/hr	Free
Des.Pressure	**20.0**	bar	Fixed
Extract.Pressure	17.5567	bar	Free
F1_in.Pressure	19.735	bar	Free
F2_in.Pressure	**20.0**	bar	Fixed
Feed.Pressure	**20.0**	bar	Fixed
Raffinate.Pressure	19.3633	bar	Free
RecPump.Pressure	21.54	bar	Free
T1_RecPump.Pressure	20.0	bar	Free
T1_TrainPump.Pressure	**20.0**	bar	Fixed
TrainPump.Pressure	**20.0**	bar	Fixed

Figure 4.89 The combined set of operating conditions.

valvs Table

	Value	Units	Spec
E1_out.Active_Specification	**0.0**	n/a	Fixed
E2_out.Active_Specification	**1.0**	n/a	Fixed
Evalve.Active_Specification	**1.0**	n/a	Fixed
FeedValve.Active_Specification	**0.0**	n/a	Fixed
RaffValve.Active_Specification	**1.0**	n/a	Fixed
RecValve.Active_Specification	**0.0**	n/a	Fixed
T1_Evalve.Active_Specification	**0.0**	n/a	Fixed
T1_FeedValve.Active_Specification	**1.0**	n/a	Fixed
T1_RaffValve.Active_Specification	**0.0**	n/a	Fixed
T1_RecValve.Active_Specification	**1.0**	n/a	Fixed

Figure 4.90 The combined valve specifications.

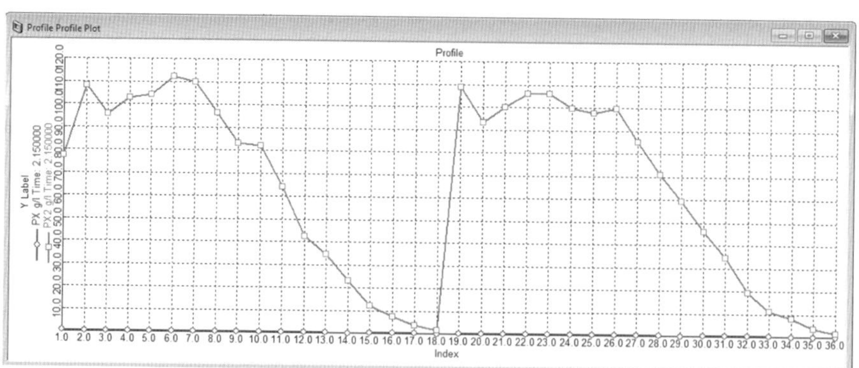

Figure 4.91 Profile behavior c. 2.5 min.

```
if Evalve.Active_Specification == 1 then
  RecValve.Active_Specification = 0;
else
  RecValve.Active_Specification = 1;
endif

if RecValve.Active_Specification == 1 then
  Din_1.Flowrate = Des.Flowrate;
else
  Din_1.Flowrate = 0;
endif

if Din_1.Flowrate == Des.Flowrate then
  Din_2.Flowrate = 0;
else
  Din_2.Flowrate = Des.Flowrate;
endif

if Din_2.Flowrate == Des.Flowrate then
  T1_RecValve.Active_Specification = 1;
else
  T1_RecValve.Active_Specification = 0;
endif

if T1_RecValve.Active_Specification == 1 then
  T1_RaffValve.Active_Specification = 0;
else
  T1_RaffValve.Active_Specification = 1;
endif

if T1_RaffValve.Active_Specification == 1 then
  RaffValve.Active_Specification = 0;
else
  RaffValve.Active_Specification = 1;
endif

if RaffValve.Active_Specification == 1 then
  FeedValve.Active_Specification = 0;
else
  FeedValve.Active_Specification = 1;
endif

if FeedValve.Active_Specification == 1 then
  F2_in.Flowrate = Feed.Flowrate;
else
  F2_in.Flowrate = 0;
endif
```

```
    if F1_in.Flowrate == Feed.Flowrate then
      T1_FeedValve.Active_Specification = 1;
    else
      T1_FeedValve.Active_Specification = 0;
    endif

    if T1_FeedValve.Active_Specification == 1 then
      T1_Evalve.Active_Specification = 0;
    else
      T1_Evalve.Active_Specification = 1;
    endif
END
```

Step 16: Create the cycle script.

With the constraints defined in step 15, all the valves are now controlled by the E1_out valve. Setting its value will completely define all the other valve states. Therefore, during our cycle, we need only to change the single variable.

Create a script with the following code:

```
Dim cycleTime
cycleTime = 2.2 * 60

for i = 1 to 20

  If E1_out.Active_Specification.Value = 1 Then
    E1_out.Active_Specification.Value = 0
  else
    E1_out.Active_Specification.Value = 1
  end if

  Simulation.RunMode = "Initialization"
  on error resume next
  Simulation.Run True

  on error resume next
  Simulation.Run True

  Simulation.RunMode = "Dynamic"

  Simulation.EndTime = cycleTime * i
  Simulation.DisplayTime = cycleTime * (i - 1)

  Simulation.Run True

next
```

This script will control the simulation run. Note the double-initialization run in each iteration. Unfortunately, when the E1_out valve state is changed, two initialization passes are required to calculate the appropriate system conditions. There is no way of forcing two initializations using tasks or the cycle organizer, which is why we are forced to write our own script.

Step 17: Run the simulation by invoking the script

Monitor the concentration profiles of the two beds during the run, and check the extract conditions afterwards. There should be a marked increase in *p*-xylene concentration in the extract.

4.8 Conclusions

In this chapter, we have learned about four different operational modes, and how to combine them. We discussed their relative pros and cons, and presented some strategies for optimizing their performance. The details of simulating these operational modes were also demonstrated.

4.9 Practice Problems

4.9.1 Simulation of a Five-Zone SMB Unit Using the ModiCon Operational Mode

Using the simulation file you created in practice problem in Section 2.6.3, create a five-zone ModiCon simulation. Vary the feed concentration between 0.15 and 4.5 $g\,l^{-1}$ to find the optimum 3-step ModiCon model.

We strongly recommend that you consider the manipulated variables in the optimization problem to be the zone lengths, and then reduce the number of search dimensions from 5 to 4 by noting that the sum of zone lengths must equal the total height of the column. See Figures 4.92 and 4.93.

4.9.2 Compare Parallel Two-Zone Results with SMB Results

Compare the performance of the parallel two-zone results with that of an ideal SMB unit of the same size. Do the following:

1) Create a four-zone SMB unit containing four adsorbent beds with the same dimensions and adsorbents as the parallel two-zone unit.
2) Ensure that the parallel two-zone unit and the SMB unit have the same flow conditions.

Run both simulations. In the parallel two-zone column, create a "raffinate report" table to see the purity and recovery of components in the raffinate stream, as shown in Figures 4.94 and 4.95.

Here, we can clearly see the benefits of the parallel two-zone separator. While the 4-column SMB unit is barely able to change the inlet concentrations of the components, the parallel two zone unit is able to increase the PX concentration that can be sent to a larger downstream SMB unit.

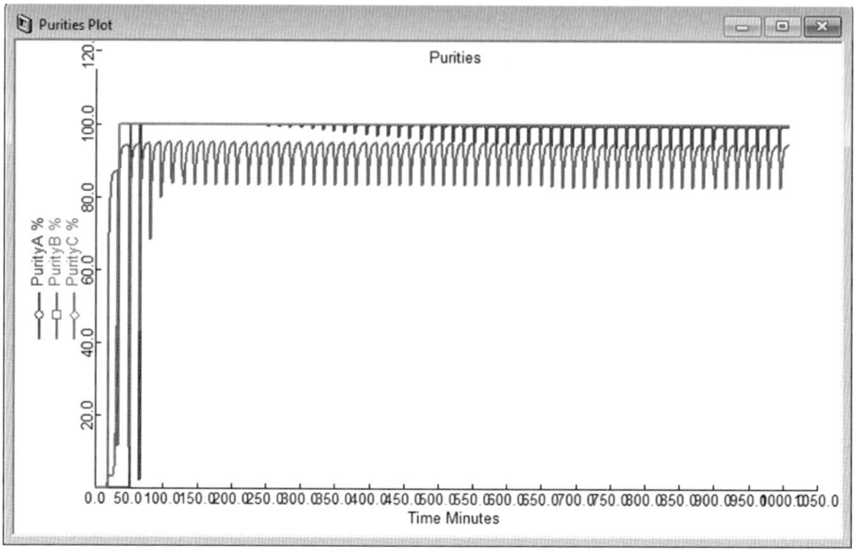

Figure 4.92 5-Zone ModiCon separation performance (purity).

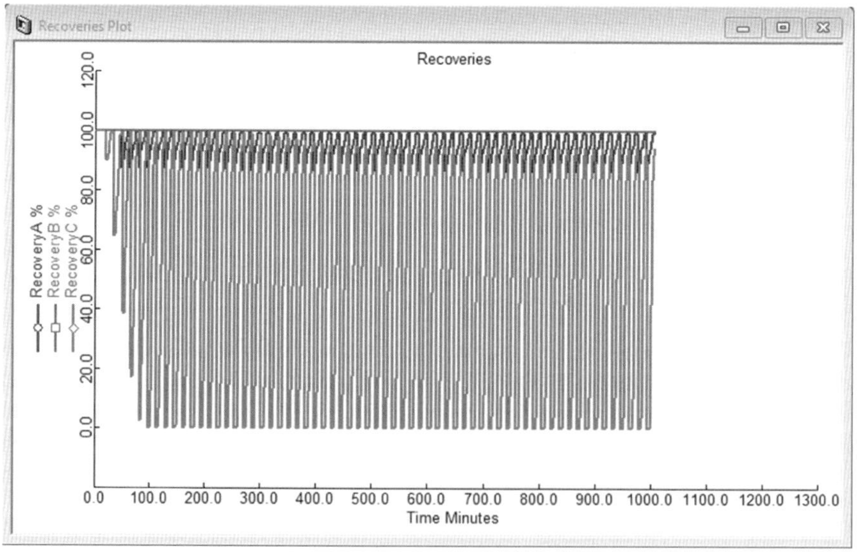

Figure 4.93 5-Zone ModiCon separation performance (recovery).

Raffinate_Report Table

	Value	Units	Description
Desorbent_Component	PDEB		Component that represents the desorbent
Raffinate_Recovery(*)			
Raffinate_Recovery("EB")	94.7336	%	Feed material recovery in raffinate
Raffinate_Recovery("MX")	96.6596	%	Feed material recovery in raffinate
Raffinate_Recovery("OX")	97.2228	%	Feed material recovery in raffinate
Raffinate_Recovery("PX")	48.2241	%	Feed material recovery in raffinate
Raffinate_Purity(*)			
Raffinate_Purity("EB")	15.5975	%	Normalized raffinate purity
Raffinate_Purity("MX")	56.497	%	Normalized raffinate purity
Raffinate_Purity("OX")	14.521	%	Normalized raffinate purity
Raffinate_Purity("PX")	13.3844	%	Normalized raffinate purity

Figure 4.94 4-Zone separation results with four beds.

Raffinate_Report Table

	Value	Units	Description
Desorbent_Component	PDEB		Component that represents the desorbent
Raffinate_Recovery(*)			
Raffinate_Recovery("EB")	94.7336	%	Feed material recovery in raffinate
Raffinate_Recovery("MX")	96.6596	%	Feed material recovery in raffinate
Raffinate_Recovery("OX")	97.2228	%	Feed material recovery in raffinate
Raffinate_Recovery("PX")	48.2241	%	Feed material recovery in raffinate
Raffinate_Purity(*)			
Raffinate_Purity("EB")	15.5975	%	Normalized raffinate purity
Raffinate_Purity("MX")	56.497	%	Normalized raffinate purity
Raffinate_Purity("OX")	14.521	%	Normalized raffinate purity
Raffinate_Purity("PX")	13.3844	%	Normalized raffinate purity

Figure 4.95 Parallel two-zone performance (recovery).

Bibliography

Huthmann, E. and Juza, M. (2001). Modification of a commercial chiral stationary phase: influences on enantiomer separations using simulated moving bed chromatography. *Journal of Chromatography A* **908** (1): 185–200.

Kawajiri, Y. and Biegler, L.T. (2006). Optimization strategies for simulated moving bed and PowerFeed processes. *AIChE Journal* **52** (4): 1343–1350.

Ludemann-Hombourger, O., Nicoud, R.M., and Bailly, M. (2000). The "VARICOL" process: a new multicolumn continuous chromatographic process. *Separation Science and Technology* **35** (12): 1829–1862.

Ludemann-Hombourger, O., Pigorini, G., Nicoud, R.M. et al. (2002). Application of the "VARICOL" process to the separation of the isomers of the SB-553261 racemate. *Journal of Chromatography A* **947** (1): 59–68.

Pais, L.S. and Rodrigues, A.E. (2003). Design of simulated moving bed and Varicol processes for preparative separations with a low number of columns. *Journal of Chromatography A* **1006** (1): 33–44.

Schramm, H., Kienle, A., Kaspereit, M., and Seidel-Morgenstern, A. (2003). Improved operation of simulated moving bed processes through cyclic modulation of feed flow and feed concentration. *Chemical Engineering Science* **58** (23–24): 5217–5227.

Subramani, H.J., Hidajat, K., and Ray, A.K. (2003). Optimization of reactive SMB and Varicol systems. *Computers & Chemical Engineering* **27** (12): 1883–1901.

Toumi, A., Hanisch, F., and Engell, S. (2002). Optimal operation of continuous chromatographic processes: mathematical optimization of the VARICOL process. *Industrial & Engineering Chemistry Research* **41** (17): 4328–4337.

Toumi, A., Engell, S., Ludemann-Hombourger, O. et al. (2003). Optimization of simulated moving bed and Varicol processes. *Journal of Chromatography A* **1006** (1): 15–31.

Xie, Y., Koo, Y.M., and Wang, N.H.L. (2001). Preparative chromatographic separation: simulated moving bed and modified chromatography methods. *Biotechnology and Bioprocess Engineering* **6** (6): 363–375.

Yu, W., Hidajat, K., and Ray, A.K. (2003). Optimal operation of reactive simulated moving bed and Varicol systems. *Journal of Chemical Technology & Biotechnology* **78** (2–3): 287–293.

Zhang, Z., Mazzotti, M., and Morbidelli, M. (2003). Multiobjective optimization of simulated moving bed and Varicol processes using a genetic algorithm. *Journal of Chromatography A* **989** (1): 95–108.

Zhang, Z., Mazzotti, M., and Morbidelli, M. (2003). PowerFeed operation of simulated moving bed units: changing flow-rates during the switching interval. *Journal of Chromatography A* **1006** (1): 87–99.

Zhang, Z., Mazzotti, M., and Morbidelli, M. (2004). Continuous chromatographic processes with a small number of columns: comparison of simulated moving bed with Varicol, PowerFeed, and ModiCon. *Korean Journal of Chemical Engineering* **21** (2): 454–464.

Zhang, Z., Morbidelli, M., and Mazzotti, M. (2004). Experimental assessment of PowerFeed chromatography. *AIChE journal* **50** (3): 625–632.

Zhang, Y., Hidajat, K., and Ray, A.K. (2007). Enantio-separation of racemic pindolol on -acid glycoprotein chiral stationary phase by SMB and Varicol. *Chemical Engineering Science* **62** (5): 1364–1375.

Zhang, Y., Hidajat, K., and Ray, A.K. (2009). Multi-objective optimization of simulated moving bed and Varicol processes for enantio-separation of racemic pindolol. *Separation and Purification Technology* **65** (3): 311–321.

5

Parameter Estimation, Regression, and Sensitivity of Adsorptive and Chromatographic Processes

This chapter discusses how to determine the parameters of adsorptive and chromatographic separations that are necessary for modeling. Section 5.1 covers a number of empirical correlations that can be used to estimate the physical properties of adsorption systems, even where no data are available. Section 5.2 contains hands-on instructions for analyzing the data collected from equilibrium experiments using three different software tools. Sections 5.3 and 5.4 provide step-by-step instructions for how to perform regression against data collected from dynamic experiments; while Section 5.3 covers a common breakthrough experiment, Section 5.4 covers the estimation of isotherm parameters from a more complicated simulated moving bed (SMB) concentration profile experiment. Section 5.5 provides the conclusions. While Sections 5.2–5.4 focus on applying the parameter estimation tools in Aspen Chromatography, essentially the same tools are included in Aspen Adsorption. We demonstrate their applications in Aspen Adsorption for parameter estimation and sensitivity analysis in two practice problems in Section 5.6. Finally, Section 5.7 provides the bibliography.

5.1 Empirical Correlations for Physical Properties

5.1.1 Axial Dispersion Coefficient

The Chung and Wen correlation [53], Eq. (5.1), was developed in 1968 by analyzing a large number of adsorption systems. It allows us to estimate the axial dispersion (E_z) of an adsorption system given the Reynolds number of the system. It is valid for Reynolds numbers between 10^{-3} and 10^3.

$$Pe = \frac{v_s d_p}{E_z} = 0.2 + 0.011(Re)^{0.48}. \tag{5.1}$$

This correlation is already implemented in Aspen, and can be chosen by selecting "Convection with Estimated Dispersion" in the Material Balance tab of any separator unit (Figure 5.1).

Design, Simulation, and Optimization of Adsorptive and Chromatographic Separations: A Hands-On Approach, First Edition. Kevin R. Wood, Y. A. Liu, and Yueying Yu.

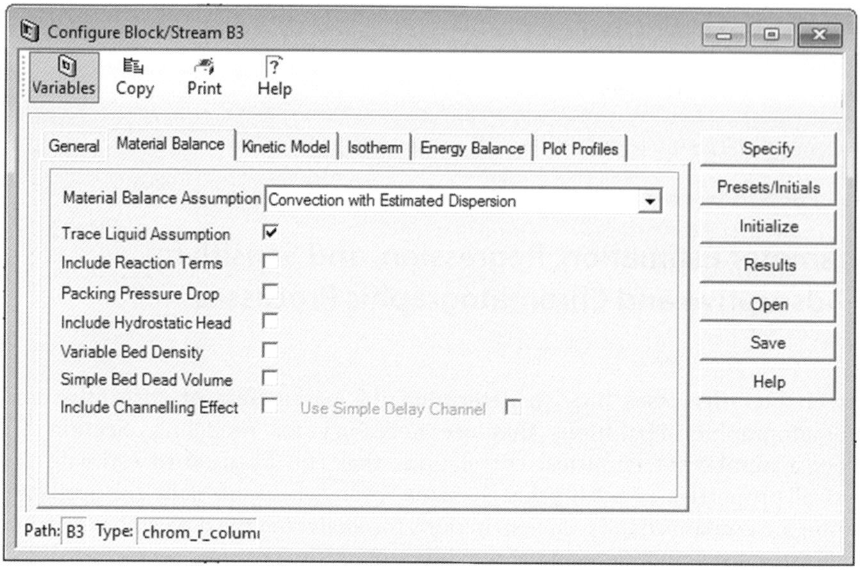

Figure 5.1 Implementing the Chung and Wen correlation in Aspen.

5.1.2 Mass Transfer Coefficient

The Geankoplis correlation [54] allows us to estimate the mass transfer coefficients. It requires the molecular diffusion coefficient of the components. It was derived from experiments where propylene glycol solutions interacted with benzoic acid, as well as from data collected by Gupta and Thados [55] and Pfeffer [56]. The correlation is only valid for Reynolds numbers below 55.

$$Sh = \frac{1.09}{\varepsilon_i} Sc^{\frac{1}{3}} Re^{\frac{1}{3}}. \tag{5.2}$$

In this equation,

$$Sh = \frac{2k_{\mathrm{f}} r_{\mathrm{p}}}{D_{\mathrm{m}}} \tag{5.3}$$

$$Re = \frac{2\rho v_{\mathrm{l}} r_{\mathrm{p}}}{\varepsilon_i \eta} \tag{5.4}$$

$$Sc = \frac{\eta}{\rho D_{\mathrm{m}}}, \tag{5.5}$$

where ρ is the density, k_{f} is the film mass transfer coefficient, v_{l} is the superficial velocity, and r_{p} is the particle radius.

This correlation is implemented in Aspen, but only in the "Particle Mass Balance" class of rate models (Figure 5.2). Therefore, in order to use this correlation without writing your own constraints, you must also supply some estimate of the tortuosity of the system. If you choose to estimate these quantities, Aspen uses the approximations given in Eqs. (5.6) through (5.8).

$$\tau_{\mathrm{tor}} = \frac{(2-\varepsilon_{\mathrm{p}})^2}{\varepsilon_{\mathrm{p}}} \tag{5.6}$$

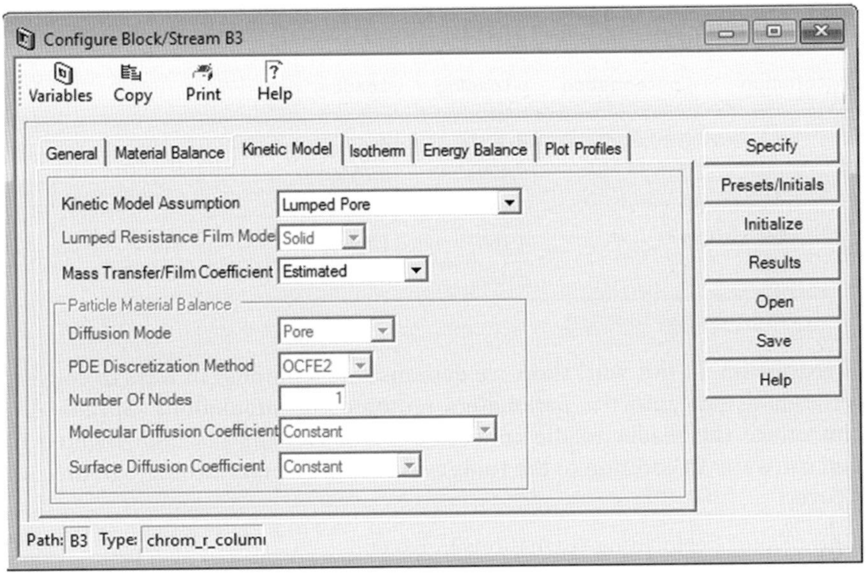

Figure 5.2 Invoking the Geankopolis correlation in Aspen.

$$D_\mathrm{p} = \frac{\varepsilon_\mathrm{p} D_\mathrm{m}}{\tau_\mathrm{tor}} \tag{5.7}$$

$$k_i = \frac{5D_\mathrm{p}}{r_\mathrm{p}}, \tag{5.8}$$

where τ_tor is a parameter describing the tortuosity, D_p is the effective diffusion coefficient within the pore, and k_i is the effective mass transfer coefficient that is internal to the liquid film.

5.1.3 Caveats

These correlations are empirical in nature, and have been developed with a somewhat restricted set of experimental data. They should serve only as a qualitative approximation or placeholder, and not as a reliable predictor of a particular system's performance.

5.2 Parameter Workshop: Regressing against Steady-State Experiments

5.2.1 Introduction to "Experiments" In Aspen Software

There are a number of tools available to fit data to steady-state experiments. You may use a programming tool, such as Excel, Mathematica, R, or Matlab. You might use a tool built specifically for statistical analysis, such as JMP or Minitab. Or you may even use the tools built into Aspen Chromatography itself to perform

Table 5.1 Equilibrium loading data.

D-Concentration	L-Concentration	D-Loading	L-Loading
0	5	0	12.5
0.5	2.5	0.8094	6.7446
1.5	1.5	2.4635	4.1058
2.5	0.5	4.1667	1.3889
5	0	8.0357	0

your regression. In this workshop, we demonstrate a number of ways to convert experimental data into the parameters required for simulations. Specifically, we introduce the reader to the concept of "Experiments" as implemented in Aspen software. In addition to the tools available in Aspen Chromatography, we will cover:

- Methods for performing regressions in Excel
- Methods for performing regressions in Mathematica
- Methods for estimating model parameters from empirical correlations.

5.2.2 Experimental Data

Let us suppose that we have run a number of tests to determine the equilibrium loading of D- and L-threonine enantiomers on an adsorbent. We vary the concentrations of both components, and then measure the maximum loading as shown in Table 5.1.

Now we want to determine the isotherm equation that fits data the best. We suspect that the Langmuir model will work well, specifically:

$$q_i = \frac{a_i b_i c_i}{1 + b_1 c_1 + b_2 c_2}, \tag{5.9}$$

where i refers to the chemical component, q is the equilibrium loading, c is the liquid concentration, and a and b are the isotherm parameters we wish to determine.

5.2.3 Parameter Regression in Excel

To begin, let us use Excel to calculate the four isotherm parameters. Excel has several tools for performing statistical analysis, but they are disabled by default, and most are somewhat limited. For example, the Data Analysis tool has a built-in Regression tool, but it only works with linear functions. We save the file for this example as **Regression.xlsx**.

Configuring the Excel add-ins

Step 1: Enable the advanced add-ins for Excel (Figure 5.3).

On the options screen, select "Manage Excel Add-ins" and press "Go" (Figure 5.4).

Figure 5.3 Setting Excel options by using File → Options.

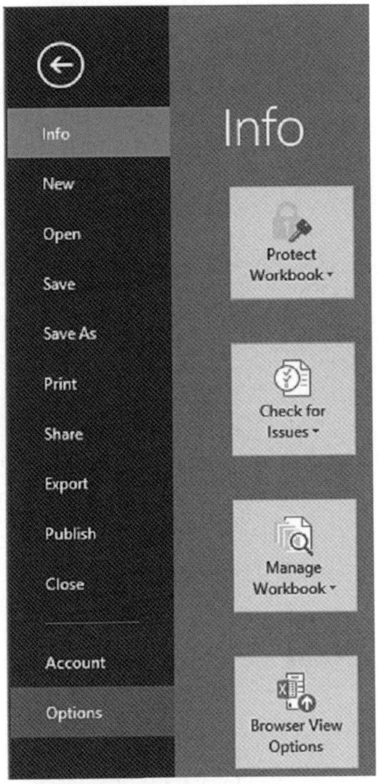

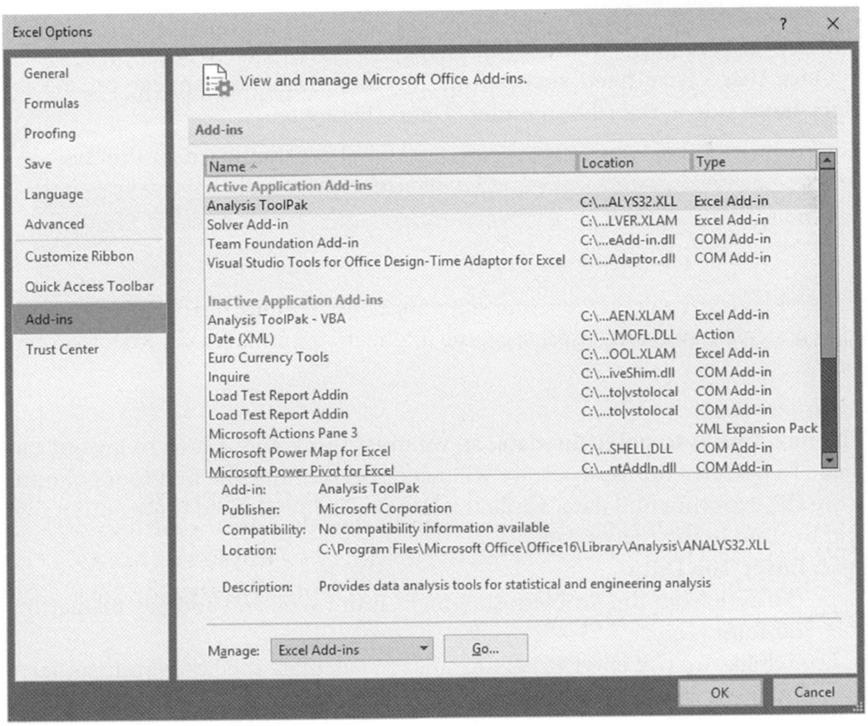

Figure 5.4 Manage Excel Add-ins.

The "Go" button will present you with a list of the built-in Excel Add-ins that are disabled by default. Place a checkmark by the Analysis and Solver Add-ins to enable them (Figure 5.5).

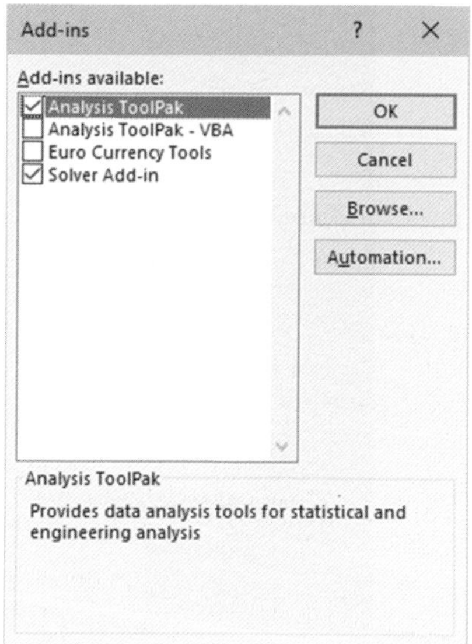

Figure 5.5 Enabling the Solver and Analysis tools.

Press "OK" to perform the installation.

Once these have been successfully installed, they will appear on the far right-hand side of the ribbon under "Data" (Figure 5.6).

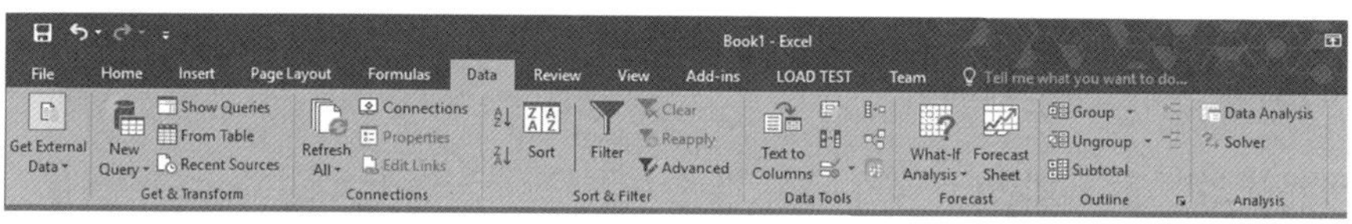

Figure 5.6 Locating the newly installed add-ins.

Spreadsheet Layout

The next task is to enter the data, so we must think about how to lay out our data. To perform our analysis, we will need six columns, the first four columns store the experimental data, while the last two columns hold information that will be used in the regression.

Step 2: Enter The Data

Type the data from the problem statement into Excel, and add any formatting that you want (Figure 5.7).

From here, we will enter Excel formulas to calculate the Langmuir isotherm given in Eq. (5.9).

D- Concentration	L- Concentration	D- Loading	L- Loading	Predicted D- Loading	Predicted L- Loading
0	5	0	12.5		
0.5	2.5	0.8094	6.7446		
1.5	1.5	2.4635	4.1058		
2.5	0.5	4.1667	1.3889		
5	0	8.0357	0		

Figure 5.7 Data entry and layout.

Defining the Functions

To define the Langmuir functions, we will use Excel's ability to name variables. But first, we need to create a place for the isotherm parameters to appear on the spreadsheet.

Step 3: Add the Isotherm Parameters to the Spreadsheet

To the right of the table we have created, add a table of isotherm parameters, as shown in Figure 5.8. The numbers represent initial guesses of the parameter values.

D- Concentration	L- Concentration	D- Loading	L- Loading	Predicted D- Loading	Predicted L- Loading		Parameters	
0	5	0	12.5				a1	1
0.5	2.5	0.8094	6.7446				a2	1
1.5	1.5	2.4635	4.1058				b1	0.1
2.5	0.5	4.1667	1.3889				b2	0.1
5	0	8.0357	0					

Figure 5.8 Adding Isotherm Parameters.

Step 4: Add Names to the Cells in the Spreadsheet

Excel has very convenient tools for naming ranges in spreadsheets. To get started, select the Parameter Table you just created, then open the "Formulas" tab on the Ribbon (Figure 5.9).

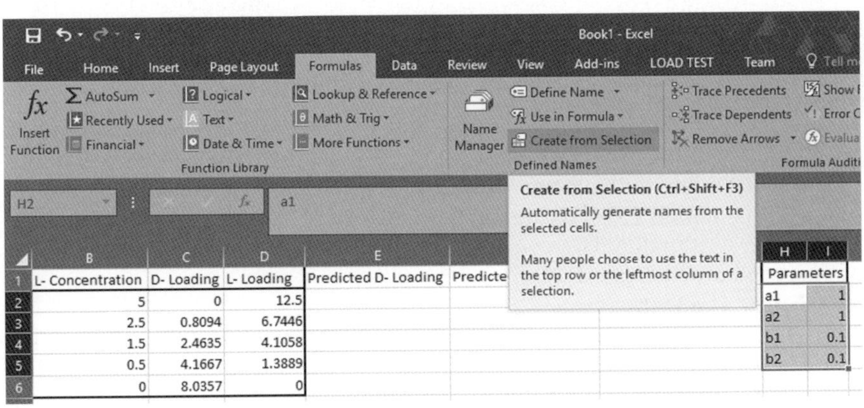

Figure 5.9 Naming the Parameters using the "Create from Selection" button.

Once you have selected the parameters as shown in Figure 5.9, click on the "create from selection" button. This will open a new window to ask where the variable names come from. Tell Excel that the left-hand column contains the names of the variables, then press "OK" (Figure 5.10).

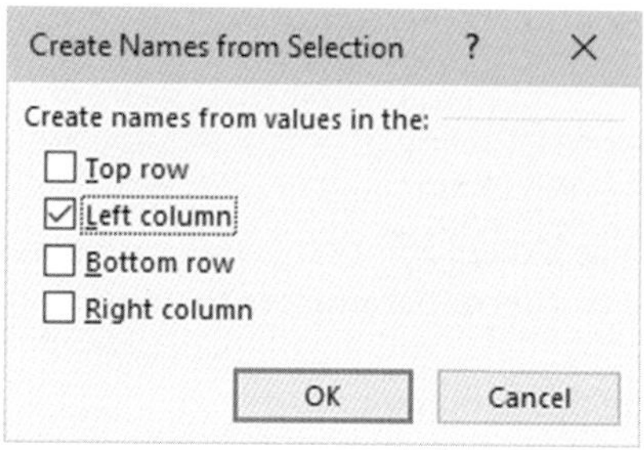

Figure 5.10 The variable names are in the Left column.

Repeat this step for the experimental data, but this time, tell Excel that the data are in the top row (Figure 5.11).

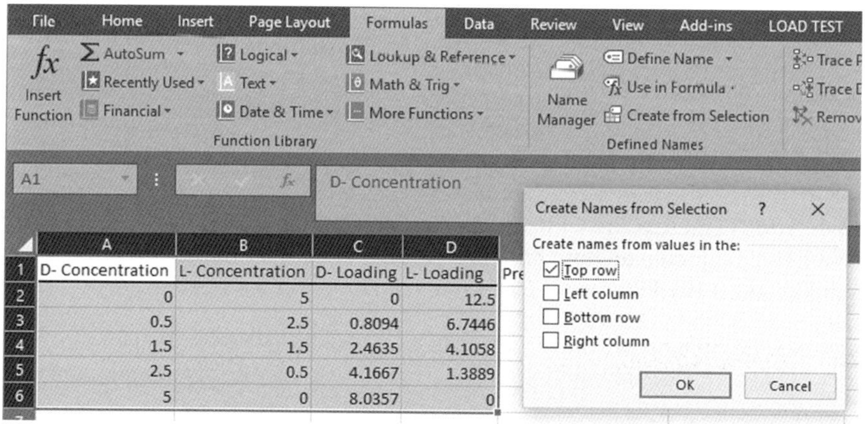

Figure 5.11 Naming the Experimental data.

To see if the names have been applied correctly, open the "Name Manager" to see all the names Excel has created (Figure 5.12):

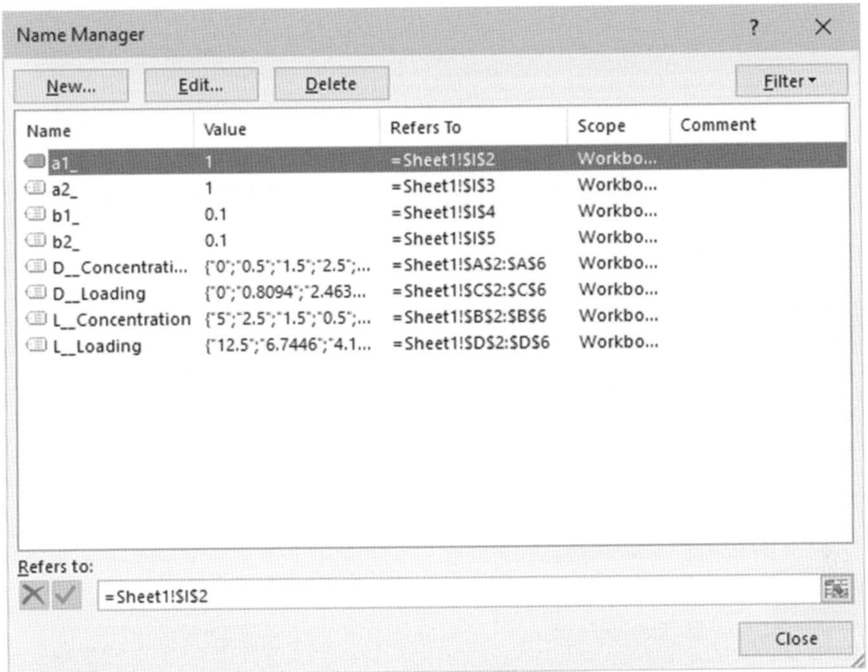

Figure 5.12 The Excel Name Manager, after our names have been created.

Step 5: Enter the isotherm equations using Excel's Names

Type in the equations using the names defined by Excel, as shown in Figures 5.13 and 5.14. The easiest way to do this is to use the "Use in Formula" dropdown, directly to the right of the Name Manager. This allows you to quickly enter a variable name into your equation (Figure 5.15)

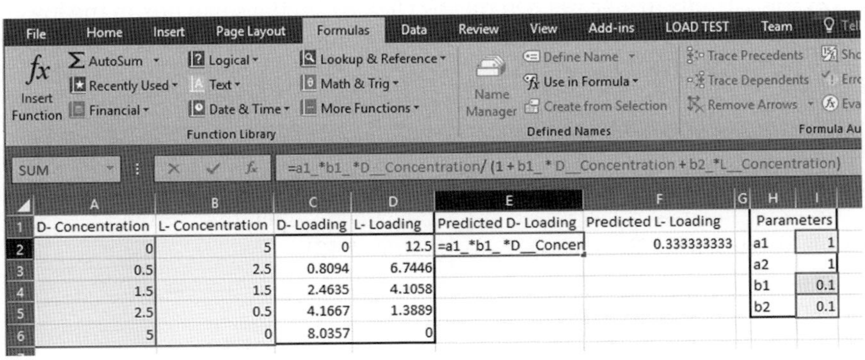

Figure 5.13 The Langmuir equation, in Excel's terms.

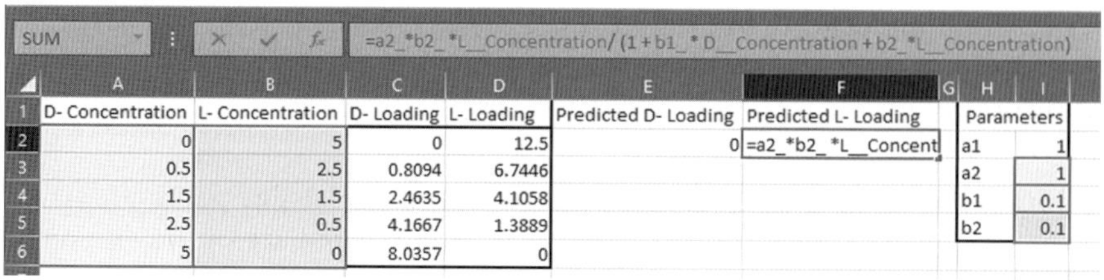

Figure 5.14 The L-Langmuir equation.

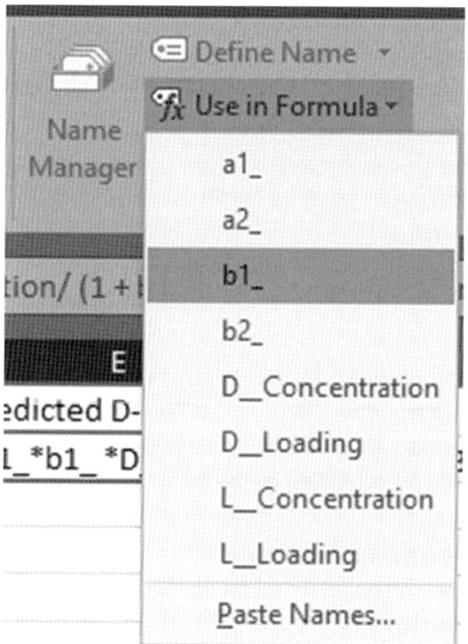

Figure 5.15 The "Use in Formula" drop down.

Once both equations have been entered, use Excel's "Fill Down" ability to apply the formula to all rows. Do this by clicking on the small box in the lower right-hand corner of your selection and dragging until all the rows have been covered. See Figure 5.16.

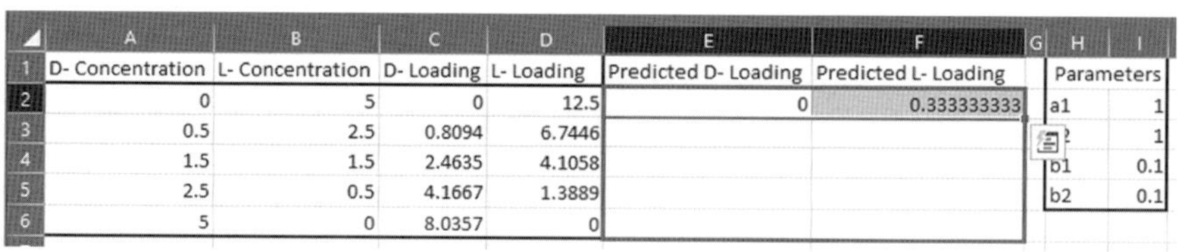

Figure 5.16 Filling in the Langmuir equations.

Step 6: Name the predicted Loadings using the same procedure as earlier.
Select the boxes and have Excel create names for the variables, just as before.

Step 7: Define the Sum Squared Error, and name it
Before we move on to the actual regression, we will first define the sum squared error. Excel has a built-in function for summing the square of the difference between two ranges, SUMXMY2. Use this function to create an entry for sum squared error on the spreadsheet. See Figure 5.17.

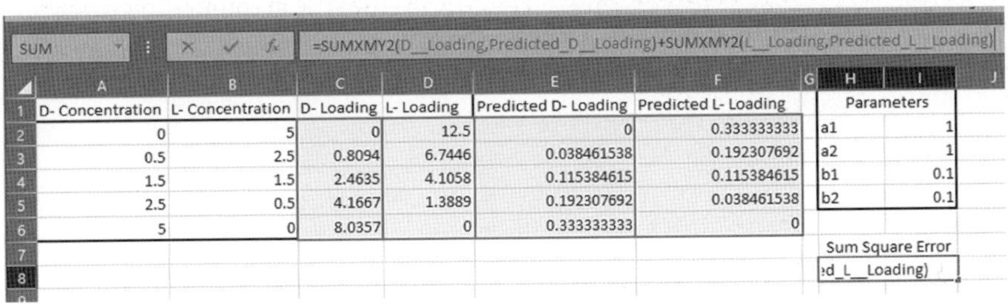

SUM =SUMXMY2(D__Loading,Predicted_D__Loading)+SUMXMY2(L__Loading,Predicted_L__Loading)

	A	B	C	D	E	F	G	H	I
1	D- Concentration	L- Concentration	D- Loading	L- Loading	Predicted D- Loading	Predicted L- Loading		Parameters	
2	0	5	0	12.5	0	0.333333333		a1	1
3	0.5	2.5	0.8094	6.7446	0.038461538	0.192307692		a2	1
4	1.5	1.5	2.4635	4.1058	0.115384615	0.115384615		b1	0.1
5	2.5	0.5	4.1667	1.3889	0.192307692	0.038461538		b2	0.1
6	5	0	8.0357	0	0.333333333	0			
7								Sum Square Error	
8								›d_L__Loading)	

Figure 5.17 Defining the Sum Squared Error.

Once you have defined the equation, be sure to give the result a name.

Regression

We now have all the pieces in place to perform our regression.

Step 8: Open the Solver
Click on the "Solver" button in the Data tab. See Figures 5.18 and 5.19.

Figure 5.18 The Solver button.

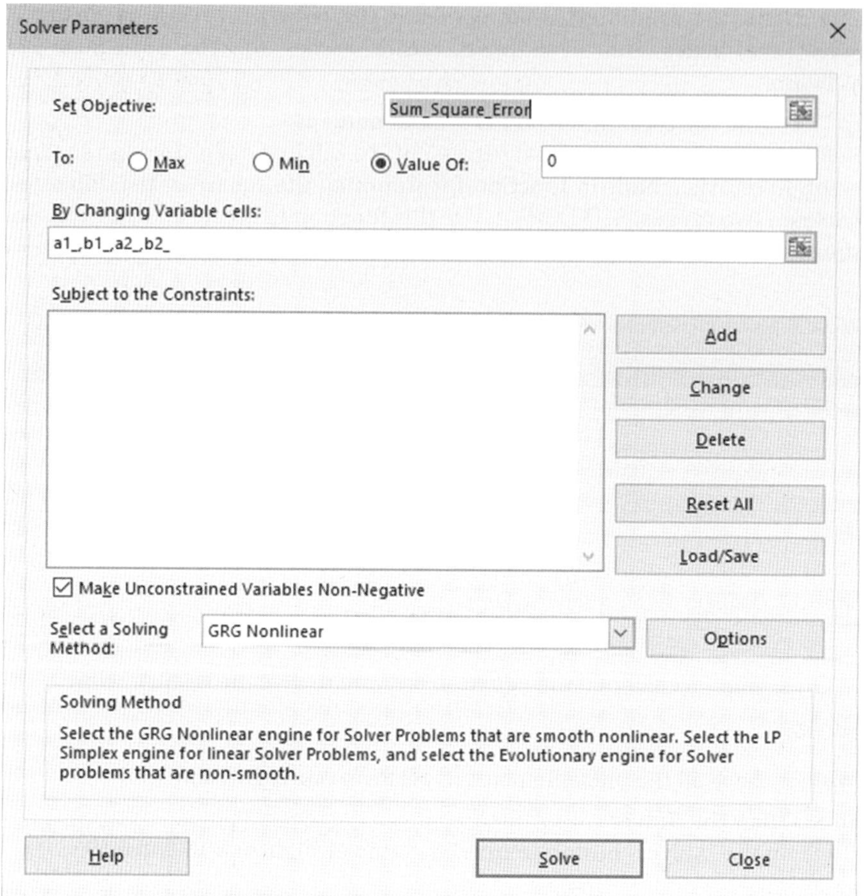

Figure 5.19 The Solver Window.

Step 9: Enter the data into the solver

To perform our least squared regression, we need to set the value of the Sum_Squared_Error to 0 by manipulating the isotherm parameters. Since we have defined names for all those quantities, you can type them directly into the solver. There is no need to specify constraints on the variables for this problem. Once you have entered the data, click on the "Solver" button.

Step 10: Consider the results

The first Solver run (using all default settings) will be completed successfully. See Figure 5.20.

Note that the solver has told us that it may have only found a local optimum (since we used the default GRG engine.) Can we do better? Put a checkmark in the "Return to Solver Parameters" box, then click "OK" (Figure 5.20).

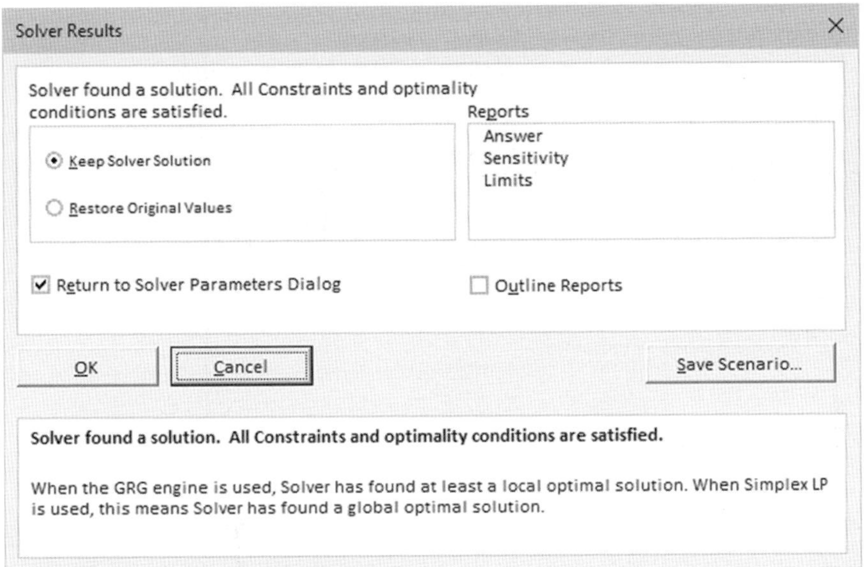

Figure 5.20 The Solver result message.

Step 11: Adjust the Solver options

This time, we will ask the solver to minimize the sum squared error, and change some of the GRG options.

Press the "Options" button next to GRG Nonlinear (Figure 5.21).

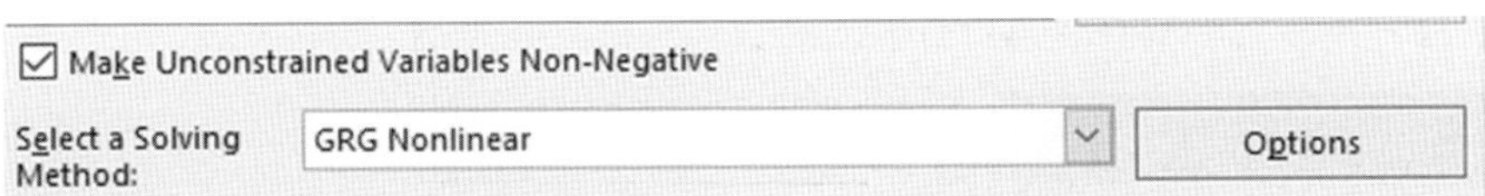

Figure 5.21 Setting Solver Options.

Open the GRG tab. Add an extra 0 to the convergence tolerance, and put a checkmark in the Multistart box. Once you have pressed "OK," solve the equations again. See Figure 5.22.

After completing the optimization run, view the results on the spreadsheet (Figure 5.23).

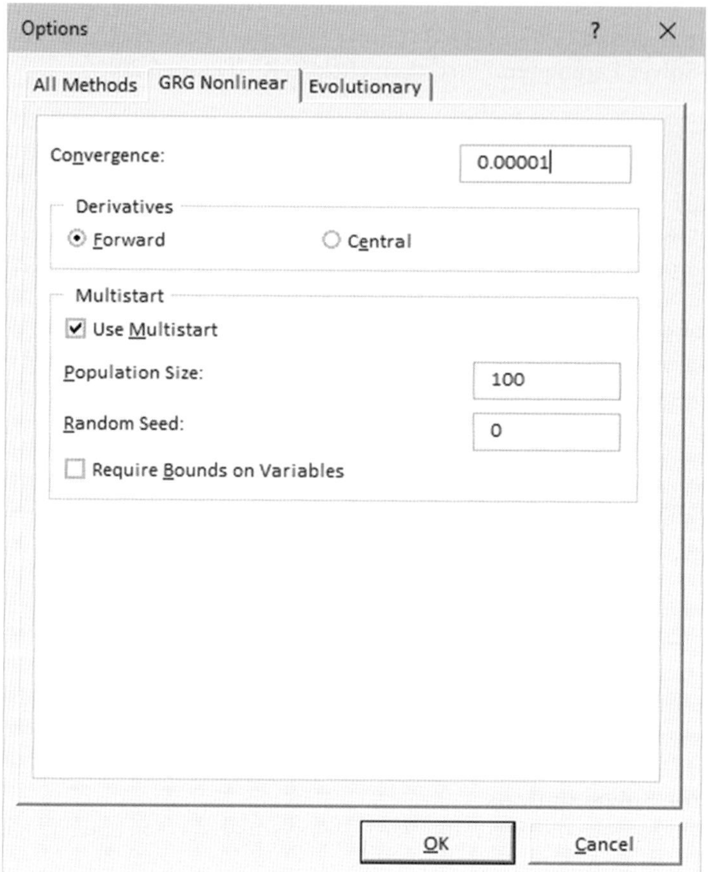

Figure 5.22 Choosing the GRG nonlinear convergence option with forward derivative and multistart options.

H	I
Parameters	
a1	74.98595
a2	74.99854
b1	0.024005
b2	0.040001
Sum Square Error	
3.27193E-09	

Figure 5.23 The optimization results.

5.2.4 Parameter Regression in Aspen Chromatography

We can repeat this regression using Aspen Chromatography itself. Aspen has several built-in tools meant to perform exactly this task, so it is more convenient. Aspen also provides more statistical information so that we can better evaluate the quality of the fit.

5.2.4.1 Defining an Estimation Flowsheet

Step 12: Add components and blocks

To prepare a flowsheet for regression, we need to convert our component list to a component set, and add our two components (Figure 5.24). We save the file for this example as **BasicRegression.cra**.

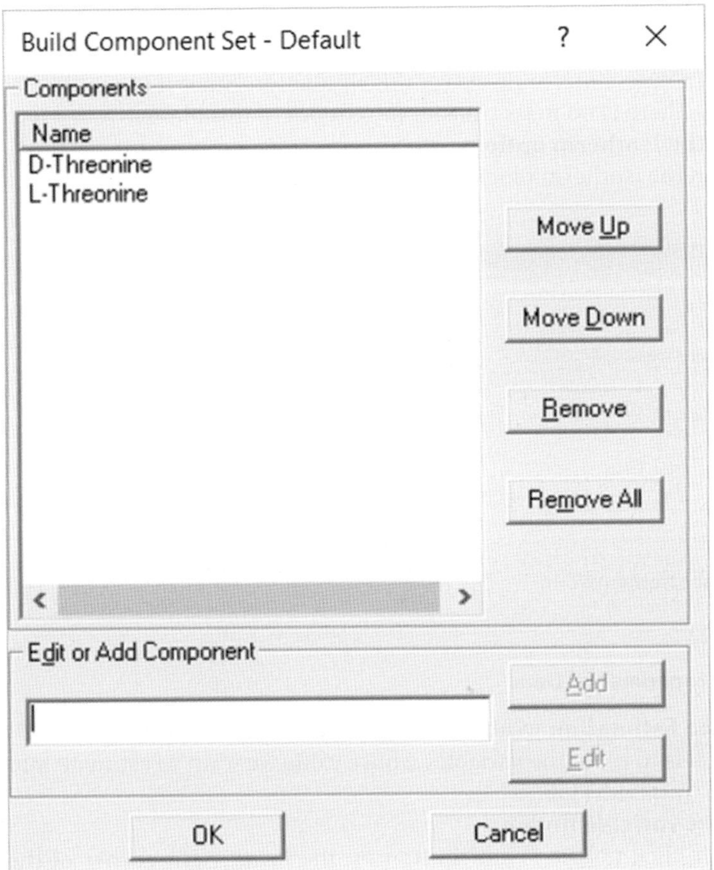

Figure 5.24 The Component Set.

We also need to add a "Static Isotherm" block from the "Utilities" tab (Figure 5.25). This block simply performs the isotherm calculation to predict equilibrium loading at a particular fluid-phase concentration.

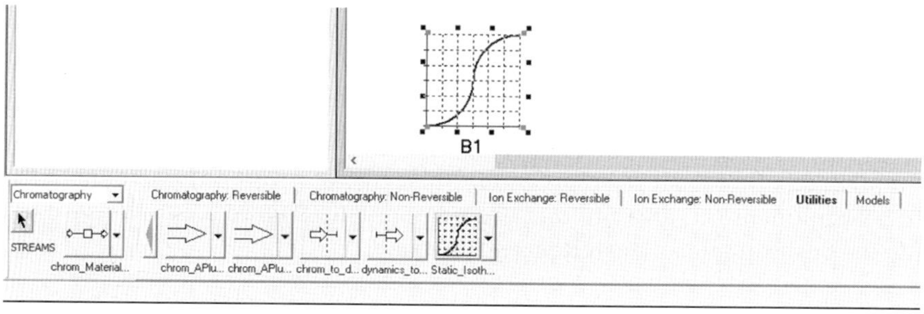

Figure 5.25 The flowsheet with a Static Isotherm.

There are no connections required for this simulation. Since it is merely predicting loading, there is no mass transfer into or out of this block.

Step 13: Define the Isotherm options

Double click on the isotherm block to open the default form (Figure 5.26).

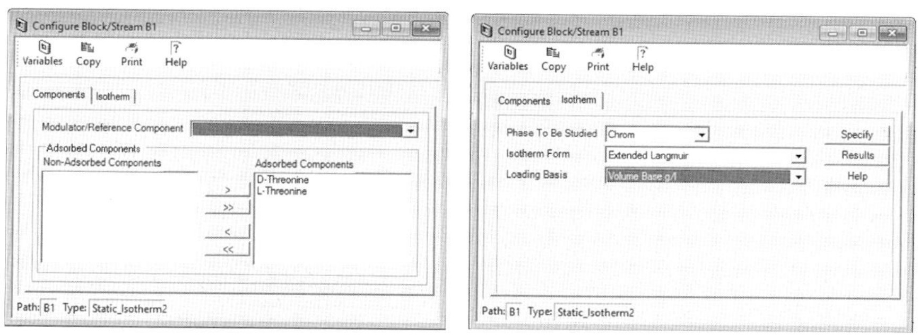

Figure 5.26 The Static Isotherm form.

5.2.4.2 Entering Experimental Data

Step 14: Open the Estimation window using Tools→Estimation (Figure 5.27)

Our next goal is to place the variables whose value we want to estimate into the "Estimated Variables" tab.

Step 15: Open the variable finder

Press ctrl+F, or click on the "Find" button on the lower right corner of the "Estimated Variables" tab. Once it is open, search for all variables in the B1 static isotherm block by typing B1 into the "Find In" box and pressing "Find" (Figure 5.28).

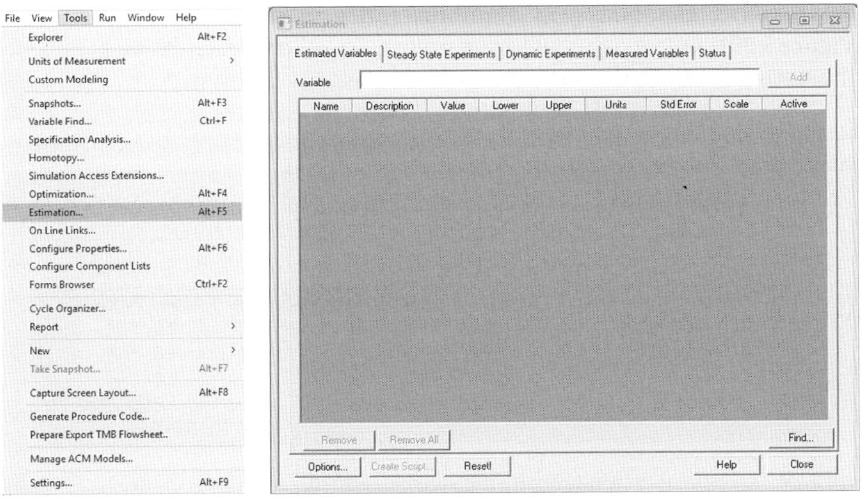

Figure 5.27 The location and appearance of the Estimation tool.

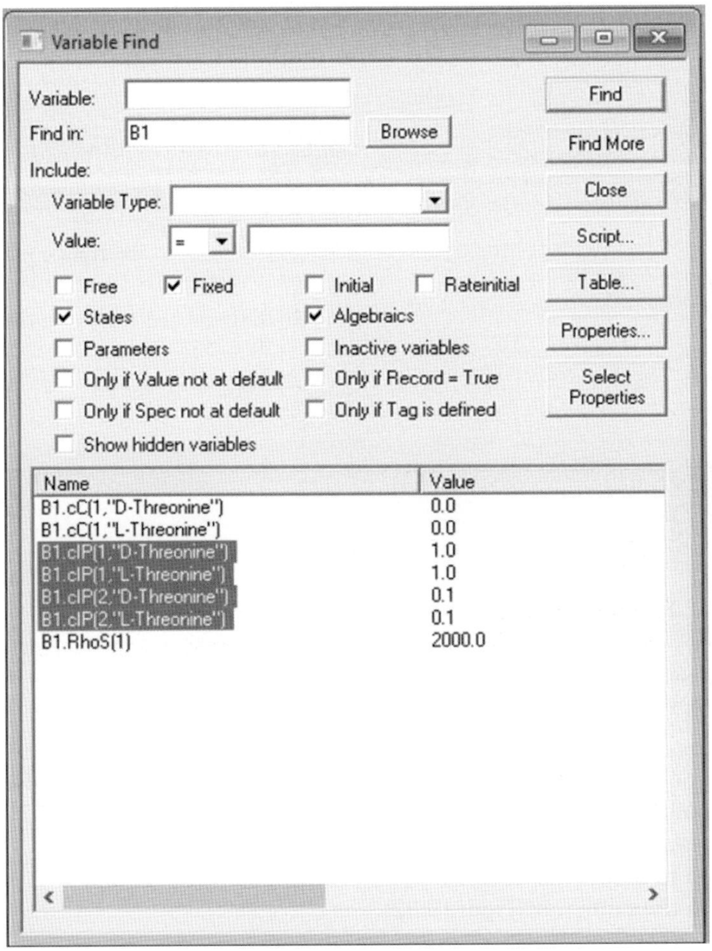

Figure 5.28 Finding the Isotherm variables.

Step 16: Drag the Isotherm variables (cIP) to the estimation window (Figure 5.29)

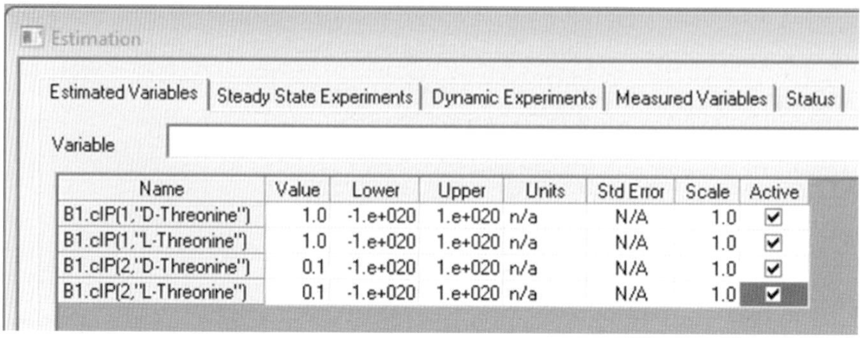

Figure 5.29 The Estimation window with the variables added.

Step 17: Enter Experimental Data

We now need to switch to the "Steady-State Experiments" tab to enter our data. Once this tab is open, press the "new" button to add experiments (Figure 5.30).

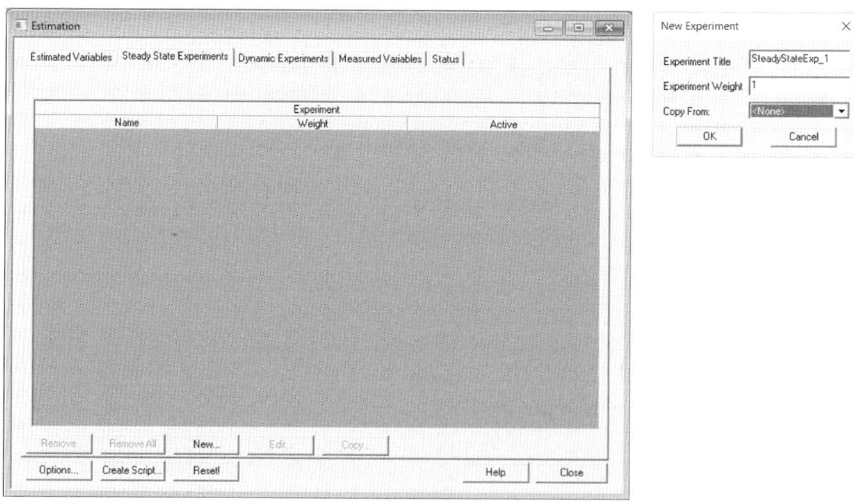

Figure 5.30 Adding new Experiments to the Steady-State Experiments tab.

Once you have created a new experiment by clicking "OK" in the new experiment dialog, you will be prompted to enter the measured variables for the experiment. You can add the variables using the "Find" tool, in the same way that we added estimated variables (Figure 5.31).

We next need to define the fixed variables on the fixed variable tab. This is accomplished in the same way (Figure 5.32).

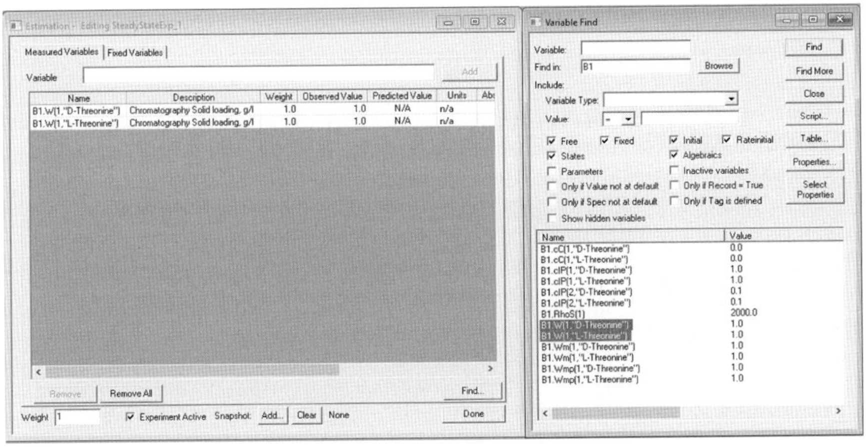

Figure 5.31 Adding Measured Variables with the "Find" Tool.

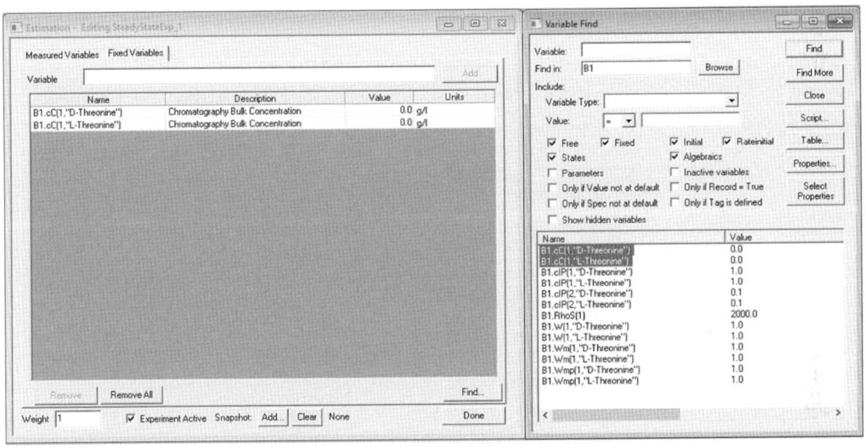

Figure 5.32 Adding Fixed Variables to the Estimation experiment.

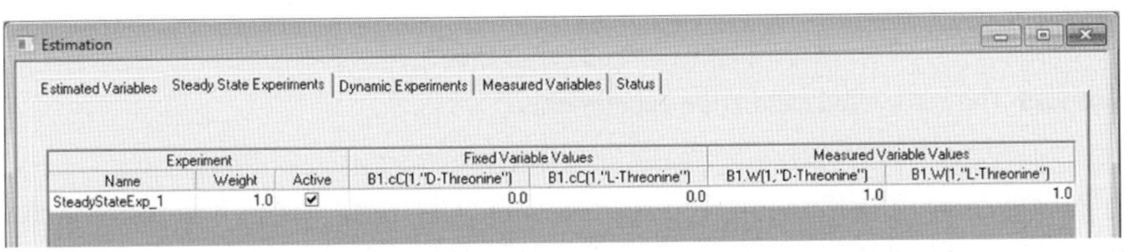

Figure 5.33 The completed experiment.

Once these variables have been entered, press the "Done" button in the lower right-hand corner. The completed experiment will look like Figure 5.33.

Now we create copies of this experiment until there are a total of five experiments. To do this, press the "Copy..." button while SteadyStateExp_1 is selected (Figure 5.34).

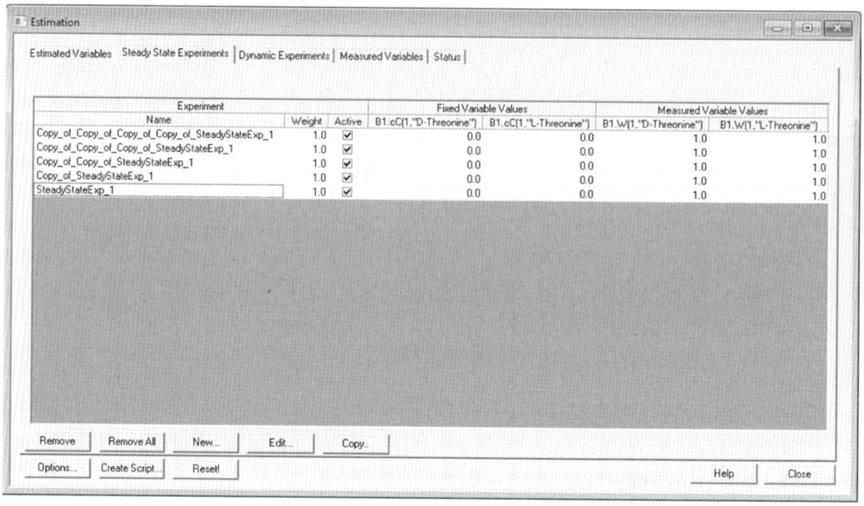

Figure 5.34 Copying the Estimation experiments.

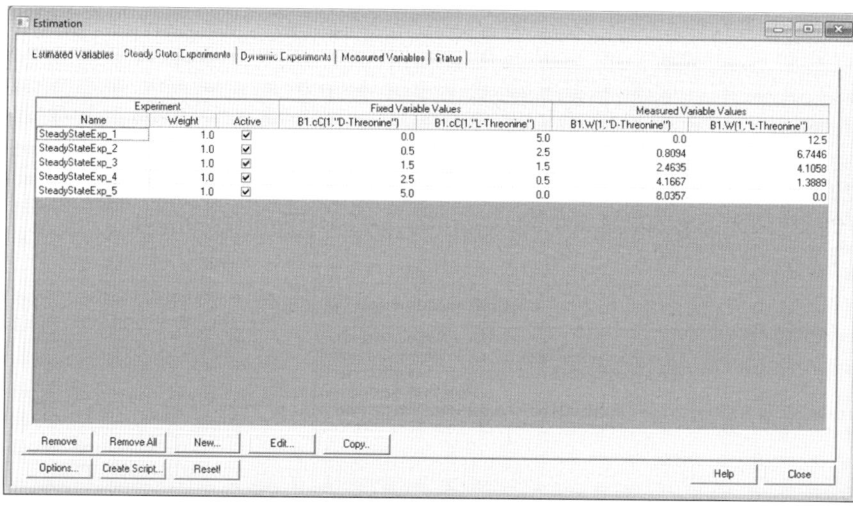

Figure 5.35 The completed Estimation experiments.

Finally, enter the data into the appropriate cells in this table, and rename the experiments in a more sensible way (Figure 5.35).

5.2.4.3 Estimation Settings

We will now investigate the options available to us in the estimation tool.

Step 18: Open the "Status" tab in the Estimation window (Figure 5.36)
From this tab, we can click on the "Options" button to see the various settings related to the estimation (Figure 5.37).

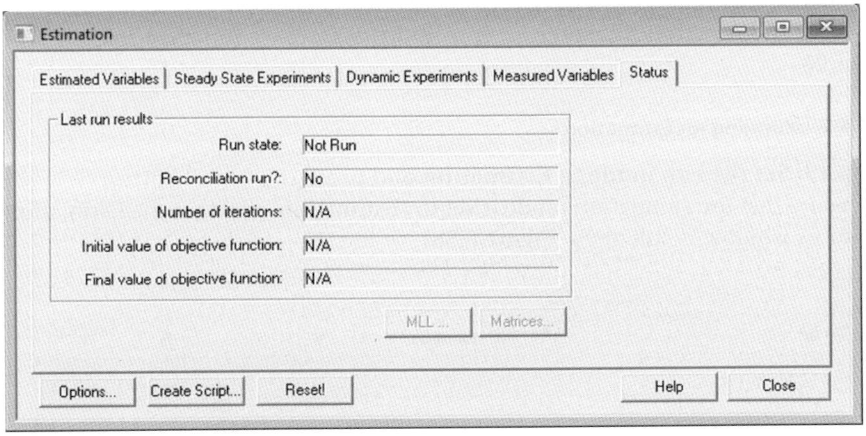

Figure 5.36 The Status of our Estimation.

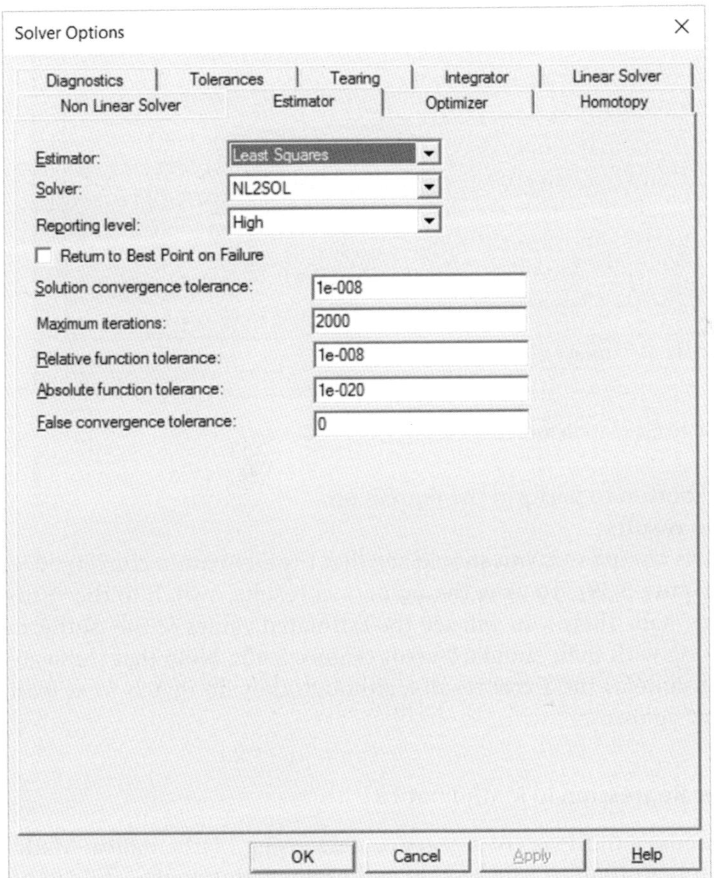

Figure 5.37 Estimation Options.

For this estimation, we set very strict convergence options. Set the convergence and function tolerance to 1e-8 and the absolute function tolerance to 1e-20.

5.2.4.4 Running an Estimation

Step 19: Set the run mode to Estimation and press the "run" button

Ensure that the simulation mode is set to "Estimation" and that the Estimation Status window is still open (Figure 5.38).

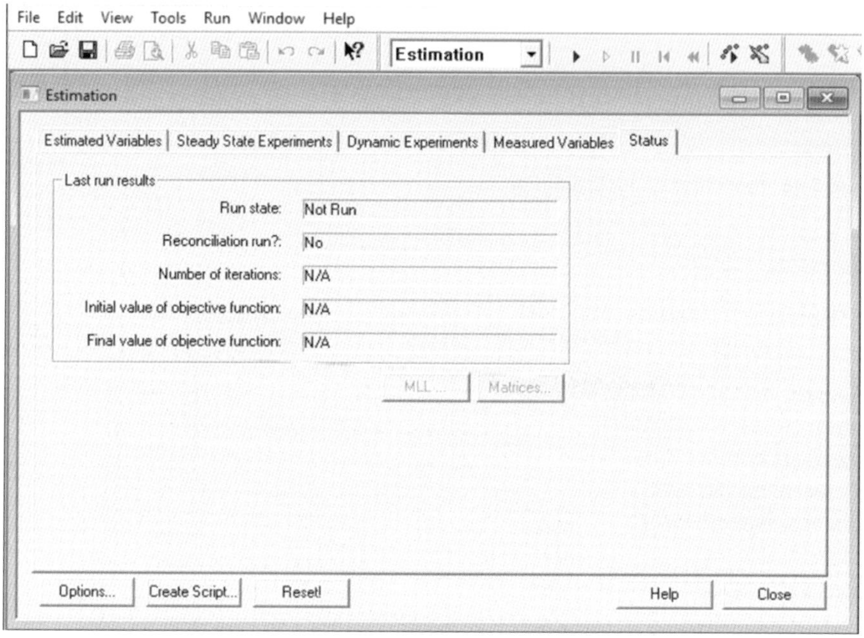

Figure 5.38 A prepared Estimation run.

Press the "run" button to perform the regression.

Step 20: View the results

Once the run has completed, you should see that the estimation converged in 29 iterations (Figure 5.39). To view the numerical results, switch to the "Estimated Variables" tab. There you will see the estimated values of the isotherm parameters, along with their standard error (Figure 5.40). Note that these are not exactly the same as the Excel results, although they do match to at least three significant figures.

5.2.5 Parameter Regression In Mathematica

We will briefly discuss a third way of regressing data: using Mathematica. Mathematica has a number of built-in regression and optimization tools; while none are perfect fits for the scenario we have here, we can still use them to get superior statistical information about our regression.

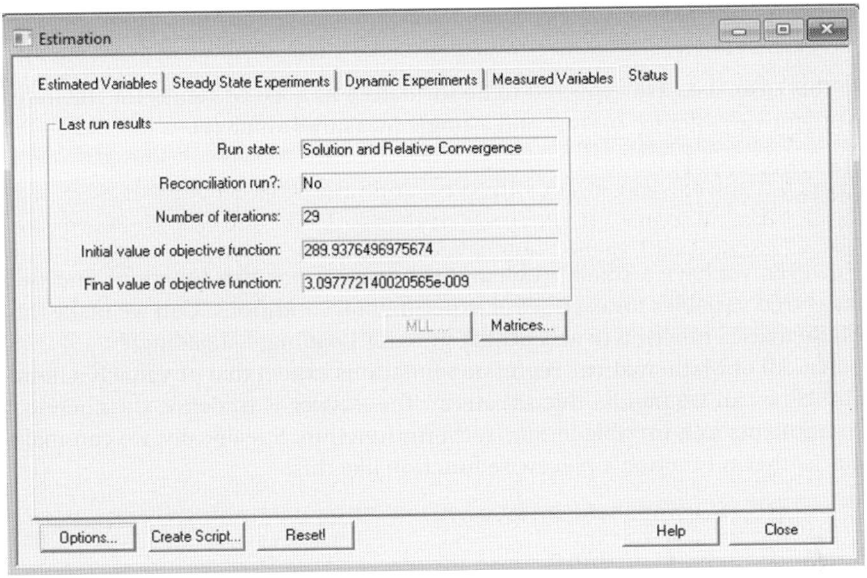

Figure 5.39 Estimation Convergence.

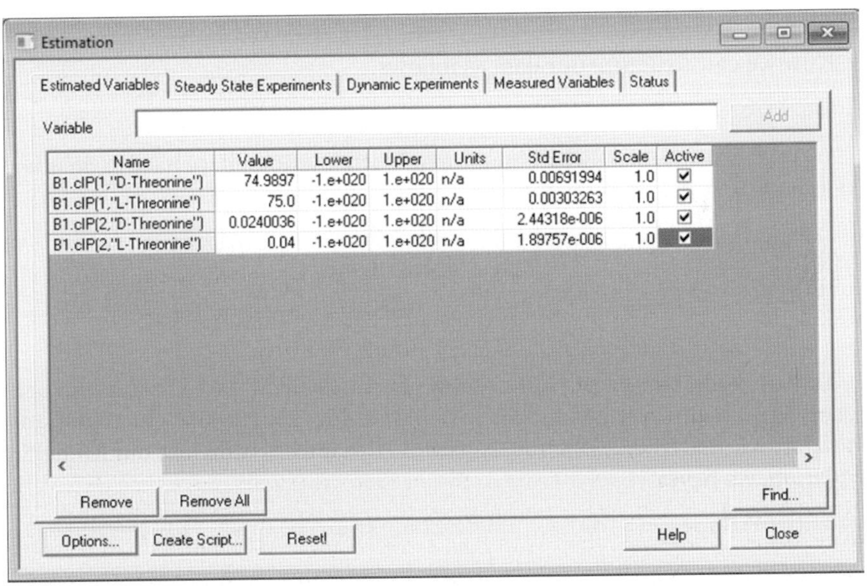

Figure 5.40 Estimation results.

5.2.5.1 Defining the Functions

Step 1: Investigate the Regression Tools in Mathematica

The two main regression tools in Mathematica are aptly named "LinearModelFit" and "NonlinearModelFit." Both work the same way:

```
NonlinearModelFit[data, function, regressionVariables,
independentVariables]
```

In this case, data are expected to be formatted as a list of values for the independent variables (a, b, etc.), and a single measured value (y).

```
{{a1,b1,…,y1},
{a2,b2,…,y2},
…}
```

However, we have a small problem. Our data set has two functions and two measured variables for each set of experimental conditions. Can we make this Mathematica function to accept that y1 = {D-Loading, L-Loading}?

No. All of Mathematica's regression functions expect that y1 value is a number. How can we handle this situation? The answer is to define the chemical components as a variable in our isotherm function. Specifically, we can make our isotherm function a piecewise function like this:

$$q(c_1, c_2, \text{component}, a_1, b_1, a_2, b_2) = \begin{cases} \dfrac{a_1 b_1 c_1}{1 + b_1 c_1 + b_2 c_2} & \text{component} = 1 \\ \dfrac{a_2 b_2 c_2}{1 + b_1 c_1 + b_2 c_2} & \text{component} = 2 \end{cases} . \tag{5.10}$$

Now we can structure our data as follows:

$$\{1, c_1, c_2, q_1\}, \{2, c_1, c_2, q_2\}.$$

So, to define this piecewise equation in Mathematica, all we have to do is use Mathematica's Piecewise function:

$$\text{isotherm} = \text{Piecewise}\left[\left\{\left\{\frac{a_1 b_1 c_1}{1 + b_1 c_1 + b_2 c_2} c == 1\right\} \times \left\{\frac{a_2 b_2 c_2}{1 + b_1 c_1 + b_2 c_2}, c == 2\right\}\right\}\right].$$

This equation has chosen our notation for the ModelFit we are about to perform. Specifically, our dependent variables are {c, $c1$, $c2$} and our regression variables are {$a1$, $b1$, $a2$, $b2$}. If we substitute this information into our ModelFit equation, we have:

```
NonlinearModelFit[data, isotherm, {a1, b1, a2, b2},
                  {c, c1, c2}]
```

5.2.5.2 Entering the Data

We now need to actually define the data. Rather than type all the numbers again, open the Excel file we created in part 1.

Step 2: Type the ImportString function into Mathematica (Figure 5.41)

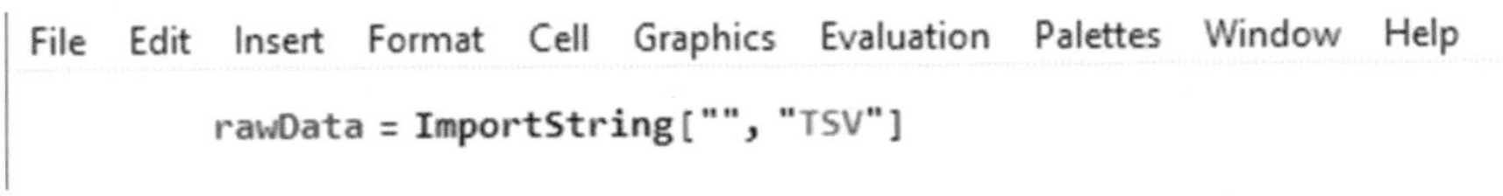

Figure 5.41 Preparing for Excel Data.

Step 3: Copy the Data from the Excel and paste it between the Quotation Marks

Copy the data shown in Figure 5.42 into Mathematica. When the dialog shown in Figure 5.43 opens, choose "Yes" to have Mathematica automatically convert the Excel data into tab-separated values.

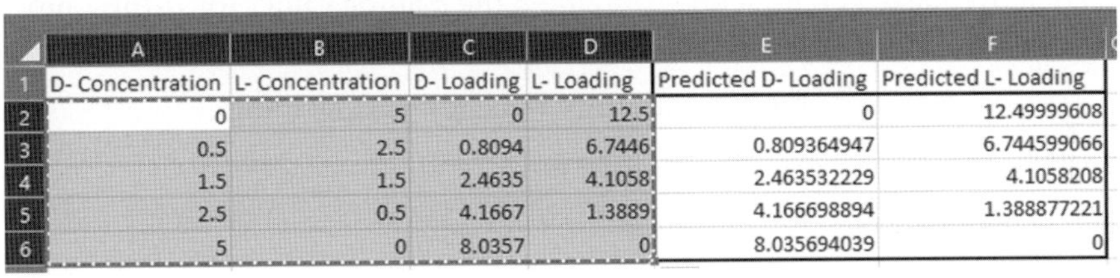

	A	B	C	D	E	F
1	D- Concentration	L- Concentration	D- Loading	L- Loading	Predicted D- Loading	Predicted L- Loading
2	0	5	0	12.5	0	12.49999608
3	0.5	2.5	0.8094	6.7446	0.809364947	6.744599066
4	1.5	1.5	2.4635	4.1058	2.463532229	4.1058208
5	2.5	0.5	4.1667	1.3889	4.166698894	1.388877221
6	5	0	8.0357	0	8.035694039	0

Figure 5.42 The Excel data to copy.

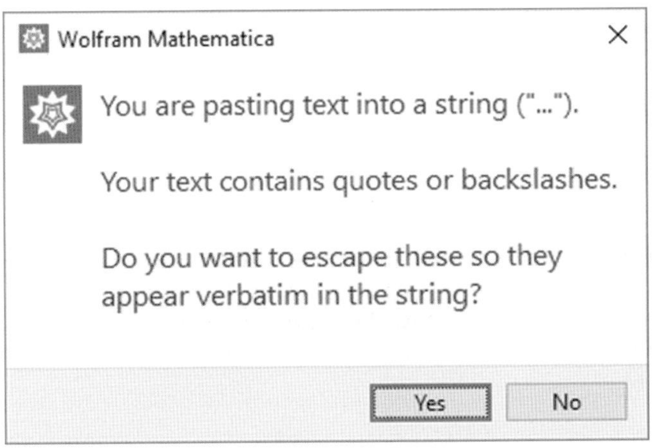

Figure 5.43 Allowing Mathematica to modify the raw Excel data.

Step 4: Execute the resulting equation (Figure 5.44)

```
File  Edit  Insert  Format  Cell  Graphics  Evaluation  Palettes  Window  Help

In[1]:= rawData = ImportString["0\t5\t0\t12.5
        0.5\t2.5\t0.8094\t6.7446
        1.5\t1.5\t2.4635\t4.1058
        2.5\t0.5\t4.1667\t1.3889
        5\t0\t8.0357\t0
        ", "TSV"]
Out[1]= {{0, 5, 0, 12.5}, {0.5, 2.5, 0.8094, 6.7446}, {1.5, 1.5, 2.4635, 4.1058}, {2.5, 0.5, 4.1667, 1.3889}, {5, 0, 8.0357, 0}}
        Assuming a matrix | Use as a list of lists instead
        matrix plot   display as   total   dimensions   more...
```

Figure 5.44 The completed Import of Excel Data.

Step 5: Format the data according to our needs

Recall that we needed our data to contain an extra "dummy" variable to indicate which component loading we were reporting. To process our data into that format, we need to make sure that only one loading value appears in each entry. To do that, we make two lists, one containing the loading values for D-threonine, and one for L-threonine. Remember that our Excel data was structured as

{D-Concentration, L-Concentration, D-Loading, L-Loading}

So, if we want to get something of the form:

{D-Concentration, L-Concentration, D-Loading}

We need to take the entries 1, 2, and 3. If we want something of the form

{D-Concentration, L-Concentration, L-Loading}

We need to take the entries 1, 2, and 4. We can accomplish this in Mathematica as explained in Figure 5.45.

```
In[11]:= rawData[[ ;; , {1, 2, 3}]]
         rawData[[ ;; , {1, 2, 4}]]
Out[11]= {{0, 5, 0}, {0.5, 2.5, 0.8094}, {1.5, 1.5, 2.4635}, {2.5, 0.5, 4.1667}, {5, 0, 8.0357}}
Out[12]= {{0, 5, 12.5}, {0.5, 2.5, 6.7446}, {1.5, 1.5, 4.1058}, {2.5, 0.5, 1.3889}, {5, 0, 0}}
```

Figure 5.45 Selecting the data we need.

So far so good, but we are still missing our dummy "component" variable. To insert that variable at the beginning, we can use Mathematica's "Prepend" function (Figure 5.46).

```
In[13]:= Prepend[1] /@ rawData[[ ;; , {1, 2, 3}]]
         Prepend[2] /@ rawData[[ ;; , {1, 2, 4}]]
Out[13]= {{1, 0, 5, 0}, {1, 0.5, 2.5, 0.8094}, {1, 1.5, 1.5, 2.4635}, {1, 2.5, 0.5, 4.1667}, {1, 5, 0, 8.0357}}
Out[14]= {{2, 0, 5, 12.5}, {2, 0.5, 2.5, 6.7446}, {2, 1.5, 1.5, 4.1058}, {2, 2.5, 0.5, 1.3889}, {2, 5, 0, 0}}
```

Figure 5.46 Prepending to the selected data.

Finally, we need to join these two lists together into a single data set, which we can do with the join function (Figure 5.47).

```
In[16]:= processedData = Join[
          Prepend[1] /@ rawData[[ ;; , {1, 2, 3}]],
          Prepend[2] /@ rawData[[ ;; , {1, 2, 4}]]
         ]

Out[16]= {{1, 0, 5, 0}, {1, 0.5, 2.5, 0.8094}, {1, 1.5, 1.5, 2.4635}, {1, 2.5, 0.5, 4.1667}, {1, 5, 0, 8.0357},
          {2, 0, 5, 12.5}, {2, 0.5, 2.5, 6.7446}, {2, 1.5, 1.5, 4.1058}, {2, 2.5, 0.5, 1.3889}, {2, 5, 0, 0}}
```

Figure 5.47 Joining the two lists we created.

5.2.5.3 Regression

With that, we are ready for our regression.

Step 6: Enter the function we described in Step 1 (Figure 5.48)

```
In[20]:= result = NonlinearModelFit[processedData, isotherm, {a1, b1, a2, b2}, {c, c1, c2}]
```

Out[20]= FittedModel[$\begin{cases} \frac{1.80003\, c1}{1+0.0240036\, c1+0.04\, c2} & c == 1 \\ \frac{3.\, c2}{1+\ll 21 \gg c1+0.04\, c2} & c == 2 \\ 0 & \text{True} \end{cases}$]

Figure 5.48 The Regression function and result.

The output of this function is now saved as "result." The output is a "Fitted-Model" object, which contains a long list of information about the regression. To see all the data available, check the "Properties" of the result (Figure 5.49).

```
In[21]:= result["Properties"]

Out[21]= {AdjustedRSquared, AIC, AICc, ANOVATable, ANOVATableDegreesOfFreedom,
          ANOVATableEntries, ANOVATableMeanSquares, ANOVATableSumsOfSquares,
          BestFit, BestFitParameters, BIC, CorrelationMatrix, CovarianceMatrix,
          CurvatureConfidenceRegion, Data, EstimatedVariance, FitCurvatureTable,
          FitCurvatureTableEntries, FitResiduals, Function, HatDiagonal,
          MaxIntrinsicCurvature, MaxParameterEffectsCurvature, MeanPredictionBands,
          MeanPredictionConfidenceIntervals, MeanPredictionConfidenceIntervalTable,
          MeanPredictionConfidenceIntervalTableEntries, MeanPredictionErrors,
          ParameterBias, ParameterConfidenceIntervals, ParameterConfidenceIntervalTable,
          ParameterConfidenceIntervalTableEntries, ParameterConfidenceRegion,
          ParameterErrors, ParameterPValues, ParameterTable, ParameterTableEntries,
          ParameterTStatistics, PredictedResponse, Properties, Response,
          RSquared, SingleDeletionVariances, SinglePredictionBands,
          SinglePredictionConfidenceIntervals, SinglePredictionConfidenceIntervalTable,
          SinglePredictionConfidenceIntervalTableEntries,
          SinglePredictionErrors, StandardizedResiduals, StudentizedResiduals}
```

Figure 5.49 The properties of our FittedModel.

We will not discuss all of these: the most important result is found in the "Parameter Table" (Figure 5.50).

```
In[22]:= result["ParameterTable"]
```

Out[22]=

	Estimate	Standard Error	t-Statistic	P-Value
a1	74.9897	0.00691994	10836.8	4.16779×10^{-23}
b1	0.0240036	2.44318×10^{-6}	9824.76	7.50537×10^{-23}
a2	75.	0.00303263	24731.	2.95018×10^{-25}
b2	0.04	1.89757×10^{-6}	21079.6	7.69367×10^{-25}

Figure 5.50 The table of parameters from the regression.

As we can see here, the results from Mathematica match exactly with the results from Aspen. However, Mathematica gives us a wealth of additional analysis much more quickly.

Mathematica can generate some nice visualizations of the fitting result (Figure 5.51):

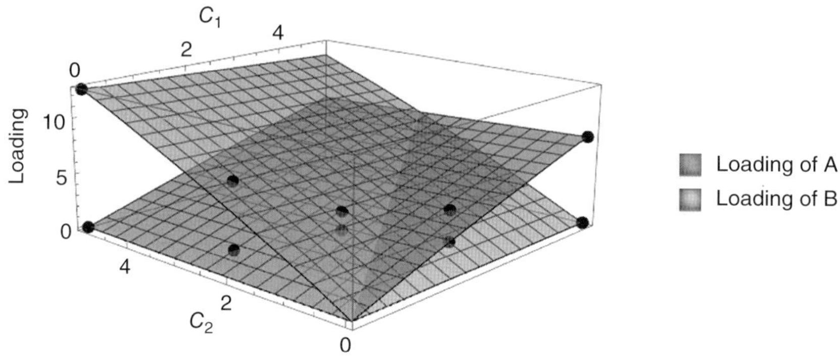

Figure 5.51 The fitted isotherms. Dots are experimental measurements.

5.3 Parameter Workshop: Regressing Against Dynamic Experiments

5.3.1 Problem Description

Aspen Chromatography is also capable of performing regression against values that are collected over time. To do this, we start out by defining an ordinary dynamic flowsheet, then creating a dynamic estimation experiment and running the simulation in estimation mode.

Once again, let us return to the separation of D- and L-threonine. This time, we have collected data from an experiment in which a racemic mixture of D- and L-threonine was fed through a column.

Step 1: Define the flowsheet using a Chrom_r_column (Figure 5.52)

Place the blocks as shown, convert the component list to a set, and add D- and L- threonine.

Step 2: Specify the values

Use the values in Table 5.2 to specify the flowsheet. Tell Aspen to estimate the dispersion through a Peclet number, Eq. (2.9). The linear lumped resistance model captures the kinetic behavior, and the isotherm is once again the extended Langmuir.

5.3.2 Dynamic Estimation Settings

It is now time to define a dynamic estimation experiment. This is analogous to the steady-state experiment we created in the previous workshop, although the values may vary with time.

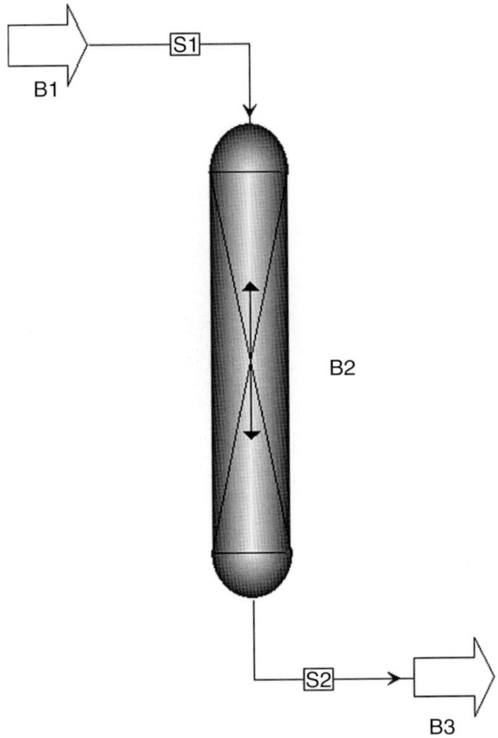

Figure 5.52 The breakthrough experimental setup

Table 5.2 Experimental conditions.

Height	25	cm
Diameter	0.46	cm
Inter-particle Voids	0.53	
Intra-particle Voids	0	
Peclet number	400	
Mass transfer coefficient	1000	1/min
Feed flow rate	5.00E-04	$l\,min^{-1}$
Feed concentration	1	$g\,l^{-1}$

Step 3: Open the Estimation Tool and add the Isotherm Parameters to the Estimated Variables
This is the same as the previous workshop, and the result should look like Figure 5.53.

Step 4: Open the "Dynamic Experiments" Tab and add a new dynamic experiment (Figure 5.54)

Step 5: Once the new experiment is created, you must add the measured variables.

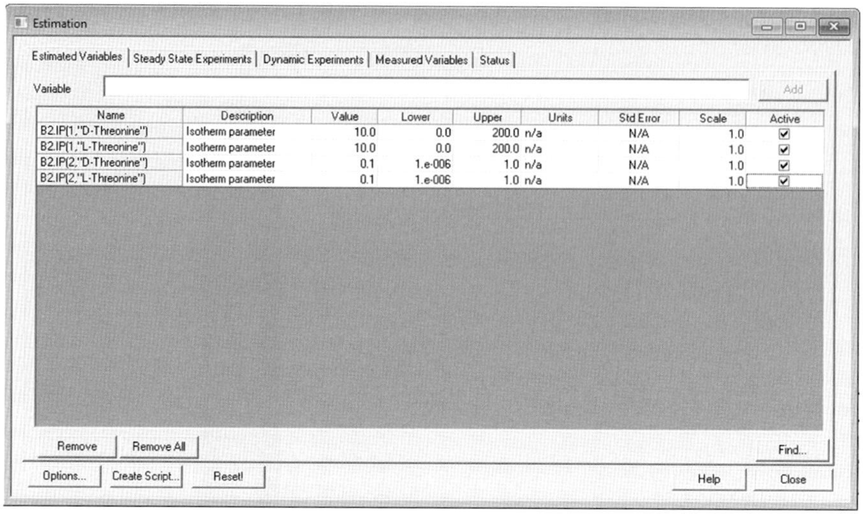

Figure 5.53 The variables to be estimated.

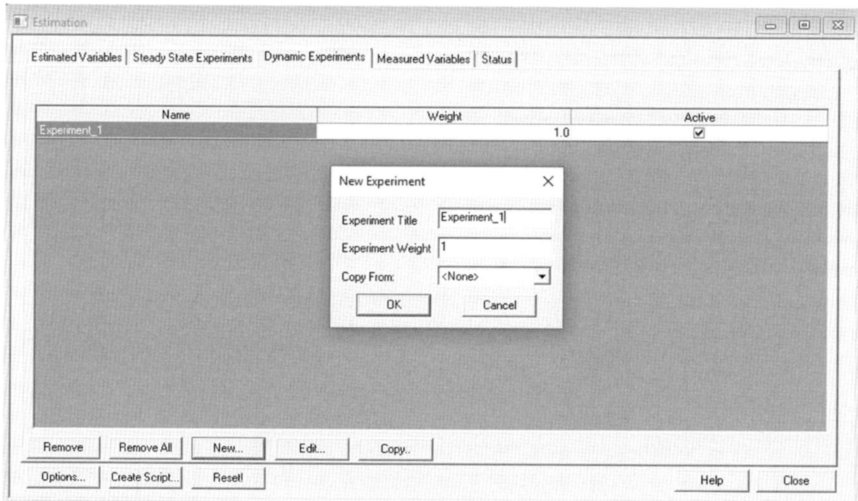

Figure 5.54 Defining a new dynamic experiment.

Just as in the previous workshop, add the outlet concentration variables to the experiment by using the "Find" tool. Search for variables in the B3 block and add the variables named "Concentration"

Once the variables have been added, insert the following experimental data. See Table 5.3.

Step 6: Run the Estimation

Once again, set strict estimation solver options, then run the estimation. This may take over a minute to complete, depending on your computer hardware (Figures 5.55 and 5.56).

Table 5.3 Experimental data.

Time	B3.Concentration("D-threonine")	B3.Concentration("L-threonine")
Min	$g\,l^{-1}$	$g\,l^{-1}$
5	0	0
7	0.010 998	8.88E-05
9	0.071 057	0.001 809
11	0.212925	0.0158
13	0.372 713	0.070 677
15	0.466 805	0.185 633
17	0.496721	0.324 477
19	0.501 003	0.42 772
21	0.500 547	0.478 018
23	0.500 142	0.494 989
25	0.500 026	0.499 129
27	0.5	0.5

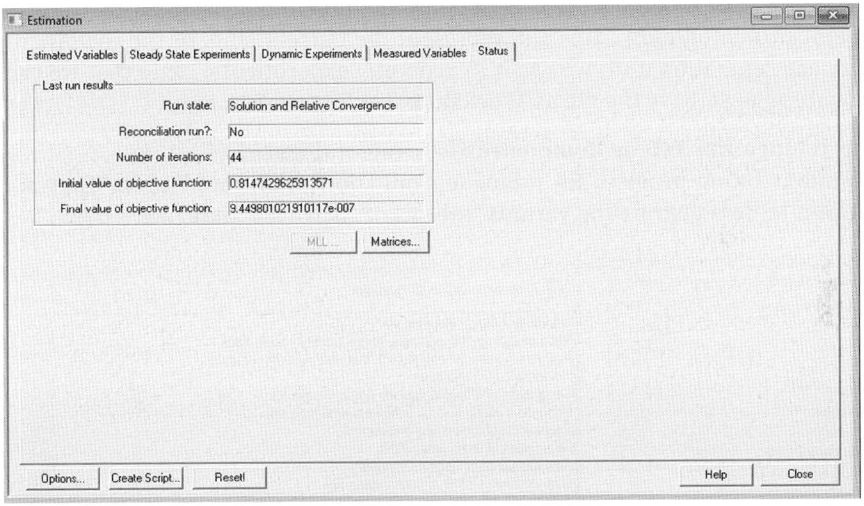

Figure 5.55 The completed Estimation.

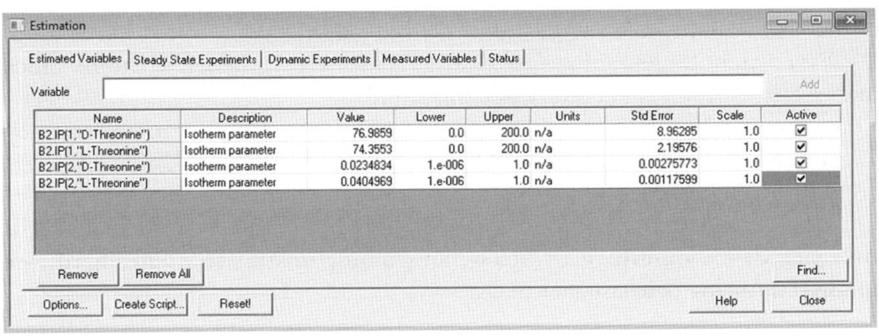

Figure 5.56 The estimated parameter values.

5.3.3 Performance Concerns

It is important to note that while we can do this kind of regression for simple breakthrough experiments, it can be very difficult to regress results from more complex experiments. In the next workshop, we will address some of these issues.

5.4 Xylene Parameter Regression

In this workshop, we suppose that you already have an SMB unit that is successfully separating xylenes. When you initially constructed the column over a year ago, you spent a lot of money to measure the isotherm parameters for your adsorbent, which would let you predict the correct operating conditions. Over time, the optimum operating conditions have slowly shifted to new values, and you suspect it is because there has been a change in the adsorption characteristics of the beds. You want to estimate a new set of isotherm parameters for your adsorbent without completely shutting down the plant. To do this, you have carefully retrieved samples from the recycle pipe in your xylene unit and had their composition analyzed.

To begin with, open the "4-zone xylene" demonstration project included in Aspen Chromatography. This will represent the current state of the hypothetical xylene separation unit; we use it to generate "experimental" data that we can compare against. Save the file as **Workshop 5-4 Regression.cra**.

Step 1: Open the Xylene demonstration project (Figure 5.57)

Demonstration projects are premade simulations built into Aspen. They are meant to demonstrate the various features of Aspen software.

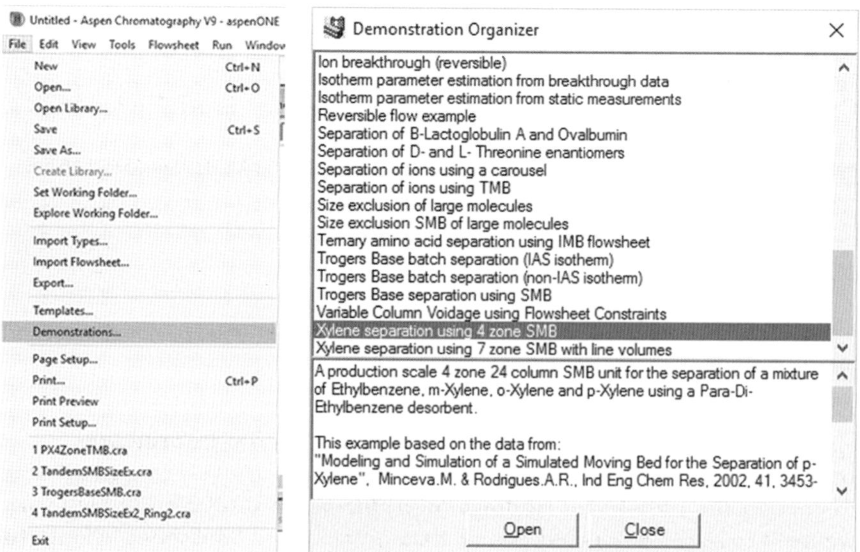

Figure 5.57 Opening a demonstration project.

Step 2: Run the Xylene demonstration in Steady-State mode

Once the file is open, ensure that the simulation is in steady-state mode, and run it.

Step 3: Convert the demonstration to dynamic SMB mode

Set the flowsheet to run in dynamic mode (do not change any column settings) and run.

Step 4: Create a History Table to record the recycle stream composition over time

Use the "new form" button to create a History Table that will record "experimental data" over time. Specifically, we want to record the recycle composition. Since Aspen does not have a literal "recycle composition" variable, we will use the composition of the outlet of the final bed. Because there are 24 beds and 15 nodes in each bed, we want the composition Chamber.Column_(24).C(15,*). See Figure 5.58.

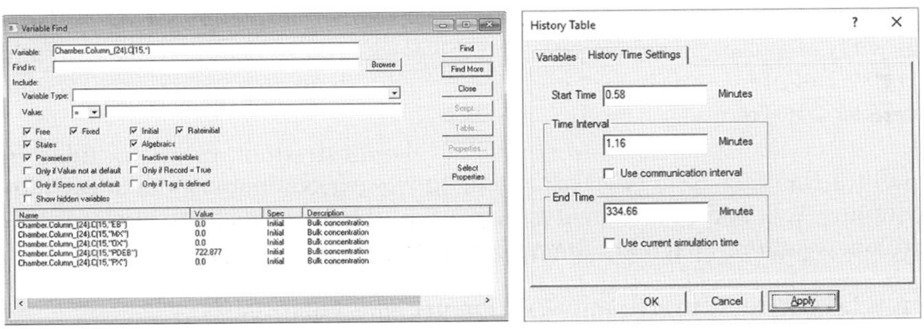

Figure 5.58 History Table configuration.

Step 5: Run the simulation

Ensure that you set the run mode to "Dynamic" then run the simulation.

Step 6: Copy the results into Excel

Once the results are available, copy the entire table into Excel (Figure 5.59).

	A	B	C	D	E	F	G
1	Time	Chamber.Column_(24).C(15,"EB")	Chamber.Column_(24).C(15,"MX")	Chamber.Column_(24).C(15,"OX")	Chamber.Column_(24).C(15,"PDEB")	Chamber.Column_(24).C(15,"PX")	Bed measure
2	Minutes	g/l	g/l	g/l	g/l	g/l	
3	0.58	2.66E-19	1.14E-18	3.22E-19	722.876	7.17E-20	24
4	1.74	2.43E-09	1.09E-08	3.14E-09	722.874	4.23E-10	23
5	2.9	1.72E-04	7.88E-04	2.31E-04	722.872	2.30E-05	22
6	4.06	0.724561	3.14678	0.896479	718.031	0.162908	21
7	5.22	2.97948	12.0903	3.3162	703.028	1.1648	20
8	6.38	6.01507	24.4242	6.7224	682.458	2.88765	19
9	7.54	17.1869	71.4468	19.858	606.872	6.17126	18
10	8.7	35.812	139.343	37.1903	491.176	15.4566	17
11	9.86	51.9733	184.467	46.7939	387.794	45.5759	16
12	11.02	43.0033	135.831	32.2378	413.943	90.4346	15

Figure 5.59 The data, as they appear in Excel.

Step 7: Convert the "History" information into "Profile" information

In order to use this simulated "experimental" data in our estimation, we need to convert the time-based readings to profile-based readings. That is to say, we need to indicate which bed each data point in our time series belongs to. Clearly, the very first data point represents data for bed 24. However, once the ports switch, the beds are effectively shifted downward. That means that the second data point must represent information about 24−1 = that is, the 23rd bed.

Rearrange the Excel spreadsheet as shown in Figure 5.60. Each column of data should have a corresponding bed number and label representing the name of the component. It is important that the component names match the names used in Aspen.

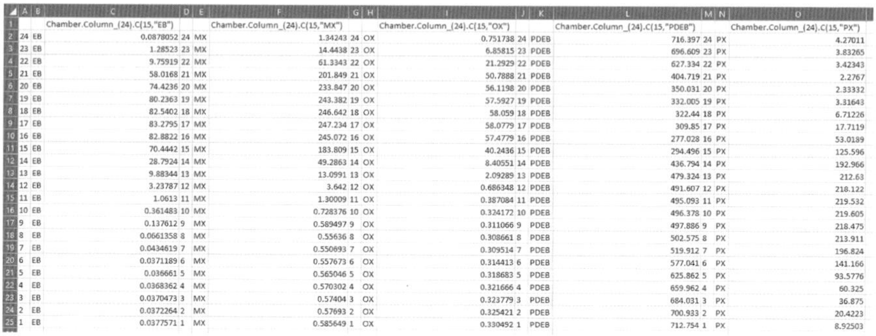

	A	B	C	D	E	F	G	H	I	J	K	L	M	N	O
1			Chamber.Column_(24).C(15,"EB")			Chamber.Column_(24).C(15,"MX")			Chamber.Column_(24).C(15,"OX")			Chamber.Column_(24).C(15,"PDEB")			Chamber.Column_(24).C(15,"PX")
2	24	EB	0.0878052	24	MX	1.34243	24	OX	0.751738	24	PDEB	716.397	24	PX	4.27011
3	23	EB	1.28523	23	MX	14.4438	23	OX	6.85815	23	PDEB	696.609	23	PX	3.83265
4	22	EB	9.75919	22	MX	61.3343	22	OX	21.2929	22	PDEB	627.334	22	PX	3.42343
5	21	EB	58.0168	21	MX	201.849	21	OX	50.7888	21	PDEB	404.719	21	PX	2.2767
6	20	EB	74.4236	20	MX	233.847	20	OX	56.1198	20	PDEB	350.031	20	PX	2.33332
7	19	EB	80.2363	19	MX	243.382	19	OX	57.5927	19	PDEB	332.005	19	PX	3.31643
8	18	EB	82.5402	18	MX	246.642	18	OX	58.059	18	PDEB	322.44	18	PX	6.71226
9	17	EB	83.2795	17	MX	247.234	17	OX	58.0779	17	PDEB	309.85	17	PX	17.7119
10	16	EB	82.8822	16	MX	245.072	16	OX	57.4779	16	PDEB	277.028	16	PX	53.0189
11	15	EB	70.4442	15	MX	183.809	15	OX	40.2436	15	PDEB	294.496	15	PX	125.596
12	14	EB	28.7924	14	MX	49.2863	14	OX	8.40551	14	PDEB	436.794	14	PX	192.966
13	13	EB	9.88344	13	MX	13.0991	13	OX	2.09289	13	PDEB	479.324	13	PX	212.63
14	12	EB	3.23787	12	MX	3.642	12	OX	0.686348	12	PDEB	491.607	12	PX	218.122
15	11	EB	1.0613	11	MX	1.30009	11	OX	0.387084	11	PDEB	495.093	11	PX	219.532
16	10	EB	0.361483	10	MX	0.728376	10	OX	0.324172	10	PDEB	496.378	10	PX	219.605
17	9	EB	0.137612	9	MX	0.589497	9	OX	0.311066	9	PDEB	497.886	9	PX	218.475
18	8	EB	0.0661358	8	MX	0.55636	8	OX	0.308661	8	PDEB	502.575	8	PX	213.911
19	7	EB	0.0434619	7	MX	0.550693	7	OX	0.309514	7	PDEB	519.912	7	PX	196.824
20	6	EB	0.0371189	6	MX	0.557673	6	OX	0.314413	6	PDEB	577.041	6	PX	141.166
21	5	EB	0.036661	5	MX	0.565046	5	OX	0.318683	5	PDEB	625.862	5	PX	93.5776
22	4	EB	0.0368362	4	MX	0.570302	4	OX	0.321666	4	PDEB	659.962	4	PX	60.325
23	3	EB	0.0370473	3	MX	0.57404	3	OX	0.323779	3	PDEB	684.031	3	PX	36.875
24	2	EB	0.0372264	2	MX	0.57693	2	OX	0.325421	2	PDEB	700.933	2	PX	20.4223
25	1	EB	0.0377571	1	MX	0.585649	1	OX	0.330492	1	PDEB	712.754	1	PX	8.92503

Figure 5.60 Data Processing in Excel.

Once the data have been arranged, copy each component's data set into columns A, B, and C, as shown in Figure 5.61. Once the data are in place, create a new column that takes the numeric bed values (1–24) and converts them into text. The Excel function for this is Text(*cell*,0). Once this column has been created, copy all the text values and paste them back to where they were using "paste value" in Excel (Figure 5.62). Finally, sort the data columns first by the string representing the bed number (Column B) and then by the component name (Column C.) The end result should look like Figure 5.63. Notice that the beds are ordered as 1, 10, 11, 12... instead of 1, 2, 3. This is actually desirable, as we will see once we create the estimation experiment.

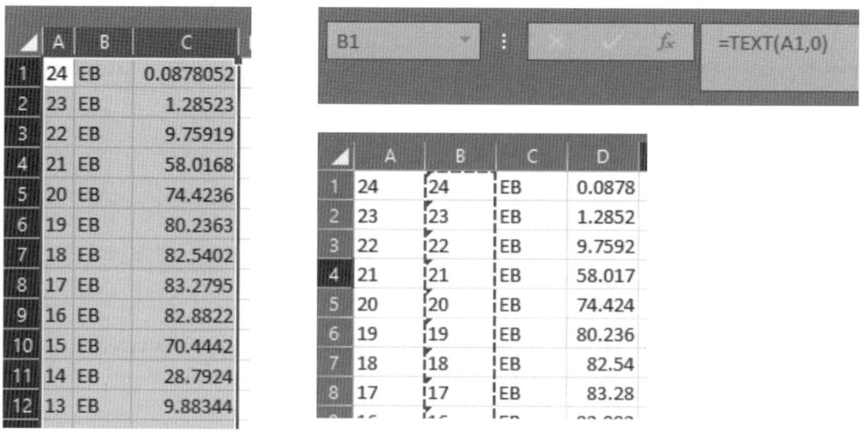

	A	B	C
1	24	EB	0.0878052
2	23	EB	1.28523
3	22	EB	9.75919
4	21	EB	58.0168
5	20	EB	74.4236
6	19	EB	80.2363
7	18	EB	82.5402
8	17	EB	83.2795
9	16	EB	82.8822
10	15	EB	70.4442
11	14	EB	28.7924
12	13	EB	9.88344

B1 =TEXT(A1,0)

	A	B	C	D
1	24	24	EB	0.0878
2	23	23	EB	1.2852
3	22	22	EB	9.7592
4	21	21	EB	58.017
5	20	20	EB	74.424
6	19	19	EB	80.236
7	18	18	EB	82.54
8	17	17	EB	83.28

Figure 5.61 Collecting the data into a single set of columns.

Figure 5.62 Pasting data by value.

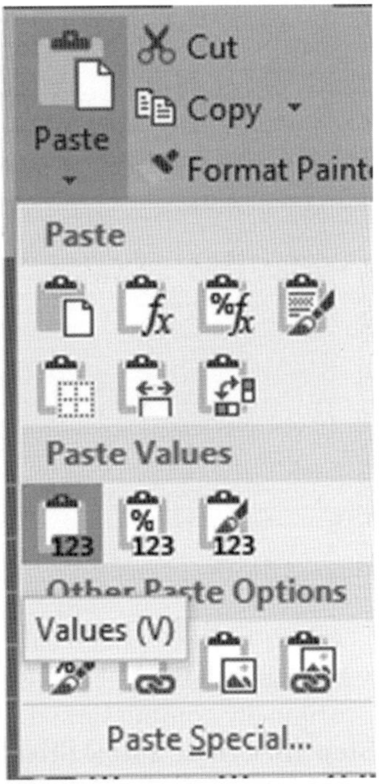

Figure 5.63 The Sorted Data.

B	C	D
1	EB	0.0378
1	MX	0.5856
1	OX	0.3305
1	PDEB	712.75
1	PX	8.925
10	EB	0.3615
10	MX	0.7284
10	OX	0.3242
10	PDEB	496.38
10	PX	219.61

Step 8: Change the demonstration project back to Steady-State mode
Once the data are fully recorded, save the demonstration project under a new name, and then reopen the original file.

Step 9: Open the Estimation Tool and enter the Estimation Variables
Use the "Tools" menu to open the estimation tool. Once you see the "Estimated Variables" tab, add the variables shown in Figure 5.64. There are two important features to note. First, we will not be estimating the PDEB isotherm parameters. That is to say, we will be assuming that the adsorbent's affinity for the solvent has not changed much over time. Second, notice that we have entered new

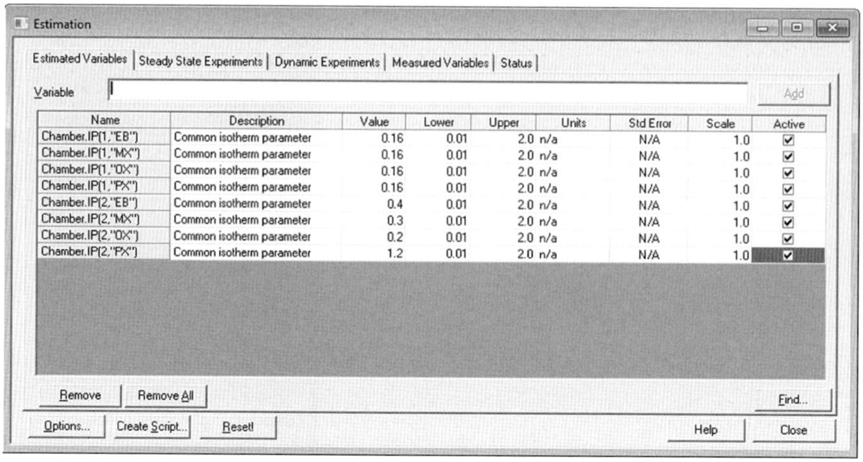

Figure 5.64 The Estimation Variables.

values in the "Value" column. Those values will represent our guess of what the values should be, based on the experiments done when the column was originally commissioned.

Step 10: Create a new Steady-State Estimation Experiment

Open the Steady-State Experiments and use the "new" button to create a new experiment (Figure 5.65).

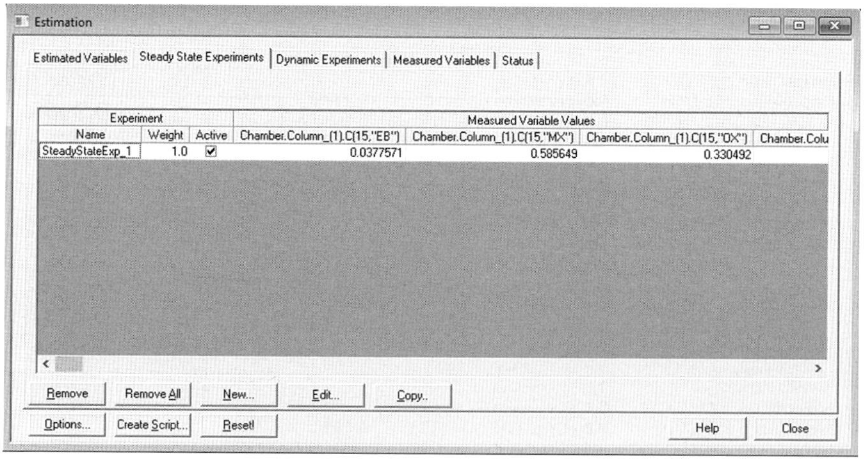

Figure 5.65 A new steady-state experiment.

Step 11: Enter the measured variables

Use the "Find" tool to locate C(7,*), this will find all the concentrations within the column at the center of the beds (Figure 5.66). Drag all these variables into the estimation experiment's measured variables tab (Figure 5.67).

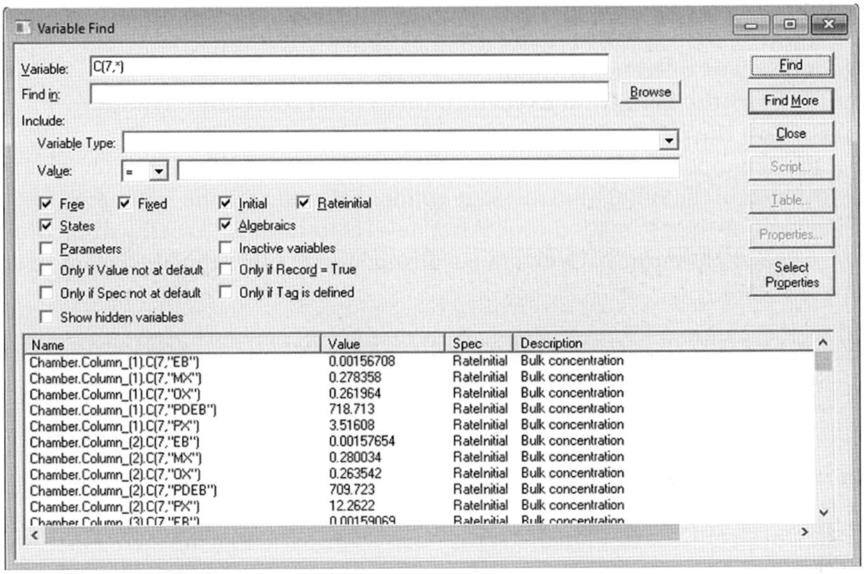

Figure 5.66 Locating the "Measured" variables.

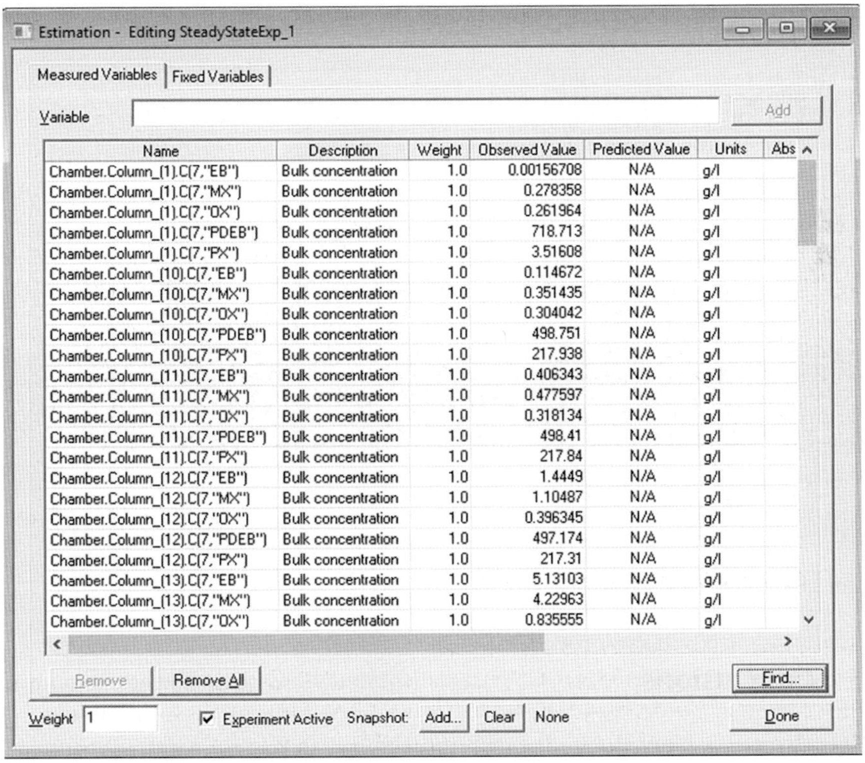

Figure 5.67 The steady-state measured values.

With this, we now see the benefit of sorting our Excel data in the unusual way. The measured variables are listed in alphabetical order, so to paste data from Excel, we need to match this ordering.

Step 12: Copy the data from Excel into the "Observed Value" column

Step 13: Run the estimation experiment

Set the run mode to "Estimation" and run the simulation. The iteration results will appear in the simulation message window, as well as in the "estimated variables" tab of the estimation tool. After several iterations, Aspen will recover the original isotherm parameters used to generate our simulated plant data (Figure 5.68).

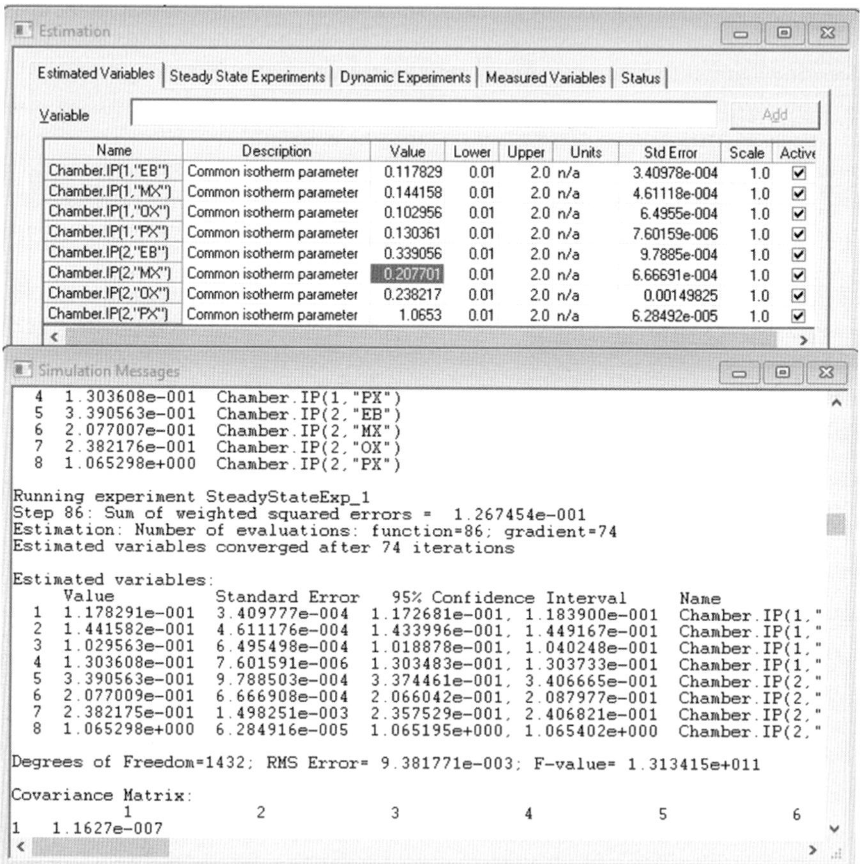

Figure 5.68 The original values, successfully recovered.

5.5 Conclusions

In this chapter, we have looked at a variety of ways to regress parameters, using Excel, Aspen, and Mathematica. We also discussed some ways of creating initial estimates of some important properties of a chromatographic separation.

5.6 Practice Problems

5.6.1 Perform Dynamic Parameter Estimation in Aspen Adsorption

The estimation tool in Aspen Adsorption is **virtually identical** to that in Aspen Chromatography. To demonstrate this, we will practice our estimation skills by performing a regression of mass transfer parameters of oxygen and nitrogen against a breakthrough curve.

Begin by creating an Aspen Adsorption flowsheet with only oxygen and nitrogen in a Component Set. Then connect a gas_bed to the Feed and Product blocks as shown in Figure 5.69.

Once the flowsheet is created, begin to set the column assumptions. The discretization can simply be a UDS1 method with 20 nodes. We are of course regressing against the extended Langmuir Isotherm (Extended Langmuir 1) as a function of concentration rather than partial pressure. The column is isothermal, and pressure drop is described by the Carman–Kozeney equation. There is no dispersion, and mass transfer is governed by linear lumped resistance based on the solid-phase concentrations.

Once those assumptions have been selected, configure the column as shown in Figure 5.70.

When you are working with data you did not collect, it is an important skill to be able to understand the implications of reported numbers. For example, someone

Figure 5.69 The breakthrough experiment setup.

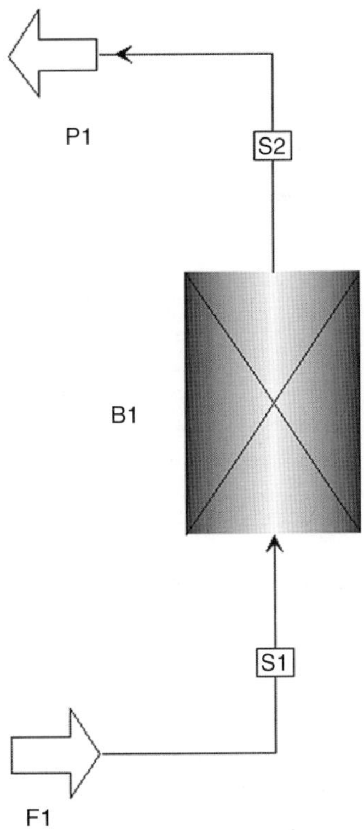

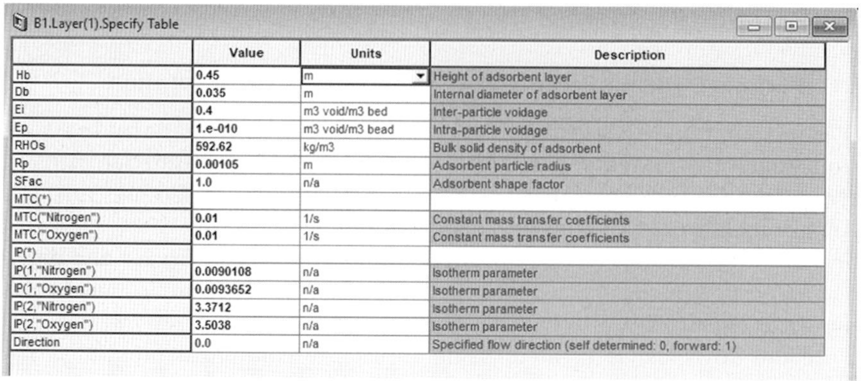
B1.Layer(1).Specify Table

	Value	Units	Description
Hb	0.45	m	Height of adsorbent layer
Db	0.035	m	Internal diameter of adsorbent layer
Ei	0.4	m3 void/m3 bed	Inter-particle voidage
Ep	1.e-010	m3 void/m3 bead	Intra-particle voidage
RHOs	592.62	kg/m3	Bulk solid density of adsorbent
Rp	0.00105	m	Adsorbent particle radius
SFac	1.0	n/a	Adsorbent shape factor
MTC(*)			
MTC("Nitrogen")	0.01	1/s	Constant mass transfer coefficients
MTC("Oxygen")	0.01	1/s	Constant mass transfer coefficients
IP(*)			
IP(1,"Nitrogen")	0.0090108	n/a	Isotherm parameter
IP(1,"Oxygen")	0.0093652	n/a	Isotherm parameter
IP(2,"Nitrogen")	3.3712	n/a	Isotherm parameter
IP(2,"Oxygen")	3.5038	n/a	Isotherm parameter
Direction	0.0	n/a	Specified flow direction (self determined: 0, forward: 1)

Figure 5.70 The column specification for the breakthrough experiment.

may not explicitly state the assumptions made in their model, and you may only determine them by analyzing which parameters are used.

Create a new dynamic experiment with the following data given in Table 5.4.

Fix the feed flow rate to 10^{-3} mol s^{-1} and then run the simulation in estimation (Figure 5.71).

From these results, we can see that the estimation was highly successful in choosing parameters that fit the data. The maximum error between the data and the predicted model was less than 1%

Table 5.4 Dynamic experiment results.

Time (s)	y_{N_2}	y_{O_2}
0	1.0000	0.0000
60	0.9955	0.0045
120	0.9683	0.0317
180	0.9201	0.0799
240	0.8737	0.1263
300	0.8413	0.1587
360	0.8216	0.1784
420	0.8100	0.1900
480	0.8031	0.1969
540	0.7989	0.2011
600	0.7961	0.2039
660	0.7943	0.2057
720	0.7930	0.2070
780	0.7921	0.2079
840	0.7914	0.2086
900	0.7910	0.2090
960	0.7907	0.2093
1020	0.7905	0.2095

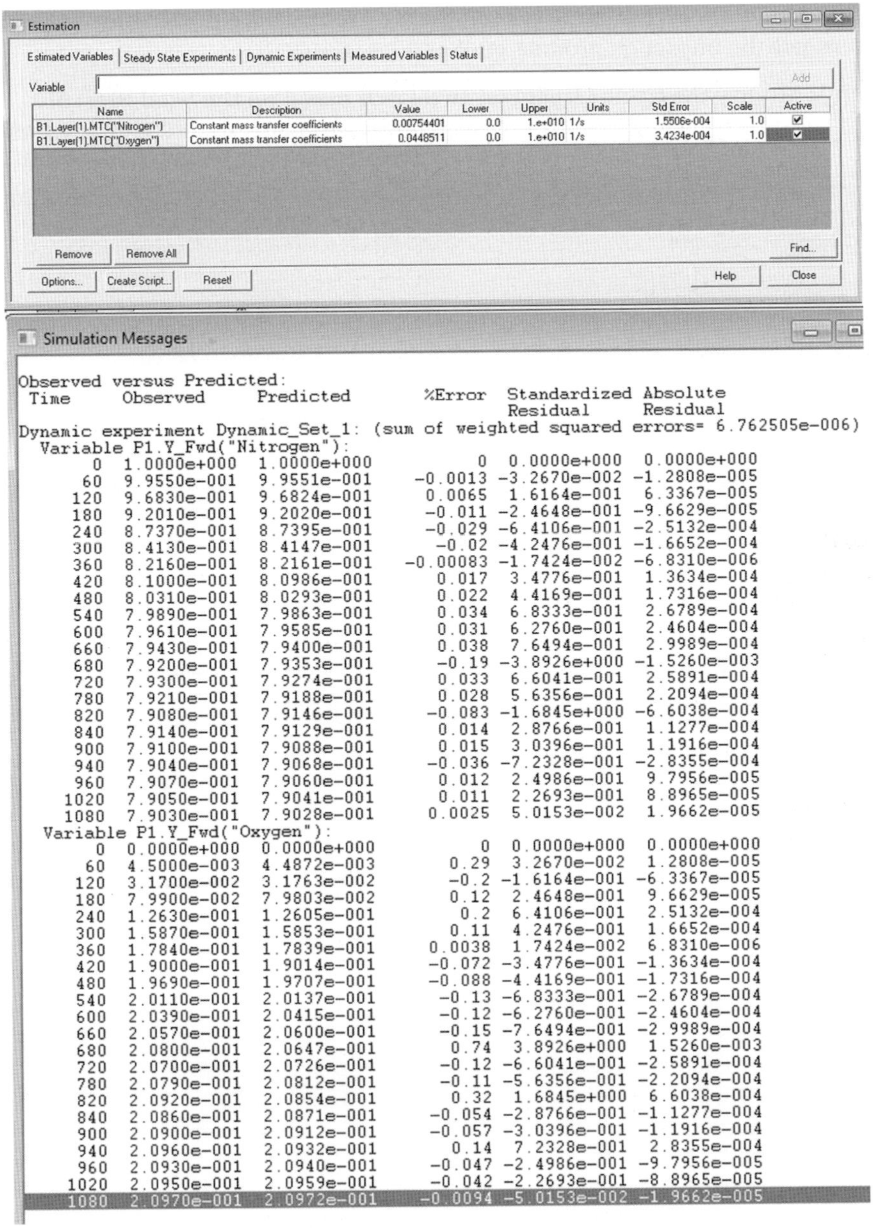

Figure 5.71 The estimation results.

5.6.2 Sensitivity Analysis Using Scripts in Aspen Adsorption

To perform a sensitivity analysis in Aspen, we will use the scripting capabilities we have already seen to automate several runs. We will also learn how to automatically export this data to Excel. To begin, open the final Aspen file from the PSA workshop in Section 1.7, where we have implemented a PSA process for the separation of nitrogen from oxygen in the air. We will vary the feed pressure to investigate the impact on product purity and recovery.

Step 1: Modify the Cycle Organizer to Switch based on the Feed Pressure

In this file, we hard coded in the switching criteria for steps 2 and 4: switch when the pressurization tank reaches 8.4 bar. However, because the feed pressure will no longer be constant, we must modify that constraint. Instead of 8.4 bar, change the criteria to 0.99*F1.P (Figure 5.72).

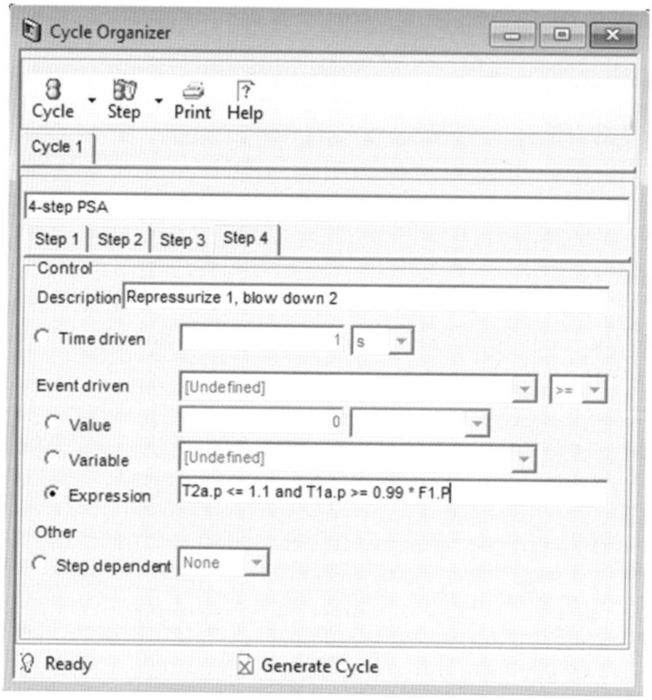

Figure 5.72 Changing the switching time constraints.

Make sure that you regenerate the task after this change.

Step 2: Add the Sensitivity Script and Add the Excel Code

We will modify this simulation file to contain a script named "Sensitivity." (Figure 5.73).

To begin with, we will write the code required to open an instance of Excel that we can write to later. Add the following code to the script and run it to ensure that it works:

```
Dim objExcel
Set objExcel = CreateObject("Excel.Application")
objExcel.Workbooks.Add
objExcel.Visible = True
```

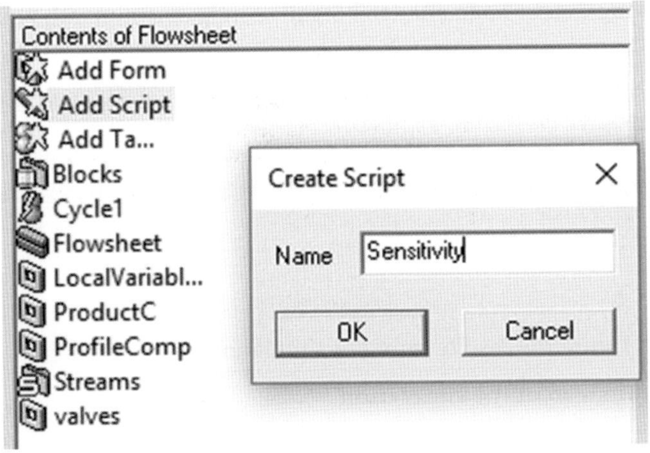

Figure 5.73 Creating the sensitivity script.

Step 3: Add the Sensitivity Logic

We will vary the feed pressure from 7 to 10 bar with a step size of 0.5. This is fairly easy to do using "for" loops in the script. Enter the following code to implement the sensitivity:

```
for p = 7 to 10 step 0.5
  F1.P.Value = p
  Simulation.Run true

  Simulation.Restart
Next
```

This code will set the value of the feed pressure, run the simulation, and then reset the simulation. We are not done yet, however, since we have not recorded any of the results. If we were to run this script, we would not have an answer stored anywhere after our simulations completed.

Step 4: Add the Excel Exporting Logic

To export data to Excel, we can write our values directly to any cells we like. Enter the following code before the "for" loop to set up some nice column headings for our Excel spreadsheet:

```
objExcel.ActiveSheet.Cells(1,1).Value = "Feed Pressure"
objExcel.ActiveSheet.Cells(1,2).Value = "Product N2 Mole
                                             Fraction"
objExcel.ActiveSheet.Cells(1,3).Value = "Product O2 Mole
                                             Fraction"
objExcel.ActiveSheet.Cells(1,4).Value = "Product Flow Rate"

Dim i
i = 2
```

Next, add the following export code after the Simulation.Run command, but before the restart:

```
objExcel.ActiveSheet.Cells(i,1).Value = p
objExcel.ActiveSheet.Cells(i,2).Value =
   P1.Y_Fwd("N2").Value
objExcel.ActiveSheet.Cells(i,3).Value =
   P1.Y_Fwd("O2").Value
objExcel.ActiveSheet.Cells(i,4).Value = P1.F.Value

i = i + 1
```

Now we will correctly store the results from our simulation.

Step 5: Run the *script* instead of the simulation

Our final script should read as follows:

```
Dim objExcel
Set objExcel = CreateObject("Excel.Application")
objExcel.Workbooks.Add
objExcel.Visible = True
objExcel.ActiveSheet.Cells(1,1).Value = "Feed Pressure"
objExcel.ActiveSheet.Cells(1,2).Value = "Product N2 Mole
   Fraction"
objExcel.ActiveSheet.Cells(1,3).Value = "Product O2 Mole
   Fraction"
objExcel.ActiveSheet.Cells(1,4).Value = "Product Flow Rate"

Dim i
i = 2

for p = 7 to 10 step 0.5
  F1.P.Value = p
  Simulation.Run true

  objExcel.ActiveSheet.Cells(i,1).Value = p
  objExcel.ActiveSheet.Cells(i,2).Value =
   P1.Y_Fwd("N2").Value
  objExcel.ActiveSheet.Cells(i,3).Value =
   P1.Y_Fwd("O2").Value
  objExcel.ActiveSheet.Cells(i,4).Value = P1.F.Value

  i = i + 1

  Simulation.Restart

next
```

Because our script will run the simulation automatically, run the script by double clicking on it, instead of using the "run" button. Note that because we are running several simulations in a row, this may take quite some time, depending on your machine. As the simulations run, the data will appear in the Excel

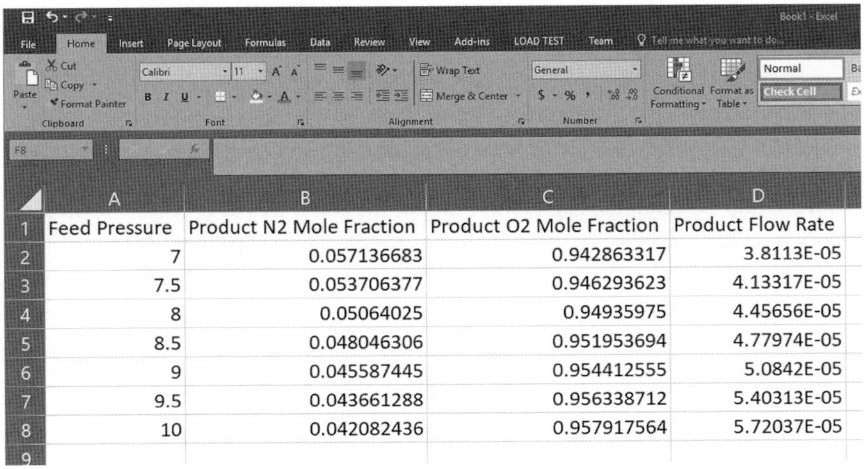

Feed Pressure	Product N2 Mole Fraction	Product O2 Mole Fraction	Product Flow Rate
7	0.057136683	0.942863317	3.8113E-05
7.5	0.053706377	0.946293623	4.13317E-05
8	0.05064025	0.94935975	4.45656E-05
8.5	0.048046306	0.951953694	4.77974E-05
9	0.045587445	0.954412555	5.0842E-05
9.5	0.043661288	0.956338712	5.40313E-05
10	0.042082436	0.957917564	5.72037E-05

Figure 5.74 Sensitivity Results.

workbook created from Aspen. Once the sensitivity analysis is complete, the Excel document should resemble Figure 5.74.

Bibliography

Ching, C.B., Arlt, W., Lisso, M., and Wozny, G. (2000). Kinetic and equilibrium study of the enantioseparation of fenoprofen in a batch setup. *Industrial & Engineering Chemistry Research* **39** (11): 4365–4369.

Chung, S.F. and Wen, C.Y. (1968). Longitudinal dispersion of liquid flowing through fixed and fluidized beds. *AIChE Journal* **14** (6): 857–866.

Feist, S.D., Hasabnis, Y., Pynnonen, B.W., and Frank, T.C. (2009). SMB chromatography design using profile advancement factors, miniplant data, and rate-based process simulation. *AIChE Journal* **55** (11): 2848–2860.

Foo, K.Y. and Hameed, B.H. (2010). Insights into the modeling of adsorption isotherm systems. *Chemical Engineering Journal* **156** (1): 2–10.

Goddard, M. and Ruthven, D. (1986). Sorption and diffusion of C8 aromatic hydrocarbons in faujasite type zeolites. II. Sorption kinetics and intracrystalline diffusivities. *Zeolites* **6** (4): 283–289.

Gupta, A.S. and Thodos, G. (1963). Direct analogy between mass and heat transfer to beds of spheres. *AIChE Journal* **9** (6): 751–754.

Pfeffer, R. and Happel, J. (1964). An analytical study of heat and mass transfer in multiparticle systems at low Reynolds numbers. *AIChE Journal* **10** (5): 605–611.

Ruthven, D. and Goddard, M. (1986). Sorption and diffusion of C8 aromatic hydrocarbons in faujasite type zeolites. I. Equilibrium isotherms and separation factors. *Zeolites* **6** (4): 275–282.

Tournier, H., Barreau, A., Tavitian, B. et al. (2001). Adsorption equilibrium of xylene isomers and *p*-diethylbenzene on a prehydrated BaX zeolite. *Industrial & Engineering Chemistry Research* **40** (25): 5983–5990.

Xiong, Y.-B., Wang, H., Wang, Y.-J., and Wang, R.-M. (2011). Novel imidazolium-based poly(ionic liquid)s: preparation, characterization, and absorption of CO_2. *Polymers for Advanced Technologies*, n/a-n/a.

Wang, G.Q., Yuan, X.G., and Yu, K.T. (2005). Review of mass-transfer correlations for packed columns*. *Industrial & Engineering Chemistry Research* **44** (23): 8715–8729.

Wilson, E.J. and Geankoplis, C.J. (1966). Liquid mass transfer at very low Reynolds numbers in packed beds. *Industrial & Engineering Chemistry Fundamentals* **5** (1): 9–14.

Literature Cited in the Text

1 Fred, K. and Paul, K.J. (1959). Separation of hydrogen sulfide and methane. US Patent 2,901,326, filed 22 December 1955 and issued 25 August 1959.
2 Milton, W.E. (1965). Method of separating oxygen from air. US Patent 3,164,454 A, filed 25 September 1959 and issued 5 January 1965.
3 Turner, N.C. (1957). Gas dehydration apparatus and process. US Patent 2,815,089 A, filed 27 June 1955 and issued 3 December 1957.
4 Hull, J.R. (1955). Dehydrator. US Patent 2,699,837 A, filed 11 July 1951 and issued 18 January 1955.
5 Skarstrom, C.W. (1960). Method and apparatus for fractionating gaseous mixtures by adsorption. US Patent 2,944,627 A, filed 12 February 1958 and issued 12 July 1960.
6 Etherington, L.D., Isidor, K., and Jonach, F.L. (1955). Separation of hydrocarbons. US Patent 2,702,826 A, filed 1 December 1950 and issued 22 February 1955.
7 Skarstrom, C.W. and Heilman, W.O. (1962). Heatless fractionator. US Patent 3,069,830 A, filed 24 March 1960 and issued 25 December 1962.
8 Langmuir, I. (1918). The adsorption of gasses on plant surfaces of glass, mica and platinum. *Journal of the American Chemical Society* **40** (9): 1361–1403.
9 Brunauer, S., Emmett, P.H., and Teller, E. (1938). Adsorption of gases in multimolecular layers. *Journal of the American Chemical Society* **60** (2): 309–319.
10 Freundlich, H. (1906). *Über die Adsorption in Lösungen*. Wilhelm Engelmann.
11 Sips, R. (1948). On the structure of a catalyst surface. *The Journal of Chemical Physics* **16** (5): 490–495.
12 Toth, J. (1971). State equations of the solid-gas Interface layers. *Acta Chim Acad Sci Hung* **69** (3): 311–328.
13 Kozeny, J., Striedieck, W.F., and Wyllie, M.R.J. (1927). *Concerning Capillary Conduction of Water in the Soil (Rise, Seepage and Use in Irrigation).* Petroleum Branch of the American Institute of Mining and Metallurgical Engineers.
14 McCabe, W. and Smith, J. (2005). *Unit Operations of Chemical Engineering*, 7the. Boston: McGraw-Hill.
15 Burke, S.P. and Plummer, W.B. (1928). Gas flow through packed columns. *Industrial and Engineering Chemistry* **20** (11): 1196–1200.

Design, Simulation, and Optimization of Adsorptive and Chromatographic Separations: A Hands-On Approach, First Edition. Kevin R. Wood, Y. A. Liu, and Yueying Yu.

16 Ergun, S. and Orning, A.A. (1949). Fluid flow through randomly packed columns and fluidized beds. *Industrial and Engineering Chemistry* **41** (6): 1179–1184.

17 Smith, J.M., Ness, H.V., and Abbott, M.M. (2004). *Introduction to Chemical Engineering Thermodynamics*, 7the. New York: McGraw-Hill.

18 Yang, R.T. (1997). *Gas Separation by Adsorption Processes.* Singapore; River Edge, NJ: World Scientific.

19 Yang, R.T. (2003). *Adsorbents: Fundamentals and Applications.* New York: Wiley.

20 Shen, D., Bülow, M., Jale, S.R. et al. (2001). Thermodynamics of nitrogen and oxygen sorption on zeolites LiLSX and CaA. *Microporous and Mesoporous Materials* **48** (1–3): 211–217.

21 Skarstrom, C.W. (1966). Oxygen concentration process. US Patent 3,237,377 A, filed 12 April 1962 and issued 1 March 1966.

22 Rege, S.U., Yang, R.T., and Buzanowski, M.A. (2000). Sorbents for air prepurification in air separation. *Chemical Engineering Science* **55** (21): 4827–4838.

23 Jayaramen, A. and Yang, R.T. Stable oxygen-selective sorbents for air separation. *Chemical Engineering Science* **60** (3): 625–634.

24 Jee, J.-G., Kim, M.B., and Lee, C.-H. (2001). Adsorption characteristics of hydrogen mixtures in a layered bed: binary, ternary, and five-component mixtures. *Industrial and Engineering Chemistry Research* **40** (3): 868–878.

25 Bárcia, P.S., Silva, J.A.C., and Rodrigues, A.E. (2010). Adsorption dynamics of C5-C6 isomerate fractions in zeolite beta for the octane improvement of gasoline. *Energy and Fuels* **24** (3): 1931–1940.

26 Kneabel, K.S. and Hill, F.B. (1985). Pressure swing adsorption: development of an equilibrium theory for gas separations. *Chemical Engineering Science* **40**: 2351.

27 Broughton, D.B., and Gerhold, C.G. (1961). Continuous sorption process employing fixed bed of sorbent and moving inlets and outlets. US Patent 2,985,589 A, filed 22 May 1957 and issued 23 May 1961.

28 Aniceto, J.P.S. and Silva, C.M. (2015). Simulated moving bed strategies and designs: from established systems to the latest developments. *Separation and Purification Reviews* **44** (1): 41–73.

29 Zhang, Z., Mazzotti, M., and Morbidelli, M. (2004). Continuous chromatographic processes with a small number of columns: comparison of simulated moving bed with Varicol, PowerFeed, and ModiCon. *Korean Journal of Chemical Engineering* **21** (2): 454–464.

30 Ludemann-Hombourger, O., Nicoud, R.M., and Bailly, M. (2000). The "VARICOL" process: a new multicolumn continuous chromatographic process. *Separation Science and Technology* **35** (12): 1829–1862.

31 Ludemann-Hombourger, O., Pigorini, G., Nicoud, R.M. et al. (2002). Application of the "VARICOL" process to the separation of the isomers of the SB-553261 Racemate. *Journal of Chromatography A* **947** (1): 59–68.

32 Pais, L.S. and Rodrigues, A.E. (2003). Design of Simulated Moving bed and Varicol Processes for preparative separations with a low number of columns. *Journal of Chromatography A* **1006** (1): 33–44.

33 Toumi, A., Engell, S., Ludemann-Hombourger, O. et al. (2003). Optimization of simulated moving bed and Varicol processes. *Journal of Chromatography A* **1006** (1): 15–31.
34 Kawajiri, Y. and Biegler, L.T. (2006). Optimization strategies for simulated moving bed and PowerFeed processes. *AIChE Journal* **52** (4): 1343–1350.
35 Zhang, Z., Mazzotti, M., and Morbidelli, M. (2003). PowerFeed operation of simulated moving bed units: changing flow-rates during the switching interval. *Journal of Chromatography A* **1006** (1): 87–99.
36 Zhang, Z., Morbidelli, M., and Mazzotti, M. (2004). Experimental assessment of PowerFeed chromatography. *AIChE Journal* **50** (3): 625–632.
37 Kurup, A.S., Hidajat, K., and Ray, A.K. (2006). Comparative study of modified simulated moving bed systems at optimal conditions for the separation of ternary mixtures under nonideal conditions. *Industrial and Engineering Chemistry Research* **45** (11): 3902–3915.
38 Yu, Y., Wood, K.R., and Liu, Y.A. (2015). Simulation and comparison of operational modes in simulated moving bed chromatography. *Industrial and Engineering Chemistry Research* **54** (46): 11576–11591.
39 Jin, W. and Wankat, P.C. (2007). Hybrid simulated moving bed processes for the purification of p-xylene. *Separation Science and Technology* **42**: 669–700.
40 Lee, J., Shin, N.C., Lim, Y., and Han, C. (2010). Modeling and simulation of a simulated moving bed for adsorptive para-xylene separation. *Korean Journal of Chemical Engineering* **27**: 609–618.
41 Wankat, P.C. (2001). Simulated moving bed cascades for ternary separations. *Industrial and Engineering Chemistry Research* **40** (26): 6185–6193.
42 Venkatesan, A. and Wankat, P.C. (2011). Simulation of ion exchange water softening pretreatment for reverse osmosis desalination of brackish water. *Desalination* **271** (1–3): 122–131.
43 Jin, W. (2007). Development and applications of two-zone SMB processes for binary separations. PhD dissertation. Purdue University. https://search.proquest.com/docview/304822154?pq-origsite=gscholar (accessed 9 December 2017)
44 Mazzotti, M. (2006). Design of simulated moving bed separations: generalized Langmuir isotherm. *Industrial and Engineering Chemistry Research* **45** (18): 6311–6324.
45 Ma, Z. and Wang, N.-H.L. (1997). Standing wave analysis of SMB chromatography: linear systems. *AIChE Journal* **43** (10): 2488–2508.
46 Mun, S., Xie, Y., and Wang, N.-H.L. (2003). Robust pinched-wave design of a size-exclusion simulated moving-bed process for insulin purification. *Industrial and Engineering Chemistry Research* **42** (13): 3129–3143.
47 Kaczmarski, K., Cavazzini, A., Szabelski, P. et al. (2002). Application of the general rate model and the generalized Maxwell–Stefan equation to the study of the mass transfer kinetics of a pair of enantiomers. *Journal of Chromatography A* **962** (1): 57.
48 Lee, K.B. and Wang, N.-H.L. (2013). *Simulated Moving Bed Chromatography for Chiral Separation: Design, Experiment, Optimization.* Lap Lambert Academic Publishing GmbH KG.

49 Ma, Z., Katti, A., Lin, B., and Guiochon, G. (1990). Simple wave effects in two-component nonlinear liquid chromatography: application to the measurement of competitive adsorption isotherms. *The Journal of Physical Chemistry* **94** (17): 6911–6922.

50 Xie, Y., Wu, D., Ma, Z., and Wang, N.-H.L. (2000). Extended standing wave design method for simulated moving bed chromatography: linear systems. *Industrial and Engineering Chemistry Research* **39** (6): 1993–2005.

51 Zang, Y. and Wankat, P.C. (2002). SMB operational strategy-partial feeding. *Industrial and Engineering Chemistry Research* **41** (10): 2504–2511.

52 Schramm, H., Kaspereit, M., Kienle, A., and Seidel-Morganstern, A. (2003). Improved operation of simulated moving bed processes through cyclic modulation of feed flow and feed concentration. *Chemical Engineering Science* **58** (23–24): 5217–5227.

53 Chung, S.F. and Wen, C.Y. (1968). Longitudinal dispersion of liquid flowing through fixed and fluidized beds. *AIChE Journal* **14** (6): 857–866.

54 Wilson, E.J. and Geankoplis, C.J. (1966). Liquid mass transfer at very low Reynolds numbers in packed beds. *Industrial and Engineering Chemistry Fundamentals* **5** (1): 9–14.

55 Gupta, A.S. and Thodos, G. (1963). Direct analogy between mass and heat transfer to beds of spheres. *AIChE Journal* **9** (6): 751–754.

56 Pfeffer, R. and Happel, J. (1964). An analytical study of heat and mass transfer in multiparticle systems at low Reynolds numbers. *AIChE Journal* **10** (5): 605–611.

Index

Design, Simulation, and Optimization of Adsorptive and Chromatographic Separations: A Hands-On Approach, First Edition. Kevin R. Wood, Y. A. Liu, and Yueying Yu.

h

X